NANOSCIENCE AND COMPUTATIONAL CHEMISTRY

Research Progress

NANOSCIENCE AND COMPUTATIONAL CHEMISTRY

Research Progress

Edited by

Andrew G. Mercader, PhD, Eduardo A. Castro, PhD and A. K. Haghi, PhD

Apple Academic Press

TORONTO NEW JERSEY

Apple Academic Press Inc.
3333 Mistwell Crescent
Oakville, ON L6L 0A2
Canada

Apple Academic Press Inc.
9 Spinnaker Way
Waretown, NJ 08758
USA

International Standard Book Number-13: 978-1-926895-59-8 (Hardcover)

Library of Congress Control Number: 2013951093

Library and Archives Canada Cataloguing in Publication

Nanoscience and computational chemistry: research progress/edited by
Andrew G. Mercader, Eduardo A. Castro, and A.K. Haghi.

Includes bibliographical references and index.
ISBN 978-1-926895-59-8
1. Nanoscience. 2. Chemical engineering--Data processing. I. Castro, E. A.
(Eduardo Alberto), 1944-, editor of compilation II. Haghi, A. K., editor of compilation
III. Mercader, Andrew G., editor of compilation

QC176.8.N35N35 2013 620'.5 C2013-906682-9

ABOUT THE EDITORS

Andrew G. Mercader, PhD

Dr. Andrew G. Mercader studied physical chemistry at the Faculty of Chemistry of La Plata National University (UNLP), Buenos Aires, Argentina, from 1995–2001. Afterwards he joined Shell Argentina to work as Luboil, Asphalts and Distillation Process Technologist as well as Safeguarding and Project Technologist, from 2001–2006. His PhD work on the development and applications of QSAR/QSPR theory was performed at the Theoretical and Applied Research Institute located at La Plata National University (INIFTA) from 2006–2009. After that he obtained a post-doctoral scholarship to work on theoretical-experimental studies of biflavonoids at IBIMOL (ex PRALIB), Faculty of Pharmacy and Biochemistry, University of Buenos Aires (UBA) from 2009–2012. He is currently a member of the Scientific Researcher Career in the Argentina National Research Council at INIFTA.

Eduardo A. Castro, PhD

Dr. Eduardo A. Castro's career was launched by studying physical chemistry at the Faculty of Chemistry of the La Plata National University of La Plata, Buenos Aires, Argentina, during 1963–1970. His diploma work to get his PhD degree was on calculation of HMO and related semi empirical methods of beta-carotene for analyze chemical reactivity and electronic spectrum. Incidentally, his only available computational resource on that time was a diagonalization subroutine for symmetric matrices and his only disposable instruction book was Andrew Streitwieser's on Theoretical Organic Chemistry. From 1971–1972 he performed his PhD work at the Physics Department of the National La Plata University, working under supervision of Manuel Sorarrain. After that he worked as a research scientist at the Theoretical and Applied Research Institute located at La Plata National University, where he founded the Group for Theoretical Chemistry in 1974. He was appointed as a member of the Scientific Researcher Career in the Argentina National Research Council, and he continues up to the present time as a Superior Researcher.

A. K. Haghi, PhD

Dr. A. K. Haghi holds a BSc in urban and environmental engineering from the University of North Carolina (USA), a MSc in mechanical engineering from North

Carolina A&T State University (USA), a DEA in applied mechanics, acoustics, and materials from the Université de Technologie de Compiègne (France), and a PhD in engineering sciences from the Université de Franche-Comté (France). He has written about 1500 original articles, 250 monographs, and 170 chapters in 40 volumes, and more than 50 academic books. It is apparent from this work that he has made valuable contributions to the theory and practice of chemical engineering, heat and mass transfer, porous media, industrial drying, polymers, nanofibers and nanocomposites.

Dr. Haghi is Editor-in-Chief of *International Journal of Chemoinformatics and Chemical Engineering* and Editor-in-Chief of *Polymers Research Journal*. He is member of many editorial boards of journals published in U.S.A.

He is Senior Editor of Apple Academic Press (Canada).

He served as associate member of University of Ottawa and was a member of Canadian Society of Mechanical Engineering. He serves as dean of faculty of Engineering at University of Guilan (Iran).

CONTENTS

List of Abbreviations .... *ix*
List of Symbols .... *xiii*
Preface .... *xv*
Acknowledgment .... *xix*

1. **Advances in Applications of Nanotechnology to Drug Delivery: A Biopharmaceutical Perspective 1**
Carolina L. Bellera, Andrea V. Enrique, Alan Talevi and Luis E. Bruno-Blanch

2. **Radiolabeled Nanoparticles Using β^+ Radionuclides as Diagnostic Agents: An Overview and a Chemotopological Approach 25**
Nancy Y. Quintero, Isaac M. Cohen and Guillermo Restrepo

3. **Carbon-Based High Aspect Ratio Polymer Nanocomposites 85**
Thomas Gkourmpis

4. **Ultrafast Laser Dynamics on Molecular Nanomagnets 125**
Georgios Lefkidis

5. **Alternant Conjugated Organic Oligomers as Templates for Sustainable Carbon Nanotube-Based Molecular Nanowire Technologies 165**
Sergio Manzetti

6. **Designing of Some Novel Molecular Templates Suitable for Hydrogen Storage Applications: A Theoretical Approach 195**
Sukanta Mondal, Arindam Chakraborty, Sudip Pan and Pratim K. Chattaraj

7. **Computational Insights of Adsorption and Catalysis Within Nanoporous Zeolites 249**
Gang Yang

8. **Cluster Bundlet Model of Single-Wall C, BC2N and BN Nanotubes, Cones and Horns 271**
Francisco Torrens and Gloria Castellano

9. **Computational Strategies for Nonlinear Optical Properties of Carbon Nano-Systems 309**
Shabbir Muhammad and Masayoshi Nakano

10. Modern Density Functional Theory a Useful Tool for the Computational Study of Nanotechnology 333

Reinaldo Pis Diez

11. Molecular Dynamics Simulations: Applicability and Scopes in Computational Biochemistry 371

Kshatresh Dutta Dubey and Rajendra Prasad Ojha

12. Structural Information From Molecular Descriptors Based on the Monte Carlo Method: Applications to Nanoscale Structures....... 401

Dan Ciubotariu, Valentin Gogonea and Ciprian Ciubotariu

Index 451

LIST OF CONTRIBUTORS

Carolina L. Bellera
Medicinal Chemistry/Biopharmacy, Department of Biological Sciences, Faculty of Exact Sciences, National University of La Plata (UNLP) – Argentinean National Council for Scientific and Technical Research (CONICET). Tel.: 542-214235333 Ext 41. 47 and 115, La Plata (B1900AJI), Buenos Aires, Argentina

Luis E. Bruno-Blanch
Medicinal Chemistry, Department of Biological Sciences, Faculty of Exact Sciences, National University of La Plata (UNLP). Tel.: 542-214235333 Ext 41. 47 and 115, La Plata (B1900AJI), Buenos Aires, Argentina.

Gloria Castellano
Facultad de Veterinaria *y* Ciencias Experimentales, Universidad Católica de Valencia *San Vicente Mártir*, Guillem de Castro-94, E-46001 València, Spain, E-mail: gloria.castellano@ucv.es, Tel.: +34-963-544-431, Fax: +34-963-543-274

Arindam Chakraborty
Department of Chemistry and Center for Theoretical Studies, Indian Institute of Technology, Kharagpur, 721302, India

Pratim K. Chattaraj
Department of Chemistry and Center for Theoretical Studies, Indian Institute of Technology, Kharagpur, 721302, India, E-mail: pkc@chem.iitkgp.ernet.in

Ciprian Ciubotariu
Department of Computer Sciences, University "Politehnica", P-ta Victoriei No. 2, 300006, Timisoara, Romania; E-mail: cheepeero@gmx.net

Dan Ciubotariu
Department of Organic Chemistry, Faculty of Pharmacy, "Victor Babes" University of Medicine and Pharmacy, P-ta Eftimie Murgu No. 2, 300041, Timisoara, Romania; E-mail: dciubotariu@mail.dnttm.ro

Isaac Marcos Cohen
Secretaría de Ciencia, Tecnología *y* Posgrado, Facultad Regional Avellaneda, Universidad Tecnológica Nacional, Argentina; Departamento de Ingeniería Química, Facultad Regional Buenos Aires, Universidad Tecnológica Nacional, Argentina. Mailing Address: 1. Avenida Mitre 750, Avellaneda (1870) Argentina; 2. Avenida Medrano 951, Ciudad Autónoma de Buenos Aires (C1179AAQ), Argentina; Tel.: 00541142221908; 00541148677562; E-mail: marcos_cohen@yahoo.com

Reinaldo Pis Diez
CEQUINOR, Center of Inorganic Chemistry, Department of Chemistry, National University of La Plata, Argentina, E-mail: pis_diez@quimica.unlp.edu.ar

Kshatresh Dutta Dubey
Biophysics Unit, Department of Physics, DDU Gorakhpur University, India, 273009. Present address: Department of Biological Sciences and Bioengineering, Indian Institute of Technology, Kanpur, India. E-mail: kshatresh@gmail.com

Andrea V. Enrique
Medicinal Chemistry, Department of Biological Sciences, Faculty of Exact Sciences, National University of La Plata (UNLP). Tel.: 542-214235333 Ext 41. 47 and 115, La Plata (B1900AJI), Buenos Aires, Argentina.

Thomas Gkourmpis
Innovation and Technology, Borealis AB, Stenungsund SE-444–86, Sweden, E-mail: thomas.gkourmpis@borealisgroup.com

Valentin Gogonea
Department of Chemistry, Cleveland State University, 2121 Euclid Avenue, SI 422, Cleveland, OH 44115; E-mail: v.gogonea@csuohio.edu

Georgios Lefkidis
Chemistry degree (1994, Aristotle University, Thessaloniki, Greece), and PhD degree in Computational Chemistry (2002, same department). Department of Physics and Research Center OPTIMAS, University of Kaiserslautern, Box 3049, 67654 Kaiserslautern, Germany. E-mail: lefkidis@physik.uni-kl.de

Sergio Manzetti
FJORDFORSK Institute of Science and Technology, Flåm, 5743 Norway / Science for Life Laboratory, Department of Cell and Molecular Biology, University of Uppsala, Sweden. E-mail: sergio.manzetti@gmx.com

Sukanta Mondal
Department of Chemistry and Center for Theoretical Studies, Indian Institute of Technology, Kharagpur, 721302, India

Shabbir Muhammad
Department of Materials Engineering Science, Graduate School of Engineering Science, Osaka University Toyonaka, Osaka 560-8531 (Japan), Fax: +81-6-6850-6268, E-mail: shabbir@cheng.es.osaka-u.ac.jp

Masayoshi Nakano
Department of Materials Engineering Science, Graduate School of Engineering Science, Osaka University Toyonaka, Osaka 560-8531 (Japan), Fax: +81-6-6850-6268, E-mail: mnaka@cheng.es.osaka-u.ac.jp

Rajendra Prasad Ojha
Biophysics Unit, Department of Physics, DDU Gorakhpur University, India, 273009. E-mail: rp_ojha@yahoo.com

Sudip Pan
Department of Chemistry and Center for Theoretical Studies, Indian Institute of Technology, Kharagpur, 721302, India

Nancy Y. Quintero
Laboratorio de Química Teórica, Universidad de Pamplona, Pamplona, Colombia. Mailing address: Avenida 6 No 6–54, Chapinero, Cúcuta, Colombia; Tel.: 005775814176; E-mail: ytrioradiac@gmail.com

Guillermo Restrepo
Laboratorio de Química Teórica, Universidad de Pamplona, Pamplona, Colombia. Mailing address: km 1 Vía Bucaramanga,Pamplona, Colombia; Phone:005775685303; / Interdisciplinary Research Institute, Universidad de Pamplona, Bogotá, Colombia. Mailing address: km 1 Vía Bucaramanga,

Pamplona, Colombia; Tel.: 005775685303; E-mail: grestrepo@unipamplona.edu.co; guillermorestrepo@gmail.com

Alan Talevi
Medicinal Chemistry/Biopharmacy, Department of Biological Sciences, Faculty of Exact Sciences, National University of La Plata (UNLP) – Argentinean National Council for Scientific and Technical Research (CONICET). Tel.: 542-214235333 Ext 41. 47 and 115, La Plata (B1900AJI), Buenos Aires, Argentina; E-mail: atalevi@biol.unlp.edu.ar

Francisco Torrens
Institut Universitari de Ciència Molecular, Universitat de València, Edifici d'Instituts de Paterna, P. O. Box 22085, E-46071 València, Spain, E-mail: francsico.torrens@uv.es, Tel.: +34-963-544-431, Fax: +34-963-543-274

Gang Yang
College of Resources and Environment & Chongqing Key Laboratory of Soil Multi-scale Interfacial Process, Southwest University, Chongqing 400715—Engineering Research Center of Forest Biopreparation, Ministry of Education, Northeast Forestry University, Harbin 150040, P. R. China, E-mail: dicpgy@yahoo.com

LIST OF ABBREVIATIONS

AFM	Atomic Force Microscopy
CDIE	Cyclopentane-D isotope Exchange
COD	Chemical Oxygen Demand
CT	Computed Tomography
DAM	Drive-Amplitude Modulation
DFT	Density Functional Theory
ECP	Effective Core Potential
EFM	Electrostatic Force Microscopy
EPR	Enhanced Permeation and Retention
FCC	Face-Centred Cubic
FEP	Free Energy Perturbation
GGA	Generalized Gradient Approximation
GIAO	Gauge-Independent Atomic Orbital
GNRs	Graphene Nanoribbons
HDL	High Density Lipoproteins
HDX	Hydrogen-Deuterium Exchange
HEDM	High-Energy Density Materials
HOMO	Highest Occupied Molecular Orbital
HOPG	Highly Oriented Pyrolytic Graphite
INS	Inelastic Neutron Scattering
IPR	Isolated Pentagon Rule
IRS	Investigated Receptor Space
KPFM	Kelvin probe force microscopy
LDA	Local Density Approximation
LIE	Linear Interaction energy
LUMO	lowest unoccupied molecular orbital
MAPE	Mean Absolute Percentage Error
MFM	Magnetic Force Microscopy
MPS	Mononuclear Phagocyte System
MRI	Magnetic Resonance Imaging
MWNTs	Multi-Walled Nanotubes
NICS	Nucleus-Independent Chemical Shift
NLO	Nonlinear Optical
PBE	Perdew-Burke-Ernzerhof
PES	Potential Energy Surface

PET	Positron Emission Tomography
QSPR	Quantitative Structure-Property Relation
RES	Reticulo-Endothelial System
RMSD	Root-Meansquared Deviation
RMSE	Root Mean Square Error
SASA	Solvent Accessible Surface Area
SFM	Scanning Force Microscopy
SPECT	Single Photon Emission Computed Tomography
SQUID	Supercomputing Quantum Interface Device
SWNC	Single-wall C-nanocones
SWNCs	Single-Wall C-Nanocones
TFD	Thomas-Fermi-Dirac
UBA	University of Buenos Aires
VASP	Vienna *abinitio* Simulation Package
VMD	Visual Molecular Dynamics
XRPD	X-ray Powder Diffraction

PREFACE

Nanoscience is the study of atoms, molecules, and objects with size on the nanometer scale. The lower limit of this scale is set by the size of atoms, and the upper limit by the highest size that allows the existence of phenomena not observed in larger structures, which by convention is 100 nm.

The phenomena that become evident as the size of the system decreases include statistical mechanical effects and quantum mechanical effects. For this reason, nanoscale materials exhibit properties that are not present in the macroscale, for example, copper becomes transparent; aluminum turns combustible; and insoluble gold becomes soluble. This enables unique and diverse applications opening a whole new spectrum in fields such as medicine, electronics, biomaterials and energy production applications.

On the other hand, nanomaterials should be regarded as new substances since as mentioned they present completely different properties. Consequently, great care must be taken in regard of their possible toxicity and negative environmental impact.

Accordingly, it is clear that being able to predict or describe the nanoscale phenomena will assist in developing safer and enhanced nanomaterials. For this purpose, computational chemistry appears as a remarkably powerful tool. Hence, advances in computational chemistry theory and applications will directly or indirectly influence nanoscience progress.

This book provides scope for academics, researchers, and engineering professionals to present their research and development work that are relevant in nanoscience and computational chemistry. Contributions cover new methodologies and novel applications of existing ones, increasing the understanding of the behavior of new and advanced systems. In addition, the book offers innovative chapters on the growth of educational, scientific, and industrial research activities, providing a medium for mutual communication between international academia and the industry.

Chapter 1 presents an overview of different nanosystems (dendrimers, lipid-based nanoparticles, liposomes, inorganic nanoparticles) and their present and potential applications as drug delivery systems. During the last 15 years, revolutionary developments in the production of nanosized particles have triggered significant expectations for their applications in medical diagnosis and drug-delivery.

Chapter 2 exhibits the properties of β^+ radionuclides and the types of nanoparticles labeled with them, the recent advances in this field, and the application of the chemotopological methodology, as a new contribution of the study of positron emitters used in PET diagnostic imaging. Radiolabeled nanoparticles designed for specific transport of therapeutic agents to diseased tissues have received attention over the last decades because they have potential as radiotracers in the visualization of tumors and metastases.

Chapter 3 reviews the properties of fillers like graphene and carbon nanotubes in a polymer matrix and their effect on the overall nanocomposite performance, illustrating their effect on properties such as: electrical and thermal conductivity, gas and liquid permeation, thermal and dimensional stability, rheology and morphology. High aspect ratio carbon-based allotropes have attracted enormous scientific and industrial attention due to their fascinating physicochemical properties and immense potential.

Chapter 4 presents *ab initio* models to describe optical, ultrafast manipulation of magnetic clusters, showing that it is possible to drive the magnetic state of a small cluster or an extended system by manipulating the spin degree of freedom using a subpicosecond laser pulse, opening exciting possibilities of translating new fascinating physics to down-to-earth computational applications, which can pave the way to move magnetism from saving information to actively processing it as well. After the discovery of laser-induced demagnetization the optical control of the magnetic state of both extended and finite materials has become an increasing field of research.

Chapter 5 reviews specific carbon-based conductive polymers as well as the development of encapsulation and molecular insertion procedures in carbon nanotubes. A brief quantum chemical analysis of five exemplary $4n/4n+2$ oligomers for future development of encapsulated alternant conjugated organic structures as tunable carbon-based conductive nanowires is also studied. This chapter shows the intriguing combination of properties of nanomaterials, the principles of organic oligomer chemistry and the method of computational quantum chemistry to provide and stimulate the knowledge of new conductive and green materials.

Chapter 6 exhibits the modeling based on the theoretical paradigm of conceptual density functional theory and its various reactivity variants of new molecular networks and aggregates, one of the most "sought after" topics in current chemical research. The utility of these molecular materials as plausible storage templates for hydrogen gas is also examined. *Ab initio* as well as classical molecular dynamics simulations are also carried out for few systems to assess their bulk properties as well as hydrogen trapping potentials.

Chapter 7 presents a series of complicated adsorption and reaction mechanisms within zeolites using computational tools. Zeolites are an important type of nanoscale materials with an ever-increasing application in a variety of fields such as separation, adsorption and heterogenous catalysis. The shown computational findings greatly help understanding the adsorption and reactivity performances within nanoporous zeolites.

Chapter 8 shows a theoretical bundlet model that describes the distribution function by size of the cluster form in organic solvents of single-wall C-nanocones (SWNCs), nanohorns (SWNHs), and BC_2N/boron nitride (BN) analogs.

Chapter 9 presents a relatively less explored feature of carbon nanosystems, which is their possible use in nonlinear optical applications. Carbon nanomaterials including zero-dimensional fullerenes, one-dimensional carbon nanotunes, and two-dimensional graphene sheets are promising to revolutionize several fields of materials sciences and are major components of nanotecnology.

Chapter 10 shows the formal evolution of the density functional theory from ancient local models to the modern hybrid meta-generalized gradient approximations and double hybrid generalized gradient approximations. In addition, the prediction of a variety of properties, such as geometries, energies and thermodynamic functions, as well as in the interpretation of spectroscopic data, using current density functional theory is explored.

Chapter 11 shows the applicability and major case studies of molecular dynamics simulations, which provide time dependent microscopic properties of biomolecules that could not be explained by experimental methods such as X-ray crystallography. In addition, an analysis is performed of the pros and cons of free energy pathways such as MM-PB (GB)/SA, LIE and other alchemical methods, frequently used to study the binding mode of several biomolecular complexes.

Chapter 12 presents molecular descriptors developed at molecular nanoscale dimensions calculated by Monte Carlo method, describing the fundamentals of the method and its general application as a stochastic integration technique for objects in an N-dimensional space. In addition, the simulation of the copolymerization reactions is showed.

— Dr. Andrew G. Mercader

INIFTA (CONICET-UNLP), Argentina

ACKNOWLEDGMENT

The editors, *Dr. Andrew G. Mercader* and *Dr. Eduardo A. Castro*, want to thank *National Research Council of Argentina (CONICET)* for their financial support.

CHAPTER 1

ADVANCES IN APPLICATIONS OF NANOTECHNOLOGY TO DRUG DELIVERY: A BIOPHARMACEUTICAL PERSPECTIVE

CAROLINA L. BELLERA, ANDREA V. ENRIQUE, ALAN TALEVI, and LUIS E. BRUNO-BLANCH

CONTENTS

Abstract ... 2
1.1 Introduction ... 2
1.2 Dendrimers ... 4
1.3 Lipid-Based Nanoparticles ... 8
1.4 Inorganic Nanosystems ... 10
 1.4.1 Gold Nanoparticles ... 10
 1.4.2 Superparamagnetic Nanoparticles ... 12
 1.4.3 Porous Inorganic Nanoparticles ... 13
1.5 Some Biopharmaceutical Considerations ... 13
 1.5.1 The Mononuclear Phagocyte System (MPS)—PEGylation ... 13
 1.5.2 The Enhanced Permeation and Retention (EPR) Effect—Size-Restriction Based Selective Distribution ... 14
 1.5.3 Alternative Routes of Administration ... 15
 1.5.4 Circumvention of ABC Transporters ... 16
 1.5.5 Active Targeting ... 17
1.6 Nanotoxicology ... 18
1.7 Conclusions ... 19
1.8 Acknowledgments and Conflict of Interest Statement ... 20
Keywords ... 20
References ... 20

ABSTRACT

During the last 15 years, revolutionary developments in the production of nano-sized particles have triggered substantial expectations for their applications in medical diagnosis and drug-delivery. This chapter will present a comprehensive overview of different nanosystems (dendrimers, lipid-based nanoparticles, liposomes, inorganic nanoparticles) and their present and potential applications as drug delivery systems. Recent reports on the advantages of those systems in relation to correction of suboptimal pharmacokinetics and distribution and location of therapeutic agents to specific cells or tissues are discussed. A final discussion on safety concerns regarding their use as pharmaceuticals is also presented.

1.1 INTRODUCTION

Multiple dose regimes are the most common drug-based therapeutic interventions. Leaving aside topical medications, conventional drug delivery systems rely on establishing a dynamic equilibrium or, more precisely, a *pseudoequilibrium* between the free drug plasmatic concentration and the free drug concentrations in all the other body tissues (let us remember that living organisms are open systems; thus, a true equilibrium is seldomly achieved due to the permanent mass exchange with the environment). After a number of doses are administered, a steady state is reached, during which plasmatic concentration will fluctuate between practically fixed maximal and minimal steady state concentrations, as long as the treatment continues. Since only the free, unbound drug can interact with its molecular target, the free drug levels at the vicinity of the site of action (generally) determine the extent of the pharmacological response (Smith et al., 2010). A non-trivial implication of the former approach is that, to attain effective concentrations of an active ingredient in its biophase or site of action, the patient is subjected to systemic exposure to the drug, which often leads to off-target related undesirable side effects. In other words, conventional drug delivery systems are characterized by nonspecific distribution: to attain therapeutic levels in the biophase, otherwise unneeded levels are accepted in the rest of the body. Therefore, patients undergoing drug therapy are exposed to: (a) large drug quantities and; (b) off-target drug effects, which could be ameliorated or avoided if targeted drug delivery systems were used. Patients receiving anticancer treatment constitute a very illustrative example of the consequences of the previous setting: the well-recognized adverse reactions to chemotherapy emerge from interactions between the drug molecules and noncancerous healthy cells.

On the other hand, back in the 1990s many drug discovery projects failed (i.e., came to an end prematurely without bringing and innovative medication to the pharmaceutical market) due to pharmacokinetic issues (which accounted for nearly 40% of failures) (Kubinyi, 2003; Schuster et al., 2005). Even though those figures have radically improved (Kola and Landis, 2003) by early recognition and weeding out of drug candidates with unfavorable disposition profiles –the "fail early fail cheap" paradigm-the previous facts reveal that a number of active ingredients cannot be fully exploited due to pharmacokinetic issues. For example, a given drug, due to its physicochemical properties, might be non-compatible with certain administration routes. Frequently, drugs with scarce aqueous solubility cannot be formulated as intravenous solution. Drugs with poor gastrointestinal absorption, high first pass metabolism or low chemical stability in the gastrointestinal media often preclude oral administration. A well-known example of this problem is insulin: due to its peptidic nature, it is readily denatured and digested in the gut; consequently, people suffering from type I diabetes are currently condemned to daily subcutaneous insulin injections. Drugs with short half-life present difficulties to built up and sustain effective levels (reducing the duration of the pharmacological effect or requiring large doses just to compensate metabolism). Specialized barrier tissues often prevent drugs from reaching particularly sensitive tissues or organs; this is especially relevant in the field of central nervous system drug development, where drugs are meant to cross the highly selective blood-brain barrier (Talevi and Bruno-Blanch, 2012; Talevi et al., 2012). Finally, interaction of the free drug with efflux transporters from the ABC superfamily (e.g., P-glycoprotein, Breast Cancer Resistance Protein) results in reduced bioavailability and is linked to multidrug resistance issues in a number of disorders such as epilepsy and cancer (Sharom, 2008; Szakàcs et al., 2006; Talevi and Bruno-Blanch, 2012).

Summarizing, we have so far described disposition and safety issues related to: unfavorable physicochemical properties (poor solubility, poor permeability); unwanted interactions with the biological systems (enzymes, transporters) and; high off-target drug levels due to nontargeted distribution. These problems, which occur regularly with traditional, small-molecule therapies, are even more frequent in the case of emerging therapies; particularly, biotherapies (gene therapy, therapeutic proteins), which owing to their intrinsic nature as complex macromolecules are far more susceptible to enzymatic cleavage, inactivation, poor permeability, slow distribution and immunogenicity (Müller and Keck, 2004; Nishikawa et al., 2005; Parra-Guillén et al., 2010; Torchilin, 2008; Vugmeyster et al., 2012). As a result, the development of adequate delivery vectors is a key issue in the field of advanced biotherapeutics. While traditional delivery systems only deal with release and absorption of the therapeutical entity, with no direct involvement on the modulation of distribution and elimination processes, advanced delivery systems should be able to retain their integrity throughout the drug distribution events, while permeating through different epithelia and endothelia, and selectively releasing the drug in the proximity of the drug target.

In this background, an ideal drug delivery device should: (a) compensate unfavorable physicochemical properties; (b) encapsulate, entrap or adsorb drug molecules; (c) conceal the drug from enzymatic cleavage, undesired biotransformations and efflux transporters recognition; (d) extravasate; (e) direct the drug to its therapeutic target; in the case of intracellular targets, promote cell uptake; (f) once in the vicinity of the target (and not before), release the drug cargo in a controlled manner; (g) present no toxicity nor accumulation within the body, preferentially, be biodegradable. Some years ago, in the light of traditional pharmaceutical technologies, a device, which gathered such a wide range of features, would have been inconceivable. Today, however, burgeoning advances on nanobiotechnology have brought us nearer and nearer to our dream delivery system. Nanosystems are currently produced from a wide diversity of materials (and materials combinations), in a wide range of tailored morphologies and sizes, and with a broad spectrum of surface coatings and functionalizations.

In this chapter, we will overview current knowledge on nano-scale drug carriers from a biopharmaceutical perspective. We will not provide details on preparation/synthetic procedures. We have focused on nonlinear polymers (dendrimers), lipidic nanosystems (liposomes, solid lipid nanoparticles (SLN), and nanostructured lipid carriers (NLC)), inorganic nanoparticles (particularly, mesoporous silica nanosystems, gold nanoparticles and superparamagnetic systems). We will present a brief outline of such devices, a separate section dedicated to general knowledge on the biopharmaceutical aspects of these nanosystems, and a final section on safety concerns related to the clinical applications of nanodevices. We would like to note that to the purposes of a comprehensive review, the profusion and exponential growth of knowledge and literature reports on nanocarriers precludes the possibility of an exhaustive review on all existent systems and advances. We will then synthetically focus on the previously listed carriers from a biopharmaceutic perspective, providing a few examples of each discussed topic.

1.2 DENDRIMERS

Dendrimers are regularly branched macromolecules with unique structural and topological features whose properties are attracting considerable interest from both scientists and technologists (Tomalia et al., 1985; Tomalia, 2005). The term dendrimer was first proposed by Tomalia in 1985. It comes from the Greek *dendron*, which means tree. They differ from traditional polymers in that they have a multibranched, 3D architecture with very low polydispersity and high functionality. A typical dendrimer comprises three different topological parts, which are: (a) a focal core; (b) building blocks with several interior layers composed of repeating units and; (c) multiple peripheral functional groups.

The first part, the focal core, can encapsulate various chemical species that exhibit unparalleled properties due to the special nano environment surrounded by extensive dendritic branching. The several interior layers, composed of repeating units, can provide a flexible space created within the voids of dendritic building blocks, which may encapsulate various small guest molecules. At last, the third part of a dendrimer is the multivalent surface, which can accommodate a large number of functionalities that can interact with the external environment, thereby defining the dendrimer's macroscopic properties (Bosman et al., 1999).

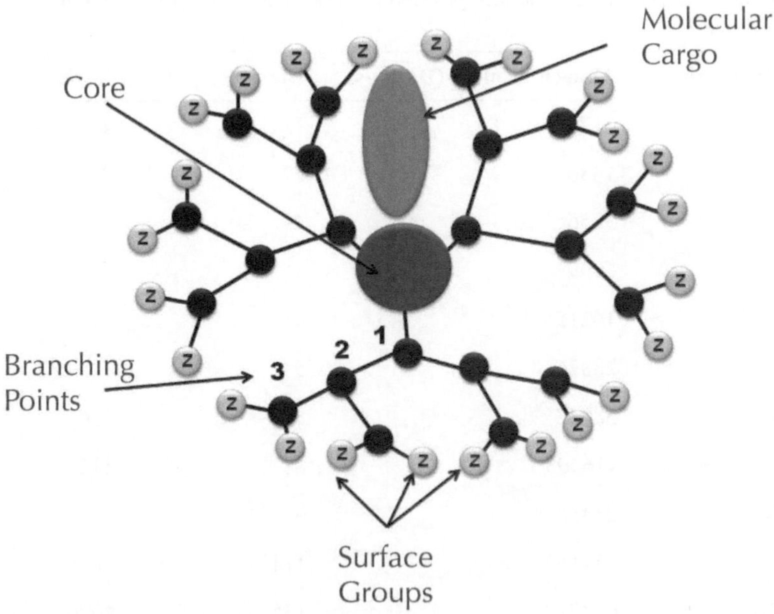

FIGURE 1 Schematic representation of a 3-generation dendrimer.

The dendrimeric structure (Fig. 1) is characterized by layers called generations (branching cycles). The number of generations corresponds to the number of focal, cascade or branching points. A fifth generation dendrimer presents five focal points between the core and the surface. The core is often called generation zero (G0). For example, in Propylene Imine (PPI) dendrimers, the core molecule is 1,4-diaminebutane; in Polyamidoamine (PAMAM) dendrimers, the initiator core is either ammonia or 1,2-etilendiamine. Dendrimers design can be based on a great diversity of functional groups, such as polyamines as in the case of PPI (Buhleier and Vögtle, 1978), a combination of amines and polyamides as in the case of PAMAM (Tomalia et al., 1985) or more hydrophobic Poly (aryl ether) dendrimers (Hawker et al., 1993). Depending on the peripheral functional groups, dendrimers can be either neutral or charged. Their physical properties vary in a regular way depending on the number of generations. The diameter of dendritic molecules increases linearly with the number of generations (Duncan and Izzo, 2005); the number of terminal groups duplicates with each generation (Klajnert and Bryszewska, 2001). Table 1 presents the relationship between the number of generations, the molecular mass, the diameter and the number of terminal groups for PAMAM dendrimers (Sampathkumar and Yarema, 2007).

TABLE 1 Variation of structural properties with the number of generations of PAMAM dendrimers.

Generation	Molecular mass (Daltons)	Diameter (A)	Surface groups
G0	517	15	4
G1	1430	22	8
G2	3256	29	16
G3	6909	36	32
G4	14215	45	64
G5	28826	54	128
G6	58048	67	256
G7	116493	81	512
G8	233383	97	1024
G9	467162	114	2048
G10	934720	135	4096

Dendrimers are mainly synthesized by means of two different approaches: divergent and convergent synthesis (Fig. 2). In divergent synthesis, the dendrimer grows from the initiator core to the surface in a stepwise fashion by iterative addition of monomer units. The high number of chemical reactions that ought to be performed on a single molecule with multiple equivalent reaction sites requires very high yielding per reaction to avoid nonideal growth events; even considering an average selectivity of 99.5% per reaction, a fifth generation PPI dendrimer will result in only 29% perfect dendrimers (Bosman et al., 1999). What is more, products are difficult to purify from structurally similar byproducts (Medina and El-Sayed, 2009). In contrast, convergent synthesis begins with the dendrimer surface units coupled to additional building blocks to from the branching structure, thus constructing dendrons from the periphery to the central, multifunctional focal point (Hawker and Fréchet, 1990; Medina and El-Sayed, 2009). Convergent reactions are easy to purify since dendrons are substantially different from the reaction byproducts, thus eliminating the need of highly efficient reactions. The convergent approach leads to improve monodispersity. Double-staged approaches coupling dendrons prepared by convergent synthesis to a *hypercore* or prefabricated lower-generation dendrimer (Wooley et al., 1991), combined divergent-convergent approaches (Kawaguchi et al., 1995) and Click chemistry (Kolb et al., 2001) have also been used for dendrimer preparation.

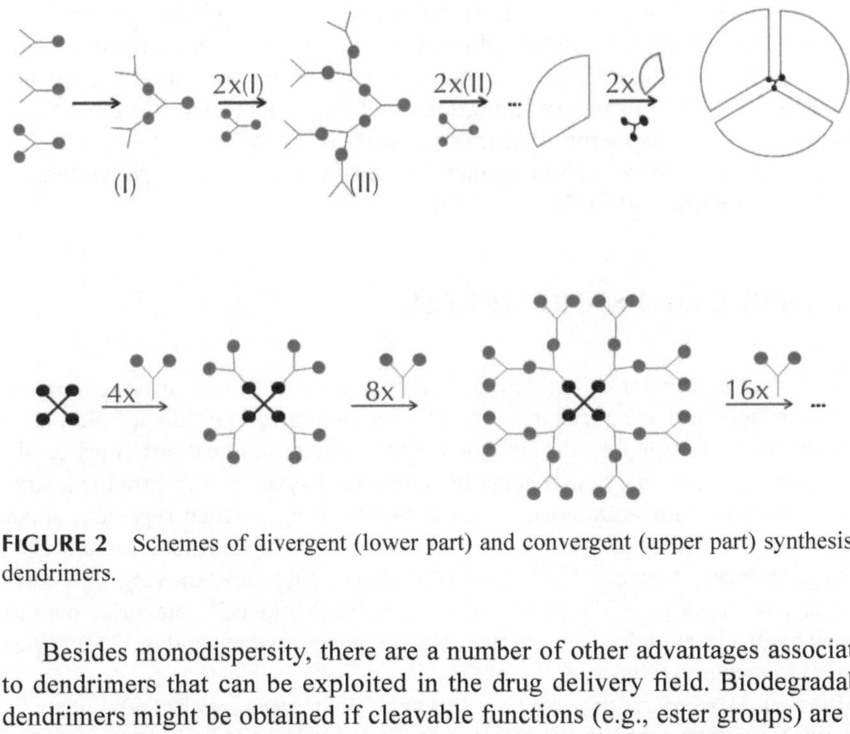

FIGURE 2 Schemes of divergent (lower part) and convergent (upper part) synthesis of dendrimers.

Besides monodispersity, there are a number of other advantages associated to dendrimers that can be exploited in the drug delivery field. Biodegradable dendrimers might be obtained if cleavable functions (e.g., ester groups) are included in the polymer backbone. What is more, degradation kinetics might be controlled by adjusting the nature of the chemical bond connecting the monomer units, the hydrophobicity of the monomer units (hydrophilic monomers result in faster degradation), the size of the dendrimer (larger dendrimers determine slower degradation due to tight packing of their surface) and the cleavage susceptibility of the peripheral and internal dendrimer structure (Medina and El-Sayed, 2009). The large number of dendrimers' surface groups and the versatility in their chemical structures allow the conjugation of different drugs, imaging agents, and/or targeting ligands while maintaining the dendrimer's compact spherical geometry in solution (Medina and El-Sayed, 2009). Asymmetric dendrimers might be prepared by coupling dendrons of different generations to a linear core, leading to "bow-tie" polyester dendrimers; the asymmetry allows for tunable structures and molecular weights, control on the number of functional groups and improved versatility in relation to attachment of diverse drugs, imaging agents and targeting moieties. Finally, the presence of many polar termini redounds in high solubility; through entrapment of guest molecules in dendrimer's voids increased solubility of poorly soluble drugs may be achieved (Morgan et al., 2003, 2006). Inclusion of nonpolar drugs might be accomplished by simple mixing of the polymer and drug solutions, where the hydrophobic drug associates with nonpolar core by hydrophobic interactions. The high specific surface and the spherical geometry confer dendrimers low intrinsic viscosity (compared to linear polymers) and high reactivity (Klajnert and Bryszewska, 2001). Among the limitations of these systems, we might highlight high production costs owing to multistep synthesis (Klajnert and Bryszewska, 2001), and difficulties to achieve controlled, sustained release in physiological conditions (Medina and El-Sayed, 2009).

1.3 LIPID-BASED NANOPARTICLES

An important parameter of the delivery vehicle pertains to low or no toxicity of the carrier itself either *in vivo* or in the environment as a byproduct; lipid-based nanoparticles are probably the least toxic for clinical applications (Puri et al., 2009). The hydrophobic constituents of lipid-based systems also provide a suitable environment for entrapment of hydrophobic drugs, which represent about 40% of newly developed drugs (Martins et al., 2007). Furthermore, around 10% of drugs approved between 1937–1997 (Proudfoot, 2005) present very high octanol-water partition coefficients and some specific therapeutic categories tend to present high lipophilicity (e.g., antineoplastic, antiparasitary and antihyperlipoproteinimic agents) (Ghose et al., 1999).

Liposomes were the first advanced drug vehicle ever proposed (Bangham et al., 1965) and thus probably the most extensively studied. As a matter of fact, a number of liposome formulations are already in clinical use (e.g., AmBisome®, liposomal amphotericin B, and Doxil®, PEGylated liposomes loaded with doxorubicin) and many others are waiting clinical trials outcomes.

They are spherical vesicles defined by a lipidic bilayer membrane structure, composed by natural or synthetic amphipathic lipids (typically, phospholipids), which encloses an inner aqueous space (Fig. 3). The former structure determines the ability of the liposome to load both aqueous soluble drugs (in their cavity) and lipophilic drugs (inside the bilayer), though the encapsulation efficiency in relation to hydrophobic drugs is rather low. They do not constitute, necessarily, nanosystems, since their size can vary from 100 nm to the low micrometer range; in fact, they can even be used to entrap other carriers (liposome-nanoparticles hybrids) (Al-Jamal and Kostarelos, 2011). Liposomes have, nevertheless, some important drawbacks beside their low capacity to encapsulate non lipophilic drugs: they are manufactured through processes that involve organic solvents, they are unstable in biological fluids and, more generally, in aqueous solutions (they cannot be commercialized as such), and they present poor batch-to-batch reproducibility and difficulties in sterilization (Beija et al., 2012; Huynh et al., 2009).

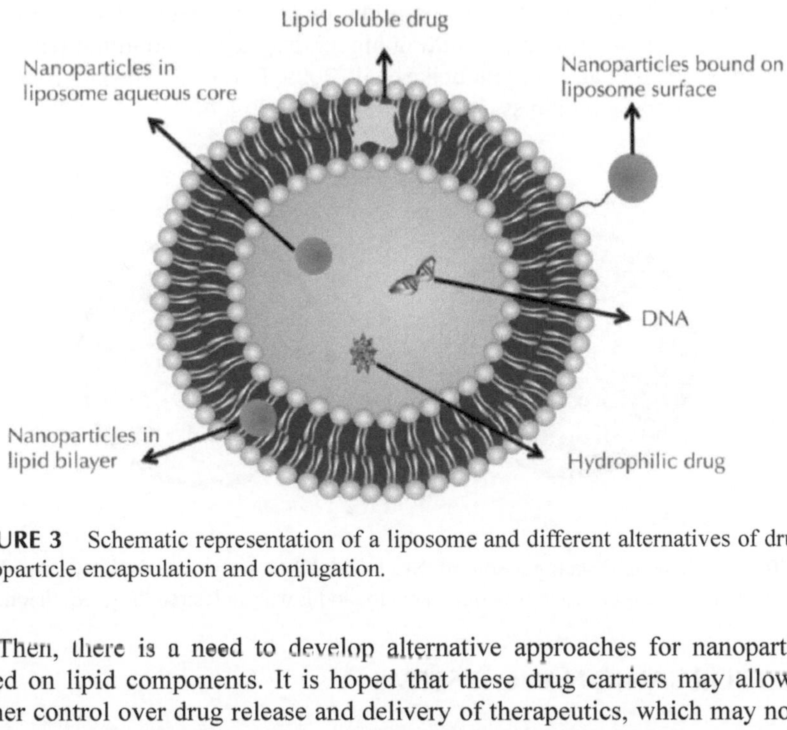

FIGURE 3 Schematic representation of a liposome and different alternatives of drug or nanoparticle encapsulation and conjugation.

Then, there is a need to develop alternative approaches for nanoparticles based on lipid components. It is hoped that these drug carriers may allow for higher control over drug release and delivery of therapeutics, which may not efficiently load into liposomes. SLN were developed at the beginning of the 1990s, by replacing the liquid lipid (oil) of an oil-in-water nanoemulsion by a solid lipid or a blend of solid lipids (Lucks and Müller, 1991). The use of solid lipids instead of oils in the emulsions preparation is a very attractive idea to achieve controlled drug release, since the drug molecule mobility in a solid matrix should be, intuitively, considerably lower compared with an oily phase (Martins et al., 2007). Moreover, large-scale manufacturing of SLN is possible (while polymeric nanoparticles have faced scaling-up issues), and solvent use can be avoided using high-pressure homogenization with extant machinery (Martins et al., 2007; Müller and Keck, 2004; Puri et al., 2009). There are, however, a couple of limitations to the use of SLN. Their loading capacity is limited because of the formation of highly ordered, perfect lipid crystal matrix (Wissing et al., 2004). After preparation, at least some of the particles crystallize in higher energy modifications that, during storage, evolve to a lower energy, more ordered modification that reduces the number of imperfections in the crystal lattice and leads to drug expulsion (Puri et al., 2009). Alternative colloidal structures may be formed in the aqueous dispersion (e.g., micelles) (Martins et al., 2007). These drawbacks have been solved by second-generation lipid nanoparticles, NLC, by blending the solid lipids with oils at certain ratios (preferably, 70:30 to 99:9:0.1). Even though the blends still present solid state at body temperature, the solid lipid matrix is less ordered compared with SLN or even amorphous, admitting higher drug loads and minimizing drug expulsion (Martins et al. 2007; Pardeike et al., 2009; Puri et al., 2009). An illustrative scheme comparing both systems is presented in Fig. 4.

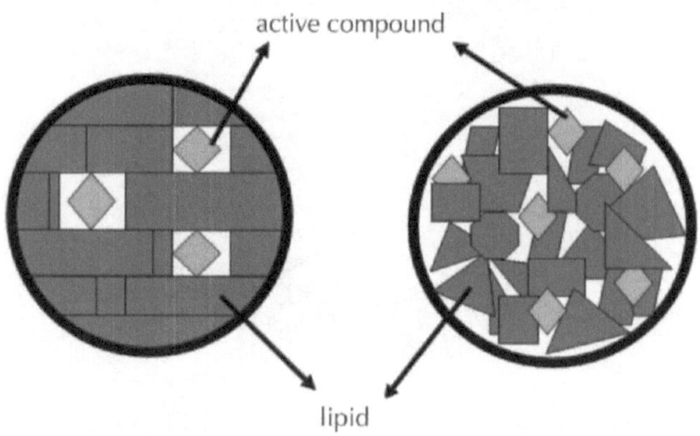

FIGURE 4 Schematic comparison of SLN and NLC structures. NLC less ordered structure results in higher encapsulation capacity and lower undesired drug expulsion.

1.4 INORGANIC NANOSYSTEMS

In this section we will overview three kind of inorganic nanosystems especially interesting in the field of drug delivery: gold nanocarriers, magnetic nanoparticles and mesoporous silica nanoparticles.

1.4.1 GOLD NANOPARTICLES

Gold nanoparticles have attracted enormous attention owing to their apparent inert nature combined with fascinating optical properties, which create a great number of potential applications in diagnostics and therapeutics.

Metals can be represented as confined plasma of positive ions and mobile conduction electrons. An interacting electromagnetic field (e.g., light) can induce a coherent oscillation against the restoring force of positive nuclei. The collective oscillations of conduction electrons upon excitation with electromagnetic radiations are known as plasmons, which give rise to a strong absorption band attributed to resonance (*surface plasmon resonance*) between the oscillating electrons and the incident radiation. Plasmon energy for bulk gold is 9 eV and it lies in the UV range; however, when conductive nanoparticles whose size is smaller than the wavelength of light are irradiated, a local surface plasmon resonance phenomenon that lies in the UV-vis range is observed (Hutter and Fendler, 2004; Kalele et al., 2006; Petryayeva and Krul, 2011; Zeng et al., 2011). Equally important, the frequency and intensity of the surface plasmon absorption bands are highly sensitive to nanoparticles composition, sizes, size-distribution, geometry, morphology, surface coating, environment and separation distance. In other words, optical properties of metallic nanoparticles are tunable. Even more, through manipulation of those variables a band shift to the infrared region can be achieved. This is interesting since biological tissues are transparent to this range of the electromagnetic spectrum (Halas, 2005). For example, Salgueiriño-Maceira et al. (Salgueiriño-Maceira et al., 2003) have constructed submicron polystirene colloids covered by multiple layers of silica-coated gold nanoparticles ($Au@SiO_2$). Thinner silica shells are linked to a red-shift compared to thicker silica shells, and for the coatings composed of Au@SiO2 nanoparticles with thin silica shells (around 8 nm), the peak position of the surface plasmon band systematically red-shifts with the number of deposited layers. When the silica shell is thick enough (larger than ca. 15 nm), the plasmon band position does not depend on the shell thickness any longer, since dipole-dipole coupling between the gold cores is fully suppressed. Similar tunable red shifts can be produced in the case of gold nanorods and gold nanoshells (Pissuwan et al., 2011; Sekhon and Kamboj, 2010). The unique and customized tailored optical properties make these systems ideal candidates for phototermal therapies (light-induced plasmonic heating), photo-triggered controlled drug release and imaging purposes, as well as integrative diagnostics and therapeutics (theranostics).

There are some other properties that turn inert metal nanoparticles attractive: unlike fluorescent materials, they experience no photobleaching; thiol-gold association allows for ready functionalization (Petryayeva and Krull, 2011); their aqueous synthesis is well established (they can be easily prepared by reducing their salts in aqueous solutions) (Kalele et al., 2006) and; they can be easily adjusted to a desirable size between 0.8 and 200 nm (Xu et al., 2006). However, they are probably, among all nanosystems, the ones that raise more safety concerns (see the specific section on Nanotoxicology).

1.4.2 SUPERPARAMAGNETIC NANOPARTICLES

Superparamagnetism is a form of magnetism that appears in small enough ferromagnetic and ferrigmagnetic nanoparticles. In such small systems, the magnetic moment of the nanoparticles is free to fluctuate in response to thermal energy, while the individual atomic moments maintain their ordered state (Colombo et al. 2012; Xu and Sun, 2009). In the absence of external magnetic field, if time used to measure magnetization is much longer than the *Néel relaxation time*, their magnetization appears to be in average zero. Under a magnetic field, however, they exhibit a magnetic signal far exceeding that of biomolecules and cells (Fig. 5). This allows identification with magnetic sensing devices (e.g., MRI scanners). Furthermore, under an alternating magnetic field, the magnetization of the superparamagnetic nanoparticles can be switched back and forth (due to either rotation of the particle or to Néel relaxation) turning the nanoparticles into local heaters (magnetic fluid hyperthermia), which seems auspicious for cancer treatment. Evidently, targeted delivery through application of an external, high-gradient magnetic field has also been proposed (Xu et al., 2006). Even though metallic magnetic nanoparticles (Fe, Co, CoFe) are ideal candidates for highly sensitive applications owing to their high magnetization value, there are also prone to fast oxidation; thus, more chemically stable magnetic oxides such as magnetite (Fe_3O_4) are often preferred for biomedical applications (Xu and Sun, 2009). Alternatively, high moment stable superparamagnetic nanoparticles can be obtained by coating Fe cores with a Fe_3O_4 shells (Peng et al., 2006) or by embedding FeCo in a carbon matrix (Seo et al., 2006). Bifunctional plasmonic and superparamagnetic systems such as dumb bell-like Au-Fe_3O_4 have also been designed (Yu et al., 2005).

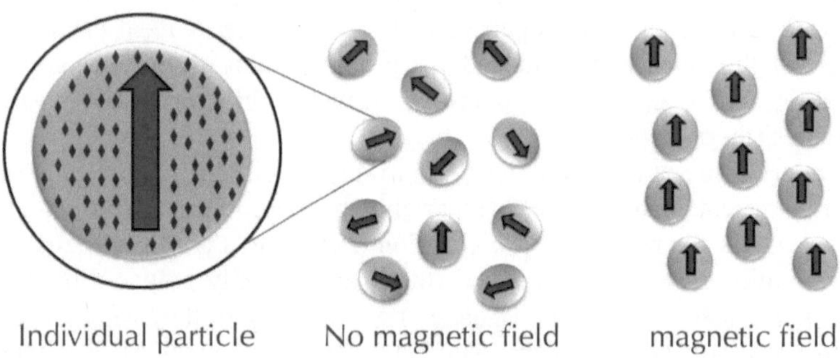

FIGURE 5 Schematic representation of the alignment of the magnetic moment of individual superparamagnetic nanoparticles in the presence of an external magnetic field.

1.4.3 POROUS INORGANIC NANOPARTICLES

Mesoporous silica nanoparticles have attracted a lot of attention for controlled delivery applications. This can be explained by considering their particular properties (Kalele et al., 2006; Liong et al., 2008; Slowing et al., 2008): they are chemically and physically stable (they are very stable against coagulation), and they present a rigid framework; they show very uniform, tunable pore size (2–6 nm), which allows loading different drugs and studying the release kinetics with high precision; they have an enormous specific surface (up to more than 1000 $m^{2/}/g$) and large pore volume (more than 0.9 cm^3/g), allowing high payloads; they present two functional surfaces (internal and external surfaces); they can be readily functionalized via silylation and; they present a unique porous structure with no interconnectivity between them that minimizes the chance of premature drug release even in the case of nonperfect capping (in other systems such as dendrimers pore encapsulated guest molecules can leak through the interconnected porous matrix when some of the pores are not capped). The major pitfalls of these kind of systems is their polydispersity and amorphous nature (a significant drawback to control their mass-transfer properties) and concerns regarding their long-term biocompatibility (a worry that is common to other inorganic nanosystems), which has not been fully studied yet.

1.5 SOME BIOPHARMACEUTICAL CONSIDERATIONS

1.5.1 THE MONONUCLEAR PHAGOCYTE SYSTEM (MPS)—PEGYLATION.

One of the most challenging problems regarding the use of nanocarriers as drug delivery systems is their rapid clearance by the mononuclear phagocyte system (MPS) (Chavanpatil et al., 2006), a part of the immune system consisting of phagocytic cells located mainly in the lymph nodes, the spleen and the liver. The clearance of nanoparticles after intravenous administration can occur already within 5 min (Wohlfart et al., 2012). It has long been recognized that surface properties and size are the major determinants of clearance and distribution of colloidal particles (Storm et al., 1995). Ancient Egyptians already known that ink colloidal particles could be stabilized by adding natural polymers. Similarly, surface attachment of surfactants and natural or synthetic polymers can be used to increase the blood circulation time. Unmodified nanoparticles rapidly adsorb plasma proteins – mainly opsonins - and as a consequence they are promptly removed from the bloodstream by the reticulo-endothelial system (RES) (Wohlfart et al., 2012). Coating the nanoparticles with hydrophilic polymers or surfactants provides an aqueous shield around the nanoparticle thus decreasing the extent of opsonization and the subsequent recognition by MPS. Poly(ethyleneglycol) (PEG)-grafted nanocarriers have been extensively used in the pharmaceutical industry for this purpose. What is more, it has been observed that PEGylation can also improve brain uptake (Calvo et al., 2001; Zara et al., 2002). The effect of particle size on *in vivo* distribution has also been studied (Yadav et al, 2011), suggesting that particle size below 100 nm tends to increase circulation lifetime. However, a number of limitations to the use of PEG have also been described (Gomes Da Silva et al., 2012). Several reports indicate that upon subsequent administrations of PEGylated systems, an immune response could be elicited, leasing to rapid blood clearance and undesirable side effects that hamper their clinical utility; this immune response includes the production of antiPEG antibodies (Dams et al., 2000; Ishida et al., 2006). PEG also hinders the cellular uptake of the colloidal entities (Gomes Da Silva et al., 2012). This later issue can be possibly overcome by the development of cleavable, stimuli responsive PEG-derivatized nanoparticles (Nie et al., 2011; Takae et al., 2008; Matsumoto et al., 2009).

1.5.2 THE ENHANCED PERMEATION AND RETENTION (EPR) EFFECT—SIZE-RESTRICTION BASED SELECTIVE DISTRIBUTION

When tumor cells multiply, angiogenesis is induced in order to cater the ever-increasing nutrition and oxygen demands of a growing tumor. This neovasculature differs greatly from that of normal tissues: the blood vessels are irregular in shape, dilated, leaky or defective; the endothelial cells are poorly aligned, with large fenestrations (Iyer et al., 2006). This determines an augmented upper limit of the size of therapeutic agents that can traverse tumor vessels. The cutoff size of tumors blood capillaries has been reported in the range from 380 to 780 nm (Hobbs et al., 1998; Unezaki et al., 1996; Yuan et al. 1995). What is more, slow venous return and poor lymphatic clearance redound in increased retention of macromolecules and colloids within the tumor. This phenomenon is known as the EPR. The differences in size-restriction of normal tissue compared to cancerous tissue have been exploited for the passive targeting of tumor cells, to increase antitumor efficacy and reduce systemic side effects. A similar passive targeting could also be conceived for other disorders that include chronic inflammation (and thus, enhanced permeability) of the affected organs or tissue (e.g., rheumatoid arthritis).

As an illustrative example, Singh et al. recently disclosed an ingenious method to avoid hair loss during paclitaxel treatment (Singh et al., 2012). The authors propose a passive targeting approach by taking advantage of the EPR effect. To this purpose, paclitaxel-loaded human serum albumin nanocapsules are prepared by means of the double emulsion method and the size distribution of the resulting nanocarriers is carefully assessed. The general idea is to achieve such a size distribution that 90% of the particles have a particle size of less than 450 nm, 10% of the particles have a particle size equal or lower than 80 nm and around 50% of the particles are about 200 nm. This size distribution responds to two purposes: limiting removal by the RES and preventing permeation from normal blood capillaries to skin (and hence hair roots) while allowing extravasation through tumor leaky vessels. When assessing tumor retentiveness and leakiness behavior of the NPs administered through intratumor route to ICRC mice carrying spontaneous mammary tumors, the authors found that two particular embodiments of the invention presented a paclitaxel tumor plasma ratio of 71.6 and 355.7, while a commercially available albumin bound paclitaxel achieved only 19.96. The performance of the proposed formulations in a murine model to study chemotherapy-induced alopecia was also assessed, finding the proposed strategy had been successful to reduce paclitaxel-associated hair loss.

There are other cutoff values regarding particle size that could be useful for passive targeting purposes. Chithrani and Chan studied the kinetics of cellular uptake of transferring-coated gold nanoparticles depending on size and shape of the particles (Chithrani et al., 2006; Chithrani and Chan, 2007). They observed spherical 50 nm particles are uptaken faster than smaller or larger ones, which relates to the balance between the endocytosis and exocytosis rates. Boyoglu et al. studied the impact of gold nanoparticles size on intracellular localization in HEp-2 cells, finding that particles below 26 nm entered the nucleus while 50 nm ones accumulated around it (Boyoglu et al., 2011). Studies like this are particularly interesting to the end of gene therapy, where a polynucleotide often needs to enter nuclei to produce the desired effect. One should be aware, anyway, that this type of cutoff values also depends on other properties such as surface characteristics and materials of the nanoparticles.

1.5.3 ALTERNATIVE ROUTES OF ADMINISTRATION

The possibility of encapsulation or grafting of a given drug to a nanocarrier of completely different physicochemical properties expands the formulation and administration opportunities. For example, insulin loaded SLN were recently proposed for pulmonary delivery of insulin as an alternative to subcutaneous injection (Liu et al., 2008). The lipid-based nanocarriers were administered to diabetic rats, finding that the insulin-SLN produced around 22% relative bioavailability using subcutaneous injection as a reference. Although evidently the formulation has to be optimized, a pulmonary efficient insulin medication would be a milestone in the treatment of diabetes. The work of Manjunath and Venkateswarlu (Manjunath and

Venkateswarlu, 2005) constitutes another good example. The authors proposed the formulation of clozapine (an antipsychotic drug with poor—below 30%—oral bioavailability due to significant first pass hepatic metabolism) as SLN in order to favor intestinal lymphatic transport (a quite unexplored absorption route) and evade portal circulation. A remarkable 91.8% relative bioavailability (compared to intravenous administration) and a significant increase in the mean residence time were achieved. Similarly, Feng et al. developed poly(lactic-coglycolic acid) (PLGA) nanoparticles, poly(lactide_)-vitamin E TPGS (PLA-TPGS) nanoparticles and PLA-TPGS montmorillonite (PLA-TPGS/MMT) nanoparticles to accomplish efficient oral delivery of anticancer agent doctaxel, avoiding first-pass extraction and Pgp efflux transport (Feng et al., 2009). The authors compared the bioavailability (in rats) of the nanoparticulated systems to that of intravenous administration. They concluded that there were some advantages linked to the nanoparticles. First, intravenous administration caused extremely high plasma concentrations, way above the maximum tolerated level[1] (thus, it would cause undesired side-effects), whereas all nanoparticle formulations showed a C_{max} within the therapeutic window. Second, the half-life of the drug was expanded from 4.5 h for intravenous administration to an astonishing half-life of 118.8 h for oral administration of the PLA-TPGS/MMT formulation. One dose of the best-performing nanoparticle formulations provided a plasma level within the therapeutic range for as long as 500 h. Finally, the oral bioavailability of the PLA-TPGS nanoparticles formulation was 91.3% compared to intravenous administration.

1.5.4 CIRCUMVENTION OF ABC TRANSPORTERS

The general strategies studied in the last 15 years to overcome ABC transporters can be synthesized as (Potschka, 2012; Talevi and Bruno-Blanch, 2012): (a) modulation of ABC transporters (i.e., reversal of multidrug resistance and down-regulation of transporters); (b) design of novel drugs which are not efflux transporter-substrates; c) bypassing drug transport (or the "Trojan horse" strategy). Regarding *transporters modulation*, the most advanced research relates to add-on therapies of specific inhibitors of ABC transporters, a strategy that was originally conceived for cancer treatment. Although preclinical and initial clinical results in the field of cancer treatment were encouraging at first, clinical trials of first, second and even third generation agents had to be stopped because of serious side- effects (Deeken and Löscher, 2007; Lhommé et al., 2008; Tiwari et al., 2011). These results have called into question the general validity of this approach, although trials continue

[1]The authors calculated the mínimum effective and the maximum tolerated concentration from *in vitro* cytotoxicity data.

in order to find more effective and safe inhibitors for Pgp and other transporters (Akhtar et al. 2011; Deeken and Löscher 2007). At this point it is important to remember that ABC transporters compose a concerted, complex efflux and influx dynamic system whose substrates are not only drugs but also endogenous compounds (e.g., waste products) and toxins. They are implicated in the inflammatory response to several stress and harmful stimuli, and, apparently, they have a role in neurodegenerative diseases such as Alzheimer's and Parkinson's disease (Hartz and Bauer, 2010). Thus, their permanent impairment or disruption is likely to result in severe side effects. The use of nanocarrier-based Trojan horses to hidden ABC transporters substrates from recognition appears then as a safer option to circumvent efflux pumps.

For illustrative purposes, propranolol, a poorly water-soluble drug that is a substrate for the P-gp efflux transporter, was conjugated to PAMAM dendrimer G3 in an attempt to enhance the transport of propranolol across Caco-2 cells. This PAMAM–propranolol prodrug indeed has shown to bypass the efflux system (D'Emanuele et al., 2004). In a related study, prodrugs between the smaller and less toxic PAMAM dendrimer G1 and the water-insoluble Pgp substrate terfenadine were synthesized using succinic acid or succinyl-diethylene glycol as linkers. All of the PAMAM–terfenadine prodrugs were more hydrophilic than the parent drug. The influence of the dendrimer prodrugs on the integrity and viability of Caco-2 cells was determined by measuring the transepithelial electrical resistance (TEER) and leakage of lactate dehydrogenase (LDH) enzyme, respectively. The LDH assay indicated that the dendrimer prodrugs had no impact on the viability of Caco-2 cells up to a concentration of 1 mM. Transport of dendrimer prodrugs across monolayers of Caco-2 cells showed an increase in the apparent permeability coefficient (Papp) of terfenadine in both apical-to-basolateral (A-B) and basolateral-to-apical (B-A) directions, with the apparent permeability coefficient for (A-B) significantly greater than (B-A) (Najlah et al., 2007). Similar results were obtained with paclitaxel and doxorubicin loaded NLC (Zhang et al., 2008), which showed 34.3 and 6.4 fold reversal powers respectively (compared to solutions of the same drugs) when tested on different multidrug resistant cell lines.

1.5.5 ACTIVE TARGETING

Active targeting often refers to the incorporation of a ligand specific for a receptor or epitope of a target tissue (Malam et al., 2009) at the nanocarrier surface. However, a stimuli-dependent distribution (e.g., guiding superparamagnetic nanoparticles to their target tissue by application of an external magnetic field) can also be conceived as a form of active targeting. Several targeting moieties have been tested so far. There is profusion of reports on the potential advantages of active targeting. For example, folate receptors are overexpressed on various types of cancer cells, and mediate endocytosis of folic acid-conjugated carriers (Leamon and Reddy, 2004; Lu and Low, 2002). Lundberg et al., produced sterically stabilized lipid drug-carriers coupled with antiCD74 antibody (LL1) (Lundberg et al., 2004) showing that LL1-liposomes show an extensive association with the Burkitt's lymphoma cells compared to nontargeted preparations (32% versus 0.6% within 24 h). Another—among numerous—suitable ligand for conjugating with nanoparticles is transferrin, since it can be specifically recognized and taken up by transferrin receptors actively expressed on the surface of various tumor cells (Yang et al., 2005). A recent patent of Lozano-López et al. (Lozano-López, 2012) introduces an easy-to-obtain nanocapsule system wherein an oily core comprizing an oil and a negative phospholipid component is surrounded by a polyarginine (PArg) shell. PArg belongs to the family known as protein transduction domains (PTD, or cell penetrating peptides) (Zahid and Robbins, in press) that are small cationic peptides able to carry other peptides, nucleic acids, viral particles and nanoparticles across the cellular membrane, into the cytoplasm. PTD engage the cellular surface through electrostatic interactions and induce their own internalization through endocytosis. Besides their proven utility in the promotion of cellular uptake of different molecules, pArg has immunostimulant properties (Mattner et al., 2002) and induces changes in paracellular transport at epithelia through tight junction disruption (Ohtake et al., 2003): all these properties have enormous potential for drug delivery in general and anticancer treatment in particular. What is more, as Lozano-López et al. state, the highly positive surface charge of the NC provides greater interaction with mucosae and, especially, tumor cells, which are often more negatively charged than healthy tissue (e.g., due to cell surface sialylation) (Marquez et al., 2004).

1.6 NANOTOXICOLOGY

Nanotoxicology is an emerging branch of toxicology dealing with the potential issues of short- and long-term-exposure to nanosystems. From our experience with micro and nano-materials such as quartz, asbestos and air-pollutants (Royal Society, 2004), we are aware that the toxicology of ultrafine materials can be radically different from that of bulk materials (Chen et al., 2009). Extensive risk assessment of nanomaterials seems particularly critical in the case of inorganic, nonbiodegradable materials.

Although a significant volume of reports assessing the potential toxicity of different types of nanoparticles have been published (using a wide diversity of experimental setups and with disparate results), we are still missing standardized procedures and coordinated research programs that allow us to reach general conclusions on this matter (Fadeel and García-Bennet, 2010; Khlebtsov and Dykman, 2011). Since the biodistribution and toxicology of nanoparticles strongly depend on a multiplicity of factors such as dose, size, shape, it has been pointed out that nanotoxicological studies without careful characterization of the physico-chemical properties of the studied systems is not meaningful (Fadeel and García-Bennet, 2010). Evaluation of the final fate of the nanoparticles and their effect on protein and gene expression levels are critical issues that to the moment have been scarcely, insufficiently investigated (Colombo et al., 2012; Fadeel and García-Bennet, 2010). Noteworthy, a study on the *in vivo* toxicity of gold nanoparticles showed that administration of 8 mg/kg/week of 8 to 37 nm particles produced, from day 14, severe side-effects such as camel-like back and crooked spine (Chen et al., 2009). Histological examination revealed various degrees of abnormality in the liver, lung and spleen of gold nanoparticle-treated mice. The median survival time was also significantly reduced. These important results underline the fact that *in vitro* assessment on a single cellular line may often not be representative of the behavior of the nanosystem in a whole organism, claiming for the development of complex cellular models. This study clearly states the cautions that should be taken regarding the clinical use of nanosystems as drug delivery devices, particularly if they are nonbiodegradable and in long-term treatment settings. It should also be kept in mind that the high specific surface of nanoparticles increases the risk of contamination with other materials (Fadeel and García-Bennet, 2010).

1.7 CONCLUSIONS

Nanosystems offer a number of features that make them tremendously attractive in the field of pharmaceutics. The wide range of materials, shapes, sizes, morphologies and surface functionalizations, and the possibility to produce nanohybrids (material combinations), along with the high dependence of physicochemical properties and biological behavior on those factors, makes targeted, tailored drug delivery systems an almost palpable reality.

Nevertheless, environmental and health risk assessment have yet to be developed. Some pieces of research suggest that some nanoparticles may have severe effects *in vivo*. Unfortunately, those types of nanoparticles that can be produced in a higher standardized manner (i.e., inorganic nanoparticles) are also the ones that potentially pose a higher health risk. In the authors' opinion, the use of biodegradable nanoparticles should be preferred as long as long-term safety studies are not completed.

On the light of the high variability of the biomedical properties of nanoparticulated systems with factors such as size, shape and surface functionality, and considering the high specific surface of nanosystems, standardization of the synthesis procedures and careful prevention of contamination with other materials are fundamental thinking of their application in the pharmaceutical sector.

1.8 ACKNOWLEDGMENTS AND CONFLICT OF INTEREST STATEMENT

A. Talevi is a member of the Scientific Research Career at CONICET. C. L. Bellera is a CONICET fellowship holder. A. Enrique and L. E. Bruno-Blanch are researchers of Facultad de Ciencias Exactas, Universidad Nacional de La Plata. The authors would like to thank UNLP (Incentivos X-597), CONICET (PIP 11220090100603) and ANPCyT (PICTs 2010–2531 and 2010–1774) for providing funds to develop our research.

The authors have no conflict of interest.

KEYWORDS

- ***hypercore***
- ***dendron***
- **lipid-based nanoparticles**
- **gold nanocarriers**
- **magnetic nanoparticles**
- **mesoporous silica nanoparticles**

REFERENCES

Akhtar, N.; Ahad, A.; Khar, R. K.; Jaggi, M.; Ágil, M.; Igbal, Z.; Ahmad, F. J.; Talegaonkar, S. *Expert Opin. Ther. Pat.* **2011,** *21,* 561–576.

Al-Jamal, W.; Kostarelos, K. *Accounts Chem. Res.* **2011,** *44,* 1094–1104.

Bangham, A. D.; Standish, M. M.; Watkins, J. C. *J. Mol. Biol.* **1965,** *13,* 238–252.

Beija, M.; Salvayre, R.; Lauth-de Viguerie, N.; Marty, J.D. *Trends Biotechnol.* **2012,** *30,* 485–496.

Bosman, A. W.; Janssen, H. M.; Meijer, E. W. *Chem. Rev.* **1999,** *99,* 1665–1688.

Boyoglu, C.; Boyoglu-Barnum, S.; Soni, S.; He, Q.; Willing, G.; Miller, M. E.; Singh, S. R. *Nanotech.* **2011,** *3,* 489–492.

Buhleier, W. W.; Vögtle F. *Synthesis* **1978,** *2,* 155–158.

Calvo, P.; Gouritin, B.; Chacun, H.; Desmaele, D.; D'Angelo, J.; Noel, J. P.; Georgin, E.; Fattal, E.; Andreux, J. P.; Couvreur, P. *Pharm. Res.* **2001,** *18,* 1157–1166.

Chavanpatil, M. D.; Khdair, A.; Panyam, J. Nanosci. *Nanotechnol.* **2006,** *6,* 2651–2663.

Chen, Y. S.; Hung, Y. C.; Liau, I.; Huang, G. S. *Nanoscale Res. Lett.* **2009,** *4,* 858–864.

Chithrani, B. D.; Chan, W. C. W. *Nano. Lett.* **2007,** *7,* 1542–1550.

Chithrani, B. D.; Ghazani, A. A.; Chan, W. C. W. *Nano Lett.* **2006,** *6,* 662–668.

Colombo, M.; Carregal-Romero, S.; Casula, M. F.; Gutiérrez, L.; Morales, M. P.; Böhm, I. B.; Heverhagen J. T.; Prosperi, D.; Parak, W. J. *Chem. Soc. Rev.* **2012,** *41,* 4306–4334.

D'Emanuele, A.; Jevprasesphant, R.; Penny, J.; Attwood, D. J. *Control. Release* **2004,** *95,* 447–453.

Dams, E. T.; Laverman, P.; Oyen, W. J.; Storm, G.; Scherphof, G. L.; van DerMeer, J.W.; Corstens, F. H.; Boerman, O. C. J. *Pharmacol. Exp. Ther.* **2000,** *292,* 1071–1079.

Deeken, J. F.; Löscher, W. Clin. Cancer Res. **2007,** *13,* 1663–1674.

Duncan, R.; Izzo, L. Adv. *Drug Deliv. Rev.* **2005,** *57,* 2215–2237.

Fadeel, B.; García-Bennet, A. E. Adv. *Drug Deliv. Rev.* **2010,** *62,* 362–374.

Fattal, E.; Andreux, J. P.; Couvreur, P. *Pharm. Res.* **2001,** *18,* 1157–1166.

Feng S. S.; Mei, L.; Anitha, P.; Gan, C. W.; Wenyou, Z. *Biomaterials* **2009,** *30,* 3297–3306.

Fréchet, J. M. J. *Science* **1994,** *263,* 1710–1715.

Ghose, A. K.; Viswanadhan, V. N.; Wndoloski, J. J. J. *Comb. Chem.* **1999,** *1,* 55–68.

Gillies, E. R.; Dy, E.; Fréchet, J. M. J.; Szoka, F. C. *Mol. Pharm.* **2005,** *2,* 129–138.

Gomes Da Silva, Fonseca, N. A.; Moura, V.; De Lima, P.; Simoes, S.; Moreira, J. *Accounts Chem. Res.* **2012,** *45,* 1163–1171.

Halas, N. J. *MRS Bull.* **2005,** *30,* 362–367.

Hartz, A. M.; Bauer, B. *Mol. Interv.* **2010,** *10,* 293–304.

Hawker, C. J. and Fréchet, J. M. *J. Am. Chem. Soc.* **1990,** *112, 21,* 7638–7647.

Hawker, C. J.; Wooley, K. L.; Fréchet J. M. J. J. *Chem. Soc. Perkin Trans* **1993,** *1,* 1287–1297.

Hobbs, S. K.; Monsky, W. L.; Yuan, F.; Roberts, W. G.; Griffith, L.; Torchilin, V. P.; Jain R. K. *Proc. Natl. Acad. Sci. USA* **1998,** *95,* 4607–4612.

Hutter, E.; Fendler J. H. *Adv. Mater.* **2004,** *16,* 1685–1706.

Huynh, N. T.; Passirani, C.; Saulnier, P.; Benoit, J. P. *Int. J. Pharmceut.* **2009,** *379,* 201–209.

Ishida, T.; Ichihara, M.; Wang, X.; Yamamoto, K.; Kimura, J.; Majima, E.; Kiwada, H. J. *Control. Release* **2006,** *112,* 15–25.

Iyer, A. K.; Khaled, G.; Fang, J.; Maeda, H. Drug Discov. Today, **2006,** *11,* 812–818.

Kalele, S.; Gosavi, S. W.; Urban, J.; Kulkarni, S. K. *Curr. Sci. India* **2006,** *91,* 1038–1052.

Kawaguchi, T.; Walker, K. L.; Wilkins, C. L.; Moore, J. S. *J. Am. Chem. Soc.* **1995,** *117,* 2159–2165.

Khlebtsov, N. G.; Dykman, L. A. *Nanotechnol. Russia* **2011,** *6,* 17–42.

Klajnert, B.; Bryszewska, M. *Acta Biochim. Pol.* **2001,** *48,* 199–208.

Kola, I.; Landis, J. Nat. Rev. *Drug Discov.* **2003,** *3,* 711–715.

Kolb, H. C.; Finn, M. G.; Sharpless, K. B. Angew. *Chem. Int. Ed. Engl.* **2001,** *40,* 2004–2021.

Kubinyi, H. *Nat. Rev. Drug Discov.* **2003,** *2,* 665–668.

Leamon, C. P.; Reddy, J. A. *Adv. Drug. Deliv. Rev.* **2004,** *56,* 1127–1141.

Lhommé, C.; Joly, F.; Walker, J.L.; Lissoni, A. A.; Nicoletto, M. O.; Manikhas, G. M.; Baekelandt, M. M.; Gordon, A. N.; Fracasso, P. M.; Mietlowski, W. L.; Jones, G. J.; Dugan M. H. *J. Clin. Oncol.,* **2008,** *26,* 2674–2682.

Liong, M.; Lu, J.; Kovochich, M.; Xia, T.; Ruehm, S. G.; Nel, A. E.; Tamanoi, F.; Zink, J. I. *ACS Nano* **2008,** *2,* 889–896.

Liu, J.; Gong, T.; Fu, H.; Wang, C.; Wang, X.; Chen, Q.; Zhang, Q.; He, Q.; Zhang, Z. *Pharm. Nanotechnol.* **2008,** *356,* 333–344.

Lozano López, V.; Alonso Fernández, J.; Torres López, D. 2012. US20120121670.

Lu, Y.; Low, P. S. Adv. *Drug Deliv. Rev.* **2002,** *54,* 675–693.

Lucks, J.S.; Müller, R. H. **1991.** EP0000605497.

Lundberg, B. B.; Griffiths, G.; Hansen, H. J. *J. Control. Release* **2004,** *94,* 155–161.

Malam, Y.; Loizidou, M.; Seifalian, A. M. Trends Pharmacol. Sci. **2009,** *30,* 592–599.

Manjunath, K.; Venkateswarlu, V. Ç. *J. Control. Release* **2005,** *107,* 215–228.

Marquez, M.; Nilsson, S.; Lennartsson, L.; Liu, Z.; Tammela, T.; Raitanen, M.; Holmberg, A. R. *Anticancer Res.* **2004,** *24,* 1347–1351.

Martins, S.; Sarmento, B.; Ferreira, D. C.; Souto, E. B. *Int. J. Nanomed.* **2007,** *2,* 595–607.

Matsumoto, S.; Christie, R. J.; Nishiyama, N.; Miyata, K.; Ishii, A.; Oba, M.; Koyama, H.; Yamasaki, Y.; Kataoka, K. *Biomacromolecules* **2009,** *10,* 119–127.

Mattner, F.; Fleitmann, J. K.; Lingnau, K.; Schmidt, W.; Egyed, A.; Fritz, J.; Zauner, W.; Wittmann, B.; Gomy, I.; Berger, M.; Kirlappos, H.; Otava, A.; Bimstiel, M. L.; Buschle, M. *Cancer Res.* **2002,** *62,* 1477–1480.

Medina, S. H.; El-Sayed, M. E. H. *Chem. Rev.* **2009,** *109,* 3141–3157.

Morgan, M. T.; Nakanishi, Y.; Kroll, D. J.; Griset, A. P.; Carnahan, M. A.; Wathier, M.; Oberlies, N. H.; Manikumar, G.; Wani, M. C.; Grinstaff, M. W. *Cancer Res.* **2006,** *66,* 11913–11921.

Morgan, T. M.; Carnahan, M. A.; Immoos, C. E.; Ribeiro, A. A.; Finkelstein, S.; Lee, S. J.; Grinstaff, M. E. *J. Am. Chem. Soc.* **2003,** *125,* 15485–15489.

Müller, R. H.; Keck, C. M. *J. Biotechnol.* **2004,** *113,* 151–170.

Najlah, M.; Freeman, S.; Attwood, D.; D'Emanuele, A. *Bioconjug. Chem.* **2007,** *18,* 937–946.

Nie, Y.; Gunther, M.; Gu, Z.; Wagner, E. *Biomaterials* **2011,** *32,* 858–869.

Nishikawa, M.; Takakura, Y.; Hashida, M. *Adv. Genet.* **2005,** *53,* 47–68.

Ohtake, K.; Maeno, T.; Ueda, H.; Natsume, H.; Morimoto, Y. *Pharm Res.* **2003,** *20,* 153–160.

Pardeike, J.; Hommoss, A.; Müller, R. H. *Int. J. Pharmaceut.* **2009,** *366,* 170–184.

Parra-Guillén, Z. P.; González-Aseguinolaza, G.; Berraondo, P.; Trocóniz, I. F. *Pharm. Res.* **2010,** *27,* 1487–1497.

Peng, S.; Wang, C.; Xie, J.; Sun, S. H. *J. Am. Chem. Soc.* **2006,** *128,* 10676–10677.

Petryayeva, E.; Krull, U. *J. Anal. Chim. Acta* **2011,** *706,* 8–24.

Pissuwan, D.; Niidome, T.; Cortie, M. B. J. *Control. Release* **2011,** *149,* 65–71.

Potschka, J. Adv. *Drug Deliv. Rev.* **2012,** *64,* 896–910.

Proudfoot, J. R. *Bioorg. Med. Chem. Lett.* **2005,** *15,* 1087–1090.

Puri, A.; Loomis, K.; Smith, B.; Lee, J. H.; Yavlovich, A.; Heldman, E.; Blumenthal, R. *Crit. Rev. Ther. Drug Carrier Syst.* **2009,** *26,* 523–580.

Royal Society. *Nanoscience and Nanotechnologies: Opportunities and Uncertainties.* **2004.**

Salgueiriño-Maceira V.; Caruso, F.; Liz-Marzán, L. M. *J. Phys. Chem. B* **2003,** *107,* 10990–10994.

Sampathkumar, S. G.; Yarema, K. J. Dendrimers in cancer treatment and diagnosis. In Nanotechnologies for the Lifesciences – Vol. 06: Nanomaterials for Cancer Diagnosis and Therapy; Kumar, C. S. S. R., Ed.; John Wiley & Sons: Hoboken, NJ, **2007,** pp. 1–17

Schuster, D.; Laggner C.; Langer, T. *Curr. Pharm. Des.* **2005,** *11,* 3545–3559.

Sekhon, B. S.; Kamboj, S. R. *Nanomed. Nanotech. Biol. Med.* **2010,** *6,* 612–618.

Seo, W. S.; Lee, J. J.; Sun, X. M.; Suzuki, Y.; Mann, D.; Liu, Z.; Terashima, M.; Yang, P. C.; McConnell, M. V.; Nishimura, D. G.; Dai, H. *Nat. Mater.* **2006,** *5,* 971–976.

Sharom, F. J. *Pharmacogenomics* **2008,** *9,* 105–127.

Singh, A.; Sarabjit, S.; Gupta, A. K.; Kulkami, M. M. **2012.** US20100166872.

Slowing, I. I.; Vivero-Escoto, J. L.; Wu, C. W.; Lin, V. S. Y. *Adv. Drug Deliv. Rev.* **2008,** *60,* 1278–1288.

Smith, D. A.; Di, L.; Kerns, E. H. *Nat. Rev. Drug Discov.* **2010,** *9,* 929–939.

Storm, G.; Belliot, S. O.; Daemen, T.; Lasic, D. D. *Adv. Drug Deliv. Rev.* **1995,** *17,* 31–48.

Szakács, G:; Paterson, J. K:; Ludwig, J. A.; Booth-Genthe, C.; Gottesman, M. M. *Nat. Rev. Drug Discov.* **2006,** *5,* 219–234.

Takae, S.; Miyata, K.; Oba, M.; Ishii, T.; Nishiyama, N.; Itaka, K.; Yamasaki, Y.; Koyama, H.; Kataoka, K. *J. Am. Chem. Soc.* **2008,** *130,* 6001–6009.

Talevi, A.; Bellera, C. L.; Di Ianni, M.; Gantner, M.; Bruno-Blanch, L. E.; Castro, E. A. *Mini-Rev. Med. Chem.* **2012,** *12,* 959–970.

Talevi, A.; Bruno-Blanch, L. E. Efflux-transporters at the blood-brain barrier: Therapeutic Opportunities. In: The blood-brain barrier: New research; Montenegro, P. and Juárez, S. M., Eds.; Nova Publishers: Hauppauge, NY, **2012,** pp. 117–144.

Tiwari, A.; Sodani, K.; Dai, C. L.; Ashby, C. R.; Chen, Z. S. *Curr. Pharm. Biotechnol.* **2011,** *12,* 570–594.

Tomalia, D. A. *Prog Polym Sci* **2005,** *30,* 294–324.

Tomalia, D. A.; Dewald, J.; Hall, M.; Kallos, G.; Martin, S.; Roeck, J.; Ryder, J.; Smith, P. *Polym. J.* **1985,** *17,* 117–132.

Torchilin, V. *Drug Discov. Today Technol.,* **2008,** *5,* e95–e103.

Unezaki, S.; Maruyama, K.; Hosoda, H. I.; Nagae, I.; Koyanagi, Y.; Nakata, M.; Ishida, O.; Iwatsuru, M.; Tsuchiya, S. *Int. J. Pharmaceut.* **1996,** *144,* 11–17.

Vugmeyster, Y.; Xu, X.; Theil, F. P.; Khawli, L. A.; Leach, M. W. *World J. Biol. Chem.* **2012,** *3,* 73–92.

Wissing, S. A.; Kayzer, O.; Müller, R. H. *Adv. Drug. Deliv. Rev.* **2004,** *56,* 1257–1272.

Wohlfart, S.; Gelperina, S.; Kreuter, J. J. *Control. Release* **2012,** *161,* 264–273.

Wooley, K. L.; Hawker, C. J.; Fréchet, J. M. *J. Am. Chem. Soc.* **1991,** *113,* 4252–4261.

Xu, Ch.; Sun, S. *Dalton Trans.* **2009,** *29,* 5583–5591.

Xu, Z. P.; Zeng, Q. H.; Lu, G. Q.; Yu, A. B. *Chem. Eng. Sci.* **2006,** *61,* 1027–1040.

Yadav, K. S.; Chuttani, K.; Mishra, A. K.; Sawant, K. K. PDA *J. Pharm. Sci. Technol.* **2011,** *65,* 131–139.

Yang, P. H.; Sun, X.; Chiu, J. F.; Sun, H.; He, Q. Y. *Bioconjug. Chem.* **2005,** *16,* 494–496.

Yu, H.; Chen, M.; Rice, P. M.; Wang, S. X.; White, R. L.; Sun, S. H. *Nano Lett.* **2005,** *4,* 379–382.

Yuan, F.; Dellian, M.; Fukumura, D.; Leunig, M.; Berk, D. A.; Torchilin, V. P.; Jain, R. K. *Cancer Res.* **1995,** *55,* 3752–3756.

Zahid, M.; Robbins, P. D. *Current. Gene Ther.* In press .

Zara, G. P.; Cavalli, R.; Bargoni, A.; Fundaro, A.; Vighetto, D.; Gasco, M. R. *J. Drug Target.* **2002,** *10,* 327–335.

Zeng, S.; Yong, K. T.; Roy. I.; Dinh, Z. Q.; Yu, X.; Luan, F. *Plasmonics* **2011,** *6,* 491–506.

Zhang, X. G.; Miao, J.; Dai, Y. Q.; Du, Y. Z.; Yuan, H.; Hu, F. Q. *Int. J. Pharmaceut.* **2008,** *361,* 239–244.

CHAPTER 2

RADIOLABELED NANOPARTICLES USING β^+ RADIONUCLIDES AS DIAGNOSTIC AGENTS: AN OVERVIEW AND A CHEMOTOPOLOGICAL APPROACH

NANCY Y. QUINTERO, ISAAC MARCOS COHEN and GUILLERMO RESTREPO

CONTENTS

Abstract 27
2.1 Introduction 27
2.2 Radionuclides Used in PET Imaging 30
2.3 Parameters For Selecting A Pet Imaging β^+ Radionuclide 31
2.3.1 Data Decay 31
2.3.2 Half-Life 32
2.3.3 Maximum Energy of the Positron 32
2.3.4 Gamma Rays 32
2.4 Types of Nanoparticles Used in Pet Imaging 32
2.4.1 Polymeric Nanoparticles 34
2.4.2 Lipid Nanoparticles 40
2.4.3 Carbon Nanoparticles 43
2.4.4 Quantum Dots (QDs) 50
2.4.5 Nanoparticles Based on Rare Earths 53
2.4.6 Other Inorganic Nanoparticles 54

2.5 Chemotopology, A Mathematical Approach For Analyzing Similarity Relationships Among β^+ Radionuclides Used For Nanoparticles Radiolabeling 62
2.5.1 Closure of X, $\overline{X}$ 67
2.5.2 Interior of X, Int (X) 67
2.5.3 Exterior of X, Ext(X) 68
2.5.4 Boundary of X, b(X) 69
2.6 Conclusions 69
Keywords 70
References 70

ABSTRACT

Nanoparticles are colloidal systems that vary in size from 10 to 1,000 nm. According to the bibliography, they have been used in biology and medicine for over 50 years. Many research laboratories around the world are now focusing on the development of diagnostic and therapeutic agents based upon radiolabeled peptides, proteins, affibodies or scaffold proteins, including nanoparticles. Radiolabeled nanoparticles designed for specific transport of therapeutic agents to diseased tissues have received attention over the last decades, because they have potential as radiotracers in the visualization of tumors and metastases. Radionuclides are bound to nanoparticles and inoculated into the human tissues, where they decay emitting particles (α, β^+, β^-, Auger electrons). Nanoparticles used as nanocarriers are characterized by their long circulation half-life and high accumulation in tumors due to their rapid penetration in human tissues.

At present, non conventional β^+ radionuclides, e.g., ^{64}Cu and ^{124}I, are studied for PET (Positron Emission Tomography) diagnostic imaging. Other radionuclides, such as ^{86}Y, ^{89}Zr, ^{62}Cu, ^{15}O and ^{68}Ga, are used in many studies for labeling nanoparticles and for tracking some biological process in animal models.

The properties of β^+ radionuclides and the types of nanoparticles labeled with these radionuclides are described. Finally, the recent advances in this field and the application of the chemotopological methodology, as a new contribution of the study of positron emitters used in PET diagnostic imaging for labeling of nanoparticles, are discussed.

2.1 INTRODUCTION

The increasing role of nuclear medicine in the clinical practice, has allowed both the *in vivo* exploration of biological systems through diagnosis methods and the treatments of many diseases, including cancer.

There are many techniques in the diagnostic field: such as ultrasound, MRI (Magnetic Resonance Imaging) or CT (Computed Tomography), among others that predominantly provide just anatomical information. Instead, molecular imaging techniques, namely SPECT (Single Photon Emission Computed Tomography) and PET (Positron Emission Tomography) supply unique functional information on the metabolic activity in physiological, biochemical or pathological processes at molecular level (Hörger, 2003) within a living subject. Additionally, these techniques are highly sensitive and specific, allowing accurate quantification and noninvasive study of cells in their natural microenvironment without limits in tissue penetration in any organ (de Barros et al., 2012; Gambhir, 2002; James and Gambhir, 2012).

At this level, it is very important to trace the movement of cells and to follow the dynamic biological and physiological processes by a visible response with high sensitivity; for this reason, nuclear techniques based on the utilization of radionuclides are commonly preferred in many research laboratories around the world. Another advantage of the use of radionuclides is that they can be bound to molecules in the form of radiopharmaceuticals, thus making viable their injection in human bodies. According to Vallabhajosula et al. (2010) the term radiopharmaceutical has several meanings, the most general one being: any radiolabeled molecule intended for human use. From a regulatory point of view, a radiopharmaceutical is pyrogen free, sterile, safe for human use, and efficacious for a specific indication. Additionally, radiopharmaceuticals must have high radionuclidic purity (fraction of the total radioactivity due to the radionuclide of interest), as well as high radiochemical purity (degree of purity of the chemical structure to which the radioisotope is bound) and *in vivo* stability, to obtain enhanced sensitivity for disease detection and improved targeting.

While many types of molecules have been used in molecular imaging, a growing area of interest is the use of peptides, proteins, affibodies or scaffold proteins, and aptamers, including nanoparticles (James and Gambhir, 2012; Silindir et al., 2012).

According to IUPAC, nanoparticles are microscopic particles whose size is measured in nanometers, often restricted to so-called nanosized particles (NPs; < 100 nm in aerodynamic diameter); NPs are also called ultrafine particles.

Nanoparticles can be made of many types of materials with diverse chemical nature (metals, metal oxides, silicates, polymers, carbon including nanodiamonds, lipids, and biomolecules) and they can have several different morphologies: spheres, cylinders, tubes, platelets, wires (de Barros et al., 2012). In contrast to atoms and molecules, nanoparticles have increased surface to volume ratio and high chemical reactivity in direct relation with the surface area. Their relatively large size, compared with small molecules and many proteins, implies that they are often taken up by the reticuloendothelial system (James and Gambhir, 2012). They can improve diagnostic imaging techniques even at the level of single cells and before the appearance of overall symptoms in many diseases (Assadi et al., 2011), because of their long blood circulation time and plasma stability (de Barros et al., 2012).

According to Welch et al. (2009), the remarkable interest for applying nanoparticles in molecular imaging has led to the development of both inorganic and organic nanoparticles, which can be functionalised through chemical bonds with radionuclides, targeting ligands, polymers such as polyethylene glycol (PEG), and liposomal agents, among other platforms. This makes possible to provide a signal for the imaging, to bring nanoparticles to the target tissue and even to modify their pharmacokinetics.

A radionuclide with widespread use in PET imaging, ^{18}F has been employed for labeling of liposomes; several studies have shown the feasibility of the use of liposome encapsulated ^{18}F-Fluorodeoxyglucose ([^{18}F]FDG), for monitoring biodistribution of liposomes, but a disadvantage is its fairly short half-life (109.7 min), which makes difficult to track liposomes distribution for more than eight hours (Goins and Phillips, 2007; Phillips et al., 2009).

Many radionuclides, can be attached by chemical methods to each nanoparticle with high yield, as it was recently demonstrated by Matson and Grubbs (2008). They synthesized nanoparticles labeled with ^{18}F for *in vivo* molecular imaging through the polimerization based on the metathesis of olephines using block copolymers, reaction known as ROMP, i.e., ring opening metathesis polymerization. So far, there are four reactions of olephine metathesis employed for the synthesis of block copolymers.

Devaraj et al. (2009) reported the synthesis and *in vivo* characterization of ^{18}F labeled nanoparticles ^{18}F-CLIO for PET-CT imaging. In addition, Sun et al. (2011) developed a simple, rapid, efficient and general strategy for synthesis of ^{18}F-labeled rareearth nanoparticles, to be used in PET imaging, through a simple inorganic reaction based on the strong bond between ^{18}F- and rare earths ions.

It can be seen that the advances in the development of new radiochemical and chemical methods, together with the improvements of labeling methods, have opened the possibilities of synthetizing functional nanoparticles with positron-emitters.

Nanoparticles show a promizing future as bioprobes, due to their special physical and chemical properties; bioprobes are typically composed of an affinity component interacting with the target tissue and a signaling component supplying intensity for imaging (Silindir et al., 2012).

Hence, a single nanoparticle can act as a bioprobe and can be tagged with various imaging agents having high selectivity with respect to the target tissue, depending on the attached ligand (Assadi et al., 2011). Many types of nanoparticles can be radiolabeled with radionuclides characterized by different decay modes, depending on their use as diagnostic or therapy reagents.

At present, there are more than 3,200 known radionuclides; more than 200 of them find different uses in science, technology and nuclear medicine (Sahoo and Sahoo, 2006); in connection with nuclear medicine, radionuclides have been used for over 70 years (Kassis, 2008). Their applications in oncologic, cardiologic and psychiatric diagnostic or therapy have increased in the last 15 years (Köster et al., 2011; Ishiwata et al., 2010); they include radiolabeling of nanoparticles with radionuclides decaying by α, β^- and β^+ emission, as well as emitters of Auger electrons.

Beta particles are more suited for solid tumors of at least 1 cm radius, because they travel relatively long distances and could impart damage to the surrounding normal cells of tumors of smaller size. Alpha particles travel a very short distance (a few mm) thus having high linear energy transfer; consequently, they are more effective in cases of micrometastases or small clusters of cancer cells, such as leukemia (but not for solid tumors). Emitters of gamma rays are seldom used, due to their high penetration, which implies an unnecessary damage to normal tissues.

Radiolabeling of nanoparticles has been performed for many purposes, including noninvasive evaluation of their biodistribution, pharmacokinetics, imaging or therapy. The selection of a radionuclide depends on various parameters; for PET imaging, the radionuclides must have appropiate decay properties conducive to detect; they emit positrons into the living subject, which interact with the electrons of the media, and after losing sufficient energy (<100 eV) annihilate rapidly, giving off in each process two 511 keV photons in opposite directions that are measured by a convenient arrangement of detectors (Agrawal et al., 2010; Smith et al., 2011; James and Gambhir, 2012).

A mathematical approach developed by Restrepo et al. (2004), chemotopology, can be applied to the study of radionuclides used for labeling of nanoparticles; results of a preliminary chemotopological study with positron emitters, recently published by Quintero et al. (2013) give the guidelines for proposing an additional case of interest related to the radiolabeled nanoparticles with β^+ radionuclides.

2.2 RADIONUCLIDES USED IN PET IMAGING

Positron emitters can be classified into three groups:

The first group includes bioisotopes: ^{11}C, ^{13}N and ^{15}O (Pagani et al., 1997) that are β^+ pure emitters, where the only electromagnetic radiations associated with their decay are the 511 keV annihilation photons. Since carbon, nitrogen and oxygen are ubiquitously found in human tissues, they can be replaced with relative easiness in the biomolecules. Moreover, their half-lives, of the order of minutes constitute an advantage for patients, because of the low radiation doses involved. However, these short half-lives preclude the accomplishment of studies involving slow biochemical processes; the radionuclides cannot be transported to long distances and must be produced *in situ* (Pagani et al., 1997).

The second group is made of radionuclides with longer half-lives, typically ^{18}F, ^{73}Se, ^{75}Br, ^{76}Br, ^{124}I and ^{120}I. ^{18}F is at present used in 80% of PET studies, due to its favorable nuclear properties, particularly its half-life (109.7 minutes) which allows labeling by complex synthesis and transport to relatively distant sites (López-Durán et al., 2007). The production of the iodine radioisotopes requires special equipment for handling high activities; in addition, some disfavorable factors affect their quality in terms of radionuclidic and radiochemical purities (Carrió et al., 2003).

The third group is composed of radioisotopes of metallic elements: ^{38}K, ^{62}Cu, ^{64}Cu, ^{55}Co, ^{68}Ga, ^{82}Rb, ^{86}Y. Most of them have convenient half-lives, which makes possible their use in places far from the location of the cyclotron (Blum, 2003). Two short-lived radionuclides belonging to this group, namely ^{68}Ga ($t_{1/2}$ = 1.135 h) and ^{82}Rb ($t_{1/2}$ = 1.27 min) can be produced in generators through their longer-lived mothers, ^{68}Ge ($t_{1/2}$ =270.8 d) and ^{82}Rb ($t_{1/2}$ = 1.27 min), respectively (Pagani et al., 1997). Nowadays, generator systems are commercially available scientific products, routinely produced. Thus, ^{68}Ga and ^{82}Rb can be accessible to most nuclear medicine departments, easily obtained by radiochemical separation from their mothers (Rösch and Baum, 2011), with high radionuclidic and radiochemical purities. Other generator systems developed are: ^{44}Ti (60.3 a)/^{44}Sc ($t_{1/2}$ = 3.927 h); ^{118}Te ($t_{1/2}$ = 6.00d)/^{118}Sb ($t_{1/2}$ = 3.6 min); ^{62}Zn (9.26 h)/^{62}Cu (9.74 min); ^{72}Se ($t_{1/2}$ = 8.4 d)/^{72}As ($t_{1/2}$= 1.083d); ^{140}Nd ($t_{1/2}$ = 3.37d)/^{140}Pr ($t_{1/2}$= 3.39 min). Although the technical feasibility of their production has been demonstrated, they have not been so far commercialized (Rösch and Baum, 2011).

Only some β^+ radionuclides, belonging to each of these three groups, have been studied in nanoparticles'radiolabeling: ^{68}Ga, ^{18}F, ^{64}Cu, ^{86}Y, ^{76}Br, ^{89}Zr and ^{124}I (Liu and Welch, 2012). Besides these radionuclides, interesting applications of ^{11}C (20.4 min) and ^{15}O (2.03 min) have been reported: despite their short half-lives, they have been employed, as well as other β^+ radionuclides, in radiolabeling of nanoparticles in animal models, at least as first approaches (Patt et al., 2010; Phillips et al., 2009). ^{15}O has also been used for studying the delivery of oxygen to the brain with liposome encapsulated hemoglobin (Goins and Phillips, 2007; Phillips et al., 2009).

2.3 PARAMETERS FOR SELECTING A PET IMAGING B+ RADIONUCLIDE

Thc most relevant parameters that guide the selection of suitable radionuclides for PET imaging are realated to their nuclear properties, namely: data decay; half-life; maximum energy of the positron and eventual presence of gamma emission.

2.3.1 DATA DECAY

It is an important property in the selection of a β^+ radionuclide for medical applications (Lambretch, 1979); the fraction of β^+ emission must be at least higher than 10% and the radionuclide should decay in a stable daughter, to avoid the generation of additional radiation (Pagani et al., 1997). Some radionuclides with relatively low β^+ percent have been studied in radiolabeling of nanoparticles, e.g., ^{64}Cu (17.60%) (Lewis et al., 2008), ^{89}Zr (22.7%) (Anderson and Ferdani, 2009; Pagani et al., 1997), ^{124}I (23%) (Koehler et al., 2010) and ^{86}Y (33%) (Walrand et al., 2003, 2011).

2.3.2 HALF-LIFE

A requisite to design a PET isotope labeled nanoparticle is the compatibility between physical half-life of the radionuclide and biological half-life of the nanoparticles (this requirement precludes, for example, the use of ^{15}O and ^{11}C). Half-lives must be long enough to allow target tissue accumulation and to assure the measurement of the activity, taken into account the duration of the clinic study. Reasonably long-lived β^+ radionuclides, such as ^{64}Cu, ^{86}Y, ^{76}Br, ^{89}Zr and ^{124}I, have been proposed in immuno-PET for radiolabeling of monoclonal antibodies (James and Gambhir, 2012; Lubberink et al., 2010) and nanoparticles.

2.3.3 MAXIMUM ENERGY OF THE POSITRON

It is a factor that affects the quality and the spatial resolution of the image; the best radionuclides are those that have low maximum β^+ energy (Laforest and Liu, 2008).

2.3.4 GAMMA RAYS

Gamma rays that eventually accompany the emission of positrons, particularly those with energies within the PET discriminator window, i.e., 350 and 700 keV (Smith et al., 2011) can cause reduction of spatial resolution and blurring of the image (Pagani et al., 1997). All of these emissions contribute to the total absorbed dose for the patient and the medical staff (Haddad et al., 2008).

2.4 TYPES OF NANOPARTICLES USED IN PET IMAGING

PET is a noninvasive nuclear imaging technique, capable of visualizing deep tissues with a high sensitivity of 0.02–0.10 cps/Bq (2–10%) (Cherry, 2001) and giving three-dimensional images of body functions (Lee et al., 2007). According to Ding and Wu (2012) PET has allowed measuring the expression of indicative molecular markers by using radiolabeled imaging agents (James and Gambhir, 2012) with radionuclides.

Silindir et al. (2012) used for PET imaging β^+ radionuclides bound to appropriate ligands in order to form the *in vivo* inoculated radiopharmaceuticals.

The half-life of β^+ radionuclides must be compatible with the time needed to achieve maximum uptake of the radiopharmaceutical in tissue, favoring their location through PET imaging (Silindir et al., 2012).

At present, combination of imaging techniques is currently employed to achieve a synergistic effect (Jarret et al., 2008); for example, PET/MRI are complementary techniques that combine high anatomical spatial resolution (MRI) and metabolic imaging at picomolar level sensitivity (PET) (Debagge and Jaschke, 2008; Ding and Wu, 2012).

In order to improve diagnosis by PET and to track physical and pathologic processes for therapy, many approaches have been developed. One of the most promizing in the latter field is the use of drug delivery systems (Li et al., 2012) especially nanoparticles in target imaging (Lian and Ho, 2001); they can be conjugated with both a specific ligand to target and a positron emitter radionuclide, by modifying surface properties (Silindir et al., 2012). The chemical or biological modification on the surface of nanoparticles ensures good biocompatibility and strong affinity between carriers and biomolecules (Jauregui-Osoro et al., 2011; Xu et al., 2006).

At present, nanoparticles with different chemical compositions have been used for biodistribution studies, *post mortem* nanoparticle determination and *in vivo* optical imaging (Balasubramanian et al., 2010).

"The information gathered helps the development of new or improved imaging probes and therapy agents. Traditional biodistribution studies using well counting require sacrifice of 50–60 animals per compound. The significant expense of lives, funds, and time limits the number of time points that can be assessed. In contrast, PET of small animals allows for serial imaging of live mice, obviates the need to sacrifice animals, and minimizes interindividual variation, resulting in better data quality" (Schipper et al., 2007).

There are many types of nanoparticles for nuclear imaging in PET (de Barros et al., 2012); the choice of a radionuclide for labeling nanoparticles depends on the imaging technique and the timing for *in vivo* evaluation (Rossin, 2011). The most current nanoparticles can be classified depending on the platform used (Fig. 1).

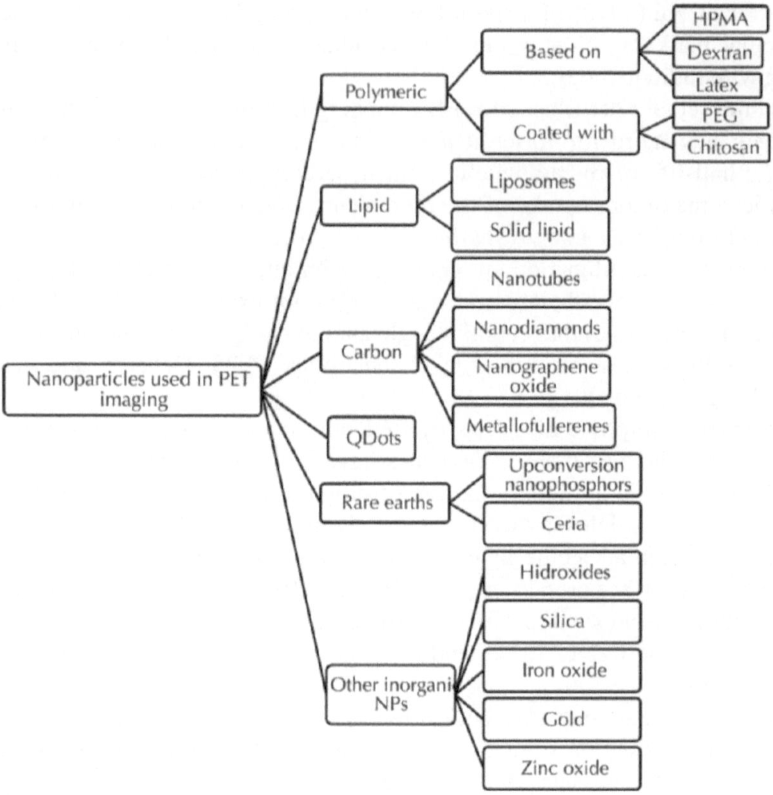

FIGURE 1 Classification of nanoparticles most used in PET imaging.

2.4.1 *POLYMERIC NANOPARTICLES*

According to Kaewprapan et al. (2012) "a number of polymers have been reported to be efficient materials for nanoparticles preparation including synthetic polymers such as poly (lactic-coglycolic acid), polylactide and polyalkylcyanoacrylates [...] and natural polymers such as albumin, gelatin, chitosan and dextran."

Important features of the polymers are the chemical flexibility and the versatility of their synthesis (Liu and Welch, 2012); polymers coating the surface of lipidic nanoparticles create an impermeable layer on the nanoparticles surface known as liposomes (Torchilin, 2005), thus improving their properties (Immordino et al., 2006).

2.4.1.1 NANOPARTICLES BASED ON HPMA

HPMA, N-(2-hydroxypropyl)-methacrylamide, is a polymeric compound, clinically approved by FDA (Food and Drug Administration) and utterly studied since many years ago (Lu, 2010); it has been used as platform for building nanoparticles and to study the mechanism of drug delivery. According to Lu (2010) these NPs can be used for "therapy of macromolecular drug conjugates. The cellular uptake, subcellular trafficking, pharmacokinetics, biodistribution and tumor targeting efficiency of HPMA copolymers have been visualized by using currently available imaging modalities, including optical imaging, γ-ray scintigraphy, single photon emission computed tomography (SPECT), positron emission tomography (PET) and magnetic resonance imaging (MRI)."

Taking into account that HPMA copolymers can stay for a long time in the bloodstream, depending on their structural design (homopolymers, random polymers or block copolymers), and that they can accumulate in tumoral tissue by the mechanism of enhanced permability and retention (Lammers et al., 2010) these interesting nanoparticles have potential for treatment of cancer in the form of conjugates (like polymer-doxorubicin) (Etrych et al., 2011; Vasey et al., 1999). "By increasing drug localization at the pathological site, and by reducing its accumulation in healthy organs and tissues, the efficacy of the intervention can often be increased, while its toxicity can be attenuated" (Lammers et al., 2010).

These NPs have also been used in some radiolabeling studies using ^{18}F (Herth et al., 2009; Moderegger et al., 2012) and ^{72}As (Herth et al., 2010) for *in vivo* PET studies.

At present, ^{72}As is considered as a promizing radionuclide because of its convenient half-life (26.0 h) that allows the study of biological processes, such as studying of targeting mechanisms of tumor therapeutics by radiolabeling of monoclonal antibodies, over several days (Jennewein et al., 2008). Nevertheless, its future in nuclear medicine, specifically in studies related to label of nanoparticles by PET imaging, is just beginning.

2.4.1.2 DEXTRAN NANOPARTICLES

Dextran is a carbohidrate of high molecular weight syntethised by bacteria taxonomically classified in the *Lactobacillaceae* family (Dhaneshwar et al., 2006); this polysaccharide has promizing future in nanomedicine, specifically as drug carrier.

Dextran is composed of α-1,6 linked D-glucopyranoside residues (Kaewprapan et al., 2012) and its convenient properties, such as non-toxicity, biodegradability and biocompatibility, have allowed the synthesis of nanoparticles based on it. This approach hold a great perspective to address the challenges related to specific site targeting and controlled drug release (Aumelas et al., 2007; Kaewprapan et al., 2012) due to their excellent physicochemical properties and physiological biocompatibility (Dhaneshwar et al., 2006).

Dextran microspheres labeled with ^{57}Co, ^{58}Co and ^{131}I have been used in biodistribution studies by Svedman et al. (1983). More recently, Keliher et al. (2011) synthesized a nanoparticle for imaging of macrophages (size = 13 nm) comprised of short, cross-linked dextrans that were modified with desferrioxamine and tethered with ^{89}Zr. The nanoparticle coated with dextran was stable, biocompatible and specifically accumulated in enriched tissues with macrophagues such as lymph nodes, liver, leukocytes in the bloodstream and splenic tissue.

Keliher et al. (2011) have also determined the ideal size of dextran NPs, its pharmacokinetics, renal clearance rates, and *in vivo* macrophage uptake; furthermore, they compared the biological behavior of linear and cross-linked dextran and found that the latter was better *in vivo* tolerated. In this opportunity, the selected β^+ radionuclide was ^{89}Zr (78.4 h), which has been used in studies of slow metabolic processes as protein synthesis, cell proliferation and labeling of antibodies, peptides and proteins (Verel et al., 2003; Walther et al., 2011).

The study performed by Keliher et al. (2011) allowed the quantification of macrophages and levels of inflammation in animal tissue, considered as important parameters in some diseases, including cancer therapy, atherosclerosis and myocardial infection. In addition, it is known that cells in inflamatories processes having high counting of macrophages can accumulate ^{18}F-FDG, so that they can be visualized on PET (Griffeth, 2005); use of ^{89}Zr for labeling of nanoparticles employed in the detection of macrophages could also give a more detailed knowledge on this physiological condition.

2.4.1.3 LATEX NANOPARTICLES

Latex is one of the highly versatile materials used for production of NPs. Latex nanoparticles designed for biomedical applications have been used in PET imaging; *in vivo* biodistribution studies have been performed in order to gauge their therapeutic potential.

Cartier et al. (2007) used nanoparticles with carboxylated latex, chemically known as poly glycidylmethacrylate (poly 2,3-epoxypropylmethacrylate = EPMA), compound used for manufactoring artificial organs or implants (Yamamuro et al., 1998). EPMA NPs (size = 144 nm) were labeled with ^{68}Ga for PET and ^{111}In for SPECT.

^{68}Ga (67.6 min) is usually bound to chelating agent known as DOTA (chemically known as 1,4,7,10-tetraazacyclododecane-1,4,7,10-tetraacetic acid) (Cantorias et al., 2008). Cartier et al. (2007) used the surface of nanoparticles based on latex for binding the radiotracer, taking into account that these surfaces have many carboxyl groups. In this case, the NPs were directly labeled with ^{68}Ga.

The results showed that ^{68}Ga mainly accumulated in organs such as the heart and the liver, and in less extent in the spleen; additionally, these studies demonstrated the biocompatibility and low toxicity of EPMA NPs. "The ability to bind different cationic tracers is provided by the carboxyl groups of the dense polymethacrylate corona. Therefore, we expect that other similar metal ions and isotopes, particularly with various half-lives could be employed, thus adapting the EPMA nanoparticles to the major biomedical imaging techniques available today. These include ^{99m}Tc, ^{67}Ga and ^{123}I used in single-photon computed tomography (SPECT) or ^{60,61,64}Cu for PET" (Cartier et al., 2007).

Another *in vivo* study using latex nanoparticles (Rossin et al., 2008) targeting the intercellular adhesion molecule 1, ICAM-1 (known as marker of inflammation that enhance the attachment of leukocytes to activated vascular endothelium) held promises for delivering therapeutics to the pulmonary endothelium. The authors determined whether it would be possible to visualize and quantify in real-time targeting of antiICAM nanocarriers to the lung endothelium by using ^{64}Cu. The possibility of obtaining *in vivo* imaging of the pulmonary endothelium through labeling of these latex beads based polymeric NPs (polymeric particles suspended in latex) with ^{64}Cu. has been demonstrated. Nevertheless, further studies are required in order to optimize the design of these NPs and to prolong their *in vivo* stability when they are labeled with the radionuclide.

2.4.1.4 NANOPARTICLES COATED WITH PEG

At present, one of the most used polymers is polyethyelene glicol (PEG), approved by the FDA for use as a vehicle, additive or excipient in foods, cosmetics and biopharmaceuticals. PEG chains can be covalently coated to other larger molecules, process known as PEGylation. Consequently, PET conjugation to protein drugs can be accomplished by PEGylation. The aim is to modify the pattern of drug distribution and to improve its therapeutic efficacy (Yamaoka et al., 1994).

PEG is chemically a linear polyether diol oligomer of ethylene oxide, with interesting properties for biomedical applications (Moghimi, 2002), which can be summarized as follows:

- Biocompatibility. They can entrap water-soluble (hydrophilic) pharmaceutical agents in their internal water compartment and water-insoluble (hydrophobic) pharmaceuticals into the membrane.
- Increased solubility in aqueous and organic media (Vert and Domurado, 2000). This feature is important for very low soluble molecules

like some anticancer drugs, e.g., taxol and camptothecin (Zacchigna et al., 2011).

- Little toxicity. PEG is eliminated from the body without any chemical change, either by the kidneys, for PEGs < 30 kDa, or in the feces, for PEGs < 20 kDa (Yamaoka et al., 1994).
- Very low antigenicity (Immordino et al., 2006; Zacchigna et al., 2011) and low immunogenicity (Moghimi, 2002).
- Half-life prolonged in the plasma, due to the increased hydrodynamic volume that reduces the kidney clearance. Since kidneys are organs composed of nephrons, which filter substances taking into account their size, the excretion of larger molecules is slower (Zacchigna et al., 2011).

PEG lessens the uptake of the nanoparticles by the reticuloendothelial system and increases their circulation time in the blood pool (de Barros et al., 2012) in comparison with nonPEGylated nanoparticles (de Barros et al., 2012; Torchilin, 2005). In addition, aggregation between nanoparticles coated with PEG and their association with serum and tissue proteins decrease, because their units have hydrophilic character, which implies a higher solubility in blood serum.

NPs coated with thick PEG layers showed longer time blood circulation and low accumulation in renal tissues, facts that suggest their posible applications as *in vivo* carriers (Welch et al., 2009).

At present, and following Wadas et al. (2007), ^{64}Cu "has emerged as an important positron emitting radionuclide that has the potential for use in diagnostic imaging and radiotherapy. However, ^{64}Cu must be delivered to the living system as a stable complex that is attached to a biological targeting molecule for effective imaging and therapy. Therefore, significant research has been devoted to the development of ligands that can stably chelate ^{64}Cu." One of these ligands is DOTA, a universal chelating agent, which is considered as a prototype for coupling di and trivalent metallic radiotracers, including ^{67,68}Ga, ^{111}In, ^{90}Y and ^{64}Cu (Al-Nahhas et al., 2007).

Rossin et al. (2005) also used ^{64}Cu for labeling nanoparticles and found that they remained for long time in the bloodstream and accumulated in tumors overexpressing folate transporters, i.e., glycozyl-phosphatidylinositols essential for normal growth and maturation. Folate transporters are linked to membrane proteins on the cells but become overexpressed in many malignant cells (Weitman et al., 1992). In the aforementioned study a high uptake in the liver was apparent, but even in such conditions it was possible to obtain tumor images.

The facts described above could open the possibility of using ^{64}Cu labeled nanoparticles in cancer therapy; those employed by Rossin et al. (2005), SCK-NPs, were obtained by cross-linking of the shell from amphiphilic diblock copolymers; then SCK-NPs were functionalised with fluorescein thiosemicarbazide

(FTSC); TETA (1,4,8,11-tetraazacyclotetradecane-*N*, *N'*, *N''*, *N'''*-tetraacetic acid) was used as chelating agent for ^{64}Cu, leading to its high *in vivo* stability. This result is in agreement with the bibliography (Sun and Anderson, 2004; Shokeen and Anderson, 2009). According to Welch et al. (2009) an advantage of the radiolabeling of polymeric nanoparticles with ^{64}Cu is to achieve high specific activities (3,700 kBq/mg [>100 μCi /μg] of polymer), which enable the administration of low doses to healthy rodents (~5 μg/animal).

Zeng et al. (2012) developed a method for radiolabeling of SCK-NPs core with ^{64}Cu using DOTA as chelating agent; this novel strategy was based on metal free "click" chemistry and increased the efficiency: "when 50 mCi of ^{64}Cu was used for radiolabeling of the SCK-NPs containing 10% in-core azides, a very high SA was obtained, up to 975 Ci/μmol (4420 Ci/μmol at EOB)."

Matson and Grubbs (2008) also used nanoparticles synthesized with block copolymers (polynorbornene) labeled with ^{18}F (61% radiochemical purity); their results give a highlight to propose these NPs as *in vivo* molecular imaging agents through PET.

1. Block copolymers are interesting platforms for obtaining nanoparticles with multifunctional capabilities, including their possible use in drug delivery (Zhang et al., 2012).
2. As it is stated by Riggio et al. (2011): "The ideal nano-carrier for drug delivery and cancer chemotherapy should (i) stabilize without altering the pharmacological activity of the drug, (ii) prevent premature metabolic degradation of the drug in the systemic circulation such that it arrives in a pristine state at the intended target, (iii) release the drug at the intended site/tumor, and (iv) exhibit similar or lower toxicity than that of the free drug."
3. Almutairi et al. (2009) reported the use of ^{76}Br and ^{124}I to radiolabel dendritic nanoprobes for PET imaging[1]; they used a core coated with polyethylene oxide (PEO) chains, thus giving a dendritic form to the nanoprobe, which gave biological stealth and allowed the pharmacokinetics in biodistribution studies.
4. This approach confirmed the utility of radiohalogens to track the angiogenesis (Almutairi et al., 2009), a process widely investigated in conjuction with the development of tumors, in particular because of the possibility of their early detection and as a new strategy of therapy.

---

[1]According to Mody (2011): "a nanoprobe can non-invasively provide valuable information about differentiate abnormalities in various body structures and organs. Therefore, the result is the determination of the extent of disease, and to assess the effectiveness of a treatment". For these reasons, a dendritic nanoprobe labelled with long half-life positron emitters (^{76}Br or ^{124}I), can be used for PET imaging.

Simone et al. (2012) illustrated in a recent study the use of polymeric nanoparticles (poly (4-vinyl)-phenol nanoparticles) labeled with ^{124}I and ^{125}I, for targeting the pulmonary endothelium, i.e., the cellular layer lining the inner surface of these blood vessels; the endothelium is considered as an organ system "involved in a multitude of physiologic functions, including the control of vasomotor tone, the trafficking of cells and nutrients, the maintenance of blood fluidity, and the growth of new blood vessels" (Aird, 2004).

The results obtained by Simone et al. (2012) showed a promizing future for improving "investigation of NPs interactions with target cells and PET imaging in small animals, which ultimately can aid in the optimization of targeted drug delivery." The use of radionuclides longer-live than ^{18}F (109.7 min.) is advantageous for evaluating *in vivo* longitudinal pharmacokinetic profiles and targeting capability of NPs characterized by long blood and tissue residence times.

2.4.1.5 NANOPARTICLES COATED WITH CHITOSAN

According to Agrawal et al. (2010) "chitosan [poly (1,4-β-D-glucopyranosamine)], an abundant natural biopolymer, is produced by the deacetylation of chitin obtained from the shells of crustaceans. It is a polycationic polymer that has one amino group and two hydroxyl groups in the repeating hexosaminide residue."

This polysaccharide has been used as coat of nanoparticles with magnetic core based on Fe_3O_4; using these NPs, Pala et al. (2012) reported the labeling of granulocytes with ^{64}Cu. Their results showed their binding capacity with Cu^{2+}; by phagocytosis, the granulocytes took up the labeled NPs, then these cells were visualized by PET imaging. The high stability of ^{64}Cu in the *in vitro* model was also confirmed. The contribution of this research was to shed light on the attainment of efficient NPs for the study of inflammatory processes characterized by high level of granulocytes.

2.4.2 LIPID NANOPARTICLES

Two types of lipid nanoparticles, liposomes and solid lipid nanoparticles, will be described.

2.4.2.1 LIPOSOMES

They can be defined as nanoparticles with spherical form or self-closed structures formed by one or several concentric lipid bilayers including an aqueous phase inside and localized between the lipid bilayers (Agrawal et al., 2010; Phillips et al., 2009; Torchilin, 2005). They are some of the most broadly used (Helbok et al., 2009; Phillips et al., 2009) and studied nanoparticles (Richardson et al., 1978; Silindir et al., 2012; Taguchi et al., 2009) with lipidic component.

Liposomes can be classified taking into consideration three of their properties (de Barros et al., 2012):

- Lipid components (natural zwitterionic phospholipids, e.g., phosphatidylcholines, phosphatidylethanolamines, stealth liposomes or long circulating nanoliposomes and targeted liposomes) (Silindir et al., 2012).
- Size (small, large or multilamellar vesicles).
- Charge (neutral and negatively or positively charged) (de Barros et al., 2012). This parameter determines how nanoliposomes interact with other molecules.

According to Silindir et al. (2012) these properties characterizing liposomes affect its vascular circulation time after intravenous administration.

To date, the use of radionuclides for labeling the liposomes is, as in other nanoparticles, well known (Li et al., 2012); these studies show insights into the pharmacokinetics of liposomes and to quantitatively measure the radioactivity distribution in *ex vivo* different tissues (Helbok et al., 2009; Laverman et al., 1999; Richardson et al., 1978).

While the internal aqueous core is suited for deliverying of hydrophilic drugs and for enclosing radionuclides, like ^{64}Cu (Petersen et al., 2011), the phospholipid bilayer is suited for encapsulating hydrophobic chemotherapeutics (Khan, 2010; Silindir et al., 2012).

The special structure of liposomes makes possible to propose them as a carrier system in cancer therapeutic applications (Riggio et al., 2011), in the same form as polymeric nanoparticles, silica nanoparticles and metallic nanoparticles can do it.

According to Torchilin (2005), there are a variety of current approaches to use targeted liposomes agents as pharmaceutical carriers: virosomes (Bhattacharya and Mazumder, 2011), magnetic liposomes (Pradhan et al., 2010), haemosomes (liposomal hemoglobin) (Taguchi et al., 2009) and ATP liposomes (Hartner et al., 2009).

In all cases, one important property characterizing the use of liposomes is the size, aspect that was evaluated years ago (Chen and Weiss, 1973; Richardson et al., 1978); for example, it has been demostrated that liposomes labeled with ^{18}F-FDG (>200 nm) accumulate in the spleen, since its size does not allow them to pass through the walls of the venous sinuses in the splenic tissue (Chen and Weiss, 1973; Oku, 1999).

In contraposition, smaller liposomes (<100 nm) can remain in the blood circulation and accumulate in tumor tissues, due to its enhanced permeability and retention (Maeda, 2001; Matsumura and Maeda, 1986; Oku et al., 2011).

If the liposomes are not functionalised, i.e., if their surfaces do not suffer any modification, they will be fastly eliminated from the blood and will be phagocytised by the cells belonging to the reticuloendothelial system, including hepatic cells. These physiological aspects do not allow their accumulation in the desired tissues or organs; for this reason, a number of developments, including the incorporation of hydrophilic PEG on the surface of liposomes, have been necessary to functionalise them.

The products obtained by coating the surface of liposomes with PEG are known as PEG-lipid conjugates; one example of them is PE-mPEG (phosphoethanolamine-methoxy (polyethylene glycol)). PEG acts as a polymeric steric stabilizer in PEG-lipid conjugates (Immordino et al., 2006; Torchilin, 2005), forming a protective layer over the liposomes surface and increasing the time of permanence in the blood circulation (Drummond et al., 1999). In this case, steric stabilization refers to the colloidal stability conferred on these NPs by hydrophilic polymers or by hidrophilic glycolipids (Drummond et al., 1999).

According to Immordino et al. (2006) these surface modifications of liposomes with PEG can occur through three processes: physical adsortion of PEG onto the surface of the liposomal vesicles, incorporation of the PEG-lipid conjugates during liposomal preparation, or covalent attachment of reactive groups onto the surface of the preformed liposomes.

It has been shown that liposomes sterically stabilized improve their antitumor therapeutic efficacy (Papahadjopoulos et al., 1991); liposomal agents containing colloidal gold inside also showed similar promizing results, i.e., they accumulate in dermal lesions resembling the malignant tumor known as Kaposi's sarcoma in transgenic mice bearing (Huang et al., 1993).

The production of aggregation by positively charged liposomes instead of neutral or negatively charged liposomes has been evaluated in presence of serum (Oku, 1999). *In vivo* tests for PET imaging with liposomes labeled with ^{18}F demonstrated that liver and spleen uptake is maximal for positively charged liposomes, while neutral liposomes have minimal uptake in these same tissues (Oku, 1999).

In some preclinical studies, liposomes labeled with radionuclides have been used for molecular imaging in pharmacokinetic studies. The following β+ radionuclides have been used up to the present time in PET imaging for labeling of liposomes: ^{64}Cu (Seo et al., 2008; Petersen et al., 2011), ^{18}F (Marik et al., 2007; Oku, 1999, 2011), ^{62}Cu, ^{15}O, ^{124}I (Phillips et al., 2009) and ^{68}Ga (Helbok et al., 2009).

Some of the single photon radionuclides employed for labeling of liposomes for SPECT, were ^{99m}Tc (de Barros et al., 2011), ^{111}In, ^{67}Ga, ^{186}Re (Marik et al., 2007) and ^{123}I (Phillips et al., 2009).

2.4.2.2 SOLID LIPID NANOPARTICLES

Solid lipid nanoparticles (SLNs) are carrier systems with potential for drug delivery to a tissue target; they are composed of physiological lipids in solid state such as triglycerides, cetyl alcohol, emulsifying wax, cholesterol, and colesterol butyrate (Loxley, 2009); SLNs were developed as alternative systems with respect to others carriers, including emulsions, liposomes, and polymeric nanoparticles (Andreozzi et al., 2011). Like other NPs, they have properties of biocompatibility, low toxicity and physical stability (Ekambaram et al., 2012) that allow their use in biological milieu.

SLNPs have other advantageous characteristics, such as their resistance against degradation, ease of scale-up and manufacturing without using organic solvents (Ridolfi et al., 2011). An additional advantage is their great potential for dermal drug delivery, since they are not cytotoxic in fibroblasts and keratinocytes cells (Ridolfi et al., 2011).

Andreozzi et al. (2011) developed a method for radiolabeling SLNs with ^{64}Cu bound to 6-[p- (bromoacetamido) benzyl]-1,4,8,11-tetraazacyclotetradecane-N, N′, N″, N‴-tetraacetic acid (BAT) conjugated to stearic acid (lipid-PEG-BAT), chelating agent that had been also tested in labeling of liposomes with the same radionuclide (Seo et al., 2008). They concluded that this radiopharmaceutical is useful for mapping the *in vivo* biodistribution using PET, and the *ex vivo* biodistribution, from gamma counting of some organs.

2.4.3 CARBON NANOPARTICLES

One interesting and promizing group of nanoparticles is formed by some carbon allotropes: nanotubes, nanodiamonds, carbon nanohorns, metallofullerenes and nanographenes. There are general approaches related to the technology of their production and the methods of characterization, and the most important feature is that the physical and chemical characterization of one of them provides the basis for similar research on the other carbon nanostructures. Due to the variety of composition, forms and properties, different nanoparticles based on carbon are better suited for different interesting applications, including PET imaging.

Carbon particles having sizes between 100 and 300 nm, labeled with ^{68}Ga (galligas) can be used in human studies of pulmonary ventilation-perfusion for determination of the V/Q ratio, where V and Q are respectively defined as the amount of air and blood reaching the alveoli (Kotzerke, et al., 2010). Similarly to what occurred with the application of ^{99m}Tc in technegas and pertechnegas aerosols, these particles have demonstrated their capability for assessment of lung function in diseases as pulmonary embolism through PET/CT imaging (Hofman et al., 2011). As these authors stated: "Compared with conventional V/Q imaging, advantages include higher-resolution, fully tomographic images with potentially better regional quantitation of lung function. The short half-life of ^{68}Ga also enables more flexible acquisition protocols with the option of performing ventilation studies selectively on patients with abnormal perfusion."

Kotzerke et al. (2010) concluded that alike technegas, galligas are easy to obtain and to apply, rendering clinically relevant information about pulmonary diseases. According to the opinion of Ament et al. (2012): "^{68}Ga aerosol (galligas) and ^{68}Ga-labeled MAA (macroaggregated albumin) are efficient substitutes for clinical use and could be an interesting alternative with high accuracy for lung V/P imaging with ^{99m}Tc-labeled radiotracers, especially in times of Mo-99 shortages and increasing use and spread of PET/CT scanners and Ga-68 generators, respectively"[2].

2.4.3.1 SINGLE WALLED CARBON NANOTUBES (SWCNTS)

They are graphene cylinders that are comprised of sp^2 bonded carbon atoms and possess highly regular structures in a hexagonal lattice (McDevitt et al., 2007).

SWCNTs are promizing transporters across cellular membrans (Kam and Dai, 2006) because they have capacity to easily pass through biological barriers.

In biologic systems, the shape of SWCNTs enables them to move into and out of the vasculature and accumulate in tumor tissues via the enhanced permeability and retention effect that characterize these tissues (James and Gambhir, 2012).

Another advantage is their high *in vivo* stability, due to their special mechanical properties; their unique electrical properties are useful for building new devices with purposes of diagnosis (Bianco et al., 2008). As liposomes, SCNTs have

[2]Pulmonary imaging using ventilation/perfusion (V/P) SPECT is a well-established diagnostic tool for pulmonary embolism.

empty internal space for encapsulation and transportation of therapeutic molecules and imaging agents, including radionuclides.

Despite these favorable qualities, SCNTs have also some disadvantages, some of them being: strong tendency to form aggregates, lack or limited availability of data on tolerance by healthy tissues (Bianco et al., 2008) and relatively high biodegradability and toxicity, once administered in the living subject. These are topics of much debate and may be the major limiting factor concerning their approval for human use (James and Gambhir, 2012). Regarding this aspect, it is necessary to conduct new studies and to obtain definitive conclusions about their toxicity (Smart et al., 2006).

From the chemical point of view, SCNTs have poor solubility in water and in all types of solvents (Bianco et al., 2008); a way to dissolve or at least to disperse them in water is to employ reagents having water-soluble side moieties, such as hydroxyl, amino, carboxyl, and poly vinyl alcohol groups (Zhang et al., 2008). In 2008, Zhang et al. evaluated the cytotoxicity of CNTs functionalised with phosphoryl choline, in order to confer them hydrophilic properties and therefore to improve their solubility in water. They used two cells lines in an *in vitro* model: clonal pheochromocytoma cells (PC12) and human colon carcinoma cell lines ($CaCo_2$); they suggested that there was not citotoxicity caused for CNTs in these conditions.

Since SWNTs are attractive nanomaterials characterized by high flexibility and compatibility with many imaging modalities (James and Gambhir, 2012), including SPECT, PET and near-infrared fluorescence imaging (NIRF)[3], the scientific world is now focused in this field, making efforts to obtain more information about their *in vivo* behavior.

At present, there are studies in preclinical phase destined to assess the behavior of CNTs previously functionalised for biocompatibility in the physiological media. For a better understanding of the physiological role of the SCNTs covalently bound to chelating agents (DOTA or desferroxamine, DFO) and to the antibody E4G10, where the new blood vessels belonging to the tumor act as target tissue, their *in vivo* renal clearance was evaluated by Ruggiero et al. (2010) using PET, microscopy and NIRF. This research showed evidences that molecular size and high aspect ratio of CNTs have an important impact on glomerular filtration

[3]PET has relatively low temporal and spatial resolutions (1–30 min, 4–10 mm) for targeted imaging in cancer (Joshi and Wang, 2010). In order to achieve better results, other modalities of imaging, including optical techniques, are often applied. Near-infrared (700–1000 nm) fluorescence imaging (NIRF) light can penetrate several centimeters in tissue and improve the spatial resolution (Xiao et al, 2012), these facts explaining its use as secondary imaging modality to complement PET in this kind of studies.

in the nephrons of renal tissue in mice; furthermore, the pK profile of these molecules allows to suggest their use for design of novel therapeutics with particular pharmacologic features (e.g., shape and aspect ratio) (Ruggiero et al., 2010) and potential use in gene delivery (Xu et al., 2006).

In the aforementioned work, Ruggiero et al. used SWCNT-([^{89}Zr] DFO) (E4G10); ^{89}Zr, the selected β^+ radionuclide for imaging of tumor vessels in murine xenograft model, has also been suggested by other authors (Holland et al., 2010; Verel et al., 2003) for *in vivo* imaging of cancer.

Looking ahead, and according to Ruggiero et al. (2010), their research shows the possibility of using SNCTs incorporating both imaging and therapeutic agents onto the same nanoplatform for targeting the blood vessels irrigating solid tumors.

Another β^+ radionuclide used for labeling CNTs is ^{86}Y; the biodistribution and pharmacokinetics of ^{86}Y-CNTs funcionalised by DOTA demonstrate that the target tissues reached are kidney, liver, spleen, and in minor grade, bone. The rapid blood clearance could suggest the feasibility of using β^+ radionuclides with short half-life in this type of studies (McDevitt et al., 2007a, 2007b).

2.4.3.2 NANODIAMONDS (NDS)

Diamond structures at nanoscale (~1 to 100 nm) including diamond films in pure phase, diamond particles, 1-D diamond nanorods and 2-D diamond nanoplatelets (Schrand et al., 2009) show great potential in bioimaging.

NDs can be spherical or prismoidal carbon particles with a truncated octahedral architecture, the cores of which are comprised of sp^3 bonded carbon atoms (Bianco et al., 2008), and can be partially coated by a shell of graphite or amorphous carbon with dangling bonds terminated by functional groups (Mochalin et al., 2012).

According to Zhu et al. (2012), NDs with sizes between 2–8 nm have some interesting properties such as chemical stability, optical transparency, biocompatibility, low cytotoxicity, extremely high hardness, stiffness and strength; due to its hability for crossing the cell membranes (Liu and Welch, 2012) they have good cell uptake (Bianco et al., 2008).

As it was found by Zhang et al. (2008) NDs have higher cellular uptake rate, in comparison with CNTs and graphene materials. This is in agreement with the results obtained by Xing and Dai (2009) who showed that NDs are much more biocompatible than to other carbon nanomaterials, including carbon blacks, fullerenes and carbon nanotubes.

NDs have additional advantages over all nanomaterials, including their large surface area and high adsorption capacity (Zhu et al., 2012). Due to their intrinsic hydrophilic surface they are unique among the carbon nanoparticles, this being one of the many reasons to be considered for biomolecular applications (Liu and

Welch, 2012). The surface of NDs contains a complex array of surface groups: carboxylic acids, esters, ethers, lactones, and amines (Schrand et al., 2009), so that they can easily bind to other molecules.

On this basis, as it was recently pointed out, NDs have a wide range of potential applications in drug delivery (Pusuluri and Kadam, 2012; Silindir et al., 2012; Zhu et al., 2012), and bioimaging (tissue engineering, and also as protein mimics and as a filler material) (Mochalin et al., 2012).

A maximum therapeutic benefit of NDs without damage to surrounding healthy tissues is expected to be reached in the future (Schrand et al., 2009). Nevertheless, some disadvantages of NDs have been recognized, the most relevant being their difficult manipulation and their insolubility in any solvent, which implies a tendency to form aggregates (Bianco et al., 2008). New developments have allowed to increase their solubility in aqueous media and to reduce particle aggregation: Martín et al. (2010) assert that NDs can be submitted to a Fenton reaction with hydrogen peroxide in order to remove the amorphous matrix that embedded them, forming large aggregates. The hydroxyl radicals produced by the Fenton reaction reduce the initial particle size from 20 nm (commercial NDs) to about 7 nm on average (Rojas et al., 2011), which represents an additional advantage. As it was mentioned in precedent paragraphs, this size is very adequate to go through the cellular membranes (Zhu et al., 2012). To date, NDs can be functionalised in a controllable manner for their further interaction with therapeutic biomolecules such as proteins and antibodies (Schrand et al., 2009).

There are studies showing the preparation of radioactive tracers using nuclides covalently bound to these nanomaterials; like in others NPs, their stability is one of the most important requisite to study their *in vivo* behavior. Biodistribution studies that have demostrated *in vivo* stability have been performed in mice using CNs and NDs with ^{99m}Tc (Wei et al., 2012). It was also proved that their behavior and fate in mice depend strongly on oxidized multiwalled carbon nanotubes (oMWCNTs) although NDs have small influence on their biodistribution and excretion.

Rojas et al. (2011) labeled nanodiamonds with ^{18}F to study their *in vivo* biodistribution by PET and found that the tissues targets were lung, spleen and liver; the excretion of NDs was checked and it was also concluded that the addition of surfactant agents induced a slight reduction in the urinary excretion rate. After the removal of higher size NDs by filtration, the uptake rate in the lung and spleen was inhibited and the uptake rate in the liver was reduced in a meaningful way (Rojas et al., 2011).

2.4.3.3 NANOGRAPHENE OXIDE (NGO)

These nanomaterials are sheets of two dimensional graphitic carbon systems comprised of sp^2 bounded carbon atoms. Conceptually, they are made by unrolling a carbon nanotube into a flat sheet; the basic material, graphene, has been considered as the rizing star of materials science and nanotechnology (Geim and Novoselov, 2007).

Like in other nanocarbons, their toxicity was studied, but no definitive conclusions have been reached (Uo et al., 2011); therefore, according to Hong et al. (2012a) their biological applications remain unexplored. However, these authors carried out two recent studies related to the use of nanographene in PET imaging by labeling with two positron emitters, ^{64}Cu and ^{66}Ga (Hong et al., 2012a, 2012b).

Hong et al. (2012b) have demonstrated that nanographene oxide sheets having PEG chains terminated in amino groups can bind to ^{66}Ga via conjugation to NOTA (1,4,7-triazacyclononane-1,4,7-triacetic acid). These nanoparticles were conjugated to monoclonal antibodies that bind to receptors overexpressed on tumor vessels. PET/CT imaging allowed to confirm the tumor uptake of these NPs, indicating the existence of affinity of nanographene for tumor vessels, the target tissue in this case.

2.4.3.4 METALLOFULLERENES

Metallofullerenes are fullerenes with components conforming a geodesic dome or carbon cage (Yamada, 2008) that can encapsulate single metallic ions (mono-, di- or trimetallofullerenes) or trimetallic nitrides.

At present, some metals belonging to groups 1 (Li, Na, K, Cs), 2 (Ca, Sr, Ba), 3 (Sc, Y, La, Ho, U) and 4 (Zr, Hf) have been successfully encapsulated into the cages (Liu and Sun, 2000). The structures containing metal atoms inside carbon cages are called endohedral metallofullerenes. Some of them have been used as imaging agents (Luo et al., 2012); for example, f-$Gd_3N@C_{80}$, is a functionalised nanoplatform containing trigadolinium that could have theranostic potential capable of delivering effective brachytherapy (Luo and Jang, 2010)[4].

Although their production is very difficult and their yields are very low (<1%), they are considered promizing materials for some applications (Liu and Sun, 2000) such as a preliminar study of xenograft brain tumor model (Shultz et al., 2011) with ^{177}Lu bound to DOTA (^{177}Lu-DOTA-f-$Gd_3N@C_{80}$).

---

[4]The @ symbol indicates that atoms to the left are encapsulated within the carbon cage written on the right (Chai et al., 1991).

Metallofullerenes labeled with ^{18}F and ^{124}I have also been used in PET imaging of rat brain tumors to estimate absorbed dose (Luo and Jang, 2010).

Recently, Luo et al. (2012) developed a nanoprobe for PET/MRI imaging using f-$Gd_3N@C_{80}$ labeled with ^{124}I for visualization of tumors and confirmation of their anatomical location in brain, by subsequent ^{18}F-FDG scanning.

"As clinical PET/MRI research advances, the current ^{124}I-f-$Gd_3N@C_{80}$ agent is promizing because it will have identical biodistribution in both imaging modalities. Additionally, the substitution of ^{124}I with ^{125}I or ^{131}I using the same synthetic methods in this work will produce a dual MR diagnostic/therapeutic (theranostic) nanoprobe" (Luo et al., 2012).

The possible applications of metallofullerenes in PET imaging are still in their beginnings. Important aspects to be studied are their functionalization and their modifications in the surface to allow the study of biologic processes.

2.4.3.5 CARBON NANOHORNS

Carbon nanohorns (CNHs), recognized as members of the fullerene family (Iijima et al., 1999) and closely related to carbon nanotubes with capacity to adsorb most types of molecules (Bianco et al., 2008) are among the most attractive nanomaterials, although their biomedical properties has not been so far satisfactorily achieved (Miyako et al., 2008). They exhibit particular structures that resemble flowers (Miyawaki et al., 2006) like dahlia or flower buds (Wang et al., 2004).

Like others nanoparticles, CNHs have an extensive surface area (Wang et al., 2004) and high number of pores, which enable the adsorption of molecules. In addition, little holes can be generated at the tips of the tubes and can be exploited to insert different therapeutic agents into their empty space (Bianco et al., 2008).

The advantage related to this feature is that the CNHs can be used as reservoirs for controlled drug release (Miyawaki et al., 2006). Although they have many desired properties, additional *in vitro* and *in vivo* studies must be carried out in order to reduce the toxicity due to their chemical composition and the process used for functionalization them. However, CNHs have shown to be nontoxic in the short-term, either by *in vitro* or *in vivo* testings (Miyawaki et al., 2006).

In vitro results provide more insights into cellular structure and function at subcellular level, whereas *in vivo* and *ex vivo* data yield information about the pharmacokinetics of these particles and their ability to reach the target tissue (Lucignani, 2009).

In this sense, the evaluation of CNHs biocompatibility for use in living cells was performed by Fan et al. (2007); their *in vitro* results showed that CNHs could readily enter the cells, probably through endocythosis.

CNHs can be produced without the use of any metal catalysts, thus implying a metal free condition and, consequently, a high purity (Wang et al., 2004).

Since they belong to the nanocarbons family, CNHs are alternative forms closely related to CNTs and appear as spherical aggregates of single-walled carbon nanotubes (Bianco et al., 2008). So far, CNHs have not been used in PET imaging; however, they show a significant potential for intracelular delivery (Fan et al., 2007) and can act as facilitators of the photothermal destruction of tumors (Whitney et al., 2011). These favorable properties, as well as their use in some bone studies (Kasai et al., 2011), make it possible to predict the development of chemical processess for functionalise them and their future utilization in other biomedical studies, including PET imaging.

2.4.4 QUANTUM DOTS (QDS)

Quantum dots are nanocrystals or fluorescent semiconductor NPs, classified as inorganic dyes, with core comprised of cadmium and selenium (de Barros et al., 2012; James and Gambhir, 2012) specifically in the form of cadmium selenide or lead selenide (CdSe or PbSe) (Bruchez et al., 1998; Murray et al., 1993). The core is coated with different inorganic semiconductor materials: zinc sulfide or zinc selenide (Kuno et al., 1997; Lee, 2007); the high bandgap energy associated with the shell coating the core limits the excitation and emission solely to the core material (Walling et al., 2009).

Briefly, a QD is a semiconductor material whose charge carriers (electrons and holes) are in all three spatial dimensions (Bertolini et al., 2008, 2009). Because of their especial composition, QDs have intrinsic optical properties that make them excellent fluorescent probes for many biological and biomedical applications (Akerman et al., 2002; Wu et al., 2002). These properties include unique bright luminescence (May et al., 2012), high quantum yields, excellent resistance to photobleaching, broad excitation spectra, narrow emission bands from UV to the near-infrared regions (Ducongé et al., 2008) and excellent photostability (Wu et al., 2002).

The hydrophobic condition of QDs is a drawback for biological applications. Like in other NPs, functionalization with secondary coatings such as mercaptopropionic acid and polyethylene glycol (PEG) improve their solubility and keep them in a non-aggregated state (Rzigalinsky and Strobl, 2009) for *in vivo* applications. Other methods for improving their biocompatibily include silanization and surface conjugation, for example, exchange with bifunctional molecules having both hydrophobic and hydrophilic sides (Azzazy et al., 2007).

PEGylation, as a process of functionalization of NPs, increases the time of *in vivo* nanoparticle circulation (Owens and Peppas, 2007). However, functionalization process of QDs by addition of PEG coatings, enlarge their size to ~20–30 nm (James and Gambhir, 2012). While QDs having a hydrodynamic diameter smaller than 5.5 nm are efficiently and fastly excreted in rodent model (Choi et al., 2007) the biological elimination process could be complicated for more voluminous particles.

The main disadvantage of QDs is the toxic core, which difficult their clinical translation, especially in the case of drug delivery (James and Gambhir, 2012) and therefore, their acceptance by regulatory agencies (Nune et al., 2011); in this respect, and taking into consideration that QDs for *in vivo* imaging offer advantages over other traditional techniques, the development of a new generation of QDs without cadmium (Allen and Bawendi, 2008; Kim et al., 2005) or by addition of silica coating the cadmium core (Botsoa et al., 2008) could avoid their toxicity (de Barros et al., 2012; James and Gambhir, 2012; Rzigalinski and Strobl, 2011). An example of the latter approach has been done by May et al. (2012); they used block copolymers (triblock copolymers, F127) encapsulating silicon quantum dots (SQDs); this enhanced method allowed to obtain cadmium free QDs, i.e., without toxicity, with hydrophilic nature and suitable for cancer imaging applications, specifically tumor targeting.

Other suitable approaches, proposed by Pradhan et al. (2007) and Xie et al. (2008) are the generation of QDs more biocompatible with less toxic core, for example based on InAs or InP for NIR imaging; but according to Cai and Hong (2012), it is necessary to perform many studies for incorporating radionuclides of As or In in these QDs. If ^{111}In has found applications in SPECT imaging (Jiang et al., 2012; Lammers et al., 2010), it would be possible to expect the development of QDs labeled with this radionuclide. Similarly, and taken into account that ^{72}As and ^{74}As are promizing radionuclides already used in some monoclonal antibodies radiolabeling studies in PET imaging (Jennewein et al., 2008) they could be employed for labeling these or others nanoparticles, as it was carried out by Herth et al. (2010).

The main question in the functionalization of QDs is how to stabilize them in aqueous solution and make them biocompatible without affecting their optical properties (May et al., 2012). Other limitations of QDs include the fact that they are prone to aggregation, due to their surface chemistry (James and Gambhir, 2012), and the application of the available methods of attaching targeting moieties to overcome this difficult can constitute a real time-consuming challenge (James and Gambhir, 2012). But even with these drawbacks, QDs have found a wide range of *in vivo* and labeling studies (Chan and Nie, 1998) including immunoassays, cell tracking in metastases (Akerman et al., 2002; Nune et al., 2011) and imaging techiques (Walling et al., 2009). Compared with small molecular dyes, QDs have intense fluorescence emission that makes it easier to track single protein molecules. According to Bertolini et al. (2008): "their large two photon cross sections allow *in vivo* imaging at greater depths"[5].

Functionalised QDs, due to their hydrophilic nature, can enter the living cells with more facility, going through cell membranes by transfection, delivery mediated by peptides, or passive uptake by endocytosis (Walling et al. 2009).

Wu et al. (2002) used QDs bound to immunoglobulin G (IgG) and streptavidin (avidin) to label a breast cancer marker, Her2, on the surface of fixed and live cancer cells; their results indicated that they can be effective in cellular imaging and offer substantial advantages over organic dyes in the detection of the target tissue.

QDs have a great potential in nuclear medicine; they have been functionalised with different tumor targeting agents including peptides, receptor ligands and antibodies; some preliminary and preclinical studies confirm their promizing future in this field (Patt et al., 2010). According to these authors, the labeling of QDs with positron emitters would allow for studying their biodistribution and pharmacokinetics by PET. In 2007, Schipper et al. evaluated by PET imaging the *in vivo* biodistribution of QDs labeled with ^{64}Cu with and without PEG coating; they found that the size of the particles had not influence on the biodistribution within the range studied (12–21 nm). Pegylated QDs showed slightly slower uptake into the liver and the spleen, and also into bone at low levels. Although the study was performed in animals, it provided useful information about the biological behavior of QDs for possible diagnostic or therapeutic application.

It has been shown that quantum dots are suitable for cell tracking (Walling et al., 2009), mapping of reticuloendothelial system and tumor targeting. Chen et al. (2008) reported the use of QDs labeled with ^{64}Cu, in dual imaging modality (NIRF/PET), for assessing *in vitro* and *in vivo* efficacy in tumor targeting, namely the vascular endothelial growth factor receptor (VEGFR). The success of this dual imaging approach may render higher degree of accuracy for NIRF imaging in deep tissues (Chen et al., 2008).

Patt et al. (2010) carried out new strategies for radiolabeling of amino QDs (functionalised nanoparticles on their surface with amino groups) using ^{18}F and ^{11}C. They studied biodistribution and pharmacokinetic by PET imaging in an animal model; their results showed that both QD tracers had rapid clearance from the blood pool and indicated a metabolically stable label.

---

[5]QDs have absorption spectra based on an optical phenomenon in which a molecule can simultaneously absorb two photons. Two photon absorption in materials such as QDs can be quantified by the two photon absorption cross section, a parameter that depends on the photon energies.

2.4.5 NANOPARTICLES BASED ON RARE EARTHS

2.4.5.1 UPCONVERSION NANOPHOSPHORS

Photon upconversion is a process where the absorption of two or more low-energy photons is followed by the emission of a single photon of higher energy, being one of the phenomena employed in bioanalytical assays through the upconverting phosphors (UCNPs), inorganic crystals composed of a transparent host lattice doped with certain trivalent lanthanide ions or transition metals, such as Yb^{3+}, Er^{3+}, Tm^{3+} and Gd^{3+} (Riuttamäki, 2011).

Like other NPs, UCPs must be modified for use in biological milieu because they are hydrophobic (Mai et al., 2006). The surface of functionalised UCPs has functional groups suitable for bioconjugation purposes, such as carboxyl, amino, and maleimide (Zhou et al., 2012); consequently, they become hidrophilic. At present, upconverting materials also can exist as upconverting nanophosphors (UCNPs) functionalised for using in bioimaging.

Like in others NPs, including QDs, there are many methods for functionalizing UCNPs, including polymer capping, surface ligand oxidation, ligand exchange, organic ligand free synthesis, cation-assisted ligand assembly (Liu et al., 2011), ligand oxidation reaction, layer by layer method, methods based on hydrophobic-hydrophobic interaction or host-guest interaction and silanization (Zhou et al., 2012).

According to Ju and Shan (2009), upconverting nanophosphors, such as phosphorescent oxide salt particles doped with rare earths, exhibit unique chemical and physical properties in comparison with their bulk materials; their applications in biolabeling and energy transfer have been emphasized by Zhang et al. (2010). The particular electronic configuration (4f) of the rare earth ions implies special optical and magnetic properties, making them ideal materials for building multifunctional bioprobes (Liu et al., 2011) to be employed in small animal imaging (Zhou, Liu and Li, 2012).

Preparation of several types of nanoparticles based on $NaYF_4$ or $NaGdF_4$ codoped with rare earth ions (Yb^{3+}, Er^{3+}/Tm^{3+}, and Gd^{3+}) having upconversion luminescence (UCL) and properties of magnetic resonance have been reported (Jing et al., 2011; Xing et al., 2012; Zhou et al., 2011); a significant progress in this field is the synthesis of a new magnetic-upconversion nanophosphors labeled with ^{18}F for usage as multimodal bioprobes through MR imaging, UCL, and PET imaging of whole-body small animals (Liu et al., 2011).

UCNPs doped with earth rare ions were modified by Gd^{3+} having amino caproic acid (AA) as ligand to functionalise them; Gd^{3+} cations were introduced on the UCNPs by ion exchange, which not only imparted magnetic resonance properties but also increased the positive charge on the surface of the UCNPs, allowing the attachment of carboxylic acid ligands (AA), characterized by a strong complexation capacity with rareearth ions.

It is pretty surprizing that these UCNPs, with unique optical properties and without autofluorescence in biological samples, are just beginning their way in nuclear medicine as effective tools in preclinical diagnostic by PET/CT imaging. The possibilities of ^{18}F-UCNPs utilization in lymphatic imaging have been investigated by Sun et al. (2011); their results showed that *in vivo* PET/CT imaging offers ultra high sensitivity for the quantitative visualization of biodistribution and lymphatic monitoring of the sentinel node. In this kind of studies, the use of $NaYF_4$: Yb, Er and $NaYF_4$: Yb, Tm UCNPs allowed assessment of the lymphatic system, a complex network of vessels, nodes and other small structures that are not directly accesible, although they are essential for the maintenance of fluid homeostasis and immunocompetence (Zhang et al., 2011).

As a result of the progress in nanotechnology, specifically in the use of nanoprobes for PET imaging, UCNPs have emerged as other novel imaging agents for small animals (Zhou et al., 2012); a reasonable expectation is that the knowledge derived of their rapid development will be extrapolated to clinical trials in few years.

2.4.5.2 CERIA NANOPARTICLES (NANO-CEO$_2$)

The potential of cerium oxide nanoparticles as antioxidant and radioprotective agents constitutes favorable advantages for their possible use in cancer therapy (Asati et al., 2010).

In a recent chapter, Rojas et al. (2012) studied the *in vivo* biodistribution of nanoparticles based on ceria (5 nm) with functionalised surface to improve their biological properties in rats and found that ^{18}F coupled with the prostetic group N-succinimidyl 4-[^{18}F] fluorobenzoate (^{18}F-SFB) (Peter et al., 2010) accumulated in some organs as lungs, spleen and liver.

2.4.6 OTHER INORGANIC NANOPARTICLES

Many solid state and inorganic materials, such as aluminum hydroxide, hydroxiapatite, silica, iron oxide and gold nanoshells (comprised of a silica core ~120 nm in diameter, and a gold shell, 8~10 nm) have been studied for their potential application in biomedical sciences (Xu et al., 2006). More recently, Hong et al. (2011) have also proposed, in a first approach, zinc oxide nanowires, with novel platform for cancer imaging and therapy.

All of these inorganic nanoparticles have been used in preclinical PET imaging, where the use of animal models in biodistribution studies allows to know and to track basic biological and pathological mechanisms related to pathogenesis, progression and treatment of many human diseases (Jauregui-Osoro et al., 2011; Nahrendorf et al., 2011; Xie et al., 2011; Zhou et al., 2012). Inorganic nanoparticles have showed low toxicity and promizing future for drug delivery (Xu et al., 2006); thus, they are a new alternative to viral carriers and cationic carriers. Inorganic nanoparticles are generally characterized by their availability, functionality, high biocompatibility, potential capability of selectively killing cancerous cells and controlled release of carried drugs (Xu et al., 2006).

2.4.6.1 HYDROXIDE NANOPARTICLES

Inorganic nanoparticles bound to ^{18}F as fluoride ion, with applications in PET imaging, were recently studied by Jauregui-Osoro et al. (2011). Hydroxyapatite, $[Ca_5(OH)(PO_4)_3]_x$–HA_x and aluminum hydroxide nanoparticles were selected among various types of materials, including silica and calcium phosphate. These selected NPs have low toxicity, efficient and stable binding of ^{18}F and high biocompatibility, relevant parameters in the studies carried out in biological tissues. The *in vivo* biodistribution of the [^{18}F]-labeled hydroxyapatite and aluminum hydroxide nanoparticles and the fate of the radiolabel were determined by PET/CT imaging; one of the most important issues for obtaining accurate diagnostic is the stability of the β^+ radionuclide after radiolabeling (Silindir et al., 2012); although [^{18}F]-labeled hydroxiapatite showed much better stability that aluminum hydroxide nanoparticles (Jaurequi-Osoro et al., 2011), both inorganic particles provide a promizing basis for the design of PET imaging agents labeled with ^{18}F.

2.4.6.2 SILICA NANOPARTICLES

Silica nanoparticles are non-toxic and biocompatible, being considered as attractive agents for nuclear medicine, specifically theranostic (Liu and Welch, 2012). Theranostic is an acronym of therapeutic and diagnostic and at present, it is considered a promise in the biomedical field (Kelkar and Reineke, 2011). According to Xiao et al. (2012), "the current focus of nanomedicine is to develop multifunctional tumor-targeting drug/agent nanocarriers, including those capable of codelivering anticancer drugs and imaging contrast agents which are termed cancer theranostics."

Theranostic and the development of related radiopharmaceuticals configure a relatively novel paradigm, a bridge towards personalized medicin, where silica nanoparticles could have a great role as drug delivery systems. "Consequently, combining noninvasive imaging with tumor-targeted drug delivery seems to hold significant potential for personalizing nanomedicine-based (chemo-) therapeutic interventions, to achieve delivery of the right drug to the right location in the right patient at the right time" (Lammers et al., 2012).

Following what stated by Vivero-escoto (2009), "these silica-based nanoparticles also offer several unique and advantageous structural properties, such as high surface area (>700 m^2/g), pore volume (>1 cm^3/g), stable mesostructure, tunable pore diameter (2–10 nm), two functional surfaces (exterior particle and interior pore surfaces), and modificable morphology (controlable particle sizes and shapes)."

Because of these reasons, silica nanoparticles have attracted particular attention in recent years and some promizing studies were performed highlighting their importance in PET imaging.

In 2011, Benezra et al., described the characterization of a multimodal silica nanoparticle known as C-dots (7 nm), or Cornell dots, because they were initially developed as optical probes at Cornell University. C-dots coated with polyethylene glycol (PEG) chains with terminal methoxy groups and labeled with ^{124}I are known as ^{124}I-cRGDY-PEG-dots; C-dots have what Benezra et al. (2011) conceived as a unique combination of structural, optical, and biological properties. Although there are many types of drugs based on nanoparticles that either have been approved or are being tested in clinic use (Yu et al., 2012) the most interesting aspect connected with C-dots is their approval by FDA for a first clinical trial in humans (de Barros et al., 2012); this clinical trial will give insights about the behavior of melanoma, one of the most aggresive and invasive tumor types; moreover, it will allow to verify the biosafety and the bioeffectiveness of C-dots in humans (Benezra et al., 2011). At this point, it is important to mention that the high sensitivity (94.2%) and the specificity (83.3%) of PET imaging using ^{18}F-FDG for the detection of melanoma has been proved since 1998, when Holder et al., concluded that PET is better than CT in the detection of melanoma metastases, and plays an important role as primary strategy in the staging[6] of this type of cancer. ^{124}I has been used as a radiopharmaceutical injectable solution in the form of ^{124}I-Beta CIT (chemically known as ^{124}I [β-carbomethoxy-3-beta-(4-iodophenyl)-tropane], a cocaine derivative that binds to dopamine and serotonin transporters) in detection of melanoma brain metastases by PET (Cascini et al., 2009).

The use of C-dots is unique and really innovative in this field, taking into consideration the growth of the nanotechnology in the last years. Furthermore, the long half-life of ^{124}I (100.22 h) will allow to follow along a significant time the kinetics of the tracer in this clinical trial.

---

[6]Cancer staging refers to a description (usually numerical) of the cancer extent through a spreading process.

A recent study of PET/MRI with NIF signal, performed by Kim et al. (2012) using magnetic silica NPs known as MNP-SiO_2(NIR797), have provided a useful tool for imaging diagnostics and *in vivo* monitoring of cells. They also developed a multimodal nanoparticle, ^{68}Ga-{MNP-SiO_2 (NIR797)}, to detect the sentinel lymph node in mice using PET/MRI/NIRF. Sentinel lymph node is the first node or groups of nodes that drain to a tisular region when a cancer occurs (Gilrendo et al., 2006) and the first one to be found by the tumoral cells in their attemps to disseminate though metastasis from a solid tumor. In comparison with other lymph nodes, the probability of finding metastatic tumor cells from the study of sentinel node becomes more favorable (Nune et al., 2011).

2.4.6.3 IRON OXIDE NANOPARTICLES

Magnetic nanoparticles offer a broad field of action in connection with the possibility of their utilization as contrast agents in MRI imaging (de Barros et al., 2012); other interesting particles, known as superparamagnetic iron oxide nanoparticles (SPIOs) are also emerging for non invasive cell tracking (Jasmin et al., 2011; Lu et al., 2007) and potential cancer therapy (hyperthemia).

These NPs can be synthetised with cores formed by superparamagnetic iron oxide (Fe_3O_4 also known as magnetite) (Xu et al., 2006); they can be coated with cross-linked dextran shell (Anzai et al., 1994), PEG (Glaus, 2008), porous silica shell (Patel et al., 2010), D-L DOPA (3,4-dihydroxy-D, L-phenylalanine) (Patel et al., 2011) or chitosan (Pala et al., 2012).

Nanoparticles with core based on magnetite have a "strong ferromagnetic behavior, low sensitivity to oxidation and relatively low toxicity compared to many other materials (e.g., iron, nickel and cobalt)" (Ünak, 2008). Additionally, magnetite nanoparticles can be easily prepared in a chemically stable form (Ünak, 2008). At present, SPIOs are used to label cells for diagnostic image through MRI (James and Gambhir, 2012; Jasmin et al., 2011); they have also been employed in the *in vivo* evaluation with positron emitters. In 2008, Flexman et al. demonstrated the use of SPIOs labeled with ^{18}F as gene delivery vehicles in a biodistribution study. Previously, Nakamura et al. (2003) had proved the possibility of using SPIOs encapsulated by the hemagglutinating virus of Japan (HVJ, also known as Sendai virus) envelope vector for *in vivo gene study.* The HVJ envelope vector (HVJ-Es) has been used because of its capacity to fuse togheter with cellular membranes and to transfer its genetic material into the cells, via viral cycle. Imaging of *in vivo* gene expression is an important strategy to elucidate the development of diseases. It is, as well, an area in which the contrast agents based on NPs can show some advantages (Hahn et al., 2011).

Devaraj et al. (2009) demonstrated that nanoparticles known as CLIO (Cross-Linked Iron Oxide) and labeled with ^{18}F showed promise for using trimodal-imaging techniques: PET, fluorescence molecular tomography and MRI. "Click" chemistry was selected in this study, because chemoselectively allows to easily conjugating a functional group to biomolecules (proteins, antibodies and peptides, for example) (Moses and Moorhouse, 2007); in this case, this procedure was used to efficiently conjugate ^{18}F to a variety of peptides and small molecules (Li et al., 2007). This research evidenced the feasibility of using targeted nanomaterials, like ^{18}F-CLIO, at lower doses (2–4 orders of magnitude) that those required for MRI imaging, an obviously important point.

Nahrendorf et al. (2011) also employed ^{18}F-CLIO with the objective of targeting macrophages and monocites for their detection in inflammation localized in mice aortic aneurysms. The nanoparticles were labeled with fluorochromes, in order to facilitate the location with optical techniques, and labeled with ^{18}F for tracking with PET imaging; their uptake into the aortic wall was quantified with PET/CT imaging.

In spite of these promizing facts, one obstacle for the potential use of CLIO Nps in clinical studies is that they are not biodegradable (Chen, 2010).

Jarret et al. (2008) synthesized two clases of nanoparticles labeled with ^{64}Cu; they developed an aliphatic amine polystirene bead and a novel SPIO nanoparticle coated with dextrane sulfate that linked to ^{64}Cu-DOTA, for targeting macrophages in vascular inflammation and evaluation by PET/MRI dual technique.

In 2010, Glaus et al. also performed the *in vivo* evaluation of ^{64}Cu labeled magnetic nanoparticles (known as micelle-coated, mSPIOs) to be used in a dual modality PET/MRI; they developed ^{64}Cu-mSPIO probes having superparamagnetic iron oxide nanoparticle core and with a micellar coating composed of DSPE-PEG-2000, a PEGylated lipid (1,2-distearoyl-sn-glycero-3-phosphoethanolamine-N-[methoxy (polyethylene glycol)-2000]). SPIOs were modified with DOTA to allow chelation of the β^+ radionuclide for PET imaging; the pharmacokinetics and the organ biodistribution were measured using PET/CT.

Nahrendorf et al. (2008) labeled NPs based on monocrystalline iron oxide (MION) and coated with cross-linked dextran and further attaching to near infrared fluorochrome and ^{64}Cu. MRI, fluorescence, and PET were possible in this way as imaging techniques. "The dextran-coated nanoparticles undergo nonspecific phagocytosis and accumulation in macrophages. Since macrophages are a common prominent constituent of atherosclerotic lesions..., we speculate that nanoparticles might be useful in imaging or even in treating atherosclerotic lesions" (Alesio et al., 2010). Their results related to atherosclerosis are in agreement with those recently obtained by Kanwar et al. (2012) by the exclusive use of MRI.

It can be seen that the use of nanoprobes for dual imaging techniques can have some additional advantages in the study of diseases, including cancer. For example, the combination of PET with CT has become the gold standard in oncologic imaging (Czernin et al., 2007), because it offers diagnostic advantages over its individual components for the major cancers (Griffeth, 2005). The results of Glaus et al. (2010) showed the stability of ^{64}Cu-mSPIO probes in mouse serum and their high initial blood retention with moderate liver cell uptake; these facts open the possibility of considering ^{64}Cu-mSPIOs as interesting molecular agents for PET imaging, with high potential for clinical applications. In this respect, it is worthwhile mentioning that the components of these NPs (iron oxide and ^{64}Cu) have been separately used in humans (Anzai et al., 1994; Anderson et al., 2001; Lewis et al., 2008); thus utilization in humans of these NPs can also be expected.

In 2010, Patel et al. reported the study of nanoparticles, for use in cell tracking by PET/MRI, having superparamagnetic iron oxide (SPIO) core encapsulated with a porous silica shell; binding with ^{64}Cu was possible through functionalization of this silica shell with appropriate ligands. The evaluation of their cell-labeling efficacy, cytotoxicity and relaxivity in comparison to Feridex, a MRI contrast agent, was performed.

Another interesting study on the use of ^{64}Cu labeled NPs with superparamagnetic iron oxide (SPIO) or manganese oxide cores coated with DL-DOPA, and DOTA as chelating agent was accomplished in 2011, by Patel et al., who evaluated cell labeling, cytotoxicity and relaxivity of these NPs concluding that these NPs had a great potential for using in cell tracking with PET/MRI and could be used for the detection of micromolar changes in copper concentration.

An important but not yet elucidated role appears to be played in Alzheimer's disease by transition metals, mainly Zn and Cu, which have been observed in large amount in patient brains with this dementia (Liu G. et al., 2010; Watt et al., 2011). Consequently, many studies related to Alzheimer's disease by PET imaging have been carried out, using different radiopharmaceuticals labeled with ^{11}C, ^{18}F and ^{64}Cu (Donelly, 2011; Fodero-Tavoleti et al., 2010; Lim et al., 2010; Vallabhajosula, 2011).

As an additional contribution to the study of neurodegenerative diseases related to the copper concentration in the human body, i.e., Alzheimer's disease, Menkes and Wilson's diseases, amyotrophic lateral sclerosis and prion diseases, Patel et al. (2011) proposed the possibility of the use of ^{64}Cu-labeled SPIOs coated with DL-DOPA.

In other interesting approach vinculated to the improvement of PET/MRI imaging, Medvedev et al. (2012) prepared nanoparticles (100 nm) with core of iron oxide labeled with ^{52}Fe; their behavior as superparamagnetic MRI contrast agent, high labeling yield (80%) and *in vivo* stability in mice confirmed the feasibility of their use. In regard to this aspect it should be mentioned that this work was mainly related with the development of instrumentation for PET/MRI rather than the properties of nanoprobes. Therefore, the accomplishment of additional studies for evaluating the *in vitro* and *in vivo* stability of these nanoparticles is neccesary.

2.4.6.4 GOLD NANOPARTICLES

Gold has been used in therapy of numerous rheumatic diseases including arthritis (Erickson and Tunell, 2009; James and Gambhir, 2012; Lockie et al., 1958; Thakor et al., 2011) and discoid lupus erythematosus, many years ago (Dalziel et al., 1986; Thakor et al., 2011). Although nanoparticles based on gold compounds have been largely superseded by newer drugs, they go on being effectively used in diagnostic methods (Thakor et al., 2011).

In consonance of the fast development of nanotechnology, gold nanoparticles have been manufactured into different shapes, such as gold nanospheres (also known as gold nanoshells), nanorods (Xiao et al., 2012), nanocages (Cobley et al., 2010), nanobelts, nanoprisms and nanostars (de Barros et al., 2012; Kumar et al., 2008; Thakor et al., 2011).

Gold nanoshells (GNS) are nanospheres consisting of a dielectric core, generally silica, and covering by a gold thin shell, where the plasmon resonance[7] frequency is determined by the relative size of the core and the metal shell (Erickson and Tunell, 2009; Xie et al., 2011). Metallic nanoparticles with plasmon resonances in the near-infrared (NIR, 750–1,300 nm) are interesting because biological tissues, including blood, display low absorption in this spectral range (Angelomé et al., 2012).

The optical resonance of gold nanoparticles can vary over hundreds of nanometers in wavelength, across the visible and into the infrared regions of the spectrum changing the relative dimensions of these two parts (dielectric core and metallic shell), (Oldenburg et al., 1998).

GNS are of special interest for cancer treatment, cancer detection and celular imaging for their unique size, composition, and physical and optical properties (Erickson and Tunell, 2009; Xie et al., 2011). As other nanoparticles, some of their advantages are biocompatibility, ability for bioconjugation with biomolecules (antibodies and biomarkers), resistance to oxidation and lack of citotoxicity, supplied by the gold shell (Erickson and Tunell, 2009); additional advantages are photostability, lack of photobleaching and possibility of detection at concentrations as low as 10–15 M in suspension (Yguerabide and Yguerabide, 2001).

At present, nanotechnology has produced composite particles of a silica particle and a gold nanoshell (Au/SiO_2) synthesized with silane cappling agents (Oldenburg et al., 1998); their surface can be functionalised by using polyethyl-

[7]Surface plasmon resonance (SPR) is a phenomenon stimulated by incident light and related to the collective oscillation of valence electrons in QDs.

ene glicol by PEGylation (Niidome et al., 2006) for increasing times of *in vivo* circulation, in the same form as it happens with liposomes; it has been shown that gold nanoshells can also easily accumulate in tumors (Xie et al., 2011) via the enhanced permeability and retention effect (Maeda, 2001). It is known that once tumors reach a certain size, new blood vessel formation, or angiogenesis, is required for their further growth, but these new vessels are poorly structured, allowing leakage of blood plasma components, including NPs, into the tumor. NPs remain inside the tumor for longer periods, due to poor lymphatic and venous drainage (Ordidge et al., 2011).

Like other NPs, gold nanoshells have been used for PET imaging by labeling with ^{64}Cu; they were also labeled with ^{111}In for SPECT imaging and the biodistribution studies results concluded that these radionuclides accumulated in rats bearing head and neck squamous cell carcinoma xenografts (Xie et al., 2011).

Nanoparticles known as gold nanorods (GNRs) have suitable size (10 nm), tunable optical properties (dependent upon the aspect ratio) and excellent chemical versatility; they have been explored in drug/gene delivery, optical imaging and combined cancer therapies (photothermal therapy and chemotherapy) as well as CT and optical imaging (Xiao et al., 2012). More recently, Xiao et al. (2012) conjugated gold nanorods with doxorubicin (a known anticancer drug) for labeling with ^{64}Cu, through NOTA, as chelating agent; they demostrated the possibility of using this multifunctional platform based on gold nanorods for targeted anticancer drug delivery by PET imaging.

The increased interest for improving the procedures in imaging and therapy inherent to cancer have allowed to perform new studies with gold nanoparticles, in exotic forms known as gold nanocages (AuNCs) (Cobley et al., 2010). "Gold nanocages were also functionalized with biological molecules to target cancer cells for early stage diagnostics and thermal therapy of tumors. A breast cancer cell line, SK-BR-3, which overexpresses epidermal growth factor receptor 2 (EGFR2 or HER2), was used to test the molecular specific binding of bioconjugated gold nanocages" (Chen et al., 2005). In this form, gold nanocages in small sizes (40 nm) were proposed as potential contrast agents for optical imaging of cancers and use as thermal therapy agent.

These results have been confirmed by Luehmann et al. (2012), who performed an *in vivo* evaluation, by PET imaging, of ^{64}Cu labeled gold nanocages (previously functionalised with thiol-PEG and then conjugated with DOTA-NHS-ester) in murine model of breast cancer. They used AuNCs prepared with controlled and optimal physicochemical properties, with emphasis in their size (55 nm and 30 nm). The conclusion was that size was determinant in the biodistribution studies: smaller NPs tend to escape through renal filtering elimination and to accumulate in the tumor after prolonged circulation.

2.4.6.5 ZINC OXIDE NANOWIRES (ZNNWS)

ZnO is an inorganic compound well characterized for its properties as fluorescent semiconductor, being a very well suited material for nanotechnology (Wang et al., 2008). Its biocompatibility, biodegradability and biosafety at cellular level, makes it very attractive for environmental or biomedical applications (Zhou et al., 2006; Li et al., 2008; Rasmussen et al., 2010). An additional advantageous feature is its solubility in water after functionalization enhanced by the presence of hydroxyl groups in the surface of NPs, which can react with other surrounding molecules (Hong et al., 2011).

Another interesting application for ZnNWs was found by Hong et al. (2011); they performed a biodistribution study using ^{64}Cu-DOTA for labeling ultrasmall ZnNWs (~20 nm in diameter and ~100–500 nm in length). For the first time, they proposed the feasibility of using these nanostructures as novel platform with promise in cancer imaging and therapy (Hong et al., 2011). "ZnO NWs can be specifically targeted to cell surface receptors *in vitro*, which opened up new avenues of future research in tumor-targeted drug delivery. The one-dimensional shape of ZnO NW is highly desirable for efficient tumor targeting since such morphology can readily take advantage of the polyvalency effect" (Hong et al., 2011).

2.5 CHEMOTOPOLOGY, A MATHEMATICAL APPROACH FOR ANALYZING SIMILARITY RELATIONSHIPS AMONG β^+ RADIONUCLIDES USED FOR NANOPARTICLES RADIOLABELING

At present, much information is available on radionuclides and nanoparticles used for both diagnostic imaging and therapy via radiolabeling through preclinical studies in animal models. Taking into account the importance of the relationship between β^+ radionuclides and labeled nanoparticles, the study of similarity relationships between radionuclides and the subsequent formulation of predictions about their use is highly relevant. There are many methodologies to achieve this goal; in particular, the chemotopological approach (Restrepo et al., 2004) will be described in this chapter.

The preceding sections consider a classification of positron emitters and an overview about the different types of nanoparticles used in PET imaging. Classifications are of two types, namely supervised and unsupervised classifications (Otto, 2007). Supervised classifications group together objects of interest by human intervention, i.e., when humans decide whether two objects belong or not to a common class. Unsupervised classifications build up classes based on the assesment of similarity among the studied objects. The aforementioned classification of positron emitters is an example of a supervised classification. An example of an unsupervised classification is that of chemical elements based upon their properties (Restrepo et al., 2004). Despite the applicability of classification techniques to make sense of a wealth of information coming from different sources, it is customary that classification studies end up with the determination of classes without further analysis. Restrepo et al. (2011) have shown that there is additional information regarding the similarity of classes and objects that can be further studied by combining classification results with topological principles. This approach has been called chemotopology.

Chemotopology has been sucessfully applied to the study of chemical elements, as well as to some chemical sets such as benzimidazoles, steroids, amino acids and monohidrides (Restrepo et al., 2004; Restrepo and Villaveces, 2005; Daza et al., 2006). More recently, the authors of the present chapter carried out a chemotopological study of positron emitters radionuclides used in PET imaging (Quintero et al., 2013).

Chemotopology does not require a specific classification method. Restrepo et al. (2004) and Quintero et al. (2013) employed hierarchical cluster analysis (HCA), whose basic assumption in HCA is that the objects belonging to the same cluster are the most similar among the objects studied. Hence, for gathering the objects on the basis of their properties the following steps are mandatory: (a) attainment of an association measure, known as similarity function; (b) building of the corresponding similarity matrix; (c) selection of a grouping methodology; and (d) Representation of the classification obtained in a bidimensional figure called dendrogram (Everitt et al., 2011).

Because the degree of similarity between objects should be assessed, after the determination of the distance matrix the following stage is to find the clusters formed by the most similar objects by the application of a grouping methodology. Several grouping methodologies, that differ in the criterion to calculate the distance between an object and the whole set of objects, can be applied; some of the methodologies currently used are those of single, complete and average linkage (Everitt et al., 2011; Hartigan, 1985; Jain et al., 1999).

HCA gathers the objects taking into acount their similarity relationships according to their properties; similar objects are grouped into the same class or cluster, distinguishing them from those belonging to another class or cluster. The final outcome of the cluster analysis is the two dimensional graphic (dendrogram), which gathers groupings or clusters that contain similarity information of the dataset. More information on HCA methodology can be found in Jain et al. (1999) and Everitt et al. (2011).

Two of the most important decisions that have to be made when applying HCA are the selection of the similarity function and of the grouping methodology.

To develop a cluster analysis of objects (in this case, β^+ radionuclides) it is necessary to characterize them by a set of properties or attributes. According to

Milligan (1996), the selected properties should provide sufficient and relevant information to obtain the correct structure of clusters. Since that the β^+ radionuclides are isotopes of elements defined by many properties, the selection performed by Quintero et al. (2013) consisted of several properties of different nature: physico-chemical, physical, nuclear, dosimetric and quantum to try to capture the different aspects of radionuclides.

After applying HCA, a further criterion has to be applied, in order to cut the dendrogram at an appropiate level. One of them is the selection number, a criterion yielding many and highly populated clusters. The maximum value of the selection number allows determining the partition of the set that offers the needed clusters (Restrepo et al., 2004). The collection of classes found by partitioning the dendrogram constitutes a topological basis $\mathcal{B}_n$ (figure 2).

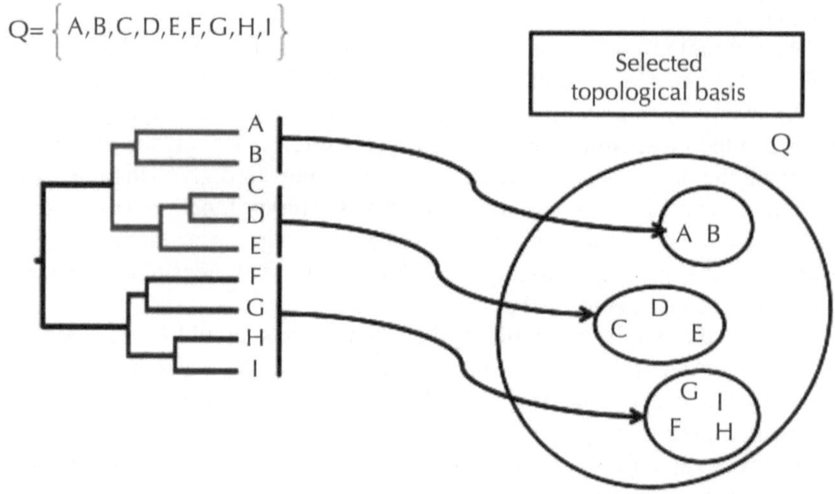

FIGURE 2 Graphical representation of the methodology applied in a chemotopological study. A dendrogram (on the left) is obtained by application of HCA over the set Q; the dashed line on the dendrogram indicates the cutting site. The topological basis is built up taking into account the sets obtained by cutting the dendrogram (Venn diagram on the right).

From this selected topological basis, different topological properties can be calculated over sets of particular interest. These properties are related to the concept of similarity, e.g., closure, interior, exterior and boundary (Restrepo and Mesa, 2011), which will be explained below.

In the first chemotopological study of positron emitters used in PET imaging (Quintero et al., 2013) two sets of β^+ radionuclides of interest were studied, the first one grouping β^+ radionuclides with known application in preclinical or clinical PET imaging and the other one grouping the same β^+ radionuclides plus 12 β^+ radionuclides with potential use for PET imaging (Table 1).

TABLE 1 β^+ Radionuclides for PET imaging.

Set I:			
Radionuclides used in preclinical and clinical trials			**Potentially useful β^+ radionuclides**
^{11}C	^{52}Mn	^{118}Sb	^{47}V
^{13}N	^{52}Fe	^{89}Zr	^{48}V
^{15}O	^{55}Co	^{43}Sc	^{70}As
^{18}F	^{66}Ga	^{44}Sc	^{115}Sb
^{60}Cu	^{68}Ga	^{45}Ti	^{116}Sb
^{61}Cu	^{122}I	^{86}Y	^{120}Sb
^{62}Cu	^{124}I	^{118}Sb	^{63}Zn
^{64}Cu	^{82}Rb	^{89}Zr	^{130}Cs
^{73}Se	^{77}Kr		^{90}Nb
^{75}Br	^{38}K		^{140}Pr
^{76}Br	^{72}As		^{106}Ag
^{51}Mn	^{86}Y		^{134}La

These latter potential radionuclides were carefully analyzed based upon their similarities with the known radionuclides (Quintero et al., 2013).

Only some cases of interest for applying topological properties were selected by Quintero et al. (2013) in each set of β^+ radionuclides; in the second set, which included the proposed β^+ radionuclides, these cases were: (a) β^+ radionuclides obtained by generator; (b) β^+ radionuclides used in peptides labeling; and (c) β^+ radionuclides proposed with possible potential in PET imaging.

Taking into account this topological background, an additional case was selected in this chapter, conformed by the β^+ radionuclides used at present for labeling various types of nanoparticles. According to Quintero et al. (2013), the topological basis $\mathfrak{B}_4$ belonging to the second set, which includes the 12 selected radionuclides with potential use in PET, is the following:

$$\mathfrak{B}_4 = \left\{ \begin{array}{c} \{^{75}Br_1, {}^{75}Br_2, {}^{75}Br_3\}, \{^{124}I_1, {}^{124}I_2\}, \{^{134}La, {}^{140}Pr\}, \{^{52}Fe_1, {}^{52}Fe_2\}, \\ \{^{115}Sb, {}^{116}Sb, {}^{118}Sb, {}^{120}Sb\}, \{^{70}As, {}^{72}As\}, \{^{61}Cu, {}^{63}Zn\}, \\ \{^{15}O_1, {}^{15}O_2, {}^{18}F_1, {}^{18}F_2\}, \{^{48}V, {}^{89}Zr, {}^{90}Nb\}, \{^{43}Sc, {}^{44}Sc, {}^{86}Y\}, \\ \{^{76}Br_1, {}^{76}Br_2\}, \{^{51}Mn_1, {}^{51}Mn_2, {}^{52}Mn_1, {}^{52}Mn_2\}, \{^{45}Ti, {}^{47}V, {}^{55}Co_1, {}^{55}Co_2\}, \\ \{^{64}Cu_1, {}^{64}Cu_2\}, \{^{130}Cs\}, \{^{38}K_1, {}^{38}K_2, {}^{82}Rb_1, {}^{82}Rb_2\}, \{^{73}Se\}, \\ \{^{60}Cu, {}^{62}Cu, {}^{106}Ag\}, \{^{66}Ga_1, {}^{66}Ga_2, {}^{68}Ga, {}^{77}Kr\}, \{^{11}C, {}^{13}N\} \end{array} \right\}$$

where each radionuclide has the form ${}^xA_y$; *A* being the β^+ radionuclide, *x* the mass number and *y* a β^+ radionuclide variant: 1, 2 or 3, only differentiated by the nuclear production parameters used in cyclotrons or accelerators, such as range of energy of projectiles, integral yield and energy threshold (Table 2).

TABLE 2 β^+ radionuclides currently used in PET and their variants.

^{11}C	^{61}Cu	$^{75}Br_3$	$^{52}Fe_1$	$^{124}I_1$	^{72}As
^{13}N	^{62}Cu	$^{76}Br_1$	$^{52}Fe_2$	$^{124}I_2$	^{86}Y
$^{15}O_1$	$^{64}Cu_1$	$^{76}Br_2$	$^{55}Co_1$	$^{82}Rb_1$	^{118}Sb
$^{15}O_2$	$^{64}Cu_2$	$^{51}Mn_1$	$^{55}CO_2$	$^{82}Rb_2$	^{89}Zr
$^{18}F_1$	^{73}Se	$^{51}Mn_2$	$^{66}Ga_1$	^{77}Kr	^{43}Sc
$^{18}F_2$	$^{75}Br_1$	$^{52}Mn_1$	$^{66}Ga_2$	$^{38}K_1$	^{44}Sc
^{60}Cu	$^{75}Br_2$	$^{52}Mn_2$	^{68}Ga	$^{38}K_2$	^{45}Ti

Based on the set of radionuclides showed in Table 2 and adding the 12 β^+ radionuclides proposed, a new set of radionuclides was built up (Quintero et al., 2013). The new case built up in this chapter gathers in the set *X* those β^+ radionuclides used at least once in PET imaging or dual techniques (PET/CT, PET/MRI) taking into account that PET tracers have often been incorporated with another modality in radiolabeling of nanoparticles.

$$X = \{^{11}C, {}^{15}O, {}^{18}F, {}^{52}Fe, {}^{62}Cu, {}^{64}Cu, {}^{66}Ga, {}^{68}Ga, {}^{72}As, {}^{76}Br, {}^{89}Zr, {}^{86}Y, {}^{124}I\}$$

Although Hahn et al. (2011) affirm in their chapter that "common isotopes that can be chelated on to or incorporated within NPs (in a analogous way to the gadolinium ions used for MRI) include ^{18}F, ^{11}C, ^{15}O, ^{13}N, ^{64}Cu, ^{124}I, ^{68}Ga, ^{82}Rb and ^{86}Y", two of these radionuclides (^{13}N and ^{82}Rb) have not been included in *X*, because there is no information about their use in radiolabeling of nanoparticles.

Then, topological properties calculated for analyzing the similarity relationships among the β^+ radionuclides belonging to X are indicated below; each property will be explained from the topological results obtained for each β^+ radionuclide in function of the radionuclides found in each topological property.

2.5.1 CLOSURE OF X, $\overline{X}$

$$\bar{X} = X \cup \{^{13}\mathrm{N}, {}^{43}\mathrm{Sc}, {}^{44}\mathrm{Sc}, {}^{48}\mathrm{V}, {}^{60}\mathrm{Cu}, {}^{70}\mathrm{As}, {}^{77}\mathrm{Kr}, {}^{90}\mathrm{Nb}, {}^{106}\mathrm{Ag}\}$$

$$\overline{X} = \left\{ \begin{matrix} ^{11}\mathrm{C}, {}^{13}\mathrm{N}, {}^{15}\mathrm{O}_1, {}^{15}\mathrm{O}_2, {}^{18}\mathrm{F}_1, {}^{18}\mathrm{F}_2, {}^{43}\mathrm{Sc}, {}^{44}\mathrm{Sc}, {}^{48}\mathrm{V}, {}^{52}\mathrm{Fe}_1, {}^{52}\mathrm{Fe}_2, {}^{60}\mathrm{Cu}, {}^{62}\mathrm{Cu}, \\ ^{64}\mathrm{Cu}_1, {}^{64}\mathrm{Cu}_2, {}^{66}\mathrm{Ga}_1, {}^{66}\mathrm{Ga}_2, {}^{68}\mathrm{Ga}, {}^{70}\mathrm{As}, {}^{72}\mathrm{As}, {}^{76}\mathrm{Br}_1, {}^{76}\mathrm{Br}_2, {}^{77}\mathrm{Kr}, \\ ^{86}\mathrm{Y}, {}^{89}\mathrm{Zr}, {}^{90}\mathrm{Nb}, {}^{106}\mathrm{Ag}, {}^{124}\mathrm{I}_1, {}^{124}\mathrm{I}_2 \end{matrix} \right\}$$

The closure of the set X contains all the β^+ radionuclides known as closure or adherence objects, i.e., those radionuclides belonging to the set of all radionuclides used in PET, which are similar to the radionuclides of X (Restrepo and Mesa, 2011). It is interesting to note that ^{44}Sc ($t_{1/2}$ = 4.0 h) and ^{90}Nb ($t_{1/2}$ = 14.6 h) are radionuclides belonging to $\overline{X}$, sharing the same cluster with ^{86}Y ($t_{1/2}$ = 14.7 h) and ^{89}Zr ($t_{1/2}$ = 78.4 h). Moderegger (2012) asserts that ^{44}Sc and ^{90}Nb are suitable radionuclides for molecular imaging of nanosized drug delivery systems by PET. In addition, Radchenko et al. (2012) recently proposed ^{90}Nb as potential PET nuclide for studying processes with slow and medium kinetics using antibodies, fragments or polymeric nanoparticles. Similarly, it is interesting to remark that ^{48}V ($t_{1/2}$ = 15.97 d), a β^+ radionuclide sharing the same cluster with ^{89}Zr and ^{90}Nb, could also be useful in radiolabeling of some nanoparticles.

Some β^+ radionuclides used for preclinical and clinical studies related to immuno-PET (Perk and Rispens, 2010) can also be employed in labeling of nanoparticles; such are the cases of ^{64}Cu, ^{86}Y, ^{76}Br, ^{89}Zr and ^{124}I, all of them found in the closure set of X.

2.5.2 INTERIOR OF X, INT (X)

$$\mathrm{Int}(X) = \left\{ \begin{matrix} ^{15}\mathrm{O}_1, {}^{15}\mathrm{O}_2, {}^{18}\mathrm{F}_1, {}^{18}\mathrm{F}_2, {}^{52}\mathrm{Fe}_1, {}^{52}\mathrm{Fe}_2, {}^{64}\mathrm{Cu}_1, \\ ^{64}\mathrm{Cu}_2, {}^{70}\mathrm{As}, {}^{72}\mathrm{As}, {}^{76}\mathrm{Br}_1, {}^{76}\mathrm{Br}_2, {}^{124}\mathrm{I}_1, {}^{124}\mathrm{I}_2 \end{matrix} \right\}$$

$$(X) = \left\{ \begin{matrix} ^{15}\mathrm{O}_1, {}^{15}\mathrm{O}_2, {}^{18}\mathrm{F}_1, {}^{18}\mathrm{F}_2, {}^{52}\mathrm{Fe}_1, {}^{52}\mathrm{Fe}_2, {}^{64}\mathrm{Cu}_1, \\ ^{64}\mathrm{Cu}_2, {}^{70}\mathrm{As}, {}^{72}\mathrm{As}, {}^{76}\mathrm{Br}_1, {}^{76}\mathrm{Br}_2, {}^{124}\mathrm{I}_1, {}^{124}\mathrm{I}_2 \end{matrix} \right\}$$

The interior set contains all the interior objects of *X*, whose neighborhoods are embedded in *X* (Restrepo and Mesa, 2011), i.e., this set only contains β^+ radionuclides belonging to *X* that have similar properties.

All the β^+ radionuclides that form part of Int (*X*) are the core of *X*; they have exclusive properties that differ to those of the others β^+ radionuclides belonging to the set of all β^+ radionuclides studied.

It is not surprizing to find ^{18}F ($t_{1/2}$ = 1.83 h) and ^{64}Cu ($t_{1/2}$ = 12.7 h) in Int (*X*), since they are two of the most used β^+ radionuclides in radiolabeling of nanoparticles for both biodistribution and pharmacokinetics through animal models. This means that these radionuclides are unique in their properties for radiolabeling nanoparticles.

It is worthwhile observing that ^{124}I ($t_{1/2}$ = 100.2 h), the first β^+ radionuclide approved for study of human melanoma by PET imaging (Benezra et al., 2011) via radiolabeling of C-dots (silica nanoparticles), also belongs to Int(*X*).

^{72}As ($t_{1/2}$ = 26.0 h) appears sharing the same cluster with ^{70}As; both are considered interior points, due to their similar properties. ^{72}As is relevant as radionuclide used in PET imaging, specifically in some studies related to labeling of monoclonal antibodies (Jennewein et al., 2008); its future in nanoparticles labeling is just beginning (Herth et al., 2010).

^{52}Fe ($t_{1/2}$ = 8.3 h) appears alone in its own cluster as another interior point; its use was investigated in a recent study, performed by Medvedev et al. (2012), destined to build up nanoparticles based on ferric oxide. Since the purpose of the work was connected with the progress in the instrumentation employed in MRI/PET, rather than the development of the nanoprobe, more studies related to *in vitro* and *in vivo* stability should be carried out in this respect.

2.5.3 EXTERIOR OF X, EXT(X)

$$\text{Ext}\ (X) = \left\{ \begin{array}{c} {}^{38}K_1, {}^{38}K_2, {}^{45}Ti, {}^{47}V, {}^{51}Mn_1, {}^{51}Mn_2, {}^{52}Mn_1, {}^{52}Mn_2, \\ {}^{55}Co_1, {}^{55}Co_2, {}^{61}Cu, {}^{63}Zn, {}^{73}Se, {}^{75}Br_1, {}^{75}Br_2, {}^{75}Br_3, {}^{82}Rb_1, {}^{82}Rb_2, \\ {}^{115}Sb, {}^{116}Sb, {}^{118}Sb, {}^{120}Sb, {}^{130}Cs, {}^{134}La, {}^{140}Pr \end{array} \right\}$$

$$(X) = \left\{ \begin{array}{c} {}^{38}K_1, {}^{38}K_2, {}^{45}Ti, {}^{47}V, {}^{51}Mn_1, {}^{51}Mn_2, {}^{52}Mn_1, {}^{52}Mn_2, \\ {}^{55}Co_1, {}^{55}Co_2, {}^{61}Cu, {}^{63}Zn, {}^{73}Se, {}^{75}Br_1, {}^{75}Br_2, {}^{75}Br_3, {}^{82}Rb_1, {}^{82}Rb_2, \\ {}^{115}Sb, {}^{116}Sb, {}^{118}Sb, {}^{120}Sb, {}^{130}Cs, {}^{134}La, {}^{140}Pr \end{array} \right\}$$

In contrast to what we commented before, the exterior set of *X* contains the β^+ radionuclides belonging to the set of all the radionuclides studied, whose neighborhoods do not contain radionuclides of *X*, i.e., those that do not have any similarity relationship with the radionuclides of *X* used in radiolabeling of nanoparticles (Restrepo and Mesa, 2011). In other words, these β^+ radionuclides could be not used in this process, taking into account the set of properties analyzed by Quintero et al. (2013).

2.5.4 BOUNDARY OF X, b(X)

$$b(X) = \left\{ \begin{matrix} {}^{11}\mathrm{C}, {}^{13}\mathrm{N}, {}^{43}\mathrm{Sc}, {}^{44}\mathrm{Sc}, {}^{48}\mathrm{V}, {}^{60}\mathrm{Cu}, {}^{62}\mathrm{Cu}, {}^{66}\mathrm{Ga}_1, {}^{66}\mathrm{Ga}_2, {}^{68}\mathrm{Ga}, \\ {}^{77}\mathrm{Kr}, {}^{86}\mathrm{Y}, {}^{89}\mathrm{Zr}, {}^{90}\mathrm{Nb}, {}^{106}\mathrm{Ag} \end{matrix} \right\}$$

The boundary set contains those radionuclides of the set of all β^+ radionuclides used in PET whose neighborhoods have radionuclides belonging to *X* and to the complement of *X* (Restrepo and Mesa, 2011), i.e., the radionuclides whose properties are similar to the ones of the radionuclides in *X* and to the rest of the radioisotopes in the set of all β^+ radionuclides studied (Table 1). The radionuclides that form part of *b*(*X*) would be similar, with respect to their properties, to the β^+ radionuclides belonging to *X* and to those that are not part of *X*.

2.6 CONCLUSIONS

To date, there are different types of nanoparticles platforms in progress for imaging, and their burgeoning use in PET imaging is mainly related to the preclinical phase. The adaptation of these studies to a regular practice in humans is not immediate, and requires the accomplishment of additional studies, in order to overcome many obstacles. The vast majority of researchers center their interest in biocompatibility, biodistribution and pharmacokinetics studies, which are the first steps in understanding the *in vivo* behavior of these nanocarriers systems, both for diagnostic by PET imaging and for therapy by drug delivery.

Many non-conventional radionuclides, e.g., ^{124}I, ^{66}Ga, ^{68}Ga, ^{89}Zr, ^{86}Y and ^{64}Cu, are being studied in radiolabeling of many types of nanoparticles. The most used β^+ radionuclide in PET imaging, ^{18}F, has proved to be very useful for this purpose, because of its suitable properties and its well-known radiochemistry. At the time of the nanotheranostic boom, the second radionuclide most used is ^{64}Cu, particularly due to their decay properties, which allows both diagnostic and therapy.

The future works will be probably centered on going in depth of the knowledge about procedures for preparation and radiolabeling, nanoparticle sizes, chemical structure and reactivity, structural design and high selectivity for targeting in both therapy and diagnostic.

As final remark, the chemotopological analysis of β^+ radionuclides for nanoparticle labeling allows predicting that the possible use of ^{90}Nb, ^{44}Sc and ^{48}V could be materialized in the future, offering the feasibility of long duration PET imaging, of the order of several hours to days.

KEYWORDS

- **β^+ radionuclides**
- **carbon nanohorns**
- **chemotopology**
- ***Lactobacillaceae* family**
- **polyethylene oxide chains**
- **quantum dots**
- **upconverting phosphors**

REFERENCES

Agrawal, P.; Strijkers, G. J.; Nicolay, K. *Advanced Drug Delivery Reviews.* **2010,** *62,* 42–58.

Aird, W. C. Crit. *Care Med.* **2004,** 32 (5 Suppl), S271–9.

Akerman, M. E.; Chan, W. C.; Laakkonen, P.; Bhatia, S. N.; Ruoslahti, E. Proc. Natl. Acad. Sci. USA. **2002,** *99,* 12617–12621.

Al-Nahhas, A.; Win, Z.; Szyszko, T.; Singh, A.; Nanni, C.; Fanti, S.; Rubello, D. *Anticancer Research.* **2007,** 27, 4087–4094.

Alesio, A. M.; Butterworth, E.; Caldwell, J. H.; Bassingthwaighte, J. B. Nano Reviews. **2010,** *1,* **5110,** DOI: 10.3402/nano.v1i0.5110. Published: Apr. *2,* 2010. http://www.ncbi.nlm.nih.gov/pmc/articles/PMC3215216/ (accessed Jan *21,* 2013).

Allen, P. M.; Bawendi, M. G. *J. Am. Chem. Soc.* **2008,** *130,* 9240–9241.

Almutairi, A.; Rossin, R.; Shokeen, M.; Hagooly, A.; Ananth A.; Capoccia, B.; Guillaudeu, S.; Abendschein, D.; Anderson, C.; Welch, M. J.; Fréchet, J. M. *PNAS.* **2009,** *106,* 685–690.

Ament, S. J.; Maus, S.; Reber, H.; Buchholz, H. G.; Bausbacher, N.; Brochhausen, C.; Graf, F.; Miederer, M.; Schreckenberger, M. *Cancer Res.* **2012,** *194,* 395–423.

Anderson, C. J.; Dehdashti, F.; Cutler, P. D.; Schwarz, S. W.; Laforest, R.; Bass L. A.; Lewis, J. S.; McCarthy, D. W. *J. Nucl. Med.* 2001; *42,* 213–221.

Anderson, C. J.; Ferdani, R. Cancer Biother. *Radiopharm.* **2009,** *24,* 379–393.

Andreozzi, E.; Seo, W. J.; Ferrara, K.; Louie, A. *Bioconjugate Chem.* **2011,** *22,* 808–*818,* 2011.

Angelomé, P. C.; Mezerji, H. H.; Goris, B.; Pastoriza-Santos, I.; Pérez-Juste, J.; Sara Bals, S.; Liz-Marzán, L. M. *Chem. Mater.* **2012,** *24,* 1393–1399.

Anzai, Y.; Blackwell, K. E.; Hirschowitz, S. L.; Rogers, J. W.; Sato, Y.; Yuh, W. T.; Runge, V. M.; Morris, M. R.; McLachlan, S. J.; Lufkin, R. B. *Radiology.* **1994,** *192,* 709–715.

Asati, A.; Santra, S.; Kaittanis, C.; Perez, J. M. *ACS Nano.* **2010,** *4,* 5321–5331.

Assadi, M.; Afrasiabi, K.; Nabipour, I.; Seyedabadi, M. *Hellenic Journal of Nuclear Medicine.* **2011,** 1–11.

Aumelas, A.; Serrero, A.; Durand, A.; Dellacherie, E.; Leonard, M. *Biointerfaces.* **2007,** *59,* 74–80.

Azzazy, M. E. H; Mansour, M. M. H.; Kazmierczak, S. C. *Clinical Biochemistry.* **2007,** *40,* 917–927.

Balasubramanian, S. K.; Jittiwat, J.; Manikandan, J.; Ong, C. N.; Yu, L. E.; Ong, W. Y. *Biomaterials.* **2010,** *31,* 2034–2042.

Benezra, M.; Penate-Medina, O.; Zanzonico, P. B.; Schaer, D.; Hooisweng, O. W.; Burns, A.; DeStanchina, E.; Longo, V.; Herz, E.; Iyer, S.; Wolchok, J.; Larson, S. M.; Wiesner, U.; Bradbury, M. S. *J. Clin. Invest.* **2011,** *121,* 2768–2780.

Bentolila, L. A.; Ebenstein, Y.; Weiss, S. J. *Nucl. Med.* **2009,** *50,* 493–496.

Bertolini, G.; Paleari, L.; Catassi, A.; Roz, L.; Cesario, A.; Sozzi, G.; Russo, P. *Current Pharmaceutical Analysis.* **2008,** *4,* 197–205.

Bhattacharya, S.; Mazumder, B. BioPharm. *International Supplements.* **2011,** *24,* s9-s14

Bianco, B.; Kostarelos, K.; Prato, M. *Expert Opin. Drug Deliv.* **2008,** *5,* 331–342.

Blum, T. Development of no-carrier-added radioselenation methods for the preparation of radiopharmaceuticals. Ph.D. Dissertation [Online], Universität zu Köln, February 2003. http://juser.fz-juelich.de/record/37411 (accessed Nov *6,* 2012).

Botsoa, J.; Lyzenko, V.; Géloën, A.; Marty, O.; Bluet, J.M.; Guillot, G. *Appl. Phys. Lett.* **2008,** *92,* 173902–173903.

Bruchez, M., Jr.; Moronne, M.; Gin, P.; Weiss, S.; Alivisatos, A. P. Semiconductor nanocrystals as fluorescent biological labels. Science. [Online] **1998,** *281,* 2013–2016. http://www.community.nsee.us/workshops/calpoly2006/Science%20v280%20Alivisatos.pdf (accessed Jan *10,* 2013).

Cai, W.; Hong, H. In a "nutshell": intrinsically radio-labeled quantum dots. Am. J. Nucl. Med. Imaging. [Online] **2012,** *282,* 136–140. http://www.ncbi.nlm.nih.gov/pmc/articles/PMC3477731/ (accessed Jan *9,* 2013).

Cantorias, M.; Smith, C.; Nanda, P.; Cutler, C. J. *Nucl. Med.* 2008; 49 (Supplement 1), 291P.

Carrió, I.; González, P.; Estorch, M.; Canessa, J.; Mitjavila, M.; Massardo, T. Medicina nuclear. Aplicaciones clínicas, Masson S.A.: Barcelona, 2003; pp 627.

Cartier, R.; Kaufner, L.; Paulke, B. R.; Wüstneck, R.; Pietschmann, S.; Michel, R.; Bruhn, H.; Pison, U. Nanotechnology. **2007,** *18,* DOI:10.1088/0957–4484/18/19/195102.

Cascini, G. L.; Ciarmiello, A.; Labate, A.; Tamburrini, S.; Quattrone, A. *Clin. Nucl. Med.* ***2009,*** *34, 698–699.*

Chai, Y.; Guo, T.; Jin, C.; Haufler, R.E.; Chibante, L. P. F.; Fure, J.; Wang, L.; Alford, J. M.; Smalley, R. E. *J. Phys. Chem.* **1991,** *95,* 7564–7568.

Chan, W. C.; Nie, S. *Science*. **1998,** *281,* 2016–2018.

Chen, J.; Saeki, F.; Wiley, B. J.; Cang, H.; Cobb, M. J.; Li, Z. Y.; Au, L.; Zhang, H.; Kimmey, M. B.; Li, X.; Xia, Y. *Nanoletters*, **2005,** *5,* 473–477.

Chen, K.; Li, Z. B.; Wang, H.; Cai, W.; Chen, X. Eur J Nucl Med Mol Imaging. **2008,** *35,* **2235,** 2244.

Chen, L. T.; Weiss, L. *Blood.* **1973,** *41,* 529–537.

Chen, S. Polymer-coated iron oxide nanoparticles for medical imaging. Ph.D. Dissertation [Online], Massachusetts Institute of Technology, 2010. http://dspace.mit.edu/handle/1721.1/59004 (accessed Jan *11,* 2013).

Cherry, S. R. *J. Clin. Pharmacol.* **2001,** *41,* 482–491.

Choi, H. S.; Liu, W.; Misra, P.; Tanaka, E.; Zimmer, J. P.; Ipe, B. I.; Bawandi, M. G.; Frangioni, J. V. Nat. Biotechnol. **2007,** *25,* 1165–1170.

Cobley, C. M.; Au, L.; Chen, J.; Xia, Y. *Expert Opin. Drug Deliv.* **2010,** *7,* 577–587.

Czernin, J.; Allen-Auerbach, M.; Schelbert, H. *Journal of Nuclear Medicine*. **2007,** *48,* 78S.

Dalziel, K.; Going, G.; Cartwright, P. H.; Marks, R.; Beveridge, G. W.; Rowell, N. R. *British Journal of Dermatology.* **1986**, *115,* 211–216.

Daza, M.C.; Restrepo, G.; Uribe, E. A.; Villaveces, J. L. *Chem. Phys. Lett.* **2006,** *428,* 55–61.

De Barros A. L. B.; Mota, L. G.; Soiares, D. C.; Coelho, M. M.; Oliveira, M. C.; Cardoso, V. N. *Biorg. Med. Chem. Lett.* **2011,** *15, 21,* 7373–7375.

De Barros, A. L. B.; Tsourkas, A.; Saboury, B.; Cardoso V. N.; Alavi, A. Emerging role of radiolabeled nanoparticles as an effective diagnostic technique. EJNMMI, [Online] **2012,** *2,* 1–15. http://www.ejnmmires.com/content/pdf/2191–219X-2–39.pdf (accessed Jan *8,* 2013).

Debbage, P.; Jaschke, W. Histochem. *Cell. Biol.* **2008,** *130,* 845–875.

Devaraj, N. K.; Keliher, E. J.; Thurber, G. M.; Nahrendorf, M.; Weissleder, R. *Bioconjugate Chem.* **2009,** *20,* 397–401.

Dhaneshwar, S. S.; Kandpal, M.; Gairola, N.; Kadam, S. S. Indian Journal of Pharmaceutical Sciences. **2006**, *68,* 705–714.

Ding, H.; Wu, F. Image Guided Biodistribution and Pharmacokinetic Studies of Theranostics. Theranostics, [Online] **2012,** *2,* 1040–1053. http://www.thno.org/v02p1040.htm (accessed Jan *11,* 2013).

Donnelly, P. S. Dalton Trans. **2011,** *40,* 999–1010.

Drummond, D. C.; Meyer, O.; Hong, K.; Kirpotin, D. B.; Papahadjopoulos, D. Pharmacological reviews. **1999,** *51,* 691–743.

Ducongé, F.; Pons, T.; Pestourie, C.; Hérin, L.; Thézé, B.; Gombert, K.; Mahler, B.; Hinnen, F.; Kühnast, B.; Dollé, F.; Dubertret, B.; Tavitian, B. Bioconjug. Chem. **2008**, *19,* 1921–1926.

Ekambaram, P.; Abdul Hasan Sathali, A.; Priyanka, K. Solid lipid nanoparticles: a review. Sci. Revs. Chem. Commun. [Online] **2012,** *2(1),* 80–102. http://www.sadgurupublications.com/ContentPaper/2012/9_117_SRCC_2(1)2012_P.pdf (accessed Jan *10,* 2013).

Erickson, T. A.; Tunnell, J. W. Gold Nanoshells in Biomedical Applications. In *Nanomaterials for the Life Sciences. Vol. 3: Mixed Metal Nanomaterials*; Challa S. S. R. Kumar Ed.; Wiley-VCH: Weinheim, Germany, **2009**; pp. 1–44.

Etrych, T.; Strohalm, J.; Chytil. P.; Říhová, B.; Ulbrich, K. J. Drug Target. **2011,** *19,* 874–889.

Everitt, B. S.; Landau, S.; Leese, M.; Stahl, D. Cluster Analysis. Wiley Series in Probability and Statistics, 5th ed.; John Wiley & Sons: United Kingdom, **2011,** pp. 1–348.

Fan, X.; Tan, J.; Zhang, G.; Zhang, F. Nanotechnology. **2007,** *18,* doi:10.1088/0957–4484/18/19/195103.

Flexman, J. A.; Cross, D. J.; Lewellen, B. L.; Miyoshi, S.; Kim, Y.; Minoshima, S. IEEE Trans. Nanobioscience. **2008,** *7,* 223–232.

Fodero-Tavoletti, M. T.; Villemagne, V. L.; Paterson, B. M.; White, A. R.; Li, Q. X.; Camakaris, J.; O'Keefe, G. J.; Cappai, R.; Barnham, K. J.; Donnelly, P. S. Journal of Alzheimer's Disease. **2010,** *20,* 49–55.

Gambhir, S. S. Nat. Rev. Cancer. **2002,** *2,* 683–693.

Geim, A. K.; Novoselov, K. S. The Rise of Graphene. Nature Mat. **2007,** 183–191.

Gil-Rendo, A.; Zornoza, G.; García-Velloso, M. J.; Regueira, F. M.; Beorlegui, C.; Cervera, M. Br. J. Surg. **2006,** *93,* 707–712.

Glaus, C. R. M. Development and analysis of radiolabeled magnetic nanoparticles for positron emission tomography and magnetic resonance imaging. Ph. D. Dissertation [Online], Georgia Institute of Technology, 2008. https://smartech.gatech.edu/handle/1853/31692 (accessed Jan *9,* 2013).

Glaus, C.; Rossin, R.; Welch, M. J.; Bao, G.; Goins, B. A.; Phillips, W. T. Methods for Tracking Radiolabeled Liposomes after Injection in the body. In *Liposome Technology,*

Volume III. Interactions of liposomes with the biological Milieu; Gregoriadis, G., Ed.; Informa Healthcare, **2007,** pp.191–206.

Griffeth, L. K. Proc (Bayl Univ Med Cent). **2005,** *18,* 321–330.

Haddad, F.; Ferrer, L.; Guertin, A. *Eur. J. Nucl. Med. Mol. Imaging.* **2008,** *35,* 1377–1387.

Hahn, M. A.; Singh, A. K.; Sharma, P.; Brown, S. C.; Moudgil, B. M. Anal. Bioanal. Chem. **2011,** *399,* 3–27.

Hartigan, J. A. *Journal of Classification.* **1985,** *2,* 63–76.

Hartner, W. C.; Verma, D. D.; Levchenko, T. S.; Bernstein, E. A.; Torchilin, V. P. Wiley Interdisciplinary Reviews: Nanomedicine and Nanotechnology. **2009,** *1,* 530–539.

Helbok, A.; Decristoforo, C.; Dobrozemsky, G.; Rangger, C.; Diederen, E.; Stark, B.; Prassl, R.; von Guggenberg, E. *Journal of Liposome Research.* **2009,** 1–*9,* DOI: 10.3109/08982100903311812.

Herth, M.; Barz, M.; Janh, M.; Zentel, R.; Rösch, F. Bioorg. Med. Chem. Letter. **2010,** *20,* 5454–5458.

Herth, M.; Braz, M.; Moderegger, D.; Allmeroth, M.; Janh, M.; Thews, O.; Zentel, R.; Rösch, F. Biomacromolecules. **2009,** *10,* 1697–1703.

Hofman, M. S.; Beauregard, J. M.; Braber, T. W.; Neels, O. C.; Eu, P.; Hicks, R. J. J. Nucl. Med. **2011,** *52,* 1513–1519.

Holder, W. D. Jr.; White, R. L. Jr; Zuger, J. H.; Easton, E. J. Jr.; Greene, F. L. Annals of Surgery. **1998,** *227,* 764–771.

Holland, J. P.; Divilov, V.; Bander, N. H.; Smith-Jones, P. M.; Larson, S. M.; Lewis, J. S. J. *Nucl. Med.* **2010,** *51,* 1293–1300.

Hong, H.; Shi, J.; Yang, Y.; Zhang, Y.; Engle, J. W.; Nickles, R. J.; Wang, X.; Cai, W. Nano Lett. **2011,** *11,* 3744–3750.

Hong, H.; Yang, K.; Zhang, Y.; Engle, J. W.; Feng, L.; Yang, Y.; Nayak, T. R.; Goel, S.; Bean, J.; Theuer, C. P.; Barnhart, T. D.; Liu, Z.; Cai, W. ACS Nano. 2012a, *6,* 2361–2370.

Hong, H.; Zhang, Y.; Engle, J. W.; Nayak, T. R.; Theuer, C. P.; Nickles, R. J.; Barnhart, T. E.; Cai, W. Biomaterials. 2012b, *33,* 4147–4156.

Hörger, I. PET-Positron Emission Tomography, an application of nuclear physics in diagnostic medicine. Dissertation [Online], Universidad Complutense de Madrid, 2003. www.gae.ucm.es/fisatom/docencia/trabajos/ines/pet.pdf (accessed Jan 10, 2013).

http://ubm.opus.hbz-nrw.de/volltexte/2012/3222/pdf/doc.pdf (accessed Jan *11,* 2013).

Huang, S. K.; Martin, F. J.; Jay, G.; Vogel, J.; Papahadjopoulos, D.; Friend, D. S. *Am. J. Pathol.* **1993,** *143,* 10–14.

Iijima, S.; Yudasaka, M.; Yamada, R.; Bandow, S.; Suenaga, K.; Kokai, F. *Chem. Phys. Lett.* **1999,** *309,* 165–170.

Immordino, M. L.; Dossio, F.; Cattel, L. International Journal of Nanomedicine. **2006,** *1,* 297–315.

Ishiwata, K.; Kimura, Y.; Oda, K.; Ishii, K.; Sakata, M.; Nariai, T.; Suzuki, Y.; Ishibashi, K.; Mishina, M.; Hashimoto, M.; Ishikawa, M.; Toyohara, J. Geriatr. Gerontol. Int. **2010,** *10,* S180-S-196.

Jain, A. K.; Murty, M. N.; Flynn, P. J. ACM Computing Surveys. **1999,** *31,* 264–323.

James, M. L.; Gambhir, S. S. Physiol. Rev. **2012,** *92,* 897–965.

Jarrett, B. R.; Gustafsson, B.; Kukis, D. L.; Louie, A. Y. Bioconjugate Chem. **2008,** *19,* 1496–1504.

Jasmin; Torres, A. L. M.; Nunes, H. M. P.; Passipieri, J. A.; Jelicks, L. A.; Gasparetto, E. L.; Spray, D. C.; Campos de Carvalho, A. C.; Mendez-Otero, R. *Journal of Nanobiotechnology.* **2011, 9**, DOI: 10.1186/1477–3155–9-4.

Jauregui-Osoro, M.; Williamson, P. A.; Glaria, A.; Sunassee, K.; Charoenphun, P.; Green, M. A.; Mullen, G. E. D.; Blower, P. J. Dalton Transactions. **2011,** *40,* 6226–6237.

Jennewein, M.; Lewis, M. A.; Zhao, D.; Tsyganov, E.; Slavine, N.; He, J.; Watkins, L.; Kodibagkar, V. D.; Rösch, F.; Mason, R. P.; Thorpe, P. E. Clin. Cancer Res. **2008,** *14,* 1377–1385.

Jiang, L.; Miao, Z.; Kimura, R. H.; Ren, G.; Liu, H. G.; Silverman, A. P.; Li, P. Y.; Gambhir, S. S.; Cochran, J. R.; Cheng, Z. Journal of Biomedicine and Biotechnology. 2012: **2012,** DOI:10.1155/2012/ 36.

Joshi, B.; Wang, T. D. Cancers (Basel). **2010,** *2,* 1251–1287.

Ju, Y.; Shan, J. Synthesis of bio-functionalized rare earth doped upconverting nanophosphors. U.S. Patent 20090121189, Apr 9, **2009**. http://patentscope.wipo.int/search/en/WO2009046392 (accessed Dec *17,* 2012).

Kaewprapan, K.; Inprakhon, P.; Marie, E.; Durand, A. *Carbohydrate Polymers*. **2012,** *88,* 875–881.

Kanwar, R. K.; Chaudhary, R.; Tsuzuki, T.; Kanwar, J. R. *Nanomedicine*. **2012,** *7,* 735–749.

Kasai, T.; Matsumura, S.; Iizuka, T.; Shiba, K.; Kanamori, T.; Yudasaka, M.; Iijima, S.; Yokoyama, A. Nanotechnology. **2011,** *22,* 065102. DOI: 10.1088/0957–4484/22/6/065102.

Kassis, A. Therapeutic radionuclides: biophysical and radiobiologic principles. *Semin. Nucl. Med.* **2008,** *38,* 358–366.

Keliher, E.; Yoo, J.; Nahrendorf, M.; Lewis, J.; Marinelli, B.; Newton, A.; Pittet, M.; Weissleder, R. *Bioconjug. Chem.* **2011,** *22,* 2383–2389.

Kelkar, S. S.; Reineke, T. M. Bioconjugate Chem. **2011,** *22,* 1879–1903.

Kham, N. W. S.; Dai, H. Physica Status Solidi (*b*). **2006,** *243,* 3561–3566.

Khan, D. R. The use of nanocarriers for drug delivery in cancer therapy. Journal of Cancer Science & therapy. [Online] **2010,** *2,* 058–062. http://www.omicsonline.org/Archive-JCST/2010/May/01/JCST-02–058.pdf (accessed Jan *11,* 2013).

Kim, J. S.; Kim, Y-H.; Kim J. H.; Kang, K. W.; Tae, E. L.; Youn, H.; Kim, D.; Kim, S-K; Kwon, J-T.; Cho, M-H.; Lee, Y-S.; Jeong, J.M.; Chung, J-K.; Lee, D. S. *Nanomedicine*. **2012,** *7,* 219–229.

Kim, S. W.; Zimmer, J. P.; Ohnishi, S.; Tracy, J. B.; Frangioni, J. V.; Bawendi, M. G. *J. Am. Chem. Soc.* **2005,** *127,* 10526–10532.

Koehler, L.; Gagnon, K.; McQuarrie, S.; Wuest, F. Molecules. **2010,** *15,* 2686–2718.

Köster, U.; Assman, W.; Barbet, G.; Chatal, F.; Fagret, D.; Haddad, F.; Hohn, A.; Jensen, M.; Miederer, M.; Pichler, B.; Polack, B.; Ratib, O.; Schibli, R.; Seimbille, Y.; Tamburella, C.; Türler, A.; Vuillez, J. P.; Wiehr, S.; Zhernosekov, K. Innovative radioisotopes for preclinical and clinical studies in nuclear medicine. CERN-INTC-2010–055/INTC-I-121-European Organization for Nuclear Research, 2011.

Kotzerke, J.; Andreeff, M.; Wunderlich, G. Eur. J. Nucl. Med. Mol. Imaging, **2010,** *37,* 175–177.

Kumar, P. S.; Pastoriza-Santos, I.; Rodríguez-González, B.; García de Abajo, F. J.; Liz-Marzán, L. M. *Nanotechnology.* **2008,** *19,* 1–12.

Kuno, M.; Lee, J. K.; Dabbousi, B. O.; Mikulec, F. V.; Bawendi, M. G. The band edge luminescence of surface modified CdSe nanocrystallites: Probing the luminescing state. J. Chem. Phys. [Online] **1997,** *106,* 9869. http://jcp.aip.org/resource/1/jcpsa6/v106/i23/p9869_s1?isAuthorized=no (accessed Dec *6,* 2012).

Laforest, R.; Liu, X. Q. J. Nucl. Med. Mol. Imaging. **2008,** 52, 151–158.

Lambretch, R. Positron emitting radionuclides. Present and future status. Brookhaven national laboratory. [Online] **1979,** 1–16 http://www.osti.gov/bridge/servlets/purl/6259031- DbfGYe/6259031.pdf (accessed Dec *5,* 2012).

Lammers, T.; Subr, V.; Ulbrich, K.; Hennink, W. E.; Storm, G.; Kiessling, F. Nano Today. **2010,** *5,* 197–212.

Lammers, T.; Yokota-Rizzo, L.; Storm, G.; Kiessling, F. Personalized nanomedicine. *Clin Cancer Res.* [Online early access]. DOI: 10.1158/1078–0432.CCR-12–1414 Published Online: July *24,* 2012; http://clincancerres.aacrjournals.org/content/early/2012/09/05/1078–0432.CCR-12–1414?cited-by=yes&legid=clincanres;1078–0432.CCR-12–1414v2 (accessed Nov *17,* 2012).

Laverman, P.; Boerman, O. C.; Oyen, W. J.; Dams, E. T.; Storm, G.; Corstens, F. H. Adv. Drug Deliv. Rev. **1999,** *37,* 225–235.

Lee, H-Y.; Li, Z.; Chen, K.; Hsu, A. R.; Xu, C.; Xie, J.; Sun, S.; Chen, X. J. Nucl. Med. **2008,** *49,* 1371–1379.

Lee, K-H. Journal of Nuclear Medicine. **2007,** *48,* 1408–1410.

Lewis, J. S.; Laforest, R.; Dehdashti, F.; Grigsby, P. W.; Welch, M. J.; Siegel, B. A. J. Nucl. Med. **2008,** *49,* 1177–1182.

Li, S.; Goins, B.; Zhang, L.; Bao, A. Bioconjug. Chem. **2012,** *23,* **1322,**1332.

Li, Z. B.; Wu, Z.; Chen, K.; Chin, F. T.; Chen, X. Bioconjugate Chem. **2007,** 1987–1994.

Li, Z.;Yang, R.; Yu, M.; Bai, F.; Li, C.; Wang, Z. L. J. Phys. Chem. C. **2008,** *112,* 20114–20117.

Lian, T.; Ho, R. J. J. Pharm. Sci. **2001,** *90,* 667–680.

Lim, S.; Paterson, B. M.; Fodero-Tavoletti, M. T.; O'Keefe, G. J.; Cappai, R.; Barnham, K. J.; Villemagne, V. L.; Donnelly, P. S. Chem. Comm. **2010,** *46,* 5381–5600.

Liu, G.; Men, P.; Perry, G.; Smith, M. A. Methods Mol. Biol. **2010,** *610,* 123–144.

Liu, S.; Sun, S. Journal of Organometallic Chemistry. **2000,** *599,* 74–86.

Liu, Y.; Welch, M. J. Bioconjug. Chem. **2012,** *23,* 671–682.

Lockie, L. M.; Norcross, B. M.; Riordan, D. J. J. *Am. Med. Assoc.* **1958,** *167,* 1204–1207.

López-Durán, F. A.; Zamora-Romo, E.; Alonso-Morales, J.; Mendoza-Vásquez, G. Revista especializada en ciencias Químicobiológicas. **2007,** *10,* 26–35.

Loxley, A. Drug Delivery Technology. **2009,** *9,* 1–5.

Lu, C. W.; Hung, Y.; Hsiao, J. K.; Yao, M.; Chung, T. H.; Lin, Y. S.; Wu, S. H.; Hsu, S. C.; Liu, H. M.; Mou, C. Y.; Yang, C. S.; Huang, D. M.; Chen, Y. C. Nano Lett. **2007,** *7,* 149–154.

Lu, Z. R. Advanced Drug Delivery Reviews. **2010,** *62,* 246–257.

Lubberink, M.; Rizv, S. N. F.; Hoekstra, O. S.; van Dongen, G. A. M. S. Medicamundi. **2010,** *54,* 41–48.

Lucignani, G. Eur. J. Nucl. Med. Mol. Imaging. **2009,** 36, 869–874.

Luehmann, H.; Wang, Y.; Xia, X.; Brown, P.; Jarreau, C.; Welch, M. J.; Xia, Y.; Liu, Y. J. *Nucl. Med.* **2012,** 53 (Supp.1), 1690.

Luo, J.; Jang, S. Estimate Absorbed Dose Using Micro PET Imaging of Rat Brain Tumor with I-124 Infusion Followed by F-18 Injection. Med. Phys. [Online] **2010,** *37,* 3176. http://online.medphys.org/resource/1/mphya6/v37/i6/p3176_s1?isAuthorized=no (accessed Dec *6,* 2012).

Luo, J.; Wilson, J. D.; Zhang, J.; Hirsch, J. I.; Dorn, H. C.; Fatouros, P. P.; Shultz, M. D. *Appl. Sci.* **2012,** *2,* 465–478.

Maeda, H. *Advan. Enzyme Regul.* **2001,** *41,* 189–207.

Mai, H. X.; Zhang, Y. W.; Si, R.; Yan, Z. G.; Sun, L-D.; You, L-P.; Yan, C-H. J. *Am. Chem. Soc.* **2006,** *128,* 6426–6436.

Marik, J.; Tartis, M. S.; Zhang H.; Fung, J. Y.; Kheirolomoom, A.; Sutchiffe, J. L.; Ferrara, K. W. *Nucl. Med. Biol.* **2007,** *34,* 165–171.

Martín, R.; Alvaro, M.; Herance, J. R.; García, H. ACS Nano, **2010,** *4,* 65–74.

Matson, J. B.; Grubbs, R. H. *J. Am. Chem. Soc.* **2008,** *130,* 6731–6733.

Matsumura, Y.; Maeda, H. *Cancer Res.* **1986,** *46,* 6387–6392.

May, J. L.; Erogbogbo, F.; Yong, K-T.; Ding, H.; Law, W-C.; Swihart, M.T.; Prasad, P. N. *Journal of Solid Tumors.* **2012,** *2,* 24–37.

McDevitt, M. R.; Chattopadhyay, D. Kappel, B. J.; Jaggi, J. S.; Schiffman, S. R.; Antczak, C.; Njardarson, J. T.; Brentjents, R.; Scheinberg, D. A. Journal of Nuclear Medicine. 2007b, *48,* 1180–1189.

McDevitt, M. R.; Chattopadhyay, D.; Jaggi, J. S.; Finn, R. D.; Zanzonico, P. B.; Villa, C.; Rey, D.; Mendenhall, J.; Batt, C. A.; Njardarson, J. T.; Scheinberg, D. A. PLoS ONE. 2007a, *2,* e907. DOI:10.1371/journal.pone.0000907.

Medvedev, D.; Mausner, L.; Meinken, G.; Maramraju, S. H; Smith, S. D.; Carroll, V.; Schlyer, D. J. Nucl. Med. **2012,** 53 (Supplement 1), 1535.

Milligan, G. W. Clustering validation: Results and implications for applied analyzes clustering and classification. In *Clustering and Classification*; Arabie, P.; Hubert, L. J.; De Soete, G., Ed.; World Scientific Publ. River Edge: Singapore, 1996; pp 341–375.

Miyako, E.; Nagata, H.; Hirano, K.; Sakamoto, K.; Makita, Y.; Nakayama, K-I.; Hirotsu, F. Nanotechnology, **2008,** *19,* 1–6.

Miyawaki, J.; Yudasaka, M.; Imai, H.; Yorimitsu, H.; Isobe, H.; Nakamura, E.; Iijima, S. Advanced Materials. **2006,** *18, 8,* 1010–1014.

Mochalin, V. N.; Shenderova, O.; Ho, D.; Gogotsi, Y. Nature Nanotechnology. **2012,** *7,* 11–23.

Moderegger, D. Radiolabeling of defined polymer architectures with fluorine-18 and iodine-131 for *ex vivo* and *in vivo* evaluation: Visualization of structureproperty relationships. Ph.D. Dissertation [Online], Johannes Gutenberg-Universität Mainz, 2012.

Moderegger, D.; Allmeroth, M.; Thews, O.; Zentel, R.; Rösch, F. In COST TD1004 MC meeting, Torino, February 17–*18,* **2012,** [Online] *In vitro* and *in vivo* evaluation of ^{18}F-labeled HPMA polymeric architectures as potential carrier system of drugs: How structure influences pharmacology or: *in vivo* veritas. http://www.cim.unito.it/website/COST/meetings/documents/Prof.%20Aime%201%2017-%202–12/ROESCH%20Torino%20COST_HPMA.pdf (accessed Dec *6,* 2012). Mody, V. V. Internet Journal of Medical Update. **2011,** *6,* 24–30.

Moghimi, S. M. Biochimica et Biophysica *Acta.* **2002, 1590,** 131–139.

Moses, J. E.; Moorhouse, A. D. *Chem. Soc. Rev.* **2007,** *36,* 1249–1262.

Murray, C. B.; Norris, D. J.; Bawendi, M. G. J. Am. Chem. Soc. **1993,** *115,* 8706–8715.

Nahrendorf, M.; Keliher, E.; Marinelli, B.; Leuschner, F.; Robbins, C. S.; Gerszten, R. E.; Pittet, M. J.; Swirski, F. K.; Weissleder, R. *Arterioscler. Thromb. Vasc. Biol.* **2011,** *31,* 750–757.

Nahrendorf, M.; Zhang, H.; Hembrador, S.; Panizzi, P.; Sosnovik, D. E.; Aikawa, E.; Libby, P.; Swirsky, F. K.; Weissleder, R. *Circulation.* **2008,** *117,* 379–387.

Nakamura, H. Kimura, T.; Ikegami, H.; Ogita, K.; Koyama, S.; Shimoya, K.; Tsujie, T.; Koyama, M.; Kaneda, Y.; Murata, Y. Molecular Human Reproduction. **2003,** *9,* 603–609.

Niidome, T.; Yamagata, M.; Okamoto, Y.; Akiyama, Y.; Takahashi, H.; Kawano, T.; Katayama, Y.; Niidome, Y. *Journal of Controlled Release.* **2006,** *114,* 343–347.

Nune, S. K.; Gunda, P.; Majeti, B. K.; Thallapally, P. K.; Forrest, M. L. Adv. Drug Deliv. Rev. **2011,** *63,* 876–885.

Oku, N. Adv. Drug Deliv. Rev. **1999,** *37,* 53–61.

Oku, N.; Yamashita, M.; Katayama, Y.; Urakami, T.; Hatanaka, K.; Shimizu, K.; Asai, T.; Tsukada, H.; Akai, S.; Kanazawa, H. Int. J. Pharm. **2011,** *403,* 170–177.

Oldenburg, S. J.; Averitt, R. D.; Wescott, S. L.; Halas, N. J. Chemical Physics Letters. **1998,** *288,* 243–247.

Ordidge, K. L.; Duffy, B. A.; Wells, J. A.; Kalber, T. L.; Janes, S. M.; Lythgoe, M. F.

Otto, M. Chemometrics. Statistics and Computer Application and Analytical chemistry, 2nd ed.; Wiley-VCH: Weinheim, Germany, **2007,** pp 121–181.

Owens III, D. E. and Peppas, N. A. International Journal of Pharmaceutics. **2006,** *307,* 93–102.

Pagani, M.; Stone-Elander, S.; Larsson, S. A. Eu. J. Nucl. Med. Mol. Imaging. **1997,** *24,* 1301–1327.

Pala, A.; Liberatore, M.; D'Elia, P.; Nepi, F.; Megna, V.; Mastantuono, M.; Al-Nahhas, A.; Rubello, D.; Barteri, M. Mol Imaging Biol. **2012,** 14:593Y*598,* 593–598.

Papahadjopoulos, D.; Allen, T. M.; Gabizon, A.; Mayhew, E.; Matthay, K.; Huang, S. K.; Lee, K-D.; Woodle, M. C.; Lasic, D. D.; Redemann, C.; Martin, F. J. Proc. Natl. Acad. Sci. USA.**1991,** *88,* 11460–11464.

Patel, D.; Kell, A.; Simard, B.; Deng, J.; Xiang, B.; Lin, H. Y.; Gruwel, M.; Tian, G. Biomaterials. **2010,** *31,* 2866–2873.

Patel, D.; Kell, A.; Simard, B.; Xiang, B.; Lin, H. Y.; Tian, G. Biomaterials. **2011,** *32,* 1167–1176.

Patt, M.; Schildan, A.; Mishchenko, O.; Habermann, B.; Patt, J.; Reischl, G.; Barthel, H.; Pichler, B.; Sabri, O. J Nucl Med. **2010**; *51* (Supplement 2), 389.

Petersen, A.; Binderup, T.; Rasmussen, P.; Henriksen, J. R.; Elema, D. R.; Kjær, A.; Andersen, T. L. Biomaterials. **2011,** *32,* 2334–2341.

Phillips, W. T.; Goins, B. A. Bao, A. Wiley Interdiscip. Rev. Nanomed. Nanobiotechnol. **2009,** *1,* 69–83.

Pradhan, N.; Battaglia, D. M.; Liu, Y.; Peng, X. Nano Lett. **2007,** *7,* 312–317.

Pradhan, P.; Banerjee, R.; Bahadur, D.; Koch, C.; Mykhaylyk, O.; Plank, C. Methods Mol. Biol. **2010,** *605,* 279–293.

Pusuluri, A.; Kadam, A. Pharma utility. **2012,** *6,* p.1–10.

Quintero, N. Y.; Restrepo G.; Cohen, I. M. Journal of Radioanalytical and Nuclear Chemistry. **2013,** *295,* 823–833.

Radchenko, V.; Hauser, H.; Eisenhut, M.; Vugts, D. J.; van Dongen, G. A. M. S.; Rösch, F. Radiochim. Acta. 2012. DOI: 10.152/ract. 2012.1971.

Rasmussen, J. W.; Martínez, E.; Louka, P.; Wingett, D. G. Expert Opin. Drug Delivery **2010,** *7,* 1063–1077.

Restrepo, G.; Mesa, H. Curr. Comput. Aided Drug Des. **2011,** *7,* 90–97.

Restrepo, G.; Mesa, H.; Llanos, E. J.; Villaveces, J. L. J. Chem. Inf. Comput. Sci. **2004,** *44,* 68–75.

Restrepo, G.; Villaveces, J. L. Croat Chem Acta. **2005,** *78,* 275–281.

Richardson, J.; Jeyasingh, K.; Jewkes, R. F.; Ryman, B. E.; Tattersall, M. H. N. Journal of Nuclear Medicine. **1978,** *19,* 1049–1054.

Ridolfi, D. M.; Marcato, P. D.; Machado, D.; Silva, R. A.; Justo, G. Z.; Durán, N. J. Phys.: Conf. Ser. **2011,** *304,* 01**2032,** DOI:10.1088/1742–6596/304/1/012032.

Riggio, C.; Pagni, E.; Raffa, V; Cuschieri, A. Journal of Nanomaterials. **2011,** 1–10.

Riuttamäki, T. Upconverting phosphor technology: exceptional photoluminescent properties light up homogeneous bioanalytical assays. Ph. D. Dissertation [Online], University of Turku, 2011. http://doria17-kk.lib.helsinki.fi/bitstream/handle/10024/71980/AnnalesAI427Riuttamaki.pdf?sequence=1 (accessed Nov *20,* 212).

Rojas, S.; Gispert, J. D.; Abad, S.; Buaki-Sogo, M.; Victor, M. V.; Garcia, H.; Herance, J. R. Mol. Pharmaceutics. **2012,** *9,* 3543–3550.

Rojas, S.; Gispert, J. D.; Martín, R.; Abad, S.; Menchón, C.; Pareto, D.; Víctor, V. M.; Alvaro, M.; García, H.; Herance, J. R. ACS Nano. **2011,** *5,* 5552–5559.

Rösch, F.; Baum, R. P. Dalton Trans. **2011,** *40,* 6104–6111.

Rossin, R. Radiolabeled nanoplatforms: imaging hot bullets hitting their target. In *Nanoplatform-based molecular imaging*; Chen, X., Ed.; John Willey & Sons: USA, 2011; pp 399–432.

Rossin, R.; Muro, S.; Welch, M. J.; Muzykantor, V. R.; Schuster, D. P. J. Nucl. Med. **2008,** *49,* 103–111.

Rossin, R.; Pan, D.; Qi, K.; Turner, J. L.; Sun, X.; Wooley, K. L.; Welch, M. J. J. Nucl. Med. **2005,** *46,* 1210–1218.

Ruggiero, A.; Villa, C. H., J. P.; Sprinkle, S. R.; May, C.; Lewis, J. S.; Scheinberg, D. A.; McDevitt, M. R. Int. J. Nanomedicine. **2010,** *5,* 783–802.

Rzigalinsky, B. A.; Strobl, J. S. Toxicol. Appl. Pharmacol. **2009,** *238,* 280–288.

Sahoo, S.; Sahoo, S. Physics Education. **2006,** 5–11.

Schipper, M. L.; Cheng, Z.; Lee, S-W.; Bentolila, L. A.; Iyer, G.; Rao, J.; Chen, X.; Wu, A. M.; Weiss, S.; Gambhir, S. S. J. Nucl. Med. **2007,** *48,* 1511–1518.

Schrand, A. M.; Ciftan Hens, S. A.; Shenderova, O. A. Critical Reviews in Solid State and Materials Sciences. **2009,** *34,* 18–74.

Seo, J.W.; Zhang, H.; Kukis, D. L.; Meares C. F.; Ferrara, K. W. Bioconjug. Chem. **2008,** *19,* 2577–2584.

Shokeen, M.; Anderson, C. J. Acc. Chem. Res. **2009,** *42,* 832–41.

Shultz, M. D.; Wilson, J. D.; Fuller, C. E.; Zhang, J.; Dorn, H. C.; Fatouros, P. P. Radiology. **2011,** *262,* 136–143.

Silindir, M.; Erdoğan, S.; Ö zer, A.Y.; Maia, S. Journal of Drug Targeting. **2012,** *20,* 401–415.

Simone, E. A.; Zern, B. J.; Chacko, A-M.; Mikitsh, J. L.; Blankenmeyer, E. R.; Muro, S.; Stan, R.V.; Muzykantov, V. R. Biomaterials. **2012,** *33,* 5406–5413.

Smart, S. K.; Cassady, A. I.; Lu, G. Q.; Martin, D. J. Carbon. **2006,** *44,* 1034–1047.

Smith, S. V.; Jones, M.; Holmes.V. Production and Selection of Metal PET Radioisotopes for Molecular Imaging, Radioisotopes-Applications. In *Bio-Medical Science*, Singh, N., Ed.; InTech: Rijeka, Croatia, 2011; pp 199–224.

Sun, X.; Anderson C. J. Methods Enzymol. **2004,** *386,* 237–*261.*

Sun, Y.; Yu, M.; Liang, S.; Zhang, Y.; Li, C.; Mou, T.; Yang, W.; Zhang, X.; Li, B.; Huang, C.; Li, F. Biomaterials. **2011,** *32,* 2999–3007.

Svedman, P.; Nosslin, B.; Rothman, U. Cardiovasc. Res. **1983,** *17,* 656–661.

Taguchi, K.; Urata, Y.; Anraku, M.; Watanabe, H.; Kadowaki, D.; Sakai, H.; Horinouchi, H.; Kobayashi, K.; Tsuchida, E.; Maruyama, T.; Otagiri, M. Drug Metabolic and Disposition. **2009,** *37,* 2197–2203.

Thakor, A. S.; Jokerst, J.; Zaveleta, C.; Massoud, T. F.; Gambhir, S. S. Nano Lett. **2011,** *11,* 4029–4036.

Torchilin, V. P. Nature reviews. **2005,** *4,* 145–160.

Ünak, P. Braz. Arch. Biol. Technol. [Online] **2008,** *51,* 31–37. http://www.scielo.br/pdf/babt/v51nspe/v51nspea06.pdf (accessed Nov *6,* 2012).

Uo, M.; Akasaka, T.; Watari, F.; Sato, Y.; Tohji, K. Dent. Mater. J. **2011,** *30,* 245–263.

Vallabhajosula, S. Semin. Nucl. Med. **2011,** *41,* 283–299.

Vallabhajosula, S.; Killeen, R. P.; Osborne, J. R. Semin. Nucl. Med. **2010,** *40,* 220–241.

Vasey, P. A.; Kaye, S. B.; Morrison, R.; Twelves, C.; Wilson, P.; Duncan R.; Thomson, A. H.; Murray, L. S.; Hilditch, T. E.; Murray, T.; Burtles, S.; Fraier, D.; Frigerio, E.; Cassidy, J. Clin. Cancer Res. 1999, *5,* 83–94.

Verel, I.; Visser, G. W.; Boellaard, R.; Stigter-van Walsum, M.; Snow, G. B.; van Dongen, G. A. J. Nucl. Med. **2003,** *44,* 1271–1281.

Vert, M.; Domurado, D. J. Biomater. Sci. Polymer Edn. **2000,** *11,* 1307–1317.

Vivero-escoto, J. L. Surface functionalized mesoporous silica nanoparticles for intracellular drug delivery. Ph.D. Dissertation [Online], Iowa State University, 2009. http://lib.dr.iastate.edu/cgi/viewcontent.cgi?article=1915&context=etd (accessed Nov *6,* 2012).

Wadas, T. J.; Wong, E. H.; Weisman, G. R.; Anderson, C. J. Curr. Pharm. Des. **2007,** *13,* 3–16.

Walling, M. A.; Novak, J. A.; Shepard, J. R. E. Int. J. Mol. Sci. **2009,** *10,* 441–491.

Walrand, S.; Flux, G. D.; Konijnenberg, M. W.; Valkema, R.; Krenning, E. P.; Lhommel, R.; Pauwels, S.; Jamar, F. Eur. J. Nucl. Med. Mol. Imaging. **2011,** *38,* S57–68.

Walrand, S.; Jamar, F.; Mathieu, I.; de Camps, J.; Lonneux, M.; Sibomana, M.; Labar, D.; Michel, C.; Pauwels, S. Eur. J. Nucl. Med. Mol. Imaging. **2003,** *30,* 354–361.

Walther, M.; Gebhardt, P.; Grosse-Gehling, P.; Würbach, L.; Irmler, I.; Preusche, S.; Khalid, M.; Opfermann, T.; Kamradt, T.; Steinbach, J.; Saluz, H. P. Appl. Radiat. Isot. **2011,** *69,* 852–857.

Wang, H.; Chhowalla, M.; Sano, N.; Jia, S.; Amaratunga, G. A. J. Nanotechnology. **2004,** *15,* 546–550.

Wang, Z. L. ACS Nano. **2008,** *2,* 1987–1992.

Watt, N. T.; Whitehouse, I. J.; Hooper, N. M. The Role of Zinc in Alzheimer's Disease. International Journal of Alzheimer's disease. [Online] **2011,** 1–10. http://www.hindawi.com/journals/ijad/2011/971021/ (accessed Jan *5,* 2013).

Wei, Q.; Zhan, L.; Juanjuan, B.; Jing, W.; Jianjun, W.; Taoli, S.; Yi'an, G.; Wangsuo, W. Nanoscale Research letters. [Online], **2012,** *7,* 1–9. http://www.nanoscalereslett.com/content/pdf/1556–276X-7–473.pdf (accessed Nov *8,* 2012).

Weitman, S. D.; Lark, R.H.; Coney, L. R.; Fort, D. W.; Frasca, V.; Zurawski, V. R. Jr.; Kamen, B. A. Cancer Research **1992,** *52,* 3396–3401.

Welch, M. J.; Hawker, C. J.; Wooley, K. L. Journal of Nuclear Medicine, **2009,** *50,* 1743–1746.

Whitney, J. R.; Sarkar, S.; Zhang, J.; Do, T.; Young, T.; Manson, M. K.; Campbell, T. A.; Puretzky, A. A.; Rouleau, C. M.; More, K. L.; Geohegan, D. B.; Rylander, C. G.; Dorn H. C.; Rylander, M. N. Lasers Surg. Med. **2011,** *43,* 43–51.

Wu, X.; Liu, H.; Liu, J.; Haley, K. N.; Treadway, J. A.; Larson, J. P.; Ge, N.; Peale, F.; Bruchez, M. P. ***Nature Biotechnology. 2002, 21,*** 41–46.

Xiao, Y.; Hong, H.; Matson, V. Z.; Javadi, A.; Xu, W.; Yang, Y.; Zhang, Y.; Engle, J. W.; Nickles, R. J.; Cai, W.; Steeber, D. A.; Gong, S. Theranostics. **2012,** *2,* 757–768.

Xie, H.; Wang, Z. J.; Bao, A.; Goins, B. A.; Phillips, W. T. Radiolabeled gold nanoshells for in vivo imaging: example of methodology for initial evaluation of biodistribution of a novel nanoparticle. In *Nanoimaging*; Goins, B; Phillips, W., Ed.; Pan Stanford Publishing, Singapore, 2011; pp 213–224.

Xie, R.; Chen, K.; Chen, X.; Peng, X. Nano Res. **2008,** *1,* 457–464.

Xing, H.; Bu, W.; Zhang, S.; Zheng, X.; Li, M.; Chen, F.; He, Q.; Zhou, L.; Peng, W.; Hua, Y.; Shi, J. Biomaterials. **2012,** *33,* 1079–1089.

Xing, Y.; Dai, L. Nanomedicine. **2009,** *4,* 207–218.

Xu, Z. P.; Zeng, Q. H.; Lu, G. Q.; Yu, A. B. Chemical Engineering Science. **2006,** *61,* 1027–1040.

Yamada, M. Studies on synthesis and characterization of endohedral metallofullerenes and derivatives. Dissertation [Online], University of Tsukuba, 2008. http://www.tulips.tsukuba.ac.jp/limedio/dlam/B27/B2754425/1.pdf (accessed Jan *13,* 2013).

Yamamuro, T.; Nakamura, T.; Iida, H.; Kawanabe, K.; Matsuda, Y.; Ido, K.; Tamura, J.; Senaha, Y. Biomaterials. 1998, *19,* 1479–1482.

Yamaoka, T.; Tabata, Y.; Ikada, Y. J. Pharm. Sci. **1994,** *83,* 601–608.

Yguerabide, J.; Yguerabide, E. E. *Journal of Cellular Biochemistry*. **2001,** *84,* 71–81.

Yu, M. K.; Park, J.; Jon, S. *Theranostics*. **2012,** *2,* 3–44.

Zacchigna, M.; Cateni, F.; Drioli, S.; Bonora, G. M. *Polymers.* **2011,** *3,* 1076–1090.

Zeng, D.; Lee, N. S.; Liu, Y.; Zhou, D.; Dence, C. S.; Wooley, K. L.; Katzenellenbogen, J. A.; Welch, M. J. *ACS Nano.* **2012,** *6,* 5209–5219.

Zhang, C.; Sun, L.; Zhang, Y.; Yan, C. Journal of Rare Earths. **2010,** 28, 807–819.

Zhang, F.; Niu, G.; Lu, G.; Chen, X. *Mol. Imaging Biol.* **2011,** *13,* 599–612.

Zhang, S.; Chan, K. H.; Prud'homme, R. K.; Link, A. J. *Mol. Pharmaceutics*. **2012,** *9,* 2228–2236.

Zhang, T.; Xu, M.; He, L.; Xi, K.; Gu, M.; Jiang, X. Carbon. **2008,** *46,* 1782–1791.

Zhou, J.; Liu, Z.; Li, F. *Chem. Soc. Rev.,* **2012,** *41,* 1323–1349.

Zhou, J.; Xu, N.; Wang, Z. L. *Advanced materials.* **2006,** *18,* 2432–2435.

Zhou, J.; Yu, M.; Sun, Y.; Zhang, X.; Zhu, X.; Wu, Z.; Wu, D.; Li, F. *Biomaterials*. **2011,** *32,* 1148–1156.

Zhu, Y.; Li, J.; Li, W.; Zhang, Y.; Yang, X.; Chen, N.; Sun, Y.; Zhao, Y.; Fan, C.; Huang, Q. *Theranostics*. **2012,** *2,* 302–312.

CHAPTER 3

CARBON-BASED HIGH ASPECT RATIO POLYMER NANOCOMPOSITES

THOMAS GKOURMPIS

CONTENTS

Abstract ... 86
3.1 Introduction ... 86
3.2 Dispersion ... 89
3.3 Melt Mixing ... 90
3.4 Solution Mixing ... 92
3.5 In-situ Polymerization ... 93
3.6 Properties of Polymer Nanocomposites ... 96
3.6.1 Electrical Conductivity ... 96
3.6.2 Thermal Conductivity ... 100
3.6.3 Mechanical Properties ... 102
3.6.4 Gas and Liquid Permeation ... 104
3.6.5 Thermal and Dimensional Stability ... 106
3.6.6 Rheological Properties ... 106
3.6.7 Morphology and Crystallization Effects ... 108
3.7 Current and Future Challenges and Applications ... 113
3.8 Conclusions ... 116
Keywords ... 117
References ... 117

ABSTRACT

High aspect ratio carbon-based allotropes have attracted enormous scientific and industrial attention due to their fascinating physicochemical properties and immense potential. In this chapter we will review the properties of fillers like graphene and carbon nanotubes once they are incorporated in a polymer matrix, and their effect on the overall nanocomposite performance. The aim is to provide some insight on the current advances by illustrating the effect such fillers have in properties like electrical and thermal conductivity, gas and liquid permeation, thermal and dimensional stability, rheology and morphology. Still while highlighting these astonishing properties one must not ignore the myriad of scientific and practical problems that have kept these systems from large-scale commercialization. For that reason we will briefly discuss these problems, paying special attention to the efficiency of the filler dispersion in the polymeric matrix, and outline the current trends, concerns and possibilities. We will conclude with a quick overview of the current and potential applications for a number of emerging technologies in the near future.

3.1 INTRODUCTION

Polymer nanocomposites based on carbon combine polymeric properties with those of the filler material (Manias, 2007). Traditionally carbon black (CB) has been the filler of choice due to the simplicity of its manufacturing methods (Kuhner, 1993) that lead to very competitive prices coupled with a very good overall performance (Wolf, 1993 and Rivin, 1971). Consequently CB-based nanocomposites have dominated the markets with the tire industry being probably the largest user, with an approximate production of 8 million metric tons per year (Rahman, 2011). Furthermore due to the inherent electrical conductivity, CB-based fillers are indispensable in all but a few electrochemical devises including fuel cells, batteries and supercapacitors (Chen, 1995). Finally it must be noted that CB offers a wide variety of structural diversity as a consequence of the different preparation methods, which leads to an array of different particle sizes, surface areas and level of activation (Herd, 1991).

Despite the advantages on performance and cost offered by the use of CB-based nanocomposites all is not perfect, as at high filler concentrations the overall mechanical performance of the system is impaired (Kong, 2000) and the subsequent viscosity increase causes a lot of processing problems (Collins, 2000). In Fig. 1 the overall increase of the composition viscosity for different types of CB-based nanocomposites as a function of the amount of CB present in the system can be seen. The viscosity is estimated by the use of the Melt Flow

Rate (MFR) index that is inversely proportional to the viscosity index (ASTM D1238-04 and ISO 1133:1997).

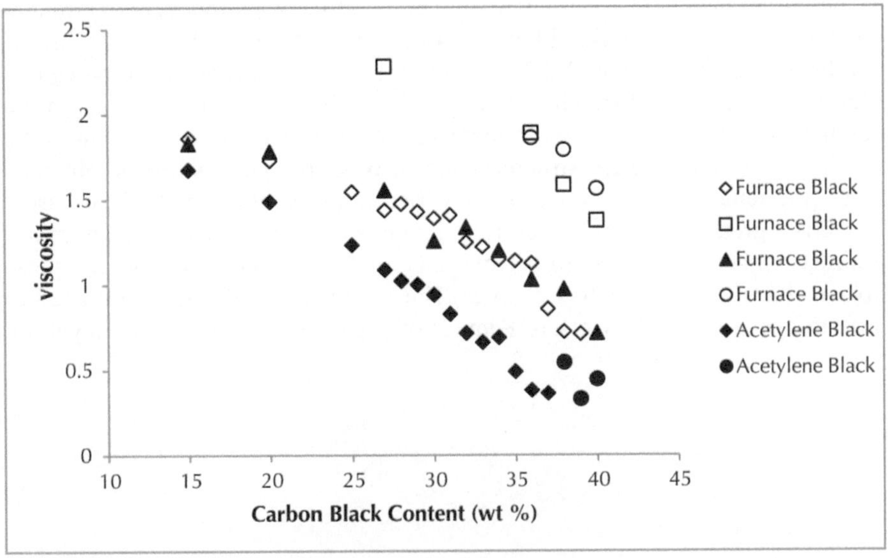

FIGURE 1 Overall viscosity increase (as seen by the MFR-Index) for different CB types in a range of polyolefin matrices (Gkourmpis, 2013). The different filler types correspond to different internal architectures and levels of internal structure that have a profound effect on the overall viscosity of the composition. Each symbol in the graph corresponds to a different polymer and a different CB type.

The results shown in Fig. 1 are representative for almost all types of Polyolefins and it can be clearly seen that there is an almost linear viscosity increase with the increase in the amount of filler introduced in the composition. Obviously different types of CBs behave in different manner as the overall rheological performance is heavily dependent on the polymeric architecture (level of polarity, existence of branches, molecular weight etc.) and on the filler's overall structure, (i.e. how easy the primary particles aggregate and coalesce and how extended the resulting agglomerates are) but the generality of the effect cannot be denied.

Alternative fillers based on carbon like graphene, nanotubes (CNT) and nanoplatelets (GNP) have been attracting attention due to their outstanding overall performance (Kim, 2010a; Paul, 2008; Potts, 2011; Stankovich, 2006), making them ideal for a wide range of industrial applications ranging from gas-barrier systems (Wu, 2012) to biomedical (Hule, 2007) and semiconductive components for electronic applications (Avouris, 2012). The main difference between these fillers and standard carbon black is the existence of high aspect ratio components that allows access to the primary carbon spatial arrangement that is lost in the CB primary particle. For example graphene can be considered as a two-dimensional (2D) sheet of sp^2 carbon atoms in a honeycomb arrangement, making it the building block for all graphitic carbon allotropes of different dimensionality (Geim, 2007). In Fig. 2, a schematic representation can be seen indicating how different allotropes like fullerenes, nanotubes and more complex structures of different dimensionality can be obtained. One has to be careful, as although these structures are in principle possible to be obtained from graphene, in reality a number of different preparation methods are used. There is a great variety of preparation methods for these systems and each has a direct impact on the overall performance of the final composition, but these are outside the scope of this chapter. The interested reader is encouraged to seek information on the work of Kim (2008; 2010a) and the references within.

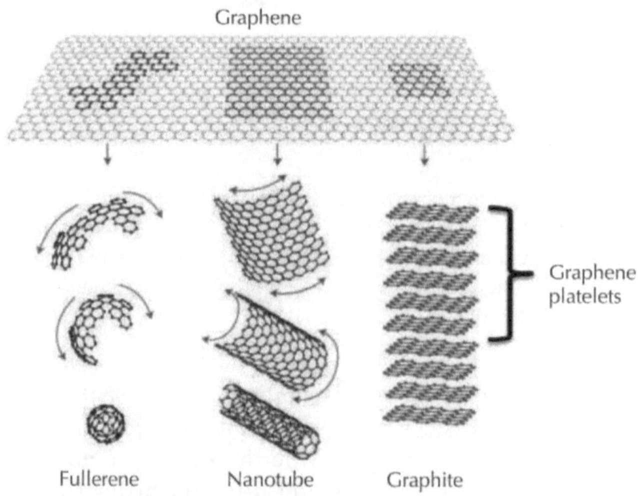

FIGURE 2 Schematic representation indicating how carbon-based allotropes like buckyballs (0D), CNTs (1D) and graphite (3D) can be obtained by wrapping the graphene sheet. Reproduced and modified from Geim, 2007.

It is precisely this possibility of gaining access to a filler with a high aspect ratio, low dimensionality and regular dimensions, that created the nanocomposites revolution initially with the discovery of fullerene in 1985 (Kroto, 1985), followed by single wall CNTs (SWCNT) in 1991 (Iijima, 1991) and more recently graphene (Geim, 2007). In principle, both graphene and CNT have similar and outstanding mechanical properties with a reported Young modulus of approximately 1TPa, thus making them much harder than traditional materials like steel,

Kevlar, high-density polyethylene (HDPE) and natural rubber (Lee, 2008). Furthermore the thermal conductivity of graphene has been reported to be of the order of 5000 W/(mK) and of CNT of the order of 3500 W/(mK). Finally these wonder materials have been reported to have very high electrical conductivity of the order of 6000 S/cm (Du, 2008). In Table 1 a summary of graphene and CNT mechanical, electrical and thermal properties can be seen in comparison with steel and other polymers.

TABLE 1 Properties of graphene and CNT in comparison with steel and polymers. aHDPE: High density polyethylene.

Material	Tensile Strength (MPa)	Thermal Conductivity (W/mk)	Electrical Conductivity (S/m)
Graphene	$(130\pm10)\times 10^3$	$4.84–5.30\times10^3$	7.2×10^3
CNT	$(60–150)\times 10^3$	3.5×10^3	$3–4\times10^3$
Steel	1.796×10^3	5–6	1.35×10^6
HDPE[a]	18–20	0.46–0.52	
Natural Rubber	$20–30\times10–^3$	0.13–0.142	
Kevlar	3.62×10^3	0.04	

In this chapter, we will review the properties of carbon-based high aspect ratio fillers like graphene, GNP and CNT and the effect they have once incorporated into a polymeric matrix. We will start by briefly reviewing the possible nanocomposite creation alternatives along with their relevant advantages and disadvantages, before we discuss in more detail the structure property relations and performance enhancement achieved with such nanofillers. We will conclude with a brief overview of the commercial situation for such systems and the future challenges. Finally, it must be noted that the area of polymer nanocomposites and nanotechnology is very diverse and impossible to cover extensively in a single book, let alone chapter. Therefore, the aim of this work is to provide a starting point for all those who are interested in the field and outline the current trends and recent advances.

3.2 DISPERSION

For a nanocomposite to exist, the filler must somehow get incorporated into the polymer matrix. The way this incorporation takes place and the final level of dispersion obtained has enormous dependence on the overall properties of the composition. Actually this is so important that a lot of the research on the field of nanocomposites has been focused on finding better methods for dispersing these high aspect ratio fillers into polymers (Alig, 2012; Moniruzzaman, 2006). One of the main problems one faces is the inability of the fillers to disperse in a random manner and the subsequent creation of agglomerates. It is believed that the main drive behind the agglomeration is the existence of weak attractive interactions between the fillers, and the effect is augmented by their sheer size that can reach dimensions up to several micrometers (Alig, 2012). This can be understood through filler–filler, filler–polymer and polymer–polymer interactions (Pötschke, 2004) as seen in the schematic representation of Fig. 3. Here we must note that polymer mobility is very important for the facilitation of the local and long-range arrangements of the filler in the matrix. With the term polymer mobility, we refer to all the length scales associated with a polymer, ranging from the local motion and spatial segmental arrangements as seen by the chain conformation (Gkourmpis, 2011) to the Rouse regime and chain reptation (Doi, 1986).

3.3 MELT MIXING

The most economically advantageous method for dispersing nanofillers into polymer matrices is melt mixing (or blending). As discussed previously, due to the attractive interactions between the filler particles, a level of clustering and agglomeration is almost always present, thus rendering good dispersion a real challenge.

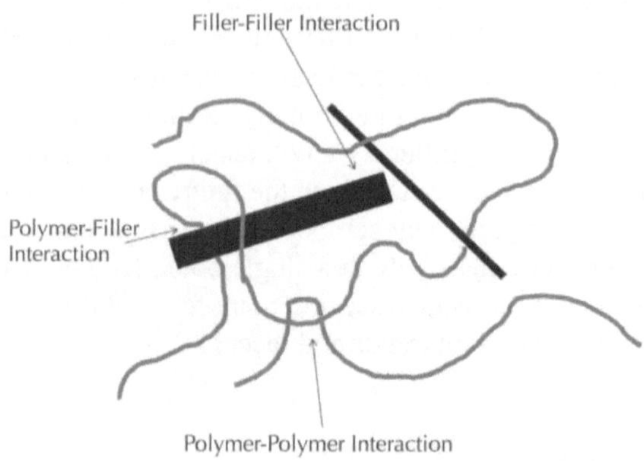

FIGURE 3 Schematic representation of the different types of interactions between an arbitrary filler and polymer.

From the technological point of view, the two main reasons for poor dispersion can be summarized to insufficient distribution of the filler during mixing, followed by secondary agglomeration due to particle diffusion and external forces (usually shear deformation) (Alig, 2012). Usually the filler particles are provided by the suppliers in bulk (powder) meaning that a degree of initial agglomeration has already taken place. These primary agglomerates have to be broken down into their individual components and then dispersed into the matrix. In the case of GNP, the clustering can be loose or extensive (see schematic in Figs. 4 and 5); whereas in the case of CNT, the nanotubes can be entangled due to structural defects originating from their preparation method and the all-present van der Waals interactions between them (Alig, 2012). In the case of industrially produced multiwalled carbon nanotubes (MWCNT), it is not uncommon that the final product is strongly deviating from the expected cylindrical shape due to the extent of curvature, intertwining and entanglements between the individual components. In Fig. 5, electron microscopy images of commercially available MWCNT and GNP as delivered can be seen. The large agglomerates of the order of a few micrometers can be clearly seen and the tight packing of the platelets indicates the enormity of the task of dispersion required.

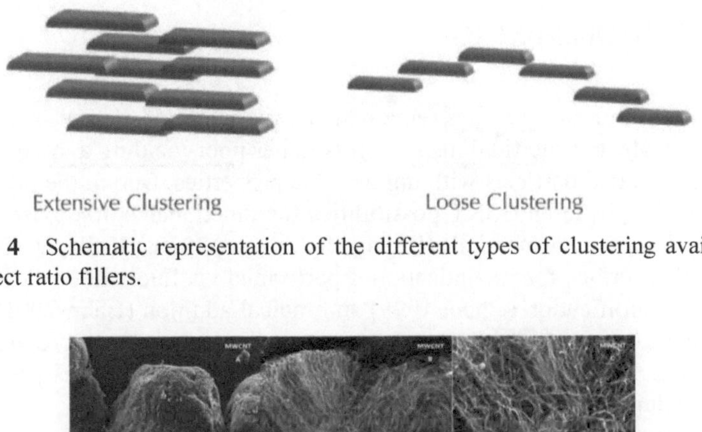

FIGURE 4 Schematic representation of the different types of clustering available for high aspect ratio fillers.

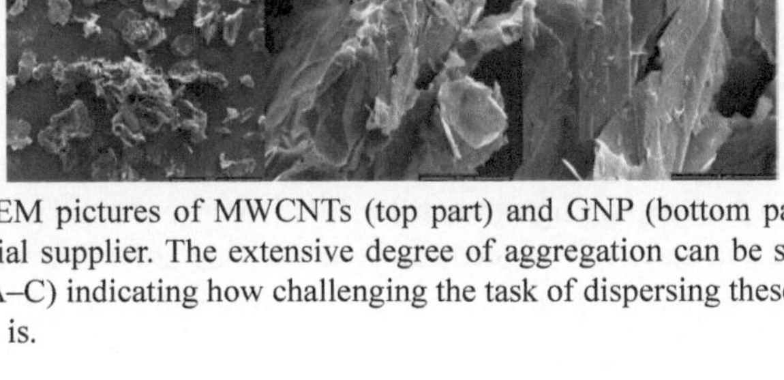

FIGURE 5 SEM pictures of MWCNTs (top part) and GNP (bottom part) as obtained from the material supplier. The extensive degree of aggregation can be seen at different length scales (A–C) indicating how challenging the task of dispersing these systems in the polymer matrix is.

It has been reported that wetting of the initial agglomerates in solution or melt can be beneficial, but a lot depends on the functional groups present on the surface of the filler as well as the level of polarity of the matrix (Alig, 2012; Barber, 2004). Polymer infiltration into the primary agglomerates has been reported to reduce their overall strength significantly (Alig, 2012). Finally the matrix viscosity, the mixing temperature and time, even the screw speed and the inserted amount of fillers into the mixer have been reported to affect the overall dispersion (Alig, 2012). In the case of CNT it has been reported that during melt dispersion the nanotubes do get fragmented and shorten (Kasaliwal, 2011; Krause, 2011; Pötschke, 2003) something that has been seen to have effect in the overall properties of the resulting nanocomposite.

Graphene melt compounding has been reported to be successful for thermally stable reduced graphene oxide (TRG) in elastomers (Kim, 2010; Liang, 2009) and glassy polymers (Kim, 2008, 2009). Despite these reports, melt mixing of graphene has been plagued by the thermal instability of the majority of chemically modified graphene, thus limiting the use in TRG.

3.4 SOLUTION MIXING

As we have seen in the previous section, melt mixing of nanocomposites in polymers is possible, but the final dispersion is rather poor yielding a system of partially agglomerated particles with unfavorable properties. Due to the preparation methods of graphene and CNT possibilities for functionalization exist, thus allowing for improved dispersions (Huang, 2012; Kim, 2010a; Kuila, 2010). In the case of CNTs surface functionalization is performed via fluorination (Mickelson, 1998), acid modification (Chen, 1998) and radical addition (Bahr, 2001). These methods have been reported to improve the solubility of the nanotubes in solvents and polymeric matrices, although the difficulty of disentangling the primary agglomerates during dispersion in a matrix is still problematic.

Traditionally graphene is synthesized from graphene oxide, a process that leaves a number of polar group on the surface that facilitate functionalization (Kim, 2010a). Due to its geometry graphene and GNP do not face the problems of entanglements seen in CNTs, but when the filler is stacked in an extensive or compact manner (see Figs. 4 and 5) the overall effectiveness is reduced. Restacking of the layered structures can be prevented by using surfactants in order to create particle suspensions (Stankovich, 2006b) or using polymers before the reduction process (Stankovich, 2006a). That way the layers can swell and be easily dispersed due to the weakening of the filler-filler forces, and the polymer is eventually absorbed and stacked between the layers during solvent evaporation (Lee, 2007). As it can be clearly seen, for such a method to work, one has to assure that the solvent is fully and properly removed, thus the choice of solvent is critical (Leroux, 2001). The process of polymer intercalation from solution is predominately driven by thermodynamics, namely the competition between the entropy gain by the desorption of the solvent molecules and the reduction in conformational energy of the intercalated polymer chains and their spatial arrangement on the graphene substrate (Ray, 2003).

3.5 IN-SITU POLYMERIZATION

Another method for ensuring adequate dispersion is the in-situ interactive polymerization; in which graphene (modified or unmodified) is mixed with liquid monomer followed by phase swelling. Depending on the type of the monomer, an appropriate initiator is allowed to diffuse in the solution and polymerization can be initiated by heat or radiation (Zheng, 2004 and Liang, 2009). A number of successful attempts on various polymers have been reported (see Kim, 2010a; Kuila, 2010 for more details), but it is important to note that so far it has been possible to polymerise monomers only in solvents. The reason is that graphene even in dilute dispersion has high viscosity, thus making bulk-phase polymerization very difficult. Surfaces have been reported to occur via chemical modification and grafting of polymer chains on graphene, but so far this has been achieved in very specific conditions and for very few systems (Potts, 2011). For CNTs similar approaches have been taken by attaching conjugating or conducting polymers on the tube surface. Again organic solvents are used and polymerization is initiated leading to a system of well-dispersed nanotubes with the polymer wrapped around them (see Xie, 2005 for more details).

As we have seen, there is no consensus among researchers on the method and approach towards dispersion of both graphene and CNT into a polymer matrix. All approaches have advantages and disadvantages and it is vital when one reviews reported results on polymer nanocomposites to take into consideration the preparation method and the sample history. Tables 2 and 3 actually prove this point as the reported properties for a number of different nanocomposites produced using different methods are presented.

Summarizing, we can say that the main advantage of melt mixing, is the elimination of the solvation phase (at least in most cases) and the consequent reduction in cost and increase in user-friendliness let alone the similarities with current industrial practices. Despite these advantages, so far the levels of dispersion of the nanoparticles on the matrix have been reported to be nowhere near the ones achieved by more advanced methods like solution mixing and in-situ polymerization. On the other hand, the solvation step and the polymerization process, are heavily dependent on the monomer/solvent/initiator combination and the filler viscosity, making such approaches very limited.

TABLE 2 Electrically conductive nanocomposites as reported in the literature.

Polymer[a]	Filler[b]	Preparation Method	Electrical percolation threshold (%vol, %wt)[c]	Ref
PS	iGO	Solvent + Hydrazine	0.1	Stankovic, 2006
PS	exfoliated graphene	Solvent + Ionic Liquid	0.13–0.17	Liu, 2008
PS	MWCNT	Solvent	0.15 (wt)	Grossiord, 2008
PS	SWCNT		0.11 (wt)	Ramasubramaniam, 2003
PS	SWCNT		0.045 (wt)	Ramasubramaniam, 2003
epoxy	partially reduced GO	In-Situ polymerization	0.52	Liang, 2009
epoxy	EG	Solvent	2.5	Celzard, 1996
PMMA	GO	In-Situ polymerization	1.3–1.6	Jang, 2009; Wang, 2004
PMMA	MWCNT		0.3 (wt)	Kim, 2004
PMMA	SWCNT		0.17 (wt)	Kim, 2004
VC/VA copolymer	GO	Solvent	0.15	Wei, 2009
EVA	GNP	Melt mixing	2.5	Stalmann, 2012
EVA	SWCNT	Solvent	0.036	Grunlan, 2004
PP	GNP	Melt mixing	7	Kalaitzidou, 2007
PP			2	Kim, 2010a
PVDF		Solvent	1.6	Ansari, 2009

TABLE 2 *(Continued)*

PVDF	EG	Solvent	5 (wt)	Ansari, 2009
PET	graphene	Melt mixing	0.47	Zhang, 2010
HDPE	EU	Melt mixing	3 (wt)	Zheng, 2004
HDPE	UG	Melt mixing	5 (wt)	Zheng, 2004
UHMWPE	SWCNT	Dry mixing	0.09 (wt)	Mierczynska, 2007
UHMWPE	MWCNT	Solution mixing	3 (wt)	Xi, 2007
LDPE	MWCNT	Milling	1–3 (wt)	Gorrasi, 2007
PC	MWCT	Melt mixing	1.44 (wt)	Pöschke, 2003a
PC	SWCNT	Solution mixing	0.11 (wt)	Ramasubramaniam, 2003

[a]PS: polystirene, PMMA: poly(methyl methacrylate), VC/VA: vinyl chloride/vinyl acetate, EVA: poly(ethyl vinyl acetate), PP: polypropylene, PVDF: poly(vinylidene fluoride), PET: poly(ethylene terepthalate), HDPE: high density polyethylene, UHMWPE: ultra high molecular weight polyethylene, LDPE: low density polyethylene, PC: polycarbonate [b]iGO: Isocyanate-treated graphite oxide, EG: expanded graphite. [c]Different groups report percolation as a function of volume or weight. To avoid transformations that require accurate densities for both polymer and filler we are reporting the values as provided by the authors. All values are in %vol except when indicated otherwise.

TABLE 3 Mechanical properties of polymer nanocomposites as reported in the literature.

Polymer[a]	Filler	Preparation Method	Filler (vol%, wt%)[b]	Modulus increase (%)	Tensile strength increase (%)	Ref
PU	SWCNT	Solution	1 (wt)	25	50	Sahoo, 2010
PU	MWCNT	Solution	1 (wt)	140	20	He, 2009
PU	MWCNT	In-situ polymerization	1 (wt)	90	90	Guo, 2008
PVA	GO	Solvent	2.5	128	70	Xu, 2009
PVA	GO	Solvent	0.49	62	76	Liang, 2009b
PVA	graphene	Solvent	1.8		150	Zhao, 2010
PMMA	EG	Solvent	21 (wt)	21		Zheng, 2003

TABLE 3 *(Continued)*

PMMA	GNP	Solvent	5 (wt)	133		Ramanathan, 2007
PMMA	GO	In-situ polymerization	1.7	54		Jang, 2009
PI	SWCNT		1 (wt)	89	9	Sahoo, 2010
PI	MWCNT		7.5	52	21	Sahoo, 2010
PP	graphite	Melt mixing	2.5		60	Kuila, 2010
PP			1.9	43		Kim, 2010a
PP	SWCNT	Melt mixing	10 (wt)	130	170	Spitalski, 2010
PP-g-MA	MWCNT	Melt mixing	8 (wt)		210	Spitalski, 2010
LLDPE	GNP	Solution	15 (wt)		200	Kuila, 2010
LLDPE	GNP	Solution	30		22	Kuila, 2010
HDPE	EG	Melt mixing	3 (wt)	100	4	Zheng, 2004
HDPE	UG	Melt mixing	3 (wt)	33		Zheng, 2004

[a]PU: polyurethane, PVA: poly(vinyl alcohol), PI: polyisoprene, PP-g-MA: polyethylene grafted maleic anhydride, LLDPE: linear low density polyethylene. [b]To avoid transformations that require accurate densities for both polymer and filler we are reporting the values as provided by the authors. All values are in %vol except when indicated otherwise.

3.6 PROPERTIES OF POLYMER NANOCOMPOSITES

As discussed earlier graphene, GNP and CNTs can be introduced in the polymer matrix in a variety of forms. In this section we will discuss the electrical, thermal, mechanical and transport properties of such systems.

3.6.1 ELECTRICAL CONDUCTIVITY

The introduction of conductive fillers into a polymer matrix can lead to a degree of electrical conductivity. The amount of filler required to achieve electron flow in the material is determined by a number of different parameters that are both filler (size, aspect ratio, polarity etc.) and polymer (architecture, polarity, molecular weight, viscosity etc.) dependant. At low filler loading conductivity is low and the composite can be considered to behave as an insulator, but as more and more filler is introduced a critical concentration called percolation threshold is reached and afterwards conductivity increases strongly until saturation is reached. In Fig. 6, a schematic representation of the percolative behavior can be seen.

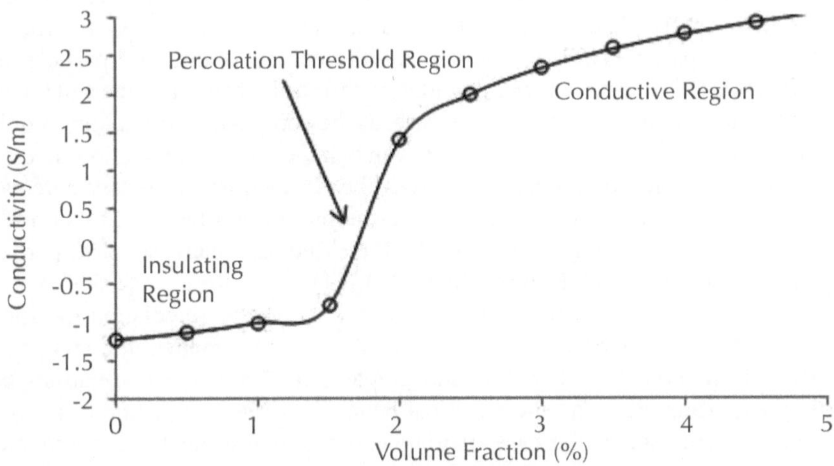

FIGURE 6 Schematic representation of the electrical percolation behavior in a polymer nanocomposite. In the insulating region the system behaves as an insulator, in the threshold region conductivity is increasing and in the conductive region saturation and maximum conductivity is achieved.

As discussed earlier, when filler is introduced in the polymer matrix different interactions between filler particles and the polymer have to be considered and since the fillers usually do not physically "touch" each other the overall dispersion of the conductive particles is of paramount importance. This type of interaction and "connectedness" (Alig, 2012) is thought of playing an important role in the electronic and thermal conduction as well as the mechanical stress of the system (Pötschke, 2004). A number of theoretical models have been suggested for the electrical percolation in polymers with most notable the pioneering work of Kirkpatrick (1973). All percolation models are assuming random dispersion of the filler in the matrix, leading to a statistical spatial distribution, direct contact between the conductive particles and random orientation for nonspherical objects (Lux, 1993). Obviously these assumptions are far from reality where due to the specific interactions between polymer and filler, a series of coalescing particles lead to agglomeration and nonrandomness of the filler's spatial distribution (Gkourmpis, 2013). The precise details of the percolation theory are beyond the scope of this discussion but the reader is encouraged to seek the work and included references of Kirkpatrick (1973) and Lux (1993).

In general, it has been seen than when high-aspect ratio fillers are introduced in a polymer a significant reduction of the percolation threshold is achieved (Kim, 2010a; Spitalsky, 2010) in comparison with more traditional fillers like carbon black. The main reason for such behavior is their aspect ratio and sheer size that affects their spatial distribution in the matrix. According to theory (Garboczi, 1995) rod-like particles percolate at one-half the volume of disk-like particles, so one expects CNTs to have a lower percolation threshold than graphene and GNPs. In reality, though this is not always the case, as the composite morphology is quite complex due to particle flexibility (CNTs can bend and coil), interactions and entanglements, leading to systems where rod-like CNT do not necessary percolate lower than disk-like graphene and GNP. Furthermore it has been seen that orientation of particles in the system affect the threshold as it increases with particle alignment (Balberg, 1984; Munson-McGee, 1991).

From Table 2, we can see that great variations on the percolation threshold values have been reported as a function of the polymer matrix, the composite preparation method, the filler type and preparation. Therefore correlations between these parameters and the electrical behavior of the compositions are very difficult to establish. In the case of graphene the lowest electrical percolation threshold reported was 0.1 vol% for a system of polystirene (PS) solvent blended with isocyanate-treated graphene oxide (iGO) followed by solution-phase reduction (Stankovich, 2006a). Similar percolation threshold onsets for single or multiwalled CNTs have been reported (Moniruzzaman, 2006). Furthermore, electrochemical exfoliation of graphite in ionic liquids has been reported to percolate at approximately 0.13–0.37vol% in a PS matrix (Liu, 2008). Systems based on solution blended poly(methyl methacrylate) (PMMA) and MWCNT have been reported to have a conductivity value of approximately 3000 S/m at 0.4 wt% for an extremely low percolation threshold of the order of 0.003 wt% (Spitalsky, 2010). MWCNTs added to polycarbonate (PC) via melt mixing lead to an increase of the maximum conductivity by 16 orders of magnitude (Pötschke, 2003b). Similarly the maximum conductivity value for a solution casted system of polyurethane (PU) and MWCNT has been reported at 2000 S/m at 27 wt% filling ratio (Spitalsky, 2010). Thermally reduced graphene has been reported to retain its high conductivity without need for any further steps and its thermal stability allows melt mixing (Kim, 2010a). Despite that it has been reported that for a system of thermoplastic polyurethane solution cased films exhibit better dispersion of the graphene sheets in comparison with similar melt mixed compositions (Kim, 2010b). Consequently, the percolation threshold in the solution-based systems was found to be significantly lower than in the melt mixed systems (0.3vol% to 0.8vol%, respectively). This has been attributed to the heavy dependence of the composites on the exfoliation conditions and compatibility with the polymer matrix.

It has also been reported (Kalaitzidou, 2007, 2008) that GNP can produce a significantly low percolation threshold of the order of 0.1–0.3vol% in a system of polypropylene (PP). Here it must be noted that GNP is not graphene but a graphene-like multilayer structure (see Fig. 2), although its high aspect ratio and layer rigidity allows the filler to remain intact during mixing, thus behaving as a high-aspect ratio nanocomposite. In the case of linear polyethylene the introduction of GNP leads to a percolation threshold of the order of 12–15 wt%, a value significantly higher than the one reported for PP (Kim, 2009b). This is quite surprizing as LLDPE and PP have similar polarity levels and such discrepancy cannot be explained easily (Kim, 2010a). A possible reasoning behind these differences has been attributed to preparation methods for improved dispersion (Kalaitzidou, 2007). For a system of poly(ethyl vinyl acetate) (EVA) and GNP prepared by melt mixing a percolation threshold of the order of 2.5 vol% has been reported (see Fig. 7), but the overall conductivity has been found to be significantly lower than expected (based on comparisons with pure GNP), something that has been attributed to the preparation method that lead to extensive aggregation of the platelets (Stalmann, 2012). Nevertheless such system has been reported to retain its conductivity upon cooling, even close to the threshold.

The high aspect ratio of the fillers discussed in this section, has been seen to have a tremendous effect on the percolation threshold values, albeit conditioned to matrix, dispersion and filler preparation methods. As discussed before the relative rigidity of the filler coupled with the high aspect ratio leads to such effects, something seen from theoretical predictions of percolation threshold for disc and rod-like fillers in comparison with spherical particles (Li, 2007 Celzard, 1996; Lu, 2005). Although nonidealized dispersion leads to experimental values that are usually higher than the equivalent theoretical predictions, it is undeniable that these fillers reduce the percolation threshold by a significant amount, even in cases where melt mixing preparation methods are used.

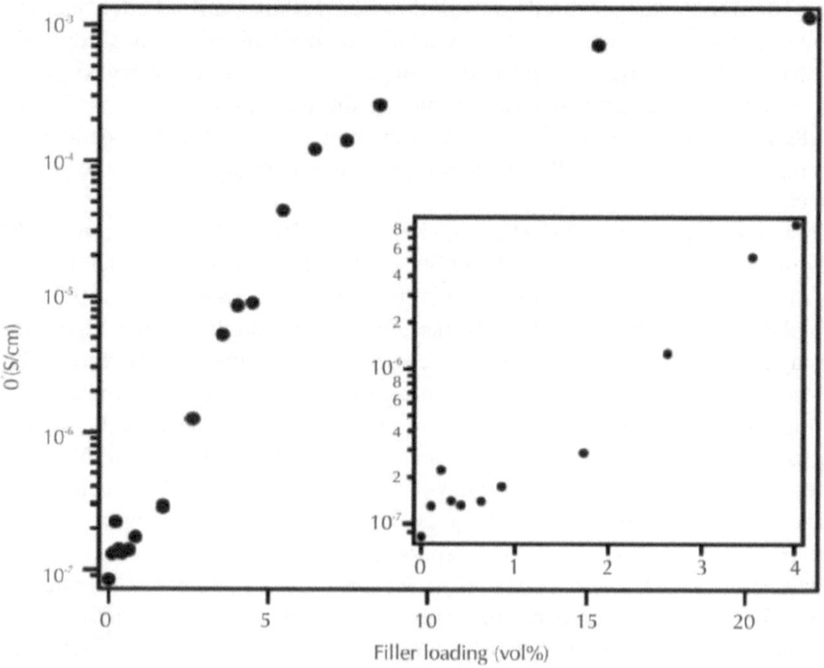

FIGURE 7 Increase in electrical conductivity as a function of the filler loading for a system of EVA/GNP measured by dielectric spectroscopy. The inlet indicates the region of the percolation threshold, reported to be approximately 2.5 vol %.

3.6.2 THERMAL CONDUCTIVITY

Thermal conductivity is affected by two main contributing carriers, electrons and phonons. In nanostructures, phonons usually dominate and the phonon properties of the structure, like group velocity, phonon scattering, heat capacity and Grüneisen parameter become vital for thermal conductivity (Hone, 2000; Ju, 2005). This is due to the finite size of such materials, leading to boundary effects and occasionally to boundary-phonon scattering phenomena that can reduce the overall thermal conductivity. The ratio of the phonon mean-free path Λ to the object size L is important, since when L>Λ the Umklapp scattering dominates reducing the thermal conductivity. In the case where L is smaller or comparable to Λ the continuous energy model that is traditionally used for bulk structures is not applicable requiring that nonlocal and nonequilibrium aspects to heat transfer be considered (Ju, 2005); under these conditions in a defectless structure phonons will propagate without being scattered and thermal conductivity becomes ballistic

(Balandin, 2002; Nikla, 2009). If one takes into account that for carbon-based nanostructures, the phonon mean free path is of the order of 1 μm, it becomes clear that these structures have all the potential to exhibit such effects and offer very high values of thermal conductivity (Ghosh, 2008). A more detailed discussion of the way phonon dispersion and mode contribution to thermal conductivity, and the 1-D quantitization of the phonon band structure influence of the overall thermal behavior of carbon-based nanocomposites, is beyond the scope of this work, but the interested reader is encouraged to seek further information in the work of Balandin (2000) and the references therein.

Superior thermal transport properties have been reported for both CNT and graphene, with values for MWCNTs of 200–3000 W/mK at room temperature (Chou, 2010) and for graphene 4840–5300 W/mK (see Table 1). Despite all this potential and the extreme thermal transport properties of these fillers, the resulting enhancement once incorporated into a polymer is not as dramatic as one would expect (Yu, 2008). This is in stark contrast to the almost exponential increase in electrical conductivity obtained with the use of these fillers as discussed previously. The reason for this huge discrepancy, has been partly attributed to the small contrast in thermal conductivity between polymers (0.1–1 W/mK) and graphitic carbons, in comparison with the differences in electrical conductivity where non-conductive polymers have values of the order of 10^{-8}-10^{-13} S/cm (Kim, 2012a). Furthermore, as thermal energy is transferred in the form of phonons (lattice vibrations) a significant amount of poor coupling in the vibrational modes at the various interfaces of the system (filler-filler and filler-polymer) can be expected (Zhong, 2006). Consequently, the poor interfacial coupling will have a significant impact on the thermal resistance (also known as Kapitza resistance), thus lowering the overall thermal conductivity of the system. It has been estimated that the thermal resistance on the CNT/polymer interface is equivalent to that of a polymer layer that is 20 nm thick (Huxtable, 2003). Here it must be said that this is an area that is not extensively studied and the basic mechanism of the Kapitza resistance in CNTs and graphene is not fully understood (Zhong, 2006). Nevertheless, ways to minimize the interfacial phonon scattering have been proposed, but excessive functionalization can reduce the intrinsic thermal conductivity of the fillers.

GNPs have been reported to improve thermal conductivity for epoxy (Yu, 2007), PP (Kalaitzidou, 2007a), PE (Fukushima, 2006) and especially in the case of epoxy a 30-fold increase in conductivity was reported under certain sample preparation conditions (Yu, 2007). As in the case of electrical conductivity systems that have been prepared by melt mixing were reported to have less significant increase in thermal conductivity (Kalaitzidou, 2007a). As always, sample preparation, degree of exfoliation, filler orientation and the interfacial interactions play a significant part on the overall thermal conductivity, something seen in the work of Yu and co-workers (2007), where the thermal conductivity was found to increase with the aspect ratio of the filler in the system. Furthermore, it has been seen that the conductivity increase was preferential along the orientation direction of the filler than in the perpendicular direction, indicating a level of macroscopic anisotropy.

3.6.3 MECHANICAL PROPERTIES

Since 2009 graphene holds the title of the world's strongest material with a breaking strength 200 times larger than steel, a tensile modulus of 1 TPa and an intrinsic strength of 130 GPa (Lee, 2008). As discussed previously, the process of graphene separation from graphite has an enormous effect on the overall mechanical performance of the resulting material, so a significant discrepancy of the reported values is to be expected depending on the composite's preparation method. Graphene is also reported to be ultra-light with a weight of approximately 0.77 mg/m^2, something being highlighted in the recent Nobel committee announcement where it was stated that one square meter of defect-free single-layered graphene hammock can support a 4 kg cat while weighing as much as one of the cat's whiskers (Nobel, 2010).

It has been reported that ripples exist upon the surface of suspended graphene layers and their existence has been attributed to thermal fluctuations (Meyer, 2007). The amplitude of these thermal and quantum fluctuations is bounded in 3D structures, but the amplitude of long wavelength fluctuations is expected to grow logarithmically with the scale of a 2D structure and in the limit of an infinite structure will become unbounded (Hohenberg, 1967). Although local deformation and elastic strain are almost unaffected by the long-range divergence in relative displacement, it is believed that a sufficiently large 2D graphene structure in the absence of any lateral tension will crumble to a fluctuating 3D structure. Consequently, the true nature of graphene's 2D structure is a hotly debated topic (Geim, 2007).

In the case of CNTs due to the cylindrical structure, the nanotube will have different properties on the axial and radial directions, with the axial being reported to be particularly strong with a Young's modulus of the order of 270–950 GPa and tensile strength of 11–63 GPa (Yu, 2000). Across the radial direction CNTs were found to be particularly soft (in comparison with the axial direction) with van der Waals forces being capable to deform adjacent nanotubes (Ruoff, 1993).

In Table 3, a brief summary of the reported results on a number of nanocomposites can be seen. In the case of graphene it must be noted that although there is a modulus increase for all polymers, the effect is more pronounced for matrices based on elastomers, something attributed to the large stiffness difference between the matrix and the filler (Kim, 2010a). For glassy polymers, a 33% improvement for a filling ratio of 0.01wt% for a PMMA system has been reported and this effect has been attributed to strong hydrogen-bonding interactions between the polymer and the filler and mechanical interlocking at the wrinkled graphene surface that immobilize the polymer segments (Ramanathan, 2008). Despite this explanation, it has been suggested that such effect, with such low filler amount is unrealistically high (Kim, 2010a) and further explanations take into consideration perfect alignment of the filler layers. Further work with functionalised graphene in a number of elastomeric matrices, indicate that the persistence length of the flexible graphene layer is the primary reason for stiffness increase and not the aspect ratio of the fully extended filler (Kim, 2010b). Furthermore, the sample preparation method has been seen to affect the final mechanical performance with compositions prepared via solution process, exhibiting higher stiffness in comparison with those prepared via melt mixing. The lower modulus of the melt mixing prepared compositions has been attributed to the reaggregation of particles during compounding (Kim, 2010a).

In the case of CNTs, similar improvements have been reported in the literature (see Table 3), with more notable the compositions produced by coagulation spinning that yield a system of CNT/PVA with modulus and strength of 80 and 1.8 GPa respectively (Vigolo, 2000 and 2002). Similarly, composite films comprised of alternating layers of polyelectrolytes and CNTs have been reported of having modulus and strength values of 11 GPa and 325 MPa respectively (Spitalsky, 2010). As in the case of graphene for compositions produced by solution methods improved mechanical properties have been reported, especially in cases where the matrix was comprised with covalently attached polymer chains (Xie, 2007). The reason for such improvement has been attributed to the enhancement of dispersion of the CNTs in the matrix achieved by grafting methods (Spitalsky, 2010). In the case of nongrafted polymers, significant improvement on the mechanical properties have been reported especially for systems prepared by solution methods, something that has been attributed to the nucleation effects of the CNT walls that lead to crystallization of the matrix (Spitalsky, 2010). In the case of melt mixed compositions, the matrix reinforcement has been relatively low if compared with alternative preparation methods, probably due to reagglomeration and CNT entanglement effects. Finally, it must be noted that systems based on ductile matrices exhibit a more efficient reinforcement from CNTs in comparison with brittle matrices where the improvement is significantly weaker. Consequently, stiff polymers tend to wrap around the nanotube creating a helical configuration (Kusner, 2006), whereas more flexible polymers with bulky aromatic groups along the backbone adopt an interchain coiling (Tallury, 2010). The chemical composition of the polymer and the chain architecture has a profound effect on the wrapping of the chain along the nanotube, with aromatic moieties forming π—π interactions with the CNT that dictate the absorption conformation. Finally, it must be noted that the geometric parameters of the polymer (chain length, size of branches etc.) and the CNT (aspect ratio, tube diameter etc.) determine the way and probability of wrapping, as when the radius of gyration of the polymer is greater than the radius of gyration of the CNT a significant improvement on the interaction and subsequent mechanical properties is expected (Mu, 2007).

3.6.4 GAS AND LIQUID PERMEATION

Permeation is the process of penetration of a gas or liquid particle (also called penetrant) in a solid, and permeability is the rate of penetration through the matrix as a function of time, subject to temperature, pressure, matrix geometry and nature of the penetrant. The process of the penetrant movement through the solid can be divided into three parts; sorption where it is absorbed by the solid's interface, diffusion where it moves through the matrix via pores or molecular gaps (free volume) and desorption where it leaves the matrix via an interface.

For defect-free graphene sheets it has been reported that they are impermeable to all gas molecules (Bunch, 2008), thus making them prime candidates for polymer-based barrier membranes. This is very important, as these membranes will have (at least in principle) superior mechanical performance as well, due to the graphene reinforcement as it was discussed previously. In practice of course things are not so simple and again preparation methods for the graphene and incorporation into the matrix play an important role as with all the other properties discussed previously. Nevertheless, in principle, graphene should possess significant advantages over alternative high aspect ratio fillers like CNTs (both single and multiwalled) and nano-fibers. For a GNP-system, it has been reported that oxygen permeability is decreased in a more efficient manner in comparison with CB, carbon nano-fibers and even organic nanofillers like modified montmorillonites (MMT) at similar loadings (Kalaitzidou, 2007a). Similar results have been presented by a number of researchers on a wide range of polymer matrices like poly(ethylene naphtalate), polyurethane, thermoplastic polyurethane, polyisoprene and polystirene and different types of graphene (Kim, 2010a). In all cases, for loadings between 1.5 and 2.2 vol% of and for a number of different gases (hydrogen, helium, nitrogen, air etc.) and preparation methods the relative reduction of the permeation reported was of the order of 30–99 % (for more details consult Table 5 in Kim, 2010a). From these results it is obvious that the type of graphene used plays a vital role in the permeation reduction, and based on results reported in the literature, incorporation of treated graphite oxide can lead to reduction of permeability by 90–99 % at loadings similar to those predicted by theoretical models for impermeable platelets of high aspect ratio (Lape, 2004). The reason why high aspect ratio fillers in general, and graphene in particular reduce the permeability so drastically has to do with the existence of tortuous paths in the system that reduce the cross sectional area that is available for permeation (Halperin, 1985).

In the case of CNTs, the situation is a little bit more complex as due to their inherent structure nanotubes can "prevent" or "facilitate" penetrant permeation and diffusion with respect to their axial or radial geometry. In general, there are two types of systems that can be used, buckypapers and isoporous membranes. Both these systems have distinctly different internal structure and porosity, with buckypapers comprizing of randomly entangles CNTs (Smajda, 2007). The preparation method of buckypapers is relatively simple and involves vacuum filtration, with the properties of the final composition depending heavily on the CNT-type and their pretreatment. Traditionally, buckypapers have a highly porous structure of large specific surface area, and they are ideal for medical and filtration applications (Sears, 2010; Vohrer, 2004). Isoporous CNT membranes, on the other hand, use the nanotube as a pore across an otherwise impermeable surface. This is a very active area of research at the moment, with a number of different suggested approaches on sample preparation (Sears, 2010), but the general consensus is that despite the small CNT diameter gas seems to permeate faster than traditional polycarbonate membranes that have cylindrical pores of the order of 10 nm diameter something achieved partially at least by the higher CNT pore density (Holt, 2006).

Finally, it must be noted that due to the way polymers tend to pack around the nanotube (see previous section on mechanical properties), the diffusion process in a CNT-based system will be heavily matrix dependant. This has been demonstrated with the comparison of oxygen diffusion in a PE/CNT and PS/CNT matrix. In the case of PS, the diffusion rate increased (Yargi, 2012), whereas for a PE matrix the diffusion rate decreased (Chrissafis, 2009). Comparative computational predictions indicate that the introduction of a SWCNT in the matrix yields a diffusion rate perpendicular to the nanotube that is approximately 95–110 % of the rate in pure PS. Similar comparisons with a PE-based system yields a rate of approximately 80 % of that in the pure matrix (Haghighatpanah, 2012). The differences in the rate of diffusion along the surface of the nanotube in comparison with the bulk polymer matrix have been attributed to the different way the polymer chains are packed, the differences in the free volume along the nanotube-polymer interface and the level of entanglement and subsequent dynamics (Gkourmpis, 2013; Karatrantos, 2012; Lim, 2007).

3.6.5 THERMAL AND DIMENSIONAL STABILITY

Graphene has been reported of having a negative thermal expansion coefficient in basal plane at room temperature (Kelly, 1972), with the thermal expansion along the thickness direction being equally small especially when compared with polymers (Nelson, 1945). Because of this, it is expected that graphene can minimize dimensional changes in the matrix, although the level of incorporation and the degree of orientation have been seen to be of great importance (Kim, 2010a). For a system of PP reinforced with properly aligned GNP the suppression of the thermal expansion of the composite was reported to be as effective as more traditional fillers like CB and nanofibers (Kalaitzidou, 2007a). Similar results have been reported for graphene oxide incorporated in epoxy, although in this case the preparation method of the filler has been seen of having a very strong effect on the overall performance of the composite (Kim, 2008, 2009a). Furthermore it must be noted that preparation methods do not alter the dimensions of the graphene layers dramatically, especially in the case of graphitic platelets (Stalmann, 2012), something that is not the case with CNTs where preparation methods can actually affect the overall size of the tubes (Alig, 2012).

Graphene has also seen to improve the thermal stability of the matrix for a number of polymers like PS (Liu, 2008) and PMMA (Villar-Rodil, 2009). Similarly thermogravimetric analysis (TGA) of CNT-based composites has shown improved thermal stability (Moniruzzaman, 2006). It has been reported that the onset of the decomposition and maximum weight loss temperatures are higher in the reinforced system in comparison with the pure polymer (Ge, 2004). The precise mechanism of this behavior is still unclear but a number of suggestions have been put forward. It has been suggested that the immobilization of the polymer in the vicinity of the filler restricts chain mobility, thus reducing the rate of decomposition (Liu, 2008). Furthermore, dispersed nanotubes have been speculated to create a dense network that can hinder the flux of degradation (Moniruzzaman, 2006). Finally, an alternative mechanism takes into account the significantly higher thermal conductivity due to the presence of the filler for the facilitation of the heat dispersion, thus limiting the degradation process (Huxtable, 2003).

3.6.6 RHEOLOGICAL PROPERTIES

Rheological measurements can be used for quantification of the overall dispersion of the filler in the polymeric matrix (Litchfield, 2006 and Vermant, 2007). In Fig. 8, the storage modulus G' obtained from small oscillatory shear versus frequency, for a wide range of concentrations of thermally expanded graphene oxide in a polycarbonate matrix (PC) can be seen (Kim, 2008). The presence of a plateau in G' is indicative of rheological percolation due to the formation of a solid-like network in the matrix (Vermant, 2007). The onset of the frequency-independent G' can coincide with other interesting phenomena (Kim, 2009a) and in this percolative regime the concentration dependence of the particle elasticity can be described by power-law scaling $G' \alpha (\phi - \phi_{perc})^{v}$ with φ_{perc} the percolation threshold and v a power-law exponent (Vermant, 2007). High aspect ratio particles with shape anisotropy (rods, discs etc.) have been seen to percolate at smaller volume fractions in comparison with spherical particles (Garboczi, 1995) but one has to remember that rheological responses are very sensitive to the level of anisotropy of the system (Krishnamoorti, 1997). Consequently, differences between theoretical predictions that assume random orientation and experimental data on laboratory prepared samples are to be expected (Kim, 2010a). These differences can be attributed, among other things to the particle alignment and subsequent reduction of the level of anisotropy induced by the preparation methods, especially in cases where mechanical mixers have been used. One way around this problem is the thermal annealing of the composite above the glass transition temperature of the matrix. In Fig. 8, the effect of annealing the nanocomposite for several hours on lowering the percolation threshold from approximately 1.5 vol% to 0.5 vol% can be observed (Kim, 2008).

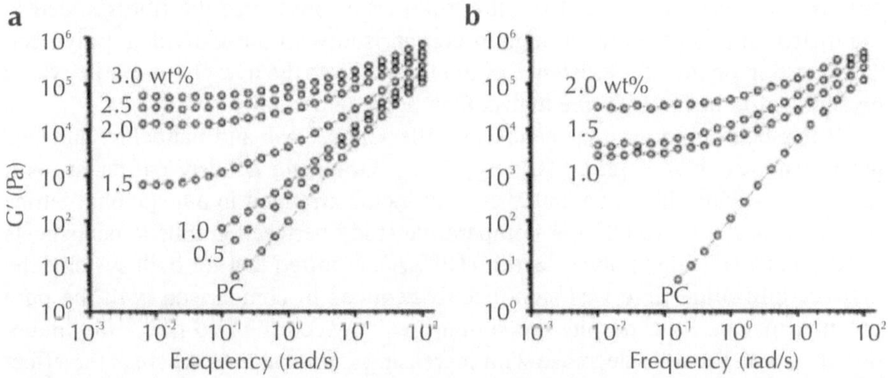

FIGURE 8 Dynamic frequency sweeps of melt-mixed TEGO1/PC indicating the low-frequency moluli changes after annealing for 10,000 s (a) and 20,000 s (b). 1TEGO: thermally expanded graphene oxide. Reproduced from Potts, 2011.

Similar rheological behavior has been observed for CNT-based composites, where at high frequencies the rheological response is not sensitive to the amount of filler in the matrix, indicating that short-range polymer dynamics are not influenced by the CNTs (Moniruzzaman, 2006). Furthermore, it has been seen that at low filler loading and in the absence of strong interfacial bonds, the glass transition temperature of the composite remains constant. Again, in a manner similar to the graphene-based systems, at low frequencies the rheological behavior moves from the liquidlike response where $G' \propto \omega^{2}$ to a solid-like response where G' is independent of the frequency. In the rheology percolative region, the superposition of the polymer entanglements and the polymer-nanotube network has been found to dominate the rheological response (Pötschke, 2004). Here it is important to make a comparison with the electrical percolation threshold where the filler-filler (the same applies for both graphene and CNTs) network is dominating the rheological response due to the necessity for electron flow through the matrix that can be achieved only by the specific arrangement of the conductive particles. In other words, a certain amount of filler-filler "contacts" (here by contacts we refer to the minimum distance required for electron flow and not necessarily physical contact) will be required to form the conductive network, therefore the electrical percolation is traditionally higher than the rheological one.

3.6.7 MORPHOLOGY AND CRYSTALLIZATION EFFECTS

So far in this discussion we treated the polymer matrix as a homogeneous system that does not interact significantly with the filler and does not alter its morphology. Nothing can be further from the truth, especially since the fillers added in the matrix in this case are so huge in comparison with the individual polymers. The question on how the existence of the filler affects the overall morphology and crystallization behavior of the matrix then pops up naturally.

It has been reported that nanoscale fillers like CNT and nanoclay, enhance heterogeneous crystallization (Cheng, 2012). Graphene is fairly particular, as it has a 2D structure like a nanoclay and sp^2 atoms arranged in a hexagonal lattice in a manner similar to CNT. A comparative study between graphene nanosheets (GNS) and CNT filled poly(L-lactide) (PLLA), reported that for both systems the half-crystallization time was significantly reduced in comparison with the pure polymer matrix. Still, despite the similarities between the two fillers the induction time was found to decrease with increasing CNT loading, whereas the effect had an inverse trend for a CNS-based system (Xu, 2010). The reason for this behavior was attributed to a mechanism of size-dependent soft epitaxy where the dimensionality of the filler plays a vital role on the crystallization kinetics. For 1D CNT, the polymer chain shows a preferential alignment along the tube axis with no lattice matching between matrix and the surface of the tube, thus simplifying the surface induction process with subsequent reduction of the induction time. Furthermore, it has been argued that the nanotube is able to offer its entire surface and surrounding area for polymer crystallization and only one nucleating site is needed in order for the crystals to grow into disc shapes on the graphitic surface.

In the case of CNS, the 2D structure makes lattice-matching the dominant force in the crystallization process. This is because the polymer chains need to be absorbed on the graphitic surface, and this will require them to undergo structural adjustments, that will require more time for the crystallization to occur. Furthermore due to the flat surface available for crystallization, the polymer crystals can grow in multiple points and with multiple orientations. Finally, it has been speculated that adjacent crystals might interfere during growth and suppress the overall crystallization kinetics of the system (Xu, 2010). On a similar study the introduction of GNS in isotactic PP (iPP), has shown to accelerate the crystallization rate under shear (Xu, 2011). In this system two nucleating origins were observed, with heterogeneous sites (two-dimensional fillers) aligning along the flow field and self-nucleating sites induced by shear (see Fig. 9). Due to this effect, it has been suggested that the existence of the filler offers a synergistic effect on promoting crystallization kinetics on the iPP, with the β phase crystals dependant on the survival of the self-nucleating sites (row-nuclei) and the α phase being heavily influenced by the presence of the filler in the system (Xu, 2011). In the case of CNT for the same iPP, the combined shear flow and the existence of the filler allowed for the creation of crystals with weak orientation at the early stages of crystallization, followed by a high degree of orientation at a later stage. This effect was attributed to the suppression of the initial orientation of the polymeric chains due to the viscoelasticity increase of the overall composition something linked to the existence of the filler. The weak orientation was stabilized at a later stage as the polymer chains were absorbed into the CNT surface (Chen, 2011). In a manner similar to the CNS/iPP system, the existence of CNT offers synergistic effects on the crystallization kinetics with an increase of crystallization rate up to 40-times in comparison with the pure polymer. Subsequent thermal analysis indicate that the iPP/CNT system (nonsheared) has a two-dimensional lamellar growth, but in the case of the sheared iPP/CNT system a mixture of two-dimensional lamellar and three-dimensional spherulitic growth was reported with β phase crystals being totally absent (Chen, 2011). Similar results were reported for SWCNT/PP system produced via melt mixing, despite the poor dispersion of the filler into the matrix (Bhattacharyya, 2003). For a system of PVDF/PMMA/graphene prepared via in-situ polymerization of the PMMA/graphene masterbatch it has been reported that the PMMA enforces β phase crystallization to PVDF but cannot stabilize them at elevated temperatures. Furthermore, the existence of the graphene sheets has been seen to restrict the *transtrans* conformations and act as nucleating agent for the crystallization of PVDF (Mohamadi, 2011). In a similar manner SWCNTs have been reported to nucleate poly(vinyl alcohol) with differences between crystallization temperatures for the filled, and unfilled systems up to 5 degrees (Probst, 2004).

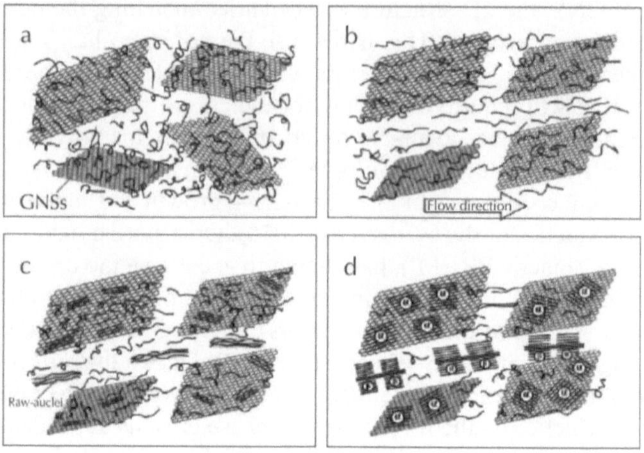

FIGURE 9 Schematic representation of the crystallization process of iPP under shear flow in the presence of graphene nanosheets (GNS). In the absence of shear, the filler constructs an isotropic network (a). During shear, the network is disturbed and transforms to an anisotropic (b). Due to geometrical confinements and polymer orientation row nuclei are created (c). Due to the nucleation ability of the filler α-phase crystals dominate with small amounts of β-phase formed (d). Reproduced from Xu, 2011.

For solution CNT-induced crystallized systems, a monohybrid shish kebab (NHSK) structure has been observed and the formation mechanism has been attributed to size-dependent soft epitaxy as we have seen in the case of iPP and PLLA (Xu, 2010 and 2011). Further studies using PE, Nylon and PVDF confirmed the effect (see Fig. 10) and for small-diameter CNTs geometric confinement has been reported to dictate the polymer orientation with exclusively orthogonal orientation between the lamellar surface and the CNT axis (Li, 2009). With increased diameter, normal epitaxy is reported to play a vital role in the multiple orientations of PE lamellar structures with the nanotube affecting the overall crystallization in a profound way in a manner similar to the iPP case discussed previously. The overall crystallization effect has to do with the CNT acting as a nucleating agent while at the same time their size and spatial arrangement creates nanoconfinement effects on the polymer chains (Li, 2009).

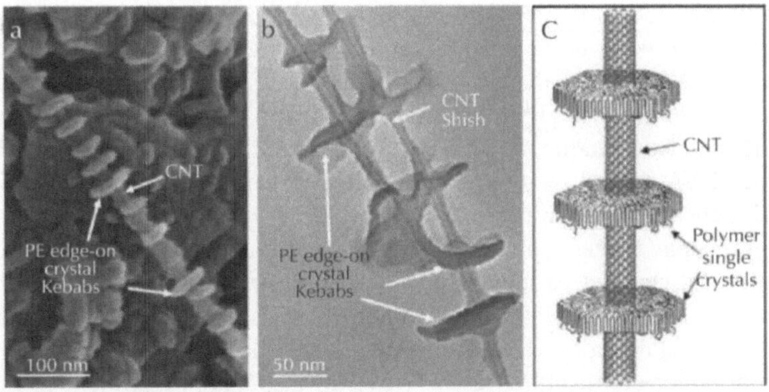

FIGURE 10 PE/MWCNT NHSK structure as seen by SEM (a), TEM (b) and a schematic representation indicating the way lamellas are growing from the CNT surface (c). Reproduced from Li, 2009

Numerical calculations have predicted that the molecular weight of the polymer does not affect the way it is absorbed on the CNT surface, with the average distance between the polymer backbone and the nanotube estimated to approximately 0.35 nm (Minoia, 2012). This geometrical arrangement allows the closest hydrogen atoms to be at a distance of 0.25–0.3 nm from the nanotube surface, thus being in range of the CH-π interactions. Due to these geometrical arrangements the polymer chain folds on the interface are predicted to act as nucleation sites for epitaxial ordering in a manner similar to the experimental predictions (Li, 2009; Xu, 2010, 2011), clearly indicating that the existence of the nanotube favors polymer crystallization. In the case of CNT folding it has been seen that the same polymer morphology is predicted to be absorbed on the surface thus not changing the energetic and morphological properties of the polymer. The nanotube diameter though, has been seen to have a major effect on the overall morphology, with increasing values leading to an increase of the polymer-CNT interactions as seen by the conformational arrangements of the chain absorbed on the nanotube surface. Detailed analysis of the polymer orientation indicates that the chains have a tendency to fold in a manner that aligns to the nanotube axis, thus minimizing the stress that will be created in the structure by forcing the lamellas to bend and follow the nanotube curvature, something predicted to be especially pronounced for small diameter tubes (Minoia, 2012). In the case of a nonstraight nanotube, it has been predicted that the polymer will align with the overall axis of the nanotube, thus leading to a situation where the chain conformation that offers a lower degree of internal stress is favorable against higher level of commensurability on the filler surface (Mionoia, 2012). This behavior is similar to the one predicted for graphene where temperature is found to play a vital role on the overall morphology with absorption and subsequent lamellar orientation being predicted to be a highly cooperative process (Wang, 2012). For a system EVA/GNP prepared by melt mixing it has been reported that the existence of the filler reduces slightly (1–2 %) the overall crystallinity and the effect on the crystallization and morphology is almost negligible, with crystallization taking place away from the graphene platelets (Stalmann, 2012). In Fig. 11, Avrami plots for pure and GNP-reinforced EVA can be seen indicating that the overall crystallization process is almost independent of the filler. When one looks at more traditional fillers like CB for the same system, prepared with the same conditions, there are distinct morphological changes as indicated by the existence of a tertiary crystallization process. This is very interesting as it comes in contradiction to the reports of GNP facilitating crystallization (He, 2011; Pang, 2012) on similar systems, indicating that the polymer architecture, level of overall dispersion, treatment of the filler and sample preparation are very important.

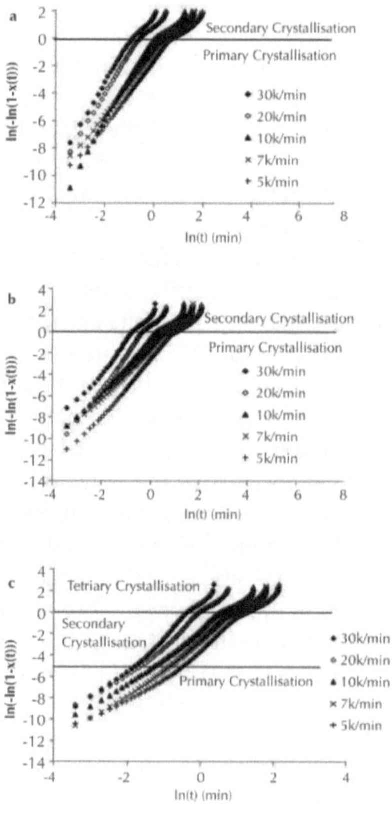

FIGURE 11 Avrami plots for pure polymer (a) GNP (b) and CB (c) based composites indicating the differences in nucleating activities.

3.7 CURRENT AND FUTURE CHALLENGES AND APPLICATIONS

So far we have discussed a wide range of properties of polymer matrices reinforced with high aspect ratio fillers like graphene and nanotubes in all their forms and variations. It is obvious that these materials offer a lot of impressive performance enhancements but at the same time one has to be cautious to the wide range of problems that restrict their large-scale incorporation in commercial applications.

Since CNTs have been available for a longer time than graphene they have been used in a wider range of applications. Still, for reasons that will discussed later and despite all the performance enhancement both these materials have not managed to find a place in large scale commercial applications. That of course does not mean that they are not used commercially today, but instead they can be found in small-scale niche applications (Endo, 2004). CNTs are heavily used in medical applications and especially in drug delivery systems and it has been suggested that they have the potential to revolutionize medicine (Lu, 2009). Due to their specific topology, surface smoothness and overall dimensions CNTs and lately graphene have been considered as potential materials for energy storage applications. In these systems the efficiency of the fuel cell is determined by the rate of electron transfer at the electrodes, and CNTs (Ajayan, 2001) and graphene electrodes (Bae, 2012) are showing impressive potential. A schematic representation of the basic working mechanism of such systems can be seen in Fig. 12. Other applications of CNTs include hydrogen storage systems and nanoscale motors but the areas where they are expected to have a huge potential is catalysis, where CNTs have been identified as potential elements for metal-free catalytic processes due to their ability of surface functionalization, (Frank, 2011) and electronics.

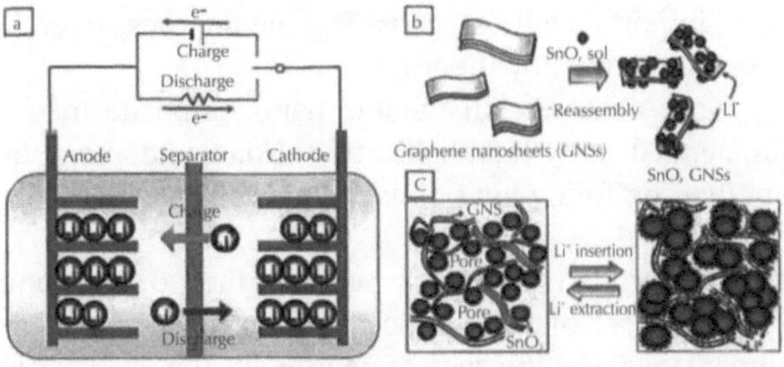

FIGURE 12 Schematic representation of the working principle of a lithium-ion battery (a), structure and synthesis of a graphene-based electrode (b) and the improvements the electrode offers in terms of storage-release cycling ability (c). In the specific diagram the porous structure with SnO_2 nanoparticles confined by graphene sheets limits the volume expansion as lithium is inserted thus improving the battery's cyclic performance. Reproduced from Liu, 2012.

Both CNTs and graphene show exceptional electron transport properties thus making them ideal candidates for electronic devices such as transistors and it has even been suggested that one day they can substitute Si-based circuits of today (Geim, 2007b). For such a system to be functional the graphene nanoribbon or the CNT has to be properly laid on the circuit with its semiconducting gap being capable of opening and closing at will due to quantum confinement of electrons. Although this sounds a lot like science fiction, recently IBM has demonstrated the possibility of using chemical self-assembly to localize individually aligned nanotubes in significant density as to form a transistor (Park, 2012). Obviously this does not mean that CNT and graphene-based computers will be available at our local shop tomorrow morning but it is an important step towards the design of totally new electronic devices that will go beyond Si (Avouris, 2012; Lin, 2010, 2011).

The most important property that fillers like CNTs and graphene bring to the design of modern nanocomposites is their sheer size and well-defined geometrical characteristics. Due to these intrinsic properties they offer superior mechanical, electrical and thermal enhancements as we have discussed previously. It is quite ironic that their biggest advantage becomes their most important liability when one thinks of large-scale commercial applications as the incorporation of these fillers into a polymeric matrix is truly challenging. Adequate dispersion of the filler into the matrix is of paramount importance and here is where the size of these fillers becomes a liability, as they have a tendency to agglomerate thus losing the potency in terms of aspect ratio the individual particle brings to the final composition. Just to be clear all fillers big or small, regular or irregular have a tendency of clustering when incorporated into a polymer matrix with CB being a very good example of such behavior. Due to the small size and irregular structure of the CB primary particles in this case, the agglomeration will not alter the overall aspect ratio in a fundamental way, thus leading to a rather mediocre decrease in properties, something of importance if one considers the immense overall amount of the final composition constituted by the filler.

In essence, every commercially viable nanocomposite must satisfy a wide range of requirements usually established by national and international standards, but it is evident that the following points have to be considered in almost all potential compositions:

1. The structureproperty relations between the polymer and the filler depend on a number of factors including purity, extent of defects and particle dimensions. For the case of nanotubes, the precise type (single wall, multiwall etc.) and its chirality has shown to play an important role on the overall performance of the final composition. Similarly for graphene, the exact type of it (single layer, nanoplatelets etc.) and its preparation method will affect the overall properties of the mixture, something that has been stressed on the previous discussion. Finally, the amount of filler

introduced in the matrix and the type and extent of polymer-filler interactions must be considered, as they will have a profound effect on the overall performance. Here it must be noted that the amount of filler present in the matrix alone is not important, but the spatial arrangement and orientation of those filler particles have to be considered, as the level of anisotropy will have an effect on electrical, thermal and mechanical properties.

2. Filler functionalization offers very attractive opportunities for improved dispersion and the modification of the interfacial properties can lead to synergistic effects that enhance orientation, overall morphology and solvent compatibility. Here it must be noted that although functionalization offers a lot of advantages it must be used in moderation as it is known to have adverse effect on the filler geometry (e.g., CNTs are known to undergo size reduction).
3. The existence of the filler in the nanocomposite offers massive potential for electrical conductivity, especially with the reduction of the percolation threshold in comparison with more traditional conductive additives. The filler network spatial arrangement and orientation has been seen to be of paramount importance for obtaining such properties. Further work on the specific interactions between polymer and filler and the effect it has on the overall morphology and performance has created a large number of design opportunities over the years.
4. The introduction of the filler in the matrix has shown to increase the overall viscosity and as a consequence the rheological, mechanical and electrical properties. The viscosity increase can also offer advantages as it slows the thermal degradation, but so far the treatment of the internal interfaces in the system and the reduction of their thermal resistance remains a challenge. For this reason, no composition has managed to take full advantage of the inherent thermal conductivity of the fillers.
5. Morphological changes and differences in crystallization due to the filler have been reported for a number of polymers. These will have an effect on the overall behavior of the final composition and the precise mechanism of the absorption of the polymer to the filler surface, the nucleation potency and the crystallization process and speed have been seen to be heavily dependent on the chain architecture and filler geometry. Recent theoretical predictions have shed some light on the mechanisms at work but a lot of unanswered questions still remain.
6. Industrial usage of these types of fillers will have to take into consideration possible health and safety risks. This is an area not properly studied at the moment and although there is a fair amount of literature no overall consensus has been reached. Public organizations like the European Commission are putting a lot of resources in the disposal of academic

and industrial researchers aiming to investigate the potential health, environmental and safety risks that are involved in the large-scale use of these materials. This is the kind of work that is constantly on-going and evolving, but until an overall consensus on how these systems can be used in a safe manner is reached, large scale usage will be limited.

7. Last but not least, the overall cost of the production and treatment of the fillers is most of the times unsustainable, especially when one needs highly purified micromechanically produced graphene monolayers or SWCNT of identical structure. A lot of alternative ways of producing graphene-based compositions and CNTs have been suggested over the years, but the overall cost is still considered high. For example, current prices of CNTs range from 1–2 $/g (MWCNT) to a few hundred $/g (SWCNT) with graphene variants reported to be equally expensive.

In conclusion, we can say that despite all their astonishing performance enhancements carbon-based high aspect ratio fillers are heavily suffering from poor dispersion and subsequent cost. In principle, all large-scale commercial applications require a certain level of simplification of the overall production cost, something that in the case of polymer nanocomposites is achieved via melt mixing. The use of more advanced preparation methods require solvents and a substantial amount of preparation steps that make the whole process unsustainable financially let alone, difficult and potentially filled with health and safety problems. The situation gets even more complicated when one has to consider that different polymers are compatible with different solvents, and different preparation methods of CNT and graphene have to be considered in terms of sustainability and overall robustness of their quality. Therefore we are currently on the verge of a potential revolution where the improvements brought by these nanocomposites can create new materials and even new applications in the future. How possible and how successful this revolution will be depends on how efficient and effective we will be in handling the challenges outlined in this section. If we succeed to produce high quality fillers at reasonable prices while we improve their overall dispersion in the matrix we will make a large step towards high volume, high quality commercial applications.

3.8 CONCLUSIONS

In this chapter we provided a quick overview of the current research on CNT and graphene-based polymer nanocomposites. Our aim was to provide some insight on the current trends of these materials and illustrate their immense potential, while highlighting the series of problems that have kept them from large-scale commercialization. It must be stressed that this is not supposed to be a comprehensive review as the area is too diverse but instead our effort was to provide some initial information and encourage the interested reader to further explore the specific areas that are of his or her particular scientific interest.

Concluding, we have to say that the area of polymer nanocomposites has been very active for a very long period of time and more and more potential applications and deeper understanding of the structure property relations of these systems have been established. It is our opinion, that all the work that has been done by all the academic and industrial research groups over the last decades on improving the deeper and more fundamental understanding of these systems points towards a very exciting future. How far these systems will get incorporated into our daily lives and how many of the issues that make their extensive commercialization possible will be solved remains to be seen.

KEYWORDS

- **carbon-based allotropes**
- **conductive fillers**
- **gas permeation**
- **liquid permeation**
- **melt mixing**
- **polymer matrix**
- **solution mixing**

REFERENCES

Ajayan, P. M.; Zhou, O. Z. *Topics Appl. Phys.* **2001,** *80,* 391–425.

Alig, I.; Pötschke, P.; Lellinger, D.; Skipa, T.; Pegel, S.; Kasaliwal, G. R.; Villmow, T. *Polymer*, **2012,** *53,* 4–28.

Ansari, S.; Giannelis, E. P. J. *Polym. Sci. Part B: Polym. Phys.* **2009,** *47,* 888–897.

ASTM D1238–04.

Avouris, P.; Xia, F. *MRS Bulletin,* **2012,** *37,* 1225–1234.

Bae, S.; Kim, S. J.; Shin, D.; Ahn, J-H.; Hong, B. H. *Phys. Scr.* **2012,** *T146,* 014024.

Bahr, J. L.; Yang, J.; Kosynkin, D. V.; Bronokowski, M. J.; Smalley, R. E.; Tour, J. M. *J. Am. Chem. Soc.,* **2001,** *123,* 6536–6542.

Balandin, A. A. *Phys. Low-Dim. Structures* **2000,** *1/2,* 1.

Balandin, A. A. Potentials, *IEEE* **2002,** *21,* 11–15.

Balberg, I.; Binenbaum, N.; Wagner, N. *Phys. Rev. Lett.* **1984,** *52,* 1465–1468.

Barger, A. H.; Cohen, S. R.; Wagner, H. D. *Phys. Rev. Lett.,* **2004,** *92,* 186103–186104.

Bhattacharyya, A. R.; Sreekumar, T. V.; Liu, T.; Kumar, S.; Ericson, L. M.; Hauge, R. H.; Smalley, R. E. *Polymer,* **2003,** *44,* 2373–2377.

Bunch, J. S.; Verbridge, S. S. Alden, J. S.; van der Zande, A. M.; Parpia, M. J.; Craighead, H. G.; McEuen, P. L. *Nano Lett.* **2008,** *8,* 2458–2462.

Celzard, A.; McRae, E.; Deleuze, C.; Dufort, M.; Furdin, G.; Mareche, J. F. *Phys. Rev. B* **1996,** *53,* 6209–6214.

Celzard, A.; McRae, E.; Mareche, J. F.; Furdin, G.; Dufort, M.; Deleuze, C. J. *Phys. Chem. Solids* **1996,** *57,* 715–718.

Chen, J.; Hamon, M. A.; Hu, H.; Chen, Y.; Rao, A. M.; Eklund, P. C.; Haddon, R. C. *Science*, **1998,** *282,* 95–98.

Chen, P. W.; Chung, D. D. L. J. *Electron. Mater.,* **1995,** *24,* 47–51.

Chen, Y-H.; Zhong, G-J.; Lei, J.; Li, Z-M.; Hsiao, B. S. *Macromolecules*, **2011,** *44,* 8080–8092.

Cheng, S.; Chen, Xi.; Hsuan, Y. G.; Li, C. Y. *Macromolecules*, **2012,** *45,* 993–1000.

Chou, T-W.; Gao, L.; Thostenson, E. T.; Zhang, Z.; Byun, J-H. *Compos. Sci. Technol.* **2010,** *70,* 1–19.

Chrissafis, K.; Paraskevopoulos, K. M.; Tsiaoussis, I.; Bikiaris, D. *J. Appl. Polym. Sci.* **2009,** *114,* 1606–1618.

Collins, P. G.; Bradley, K.; Ishigami, M.; Zettl, A. *Science*, **2000,** *287,* 1801–1804.

Doi, M.; Edwards, S. F. *The Theory of Polymer Dynamics*, Clarendon, Oxford, 1986.

Du, X.; Skachko, I.; Barker, A.; Andrei, E. Y. *Nature Nanotechnol.*, **2008,** *3,* 491–495.

Endo, M.; Hayashi, T.; Kim, Y, A.; Terrones, M.; Dresselhaus, M. S. *Phil. Trans. R. Soc. Lond. A*, **2004,** *362,* 2223–2238.

Frank, B.; Blume, R.; Rinaldi, A.; Trunschke, A.; Schlögl, R. Angew. *Chem. Int. Ed.* **2011,** *50,* 10226–10230.

Fukushima, H.; Drzal, L. T.; Rook, B. P.; Rich, M. J. J. *Therm. Anal. Calorim.* **2006,** *85,* 235–238.

Garboczi, E. J.; Snyder, K. A.; Douglas, J. F.; Thorpe, M. F. *Phys. Rev. E* **1995,** *52,* 819–828.

Ge, J. J.; Hou, H.; Li, Q.; Graham, M. J.; Greiner, A.; Reneker, D. H.; Harris, F. W.; Cheng, S. Z. D. *J. Am. Chem. Soc.* **2004,** *126,* 15754–15761.

Geim, A. K.; MacDonald, A. H. Phys. *Today*, **2007,** *60,* 35–41 (b).

Geim, A. K.; Novoselov, K. S. *Nature Mater.* **2007,** *6,* 183–191.

Ghosh, S.; Calizo, I.; Teweldebrhan, D.; Polikatilov, E. P.; Nikla, D. L.; Balandin, A. A.; Bao, W.; Miao, F.; Nau, C. N. *Appl. Phys. Lett.* **2008,** *92,* 151911.

Gkourmpis, T. *Phys. Rev. B* **2012** (submitted).

Gkourmpis, T.; Mitchell, G. R. *Macromolecules*, **2011,** *44*, 3140–3148.

Gkourmpis, T.; Svanberg, C.; Kaliappan, S. K.; Schaffer, W.; Obadal, M.; Tranchida, D. Eur. *Polym. J.* **2013,** 49, 1975–1983.

Gorrasi, J.; Sarno, M.; Di Bartolomeo, A.; Sannino, D.; Ciambelli, P.; Vittoria, V. *J. Polym. Sci. B* **2007,** *45*, 597–606.

Grossiord, N.; Loos, J.; Laake, L. V.; Maugey, M.; Zakri, C.; Koning, C. E.; Hart, A. *J. Adv. Funct. Mater.* **2008,** *18*, 3226–3234.

Grunlan, J. C.; Mehrabi, A. R.; Bannon, M. V.; Bahr, J. L. *Adv. Mater.* **2004,** *16,* 150–153.

Guo, S. Z.; Zhang, C.; Wang, W. Z.; Liu, T. X.; Tjiu, W. C.; He, C. B. *Polym. Compos.* **2008,** *16*, 501–507.

Haghighatpanah, S.; Bolton, K., Phys. Rev. B 2012 (in print).

Halperin, B. I.; Feng, S.; Pen, P. N. Phys. Rev. Lett. **1985,** *54,* 2391–2394.

He, F.; Fan, J.; Lau, S.; Chan, L. H. J. Appl. Polym. Sci. **2011,** *119,* 1166–1175.

He, H.; Zhang, Y.; Gao, C.; Wu, J. Chem. Commun. **2009,** 1655–1657.

Herd, C. R.; McDonald, G. C.; Hess, W. M. *Rubber Chem. Technol.*, **1991,** *65,* 107–129.

Hohenberg, P. C. *Phys. Rev. B* **1967,** *158,* 383–386.

Holt, J. K.; Park, H. G.; Wang, Y.; Staderman, M.; Artyukhin, A. B.; Grigoropoulos, C. P.; Noy, A.; Bakajin, O. *Science*, **2006,** *312,* 1034–1037.

Hone, J.; Liaguno, M. C.; Nemes, N. M.; Johnson, A. T.; Fisher, J. E.; Walters, D. A.; Casavant, M. J.; Schmidt, J.; Smalley, R. E. *Appl. Phys. Lett.* **2000,** *77,* 666–668.

Huang, Y. Y.; Terentjev, E. M. *Polymers*, **2012,** *4,* 275–295.

Hule, R. A.; Pochan, D. J. *MRS Bulletin*, **2007,** *32,* 354–358.

Huxtable, S. T.; Cahill, D. G.; Shenogin, S.; Xue, L.; Ozisik, R.; Barone, P.; Usrey, M.; Strano, M. S.; Siddons, G.; Shim, M.; Keblinski, P. *Nature Mater.* **2003,** *2,* 731–734.

Iijima, S. *Nature*, **1991,** *354,* 56–58.

ISO 1133:1997.

Jang, J. Y.; Kim, M. S.; Jeong, H. M.; Shin, C. M. *Compos. Sci. Technol.* **2009,** *69,* 186–191.

Ju, Y. S. Appl. Phys. Lett. **2005,** *87,* 153106.

Kalaitzidou, K. Fukushima, H.; Askeland, P.; Drzal, L. T. *J. Mater. Sci.* **2008,** *43,* 2895–2907.

Kalaitzidou, K. Fukushima, H.; Drzal, L. T. *Compos. Sci. Technol.* **2007,** *67,* 2045–2051.

Kalaitzidou, K.; Fukushima, H.; Drzal, L. T. *Carbon*, **2007,** *45,* 1446–1452 (a).

Karatrantos, A.; Composto, R. J.; Winey, K. I.; Kröger, M.; Clarke, N. *Macromolecules*, **2012,** *45,* 7224–7281.

Kasaliwal, G. R.; Pötschke, P.; Göldel, A.; Heinrich, G. *Polymer*, **2011,** *52,* 1027–1036.

Kelly, B. T. Carbon, **1972,** *10,* 429–433.

Kim, H. M.; Kim, K.; Lee, C. Y.; Woo, J.; Cho, S. J.; Yoon, H. S. *Appl. Phys. Lett.* **2004,** *84,* 589.

Kim, H. Macosko, C. W. *Macromolecules*, **2008,** *41,* 3317–3327.

Kim, H. Macosko, C. W. *Polymer*, **2009,** *50,* 3797–3809 (a).

Kim, H.; Abdala, A. A.; Macosko, C. W. *Macromolecules,* **2010,** *43,* 6515–6922 (a).

Kim, H.; Miura, Y.; Macosko, C. W.; *Chem. Mater.* **2010,** *22,* 3441–3450 (b).

Kim, S.; Do, I.; Drzal, L. T. *Macromol. Mater. Eng.* **2009,** *294,* 196–205 (b).

Kirkpatrick, S. *Rev. Mod. Phys.* **1973,** *45,* 574–588.

Kong, J.; Franklin, N. R.; Zhou, C.; Chapline, M. G.; Peng, S.; Kyeongjae, C. D. H. *Science*, **2000,** *28,* 622–625.

Krause, B. Boldt, R.; Pötschke, P. *Carbon*, **2011,** *49,* 1243–1247.

Krishnamoorti, R.; Giannelis, E. P. *Macromolecules*, **1997,** *30,* 4097–4102.

Kroto, H. W.; Heath, J. R.; O'Brien, S. C.; Curl, R. F.; Smalley, R. E. *Nature*, **1985,** *318,* 162–163.

Kuhner, G.; Voll, M., *Manufacture of Carbon Black in Carbon Black Second Edition*, Eds. Donnet, J-B.; Bansal, R. C.; Wang, M. J.; Taylor and Francis, London 1993.

Kuilla, T.; Bhadra, S.; Yao, D.; Kim, N. H.; Bose, S.; Lee, J. H. *Prog. Polym. Sci.,* **2010,** *35,* 1350–1375.

Kusner, I.; Srebnik, S. *Chem. Phys. Lett.* **2006,** *430,* 84–88.

Lape, N. K.; Nuxoll, E. E.; Cussler, E. L. *J. Membr. Sci.* **2004,** *236,* 29–37.

Lee, C.; Wei, X.; Kysar, J. W.; Hone, J. *Science*, **2008,** *321,* 385–388.

Lee, W. D.; Im, S. S. *J. Polym. Sci. Part B Polym. Phys.* **2007,** *45,* 28–40.

Leroux, F.; Besse, J. P. *Chem. Mater.* **2001,** *13,* 3507–3515.

Li, J.; Kim, J-K. *Compos. Sci. Technol.* **2007,** *67,* 2114–2120.

Li, L.; Li, B.; Hood, M. A.; Li, C. Y. *Polymer*, **2009,** *50,* 953–965.

Liang, J.; Huang, Y.; Zhang, L.; Wang, Y.; Ma, Y.; Guo, T.; Chen, Y. *Adv. Funct. Mater.* **2009,** *19,* 2297–2302 (b).

Liang, J.; Wang, Y.; Huang, Y.; Ma, Y.; Liu, Z.; Cai. J.; Zhang, C.; Gao, H.; Chen, Y. *Carbon*, **2009,** *47,* 922–925.

Liang, J.; Xu, Y.; Huang, Y.; Zhang, L.; Wang, Y.; Ma, Y.; Li, F.; Guo, T.; Chen, Y. J. *Phys. Chem. C*, **2009,** *113,* 9921–9927.

Lim, S. Y.; Sahimi, M.; Tsotsis, T. T.; Kim, N. *Phys. Rev. E* **2007,** *76,* 011810–011825.

Lin, Y-M.; Dimitrakopoulos, C.; Jenkins, K. A.; Farmer, D. B.; Chiu, H-Y.; Grill, A.; Avouris, P. *Science*, **2010,** *327,* 662.

Lin, Y-M.; Valdes-Garcia, A.; Han, S-J.; Farmer, D. B.; Meric, I.; Sun, Y.; Wu, Y.; Dimitrakopoulos, C.; Grill, A.; Avouris, P.; Jenkins, K. A. *Science*, **2011,** *332,* 1294–1297.

Litchfield, D. W.; Baird, D. G. *Rheol. Rev.* **2006,** 1–60.

Liu, J.; Xue, Y.; Zhang, M.; Dai, L. *MRS Bulletin*, **2012,** *37,* 1265–1272.

Liu, N.; Luo, F.; Wu, H.; Liu, Y.; Zhang, C.; Chen, J. *Adv. Funct. Mater.* **2008,** *18,* 1518–1525.

Lu, C.; Mai, Y-W. *Phys. Rev. Lett.* **2005,** *95,* 088303.

Lu, F.; Gu, L.; Meziani, M. J.; Wang, X.; P. G.; Luo, P. G.; Veca, L. M.; L.; Cao, L.; Sun, Y. *Adv. Mater.* **2009,** *21,* 139–152.

Lux, F. J. Mater. Sci., **1993,** *28,* 285–301.

Manias, E, *Nature Mater.,* **2007,** *6,* 9–11.

Meyer, J. C.; Geim, A. K.; Katsnelson, M. I.; Novoselov, K. S.; Booth, T. J.; Roth, S. *Nature*, **2007,** *466,* 60–63.

Mickelson, E. T.; Huffman, C. B.; Rinzler, A. G.; Smalley, R. E.; Hauge, R. H.; Marfrave, J. L. *Chem. Phys. Lett.*, **1998,** *296,* 188–194.

Mierczynska, A.; L'Hermite, M.; Boiteux, G.; Jeszja, J. K. J. *Appl. Polym. Sci.* **2007,** *105,* 158–168.

Minoia, A.; Chen, L.; Beljonne, D.; Lazzaroni, R. *Polymer*, **2012,** *53,* 5480–5490.

Mohamadi, S.; Sharifi-Sanjani, N. *Polym. Comp.* **2011,** *32,* 1451–1460.

Moniruzzaman, M.; Winey, K. I. *Macromolecules*, **2006,** *39,* 5194–5205.

Mu, M.; Winey, K. I. J. *Phys. Chem. C* **2007,** *111,* 17923–17927.

Munson-McGee, S. H. *Phys. Rev. B* **1991,** *43,* 3331–3336.

Nelson, J. B.; Riley, D. P. *Proc. Phys. Soc.* London, **1945,** *57,* 477.

Nikla, D. L.; Pokatilov, E. P.; Askerov, A. S.; Balandin, A. A. *Phys. Rev. B* **2009,** *79,* 155413.

Nobel committee announcement, The Nobel Prize in Physics 2010–29 Oct 2012 URL: http://www.nobelprize.org/nobel_prizes/physics/laureates/2010/.

Pang, H.; Zhong, G.; Xu, J.; Yan, D-X.; Ji, X.; Li, Z-M.; Chen, C. Chin. *J. Polym. Sci.* **2012,** *30,* 879–892.

Park, H.; Afzali, A.; Han, S-J.; Tulevski, G. S.; Franklin, A. D.; Tersoff, J.; Hannon, J. B.; Haensh, W. *Nature Nanotech.* **2012,** *12,* 787–791.

Paul, D. R.; Robeson, L. M. *Polymer*, **2008,** *49,* 3187–3204.

Pötschke, P.; Abdel-Goad, M.; Alig, I.; Dudkin, S.; Lellinger, D. *Polymer*, **2004,** *45,* 8863–8870.

Pötschke, P.; Bhattacharyya, A. R.; Janke, A.; Goering, H. *Compos. Interf.* **2003,** *45,* 8863–8870 (*b*).

Pötschke, P.; Dudkin, S. M.; Alig, I. *Polymer*, **2003,** *44,* 5023–5030 (a).

Potts, J. R.; Dreyer, D. R.; Bielawski, C. W.; Ruoff, R. S. *Polymer*, **2011,** *52,* 5–25.

Probst, O.; Moore, E. M.; Reasco, D. E.; Grady, B. P. *Polymer*, **2004,** 4437–4443.

Rahman, A.; Ali, I.; Zahrani, S. M. A. *Nano*, **2011,** *6,* 185–203.

Ramanathan, T.; Abdala, A. A. Stankovich, S.; Dikin, D. A.; Herrera-Alonso, M.; Piner, R. D.; Adamson, D. H.; Schniepp, H. C.; Chen, X.; Ruoff, R. S.; Nguyen, S. T.; Aksay, I. A.; Prud'Homme, R. K.; Brinson, L. C. *Nature Nanotechnol.* **2008,** *3,* 327–331.

Ramanathan, T.; Stankovich, S.; Dikin, D. A.; Liu, H.; Shen, H.; Nguyen, S. T. *J. Polym. Sci. Part B: Polym. Phys.* **2007,** *45,* 2097–2112.

Ramasubramaniam, R.; Chen, J.; Liu, H. Appl. Phys. Lett. **2003,** *83,* 2928–2930.

Ray, S. S.; Okamoto, M. *Prog. Polym. Sci.* **2003,** *28,* 1539–1641.

Rivin, D. Rubber Chem. Technol., **1971,** *44,* 307–343.

Ruoff, R. S.; Tersoff, J.; Lorents, D. C.; Subramoney, S.; Chan, B. *Nature*, **1993,** *364,* 514–516.

Sears, K.; Dumee, L.; Schutz, J.; She, M.; Huynh, C.; Hawkins, S.; Duke, M.; Gray, S. *Materials*, **2010,** *3,* 127–149.

Smajda, R.; Kukovecz, A.; Konya, Z.; Kiricsi, I. *Carbon*, **2007,** *45,* 1176–1184.

Spitalsky, Z.; Tasis, D.; Papagelis, K.; Galiotis, K. *Prog. Polym. Sci.,* **2010,** *35,* 357–401.

Stalmann, G. MSc Dissertation, University of Gothenburg, 2012.

Stankovich, S.; Dikin, D. A.; Dommett, G. H. B.; Kohlhaas, K. M.; Zimney, E. J.; Stach, E. A.; Piner, R. D.; Nguyen, S. T.; Ruoff, R. S. *Nature*, **2006,** *442,* 282–286 (a).

Stankovich, S.; Piner, R. D.; Chen, X.; Wu, N.; Nguyen, S. T.; Ruoff, R. S. *J. Mater. Chem.* **2006,** *16,* 155–158 (b).

Tallury, S. S.; Pasquinelli, M. A. *J. Phys. Chem. B* **2010,** *114,* 4122–4129.

Vermant, J.; Ceccia, S.; Dolgovskij, M. K.; Maffettone, P. L.; Macosko, C. W. *J. Rheol.* **2007,** *51,* 429–450.

Vigolo, B.; Penicaud, A.; Coulon, C.; Sauder, C.; Pailler, R.; Journet, C.; Bernier, P.; Poulin, P. *Science*, **2000,** *290,* 1331–1334.

Vigolo, B.; Poulin, P.; Lucas, M.; Launois, P.; Bernier, P. *Appl. Phys. Lett.* **2002,** *81,* 1210–1212.

Villar-Rodil, S.; Paredes, J. I.; Martinez-Alonso, A. Tascon, J. M. D. *J. Mater. Chem.* **2009,** *19,* 3591–3593.

Vohrer, U.; Kolaric, I.; Haque, M. H.; Roth, S.; Detlaff-Weglikowska, U. *Carbon*, **2004,** *42,* 1159–1164.

Wang, L.; Duan, L-L. *Comput. Theor. Chem.* **2012,** (in press).

Wang, W-P.; Pan, C-Y. *Polym. Eng. Sci.* ***2004*****,** *44,* 2335–2339.

Wei, T.; Luo, G.; Fan, Z.; Zheng, C.; Yan, J.; Yao, C.; Li, W.; Zhang, C. *Carbon*, **2009,** *47,* 2296–2299.

Wolf, S.; Wang M-J.; In *Carbon Black Reinforcement in Elastomers in Carbon Black* Second Edition, Eds. Donnet, J-B.; Bansal, R. C.; Wang, M. J.; Taylor and Francis, London 1993.

Wu, H.; Drzal, L. T. Carbon, **2012,** *50,* 1135–1145.

Xi, Y.; Yamanaka, A;. Bin, Y.; Matsuo, M. J. *Appl. Polym. Sci.* **2007,** *105,* 2868–2876.

Xie, L.; Xu, F.; Qiu, F.; Lu, H.; Yang, Y. *Macromolecules*, **2007,** *40,* 3296–3305.

Xie, X-L.; Mai, Y-W.; Zhou, X-P. *Mater. Sci. Eng.* **2005,** *49,* 89–112.

Xu, J. Z.; Chen, T.; Wang, Y.; Tang, H.; Li, Z-M.; Hsiao, B. S. *Macromolecules* **2011,** *44,* 2808–2818.

Xu, J. Z.; Chen, T.; Yang, C. L.; Li, Z. M.; Mao, Y. M.; Zeng, B. Q.; Hsiao, B. S. *Macromolecules* **2010,** *43,* 5000–5008.

Xu, Y.; Hong, W.; Bai. H.; Li. C.; Sh, G. *Carbon,* **2009,** *47,* 3538–3543.

Yargi, Ö.; Ugur, S.; Pekcan, Ö. *Polym. Eng. Sci.* **2012,** *52,* 172–179.

Yu, A.; Ramesh, P.; Itkis, M. E.; Bekyarova, E.; Haddon, R. C. *J. Phys. Chem. C* **2007,** *111,* 7565–7569.

Yu, A.; Ramesh, P.; Sun, X.; Bekyarova, E.; Itkis, M. E.; Haddon, R. C. *Adv. Mater.* **2008,** *20,* 4740–4744.

Yu, F-M.; Lourie, O.; Dyer, M. J.; Moloni, K.; Kelly, T. F.; Ruoff, R. S. *Science,* **2000,** *287,* 637–640.

Zhang, H-B.; Zheng, W-G.; Yan, Q.; Yang, Y.; Wang, J-W.; Lu, Z-H.; Ji, G-Y.; Yu, Z. Z. *Polymer,* **2010,** *51,* 1191–1196.

Zhao, X.; Zhang, Q.; Chen, D. *Macromolecules* **2010,** *43,* 2357–2363.

Zheng, W.; Lu, X.; Wong, S. C. *J. Appl. Polym. Sci.* **2004,** *91,* 2781–2788.

Zheng, W.; Wong, S. C. *Compos. Sci. Technol.* **2003,** *63,* 225–235.

Zhong, H.; Lukes, J. R. *Phys. Rev. B* **2006,** *74,* 125403/1–125403/10.

CHAPTER 4

ULTRAFAST LASER DYNAMICS ON MOLECULAR NANOMAGNETS

GEORGIOS LEFKIDIS

CONTENTS

Abstract 126
4.1 Introduction 126
4.2 Computational Methods 128
4.3 The Λ Process as a Spin-flipping Mechanism 129
4.4 Genetic Algorithm 131
4.5 Clusters with One Magnetic Center 133
4.6 Finite Clusters with More Than One Magnetic Centers 137
4.7 Phononic Degrees of Freedom 145
4.8 Substrate 153
4.9 Functionalization 155
4.10 Conclusions 160
Acknowledgments 161
Keywords 161
References 161

ABSTRACT

This chapter is dedicated to *ab initio* model to describe optical, ultrafast manipulation of magnetic clusters. After the discovery of laser-induced demagnetization the optical control of the magnetic state of both extended and finite materials has become an increasing field of research. The investigations mainly follow two paths. The first one is to try and understand the microscopic processes behind this exciting phenomenon while the other one aim at using it to coherently manipulate magnetic materials and, ultimately, design spintronic applications. Here, after a short historic review we will develop a first-principles model, which can explain ultrafast spin-switching on single magnetic centers. Then we will extend the system to include more magnetic centers. Doing so will allow, beside spin flip, also for spin transfer and thus will enable us to implement magnetic logic. As side considerations, we will discuss the effects of phonons, as an additional degree of freedom also connected to temperature, and of a metallic substrate, since they both can affect the optical selection rules in the clusters. We will close the chapter with a short assessment of the importance of our theoretical findings.

4.1 INTRODUCTION

Today's computer and electronics application technology is (at least to a certain extent) guided by the ongoing pressure of the market for faster information processing and higher speed for information storage. Until last decade this demand has been mainly fulfilled by designing and fabricating denser CPUs and by producing magnetic hard discs with higher storage capacity. At present this trail seems to become saturated mainly because the used technologies are gradually reaching their functional limits. Sizes are becoming so small and functional elements so closely packed that quantum effects can no longer be neglected. Magnetic switching of domains in ferromagnetic materials (the technology mainly invoked for magnetic hard discs nowadays) seems to reach a fundamental speed limit of the picosecond regime. Of course many novel ideas have been proposed and successfully tested, e.g., thermally assisted switching perpendicular double magnetic-field pulses, but they either pose some additional problems, like disposing of the additional energy supplied to the system due to the heat for repeated use, or only slightly push the speed limits further (up to a factor of ten). Clearly what the market needs on the long run are some fundamentally new ideas which can speed up processes by several orders of magnitude.

If one considers the time scales of the main scattering mechanism on the nanoscale, i.e., electron-electron, electron-photon, electron-phonon, spin-electron, spin-spin, spin-phonon, one clearly sees that the use of photons as means of exerting coherent magnetic control is among the favorite candidates, among other reasons due to the progress of laser technology: laser pulses with energies in the optical and/or X-ray regime and subpicosend or even subfemtosecond time scales are becoming standard in many laboratories around the globe.

The pioneering work of Beaurepaire et al. (1996), who for the first time demonstrated the possibility of demagnetizing a metallic ferromagnetic sheet of Ni using only laser pulses, has inspired a very wide range of experimental and theoretical work trying to both elucidate the mechanisms involved in the process and improve it (Chovan et al., 2006; Hübner et al., 2009; Kirilyuk et al., 2010; Koopmans et al., 2005; Zhang et al., 2009). After a long period of controversy with respect to fundamental processes involved and their order, today there seems to be a consensus that, at least to a certain extent, the laser pulse excites the electrons in a coherent manner during the first few hundred femtoseconds, which, in turn, scatter with the spins. This is followed by an incoherent, thermalizing process, in which the phonons play a major role (Bigot et al., 2009; Bovensiepen, 2009; Lefkidis et al., 2009).

At the same time the size of the materials in question is becoming smaller, a fact that points towards miniaturization of magnetic elements. More than simply devizing magnetic elements solely for storing information, new, alternative ways of magnetically processing information are being sought as well. As examples we mention the logical NOT, logical AND, signal fan-out, and signal cross-over elements constructed out of planar magnetic wires that are less than a micrometer in width (Allwood et al., 2005), and the three-input majority logic gate based on a network of physically coupled, nanometer-scale magnets (Imre et al., 2006). With respect to miniaturization one can mention two capturing successes. One is represented by the "moleculators", for which changing the concentration of cations in a solution of conjugated molecules leads to different absorption spectra and thus allows for logic operations (de Silva and Uchiyama, 2007), and the other by the recent "publication" of Church et al. (2012), who stored an html-coded draft of a book that included 53,426 words, 11 JPG images and 1 JavaScript program into a 5.27 megabit bitstream using next-generation DNA synthesis and sequencing technologies (an amazing storage density of 5.5 petabits/mm^3). Despite their apparent lacking behind with respect to the time scales involved, the last two examples perfectly put forth the vision in computer and storage technology, also reflected in the new microSDXC standard of IEEE for flash RAMs, capable of handling up to 2 terabits per card (a storage density of about one bit every few hundred atoms).

All these are tokens of the fact that application advances in the foreseeable future will most probably be based not only on improving existing technologies but also on fundamentally new and therefore exciting physics. The latter includes, further than just device miniaturization, also speeding them up. Inevitably the investigation of the interaction of subpicosecond laser pulses with magnetically ordered materials (both extended and finite) is becoming one of the most timely fields of active scientific research. Some additional information can be found in Baletto and Ferrando (2005) as well as in Kirilyuk et al. (2010).

4.2 COMPUTATIONAL METHODS

Since we are interested in describing magnetic phenomena, correlations play always a major role in our systems. This, in fact, answers the first somehow more technical question: which methods should be used for the quantum mechanical description of the systems. One is usually faced with two choices, namely quantum chemistry in real space and density functional theory (DFT). Although there are numerous arguments speaking for and against both approaches, it seems that the pure quantum mechanical one has some definite advantages. The main one is the possibility to systematically increase the level of theory, while DFT encloses a certain amount of empirical parameters hidden in the exchange potential. Two additional components are the need for a good description of many excited states as well as the ground state, and the desire for accurate wavefunctions. As it turns out for the systems discussed here (mainly including atoms up to $3d$ transition elements, i.e., Fe, Co and Ni) the inclusion of correlations at a high post-Hartree-Fock level and the use of relativistic, effective core potentials prove adequate enough. Therefore it is our method of choice (for a further discussion on the several methods on some of the systems discussed here see also Aryasetiawan and Gunnarsson, 1995; de Graaf et al., 1996; Faleev et al., 2004; Lichtenstein and Katsnelson, 1998; Satitkovitchai et al., 2003; Sawatzky and Allen, 1984; and references therein).

As a first step, we typically calculate the nonrelativistic electronic states of our system with a commercial quantum chemistry package. The post-Hartree-Fock methods used vary, depending on the system from configuration interaction (CI) and complete-active-space selfconsistent field (CAS-SCF) to symmetry-adapted-cluster configuration interaction (SAC-CI) and equation-of-motion-coupled-cluster (EOM-CC). It is important for the subsequent steps that the wavefunction of all electronic states involved be expressed (or convertible) to some determinental expansion (CI expansion). The Hartree-Fock molecular orbitals have to be of restricted form (i.e., identical set of orbitals for alpha and beta electrons), so that we can construct spin-eigenfunctions and build all sublevels of any given multiplicity, i.e., for a given triplet we can build the wavefunction for all three s_z projections. This is essential for the subsequent inclusion of spin-orbit coupling (SOC).

In a second step, we include in a perturbative way the SOC operator $\widehat{H}_{SOC}$ and an external, static magnetic field **B**. For $\widehat{H}_{SOC}$ we restrict ourselves to the contributions from one-electron integrals $\left\langle \psi_i \left| \frac{\mathbf{r}\times\mathbf{p}}{r^3} \right| \psi_j \right\rangle$ while the two-electron contributions are estimated using effective nuclear charges (Koseki et al., 1998). The existence of the infinitesimal **B** field ensures the separation of the spin-up and spin-down states (quantum mechanically we cannot unambiguously separate degenerate states for any linear combination of them is again an eigenstate of the Hamiltonian).

Finally, with the magnetic states at hand we propagate the total wavefunction under the influence of proper laser pulses in order to push the electronic population from an initial state $|i\rangle$ to a desired, target state $|f\rangle$. The states are chosen from their properties (expectation values and localization of the spin and charge densities – the latter is performed via Mulliken population analysis). The laser pulse is optimized with a genetic algorithm (Schwefel, 1981), which was especially developed for this purpose (Hartenstein et al., 2008). The algorithm encodes the geometry of incidence, helicity, duration and amplitude of a subpicosecond laser pulse in a single binary arithmetic sequence. Mimicking nature the algorithm sorts out many such sequences (individuals), which are the most promizing ones, and mixes their genetic codes. The procedure continues until one individual reaches the sought fitness level, that is it performs the electronic population transfer we are after.

The time evolution is described by time-dependent perturbation theory within the interaction picture and is performed with an embedded sixth order Runge-Kutta method and adaptive Cash-Karp integration steps (1990). The actual propagation is performed by solving the system of differential equations:

$$\frac{\partial c_\alpha(t)}{\partial t} = \frac{-i}{\hbar} \sum_\beta \left\langle \alpha \left| \widehat{H}'(t) \right| \beta \right\rangle c_\beta(t) e^{-1} \left(E_\beta - E_\alpha \right) t/\hbar \quad (1)$$

where $|\alpha\rangle$ and $|\beta\rangle$ are the unperturbed eigenstates, $\widehat{H}'(t)$ is the perturbation term in the Hamiltonian (i.e., the interaction with the laser pulse), c_α the coefficients of state $|\alpha\rangle$ in the total wavefunction $\Psi(t) = \sum_\alpha c_\alpha(t) e^{-1E_\alpha t/\hbar} |\alpha\rangle$, and E_α its energy.

4.3 THE Λ PROCESS AS A SPIN-FLIPPING MECHANISM

Magnetic storing of information is mostly based on flipping the magnetization vector of domains or clusters of magnetically ordered materials using different techniques and magnetic fields or pulses. On the atomic or molecular scale this amounts to changing the projection of the spin expectation value from a positive to a negative value (provided we are dealing with a spin multiplicity higher than a singlet). Borrowing the idea of the brachystochrone from classical mechanics, where the fastest path between two points is not necessarily the shortest one, we induce spin flipping via Λ processes (Fig. 1). There are two main reasons for this: (i) the energy difference between the initial (spin-up) and the final (spin-down) state is usually in the order of a few meV rendering the process slow, and (ii) the direct transition between the states is forbidden within the electric-dipole (ED) approximation, so one would have to rely on magnetic-dipole (MD) transitions which are about six orders of magnitude weaker.

In our process we substitute two fast laser-induced ED transitions (from the initial state to an intermediate state with higher energies, and from the intermediate state to the final state) for the direct MD transition. Obviously the prerequisite is that the intermediate level(s) be optically coupled to both the initial and the target states. This prerequisite is fulfilled by the action of SOC, which can mix spin-up with spin-down states as well as different multiplicities. Of course if one includes SOC the spin is not a good quantum number anymore but still one can calculate its expectation value and often classify the states as singlets or triplets in a loose manner.

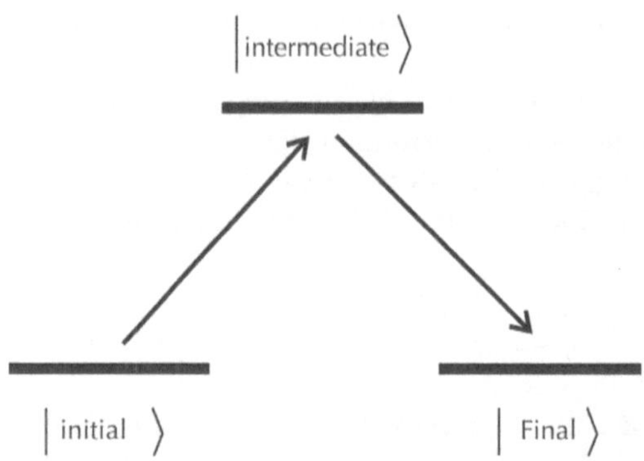

FIGURE 1 The spin-flipping Λ process. The initial state and the final state have different spin-orientations (up and down, respectively) or different spin-localization, while the intermediate state is a SOC mixed state.

Although SOC affects all multiplicities higher than 1, we prefer triplets to doublets. The reason for this is mainly that in the case of doublets the actual spin and charge carrier is the same (one electron) while in the case of a triplet state a combination of at least two electrons is needed, which allows for separation of local spin and charge dynamics while presenting a minimum of correlations (Hartenstein et al., 2008; Hübner et al., 2009; Lefkidis et al., 2009; Zhang et al., 2009). A large number of magnetic clusters have been investigated using this model and several spin-flipping scenarios have been derived. These clusters have varying number of active magnetic centers (from one to three).

One major challenge is to find a laser pulse capable of driving the Λ process. Although in principle we have many parameters at hand, the ones we actually use are the intensity, the width, the frequency, the helicity and incidence geometry of the pulse. The envelope of the pulse is always a sech^2 function (it has a shape quite similar to a Gaussian curve but is computationally easier to handle). Additional degrees of freedom, like changing the geometry or the frequency during the pulse are not needed, although theoretically they might further improve the efficiencies. (There is one exception to this when we realize the ERASE function discussed later in this chapter.)

4.4 GENETIC ALGORITHM

Since the dependence of the outcome of the laser pulse is a very complicated function of the laser parameters, we use a genetic algorithm to optimize those (Hartenstein et al., 2008). To this end we first encode in a 64–bit integer i the frequency, the intensity, the width, and the three angles of incidence θ, φ and γ (θ and φ give the direction of the propagation of the light in spherical coordinates, while γ is the angle of the polarization of the light with the optical plane in the case of linearly polarized light – for circularly polarized light it is obviously irrelevant). This integer is called individual and represents a single laser pulse. We arbitrarily generate a generation of individuals (typically 1000) and propagate the investigated system under the influence of all of them. We also need to define a fitness function $f(i)$ which describes the efficiency of the laser pulse. Intuitively one could simply use as $f(i)$ the population of the desired final state after the laser pulse. This definition however would not take into account laser pulses for which the final state is unpopulated at the end but gets some significant population during the laser pulse (which means that although per se unsuccessful, they might be close to a successful one since they obviously shake things up!). Therefore we use a weighted function of the end population p and the maximal population p_{max} of the final state, using as weight the population of the final state itself

$$f(i) = p^2 + 0.7p_{\max}(1-p) \quad (2)$$

From every generation we keep 25% of the most successful individuals and combine them in two different ways to create 50% of the individuals of the next generation: we arbitrarily cross two of them (parents) to create two new ones (children), and we arbitrarily toggle one or two bits (mutants). In order to avoid genetic drift each following generation still consists of 25% completely random individuals. Fig. 2 shows the flowchart of the genetic algorithm.

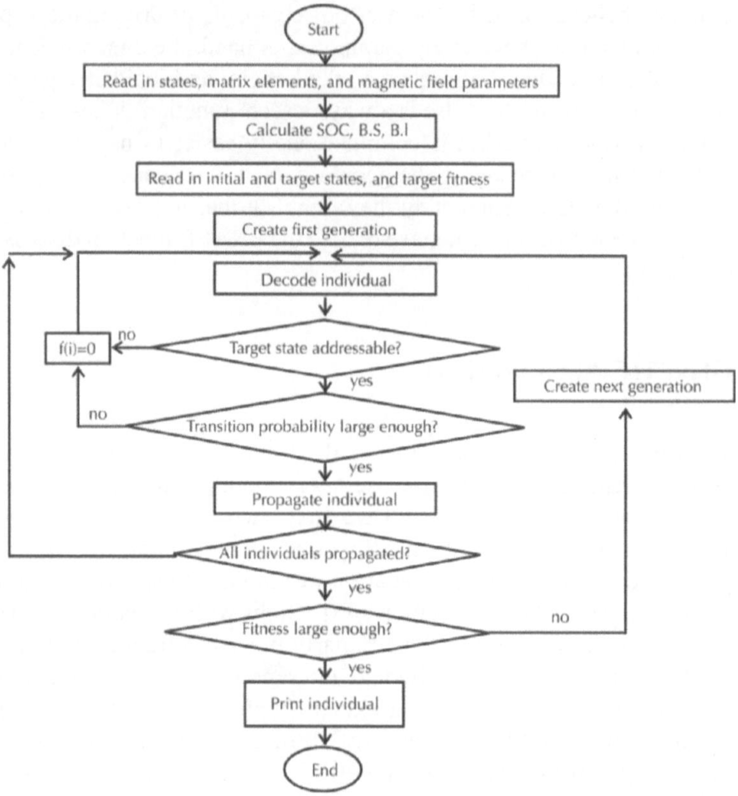

FIGURE 2 Flowchart of the genetic algorithm. (Hartenstein et al., 2008).

Since we typically propagate 1000 individuals in every generation (often we need several hundred generations) this is the computationally most intensive step and, although highly parallelizable, it still takes the longest time to complete. Therefore we have incorporated in the algorithm some control mechanisms. One checks in advance which states are addressable with the laser pulse at hand (selection rules); the unnecessary ones are left out of the propagation. Another one skips propagations for which we know the outcome will not be satisfactory enough. So cases for which a single Rabi oscillation would take more than 1000 times longer than the laser duration itself are left out – same holds for cases where the two transition matrix elements (initial to intermediate state, and intermediate to final state) differ by more than one order of magnitude (since we analytically know from model Λ systems that for a 90% population transfer the two transition-matrix elements cannot differ by more than about 30%).

4.5 CLUSTERS WITH ONE MAGNETIC CENTER

The first realistic substance we applied the Λ process was NiO (both bulk and the 001 surface) modeled with the doubly embedded NiO_6^{-10} and NiO_5^{-8} clusters, respectively (Fig. 3). The clusters were first embedded in a shell of effective core potentials describing the Ni atoms to be found in the immediate vicinity and a charge-point field accounting for the Madelung potential of the extended system (Gómez-Abal et al., 2004; Lefkidis and Hübner, 2007). This cluster model has previously been successful in accurately describing the intragap, *d*-character energy states of NiO at various quantum-theory levels, and gives very good agreement with experimental results, especially since these states exhibit a very strong spin localization on the Ni atom (for further discussion on theory and experiment see also Faleev et al., 2004; Fiebig et al., 2001; Geleijns et al., 1999; Gorschlüter and Merz, 1994; Lefkidis and Hübner, 2005, Rubano et al., 2010).

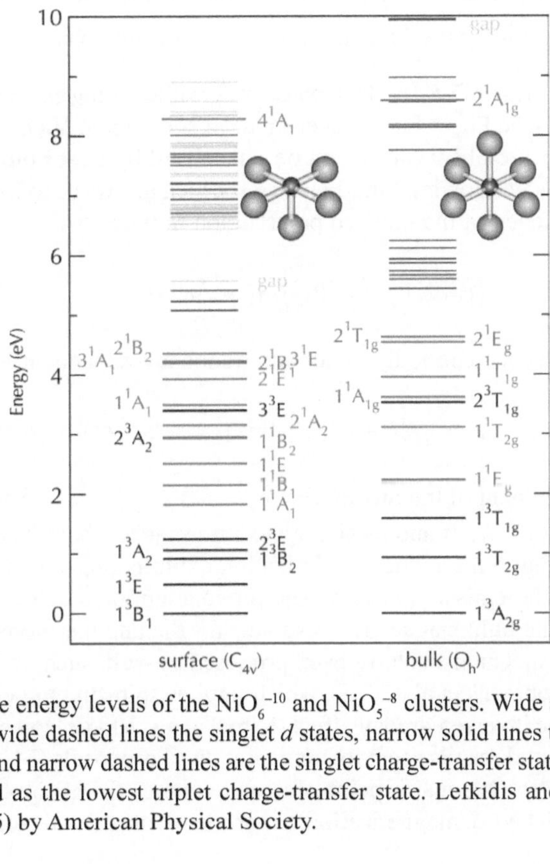

FIGURE 3 The energy levels of the NiO_6^{-10} and NiO_5^{-8} clusters. Wide solid lines are the triplet *d* states, wide dashed lines the singlet *d* states, narrow solid lines the triplet charge-transfer states, and narrow dashed lines are the singlet charge-transfer states. The electronic gap is identified as the lowest triplet charge-transfer state. Lefkidis and Hübner (2005). Copyright (2005) by American Physical Society.

One major finding is that the process is possible within 300 femtoseconds only with the application of linearly polarized laser pulses. The reason becomes apparent if one considers the selection rules governing the transitions. We identify state $|i\rangle$ with the $\langle s_z\rangle = 1$ sublevel of the ground-state triplet ($^3B_1$ for the surface and $^3A_{2g}$ for the bulk) and state $|f\rangle$ with the respective $\langle s_z\rangle = -1$ sublevel. As intermediate excited states we use a set of spin-mixed states situated at about 0.4 eV with $\langle s_z\rangle = 0$ (remember that these are not good quantum numbers since we include SOC and their values are only approximate). The process takes place in two phases. During the excitation phase a photon with helicity σ_- is absorbed while during the deexcitation phase a photon with the opposite helicity σ_+ is emitted. Since both transitions are induced by one single laser pulse the simultaneous presence of both helicities is necessary, which is tantamount to linearly polarized light (Lefkidis and Hübner, 2007). This conclusion, which stems only from the conservation of angular momentum, was also confirmed by numerical calculations. To this goal we investigated the induced, time-dependent polarization $\langle P(t)\rangle$ in the material (Lefkidis et al., 2009). If one draws the components of $\langle P(t)\rangle$ vs. time one sees that the components $\langle P_x(t)\rangle$ and $\langle P_y(t)\rangle$ are not in-phase. In fact, during the absorption phase the oscillations of $\langle P_y(t)\rangle$ precede those of $\langle P_x(t)\rangle$ by a phase of $\frac{\pi}{2}$ while during emission the reversed order is the case (see Fig. 4 for the laser-induced spin flip on NiO_5^{-8}).

This qualitative observation can be quantified by first Fourier-transforming <P(t)> with a windowing function *h*(*t*–*s*), which allows us to follow the evolution of the frequency of the induced polarization in time

$$\left\langle \tilde{\mathbf{P}}(\omega,t)\right\rangle = \int_{-\infty}^{+\infty} \langle \mathbf{P}(s)\rangle h(t-s) e^{i\omega s} ds \qquad (3)$$

and then use $\left\langle \tilde{\mathbf{P}}(\omega,t)\right\rangle$ to build the time- and frequency-dependent Stokes vector:

$$\mathbf{s}(\omega,t) = (I,Q,U,V) = \left(|P_x|^2 + |P_y|^2, |P_x|^2 - |P_y|^2, 2\Re(P_x P_y^*), 2\Im(P_x P_y^*) \right). \qquad (4)$$

The fourth component of the Stokes vector $V = 2\Im(P_x P_y^*)$ is connected to the circular polarization of $\left\langle \tilde{\mathbf{P}}(\omega,t)\right\rangle$ and its sign gives information about its helicity. As one clearly sees in Fig. 5 the induced polarization exhibits opposite helicities for the absorption and the emission phases. The obvious importance of the finding lies in the fact that the light can act as a reservoir for the angular momentum surplus. Of course other mechanisms have been proposed as well, such as the Elliot-Yafet mechanism (Koopmans et al., 2005), which implies spin-flipping electron-phonon coupling, and the inverse Faraday effect (Kimel et al., 2005). However some more recent experimental results seem to contradict these proposals (Meier et al., 2009; Radu et al., 2009). One major point of dispute comes from Bigot et al., who repeated laser-induced demagnetization experiments on thin, ferromagnetic films

(Ni) varying their thickness (2009). The argument was that if phonons played an important role, then moving to thinner films would speed up the recovery of the magnetization, since thinner films induce more surface phonons; however the results showed no such dependence. In fact Bigot et al. went even further proposing that relativistic effects (SOC) stemming from the electric field of the laser itself and not just of the nucleus might represent a significant contribution to the magnetization dynamics.

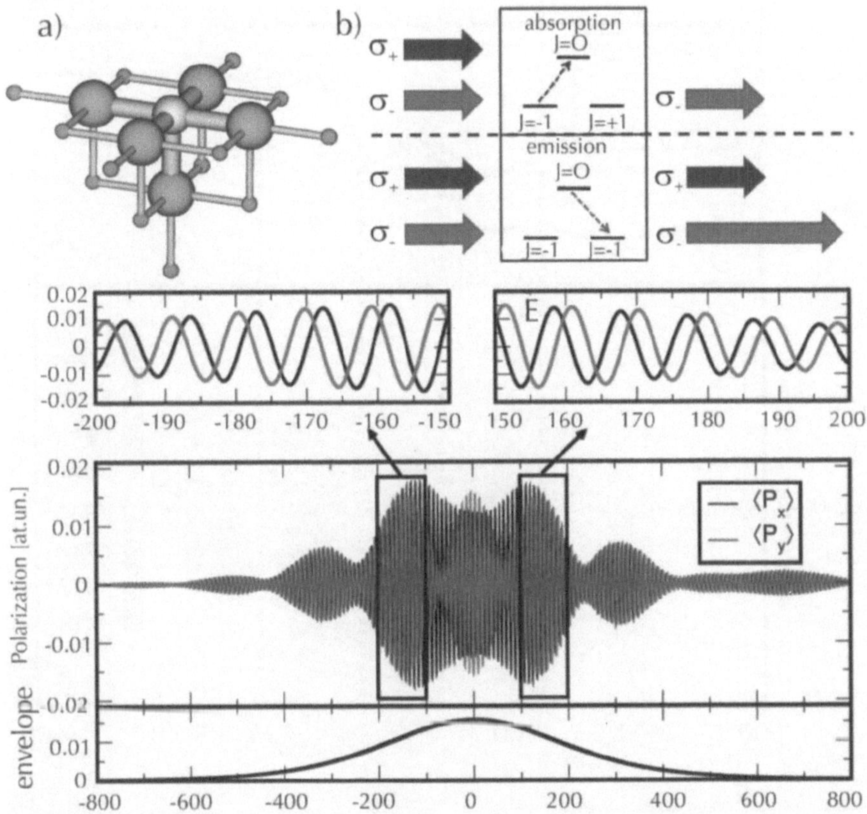

FIGURE 4 Time-dependent light-induced polarization in the material. Panel (a) shows the NiO_5^{-8} cluster and panel (e) its induced polarization throughout the whole Λ process. In panels (c) and (d) two time windows are magnified and one can see that the phase difference between $\langle P_x(t)\rangle$ and $\langle P_y(t)\rangle$ changes sign, which corresponds to different helicities of the participating light. Panel (f) shows the envelope of the laser pulse. Panel (*b*) is a schematic view of the angular-momentum flow during absorption and emission; the cluster has always a net change of $\Delta J = +1$ and the reflected light a net total helicity of σ_- (Lefkidis et al., 2009). Copyright (2009) by American Physical Society.

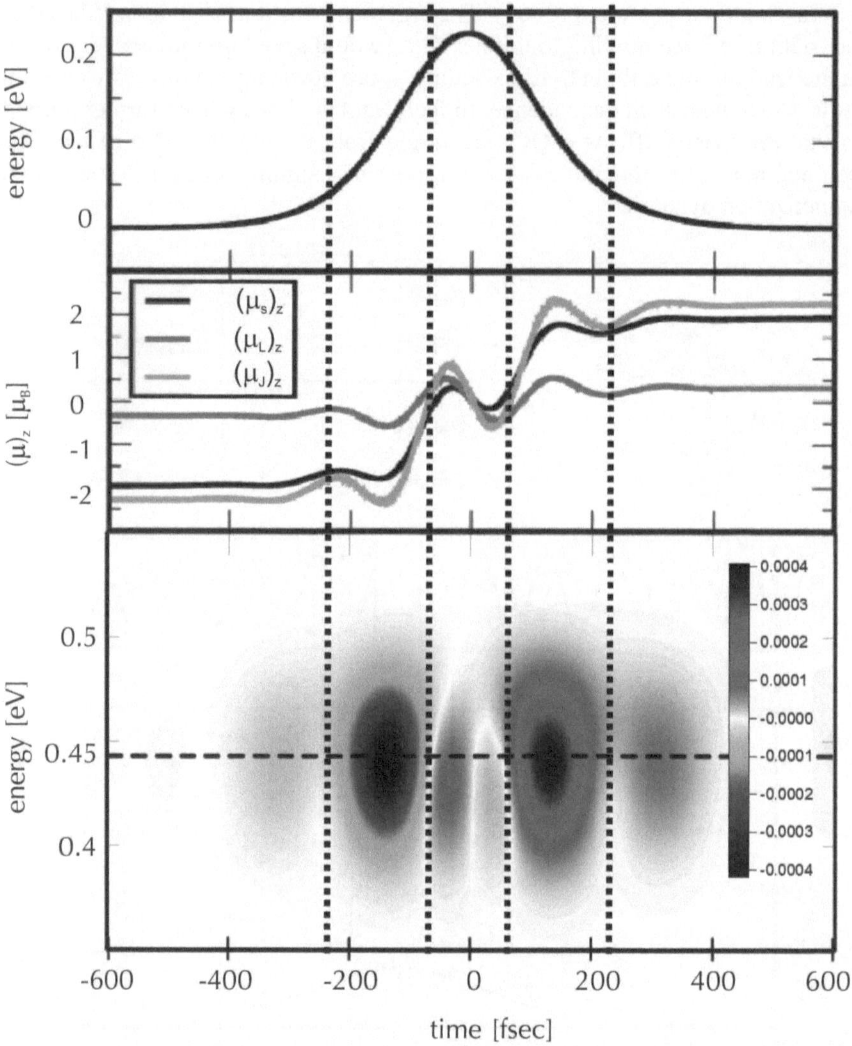

FIGURE 5 Upper panel: Expectation value of the energy of the cluster. Middle panel: Expectation values of the spin, orbital angular momentum, and total angular momentum. Lower panel: Contour plot of the polarization of the emitted light at frequencies around that of the incident light as a function of energy and time. Blue [black in phase (i)] means σ_+ and red [gray in phase (iii)] σ_- light (color code in atomic units). The horizontal dashed line indicates the energy (0.443 eV) of the incident σ_0 light. One can clearly distinguish the change of the polarization when going from absorption to emission (compare upper and lower panel). The four vertical dashed lines roughly indicate phases (i), (ii), and (iii) (see text). (Lefkidis et al., 2009.) Copyright (2009) by American Physical Society.

4.6 FINITE CLUSTERS WITH MORE THAN ONE MAGNETIC CENTERS

One major point of interest is to which extent the behavior of the previous systems change when moving over to actual nanoclusters. Please bear in mind that both NiO clusters, although finite and consisting of one Ni atom where intended to describe extended systems, i.e., infinite in real space but confined in reciprocal space. It is natural to expect that scaling down to molecular clusters the situation might drastically change.

One family of the smallest and easiest clusters one can come up with, are linear chains with two magnetic centers. These chains can be thought of as deposited on an inert surface such as a Cu(001) surface, which fixes their geometry, or in the gas phase. Clearly a substrate can mediate interaction as well (Pal et al., 2010), but for the sake of simplifying physics we can neglect those effect for the moment. For our investigations we first calculated a series of chains, in which two magnetic atoms (Co, Fe, or Ni) are connected through Na atoms. The choice was driven by the fact that Na atoms are metallic but not magnetic (which can lead to separation of spin and charge dynamics), have an odd number of electrons (so that by varying their number we can ensure that we end up with a cluster with an even number of electrons, which means that we are dealing with singlets and triplets, instead of doublets and quartets), and, last but not least, are computationally cheap. The first surprizing result, which turned out to be an almost universal feature in nonsymmetric molecules) was that the spin density for every state is mainly localized on one magnetic atom for the electronic low-lying states (one magnetic atom carries typically more than 80% of the total spin density). This fact gave rise to a new possibility, namely not only to flip the spin but also to transfer it from one center to the other. The underlying mechanism still remains the same (a Λ process) but depending on the choice of initial and final state (different spin orientations or different spin localizations) one can realize spin flip or spin transfer. However, for the linear chains the transfer turns out to be impossible, the reason being that the high-symmetry rotation axis (point groups $C_{\infty v}$ or $D_{\infty h}$ for two identical or two different magnetic atoms, respectively) forbids the necessary transitions (Li et al., 2012). Local spin flip is practically never a problem.

The behavior of Co, Fe and Ni with respect to local spin flip is somehow counterintuitive. Since the latter one has the only two holes in the 3*d* shell, and hence the simplest level scheme, one would expect it to exhibit the most straightforward spin-flip scenarios, followed by Fe and Co with three and four holes, respectively. As it turns out this is not the case. Fe is the most well behaved metal, for which the electronic-population transfer from the spin up to the spin down state very much resembles a simple Rabi oscillation cycle. Ni, on the other hand, exhibits an almost chaotic behavior (however, by slightly changing the initial conditions and repeating all the propagations, one sees that the processes are not really chaotic). Finally Co is somehow in between and can flip its spin only after a finite number of cycles (usually between five and eight cycles). This local behavior persists regardless of the nature of bridging atoms (see Fig. 6 for the case of a bridging oxygen).

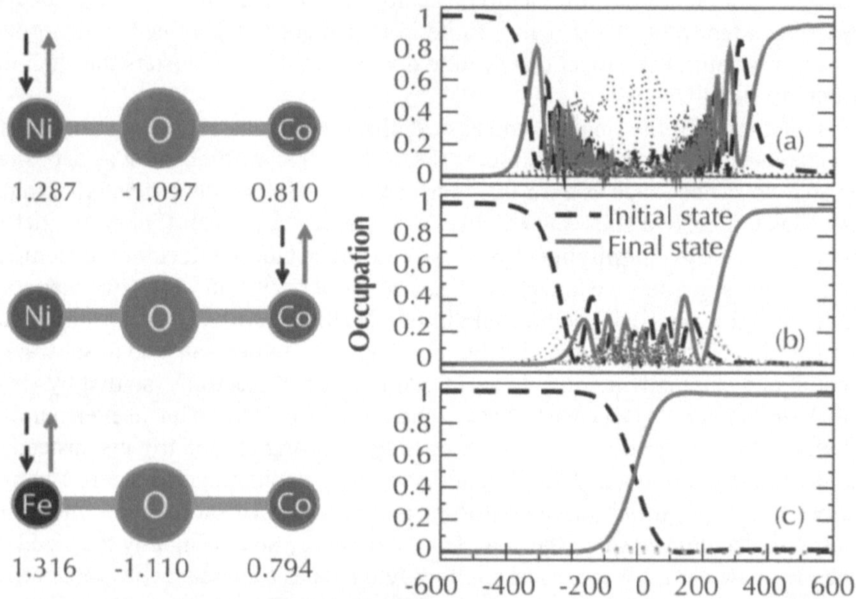

FIGURE 6 Right panel: Local spin flips at the Ni, Co, and Fe ends of the linear structures via the Λ process. (a) Spin flip at the Ni end of $[Ni\text{-}O\text{-}Co]^+$. (b) Spin flip at the Co end of $[Ni\text{-}O\text{-}Co]^+$. (c) Spin flip at the Fe end of $[Fe\text{-}O\text{-}Co]^+$. Left panel: Corresponding sketches of the optimized linear structures with arrows indicating the atoms where spin flip occurs. The numbers below the atoms give the atomic charge densities. (Li et al., 2011). Copyright (2011) by American Physical Society.

A second family consists of ligand-stabilized clusters, like the synthesized homodinuclear complex $[Ni^{II}_2(L\text{-}N_4Me_2)(emb)]$ (Jin et al., 2012; Lefkidis et al., 2011). The complex consists of two Ni cations in different local geometries: One in square planar geometry ($Ni_{sqpl.}$) leading to a singlet state and one in a slightly distorted octahedral geometry ($Ni_{oct.}$) leading to a triplet ground state (Fig. 7). The latter leads to the spin density being, to a very great extent, localized on one Ni atom only. Table 1 gives the relevant information for some of the low-lying energy levels.

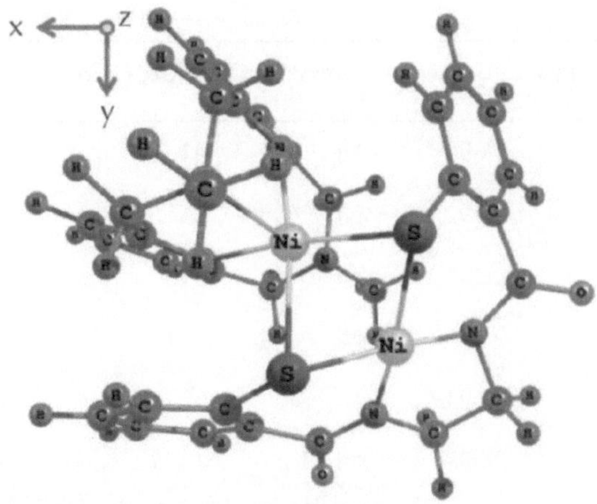

FIGURE 7 The homodinuclear complex $[Ni^{II}_2(L\text{-}N_4Me_2)(emb)]$. Reprinted with permission from Lefkidis et al. (2011). Copyright (2011) by American Chemical Society.

TABLE 1 Some of the low-lying many-body states and their spin localization (triplets only). For the exact definition of the polar angles θ and φ, see Fig. 6. The calculations include a static external B field of 10^{-5} a.u., pointing at $\theta = 35°$ and $\varphi = 0°$. States 2, 3, 16, 17, 18, and 19 (in bold) are used for the spin-flip process as intermediate states. States 18 and 19 have no appreciable spin expectation value along the *z* axis. (Lefkidis et al., 2011). Copyright (2011) Americal Chemical Society.

State	Energy (eV)	$\langle S \rangle$	Atom	θ (degrees)	φ (degrees)
35	5.4449	0.197	$Ni_{sqpl.}$	127.6	-111.1
30	5.1690	0.174	$Ni_{sqpl.}$	86.8	74.4
29	5.1666	0.183	$Ni_{sqpl.}$	116.8	-102
28	4.2285	0.811	$Ni_{oct.}$	40.3	105.5
27	4.2280	0.774	$Ni_{oct.}$	117.3	-61.6
24	4.0783	0.614	$Ni_{oct.}$	89.2	123.8
23	4.0775	0.782	$Ni_{oct.}$	138.9	-75.9
19	2.3015	0.000			
18	2.1853	0.002			

TABLE 1 *(Continued)*

State	Energy (eV)	$\langle S \rangle$	Atom	θ (degrees)	φ (degrees)
17	2.1842	0.090	$Ni_{oct.}$	41.7	120.1
16	2.1802	0.092	$Ni_{oct.}$	131	-48.9
12	1.1287	0.397	$Ni_{oct.}$	41.7	120.1
11	1.1267	0.395	$Ni_{oct.}$	131	-48.9
5	0.7361	0.094	$Ni_{oct.}$	33.6	143.1
4	0.7265	0.098	$Ni_{oct.}$	149.6	-77.9
3	0.0029	0.813	$Ni_{oct.}$	27.9	103.8
2	0.0016	0.819	$Ni_{oct.}$	156.7	-57.7
1	0.0000	0.131	$Ni_{oct.}$	0.131	-130.7

The best, theoretical spin-flip scenario is found for a laser pulse with $\theta = 49.9°$, $\gamma = 74.4°$, and $\varphi = 322.1°$ (θ and γ denote the angles of incidence in spherical coordinates, and φ is the angle between the polarization of the light and the optical plane with respect to the Cartesian coordinate system as in Fig. 6). The $sech^2$-shaped laser pulse has an energy of 2.29 eV and a duration (full-width-at-half-maximum) of 53 femtoseconds.

One extension of the spin-flipping scenarios on the complex, was that for the first time charge-transfer states were used as intermediate states as well. The advantage of this is that the transitions from the initial to the intermediate, and from the intermediate to the final state are no more Laporte-forbidden (as a reminder, *d–d* transitions are strictly speaking only in spherical symmetry, so the slight distortion in the real molecules makes them allowed, nonetheless they remain very weak). Therefore the relevant dipole-transition matrix elements are about one order of magnitude larger, allowing a much weaker laser pulse (0.25 $Js^{-1} m^{-1}$, about 100 times weaker than using *d* states as intermediate states) to achieve the same result (Jin et al., 2012).

An additional effect discovered in this complex is the fact that the intermediate states do not necessarily have to be populated during the process, although they still need to exist. This can be checked by selectively removing them from the time propagation – we find that the whole process gets blocked (Fig. 8). In principle this means that the intermediate states sometimes do not retain the electronic population but instantaneously deflect it to the final state.

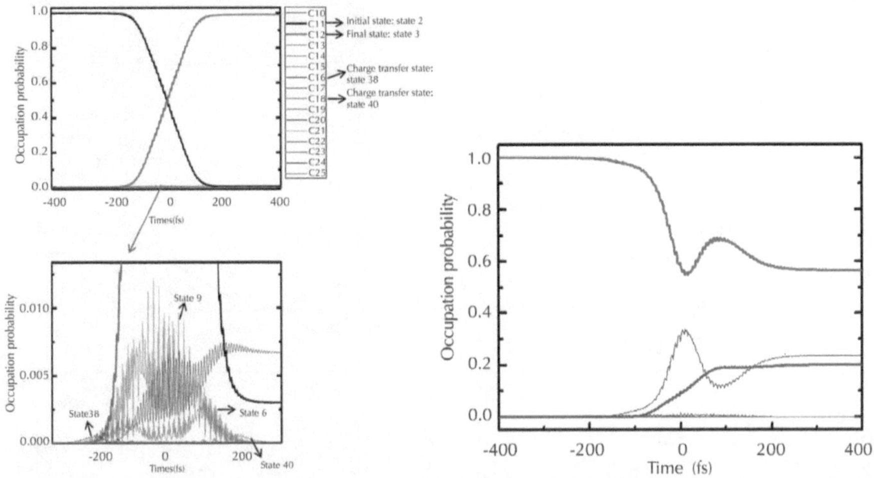

FIGURE 8 Partial blocking of the spin switching in the homodinuclear complex $[Ni^{II}_2(L\text{-}N_4Me_2)(emb)]$. Upper left panel: population transfer between initial and final state. There is no significant population at the intermediate states. Lower left panel: magnification of the upper panel. Right panel: The final state gets a maximum population but loses it again, if the transition matrix elements towards one of the intermediate excited states are set to zero.

This complex was the first one on which magnetic dynamics was experimentally investigated as well. Fig. 9(a) shows the UV/Vis difference absorption spectra of the complex in methanol obtained after excitation at 2.53 eV between 1.77 eV and 2.70 eV for delay times up to 35 ps. By a global analysis of the transient absorption differences $\Delta[OD(\lambda, t)]$ we find that there are two different transient states Γ_1 and Γ_2 with corresponding relaxation time $\tau_1 = 1.6$ ps and $\tau_1 = 4.4$ ps, respectively [Fig. 9(b)]. Fig. 9(c) shows the absorption spectra of the two transient states (Γ_1 and Γ_2) fitted with four Gaussians each (the arrows indicate the center of the Gaussians). Figure 9(d) shows the theoretically calculated spectra for the transient states Γ_1 and Γ_2. This combined experimental and theoretical effort in fact proves that the laser induced coherent control of the magnetic states of small molecules can indeed go beyond the mere proof-of-principle.

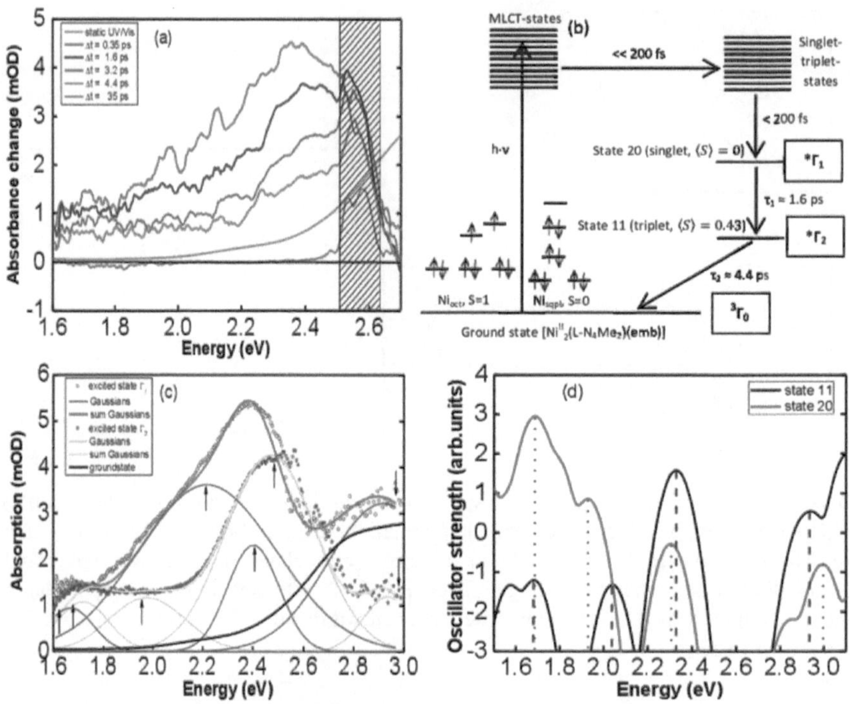

FIGURE 9 Transient absorption spectra. (a) Transient difference absorption spectra of Ni-complex/methanol at selected delay times after excitation at 2.53 eV (490 nm). Shaded region: Contribution of scattering artifact by pump pulse, which at no times relates to relaxation dynamics. At 35 ps this artifact is the only contribution. (b) Proposed reaction dynamics of the two excited-state processes according to the global analysis of the data in (a). (c) Absorption spectra of the two transient states Γ_1 (circles) and Γ_2 (stars). The arrows indicate the peak positions. (d) Theoretical spectra of states Γ_1 and Γ_2, and corresponding experimental energies (blue dashed and dotted lines). (Jin et al., 2012.) Copyright (2012) by American Physical Society.

Another family of clusters investigated was bare clusters in the gas phase. These can be easily synthesized and handled, e.g., by nozzle expansion or laser ablation followed by mass-spectroscopy selection (Bialach et al., 2009, 2010; Sebetci, 2008). From a computational point of view they are most interesting because they carry all the interesting physics, i.e., the metallic/magnetic centers, without the side effects of the surrounding chemistry (ligands), which can obscure or even hide the basic physics. We mainly investigated such families in order to unveil the role of the structural parameters. One important issue is the role of the bridging atoms. Fig. 10 depicts the delicate balance needed for the two energy differences in order to have a successful Λ process: The energy difference between initial and final state has to be small enough, so that a single laser pulse can be resonant to both the excitation and deexcitation frequencies of the system, but far enough so that they can be distinguishable (i.e., not equally populated at finite temperatures). The energy of the intermediate state must be far enough from the ground (initial and final) states but then again not too high, since otherwise the same pulse can address neighboring excited states as well (because the relative detuning becomes very small) leading to additional undesired transitions. The challenge thus lies in designing systems with the proper energy differences.

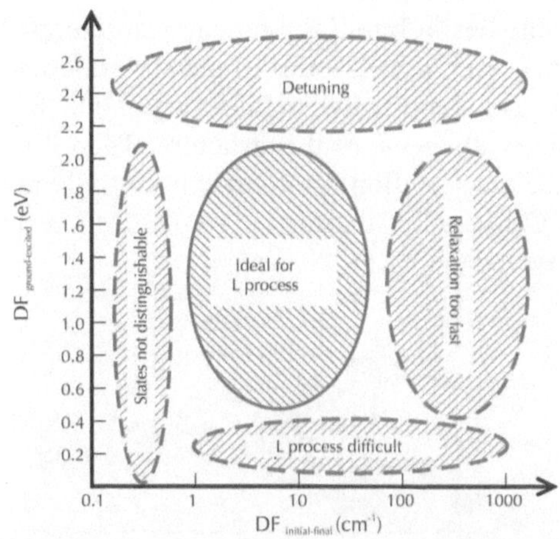

FIGURE 10 Sketch of the required energy differences between initial and final states (abscissa axis in logarithmic scale) and between the initial state and the intermediate excited state (ordinate axis) in a Λ process. (Li et al., 2011.) Copyright (2011) by American Physical Society.

The experience collected so far suggests that the spin flip is relatively easy to achieve. The reason is that the initial and the final state are typically separated only by the Zeeman splitting (induced by an external field or the magnetocrystaline anisotropy of the system) while the excited state can be appropriately chosen. Spin transfer, on the other hand, proves to be much more difficult. This can be understood if one takes into account two different tendencies when designing a cluster. If the cluster is very symmetric (as is the case for the homonuclear metallic chains Ni-Na-Ni or Co-Na-Na-Co) the spin density is not localized, but equally distributed among the magnetic centers. If the symmetry is very strongly broken (for instance Ni-Na-Co) then the two magnetic centers have so different level schemes that their respective magnetic behavior remains isolated and no

Λ process can be found to transfer the spin from the one to the other. Therefore, linear structures are still not suitable candidates. This time the selection rules are the reason. A detailed group-theory analysis reveals that due to the rotational symmetry of the structure (point groups $D_{\infty h}$ and $C_{\infty v}$ for homo- and heterodinuclear chains, respectively) no excited state can be found which is optically addressable from both initial and final state ($\Delta_u \leftrightarrow \Phi_g$ and $\Delta_g \leftrightarrow \Phi_u$ for point group $D_{\infty h}$, and $\Delta \leftrightarrow \Phi$ for point group $C_{\infty v}$), unless higher-angular momentum states are involved (*f* states), which, however, are not relevant for our 3*d* transition metals (Li et al., 2012). As it turns out, using the same atoms as magnetic centers and inducing the asymmetry through geometrical structure is the best way to keep the necessary symmetry breaking small enough.

The solution to this lies in building branched structures. If we add one more metallic atom (not necessarily magnetic) to the structure and reoptimize the geometry we typically get a distorted triangular shape, which removes the rotation axis from the symmetry elements of the structure. Then it is possible to transfer the spin as well, since the selection rules get substantially relaxd. As an example we mention the $[Ni\text{-}O(Mg)\text{-}Co]^+$ cluster for which, even a simultaneous spin flip and transfer was achieved (Fig. 11).

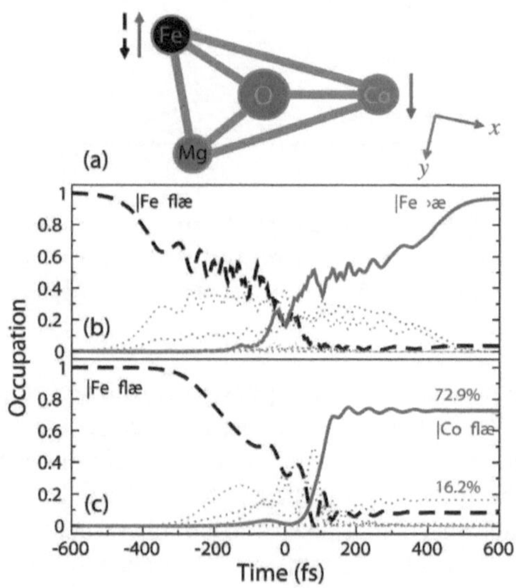

FIGURE 11 Spin flip and spin transfer processes achieved in the structure $[Fe\text{-}O(Mg)\text{-}Co]^+$. (a) Sketch of the optimized geometry with arrows indicating the initial and the final spin localization and direction for the spin flip and spin transfer processes. (b) Local spin flip at the Fe site. (c) Spin transfer from Fe to Co. (Li et al., 2011.) Copyright (2011) by American Physical Society.

4.7 PHONONIC DEGREES OF FREEDOM

In reality what one wants to ultimately achieve is coherent control at finite temperatures (if possible even at room temperature) in order to have handy spintronics applications. This inevitably imposes consideration of the molecular vibrations as well. The first substance on which we investigated this was again NiO, more specifically we analyzed the effect of the distortion of the lattice on the optical transitions possible by looking into the second order susceptibility tensor $\chi^{(2)}$ (Lefkidis and Hübner, 2006). The reason for choosing $\chi^{(2)}$ over the first order susceptibility tensor $\chi^{(1)}$ is twofold: first it is much more sensitive to small perturbations and hence it can reveal more information, and second there are ample experimental and theoretical data for NiO which allow us to validate the quality of our calculation method (as examples see de Graaf et al., 1996; Fiebig et al., 2001; Geleijns et al., 1999). In this particular case the method of choice is the complete-active-space configuration-interaction (CAS-SCF), since it correctly accounts for the static correlations needed to accurately estimate the intragap *d*-states of NiO (both 001 surface and bulk). What we do is we manually dislocate all atoms and by calculating the total energy of the cluster we derive the force matrix, the diagonalization of which yields the phononic modes. We then choose the Γ point of the lowest optical phonon and take snapshots of the second harmonic signal at different times throughout the phononic period (i.e., for different transient geometries – frozen phonon approximation). The susceptibility tensor is calculated as

$$\chi^{(2)}_{ijk} \propto \sum\nolimits_{\alpha\beta\gamma} \left[V^{i}_{\gamma\alpha} \overline{V^{j}_{\gamma\alpha} V^{k}_{\beta\alpha}} \frac{\frac{f(E_\gamma)-(E_\beta)}{E_\gamma - E_\beta - h\omega + ih\delta} - \frac{f(E_\beta)-(E_\alpha)}{E_\beta - E_\alpha - h\omega + ih\delta}}{E_\gamma - E_\alpha - 2h\omega + 2ih\delta} \right] \qquad (5)$$

where $V^{i}_{\alpha\beta}$ is the *i*-th component of the electric-dipole transition matrix element between states $|\alpha\rangle$ and $|\beta\rangle$, E_α and $f(E_\alpha)$ are the energy and population of state $|\alpha\rangle$, ω is the frequency of the probing light, and δ an empirical broadening parameter. The overbar in $\overline{V^{J}_{\alpha\beta} V^{K}_{\beta\alpha}}$ stands for symmetrization (since we do not know which of the two photons gets absorbed first). Clearly $\chi^{(2)}_{ijk}$ is a function of the phononic coordinate (q). When $q \neq 0$, which means away from the equilibrium geometry, the cluster has a lower symmetry. So while the bulk of NiO ideally has O_h symmetry in equilibrium, the point group becomes C_{4v} at the Γ point of the optical phonon (since all Ni atoms move in one direction and all O atoms in the opposite) and even C_{2v} at the *X* point. Obviously this affects the optical signal in two ways. First the degeneracy of the electronic levels gets lifted (Fig. 12) and the respective peaks in the spectra split and shift, and second it relaxes the selection rules. The latter effect is in fact more interesting than one might suspect because it is not linear. Thus one of the most exciting results is that second harmonic generation, which is symmetry-forbidden in the O_h symmetry, becomes allowed albeit weak. In a classical picture the amplitude of the phononic oscillations is temperature dependent, so more elevated temperatures render the signal stronger (Fig. 13, upper panel). It should also be noted here that integrating over the whole phononic period does not cancel out the signal (Fig. 13, lower panel).

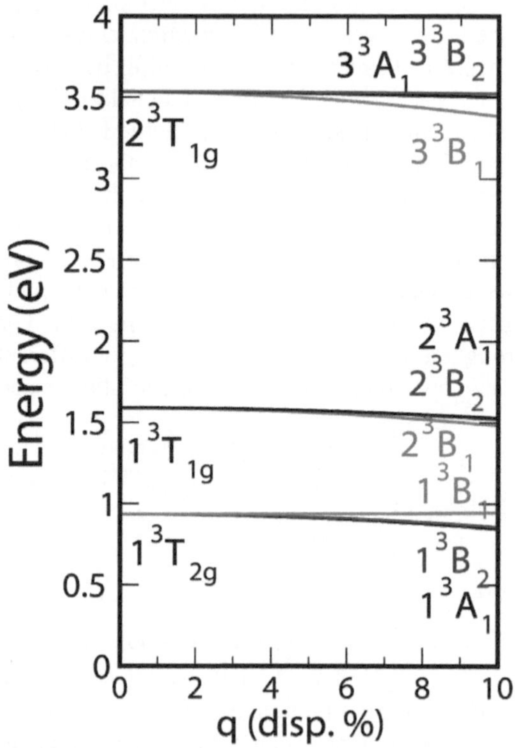

FIGURE 12 Splitting of the levels of the NiO_6^{-10} cluster vs. phononic displacement q in the case of the transversal plane wave along the z direction (C_{2v} symmetry). The splitting is an even function of q. (Lefkidis and Hübner, 2006.) Copyright (2006) by American Physical Society.

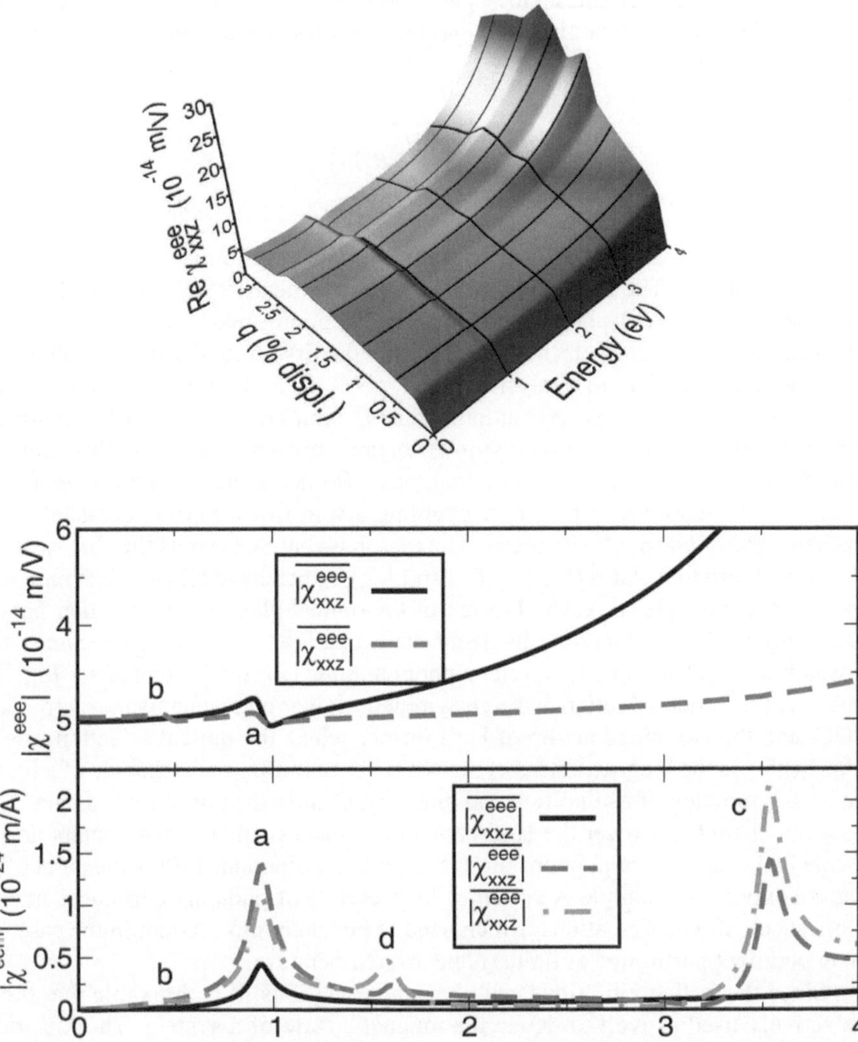

FIGURE 13 Second order susceptibility tensor. Up: $\chi^{(2)}_{xxz}$ vs. photon energy and phononic coordinate. A peak starts to appear after $q = 0.25\%$ at energy around 0.95 eV. Bottom: Non-vanishing time-averaged $\left|\chi^{(2)}_{eee}\right|$ and $\left|\chi^{(2)}_{eee}\right|$ tensor elements. Peaks a and c are split by 5 meV due to the lifting of the phononinduced electronic level degeneracy. (Lefkidis and Hübner, 2006.) Copyright (2006) by American Physical Society.

One can go further and include phononelectron scattering terms in the Hamiltonian. To do so we quantized the phononic modes and calculated the electron-phonon coupling

$$\lambda^{i}_{\alpha,\beta} = \left\langle \alpha, 0_i \left| \frac{\partial \widehat{H}}{\partial q} \right| \beta, 1_i \right\rangle \tag{6}$$

where $|\alpha\rangle$ and $|\beta\rangle$ are the electronic states, $\widehat{H}$ the Hamiltonian, and q the phononic coordinate. The phonondressed state $|\alpha, 0_i\rangle$ means electronic state α and the ground state (zero phonons) of the i-th phononic mode, while the phonon dressed state $|\beta, 1_i\rangle$ means electronic state β and the first excited state (one phonon) of mode i. We restrict our analysis to the Γ point of the first optical mode of the crystal (T_{1g} irreducible representation in the O_h point group of the rock-salt structure of NiO), since it is the most symmetric, optically active one, and thus mainly affects the selection rules of the transitions within the electric-dipole approximation. In the absence of phononic coupling a spin flip scenario necessitates an excited state of *ungerade* symmetry. The reason is that SOC splits the first excited state ($^3T_{2g}$) into four states (E_g, T_{1g}, T_{2g}, and A_{1g}), which have all *gerade* symmetry like the ground state (A_g). The Γ point of the optical phonon, on the other hand, has is *ungerade* (T_{1u}) and thus an electronic excitation (e.g., $A_g \leftrightarrow E_g$) becomes allowed if at the same time we excite a phonon mode (for the Γ point $A_g \leftrightarrow T_{1u}$). In this case the lattice vibration acts as a symmetry lowering mechanism exactly like SOC, and the combined action of both further relaxs the optical selection rules. Fig. 14 shows the magnetic state of the NiO_5^{-8} cluster after the influence of a laser pulse as a function of its duration and intensity. Clearly the possibility of spin-flip does not change, however the landscape of the successful laser parameters does. While in the absence of phonons the successful combinations follow the parabolic pattern expected in simple Λ systems, the presence of phonons introduces many more successful combinations, which need to be taken into account in the case of an experiment performed at finite (nonzero) temperature.

Since the lattice vibrations can alter the transitions, it is thinkable that they be actively used as well to detect the magnetic state of a system. The idea behind is that in order to facilitate the experimental detection of the spin state of a small cluster, one can use the vibronin/phononic coupling to the electronic/spin state and hope that changing the latter (with some coherent spin manipulation scheme) will affect the vibronic spectrum (infrared spectrum) enough to be detected.

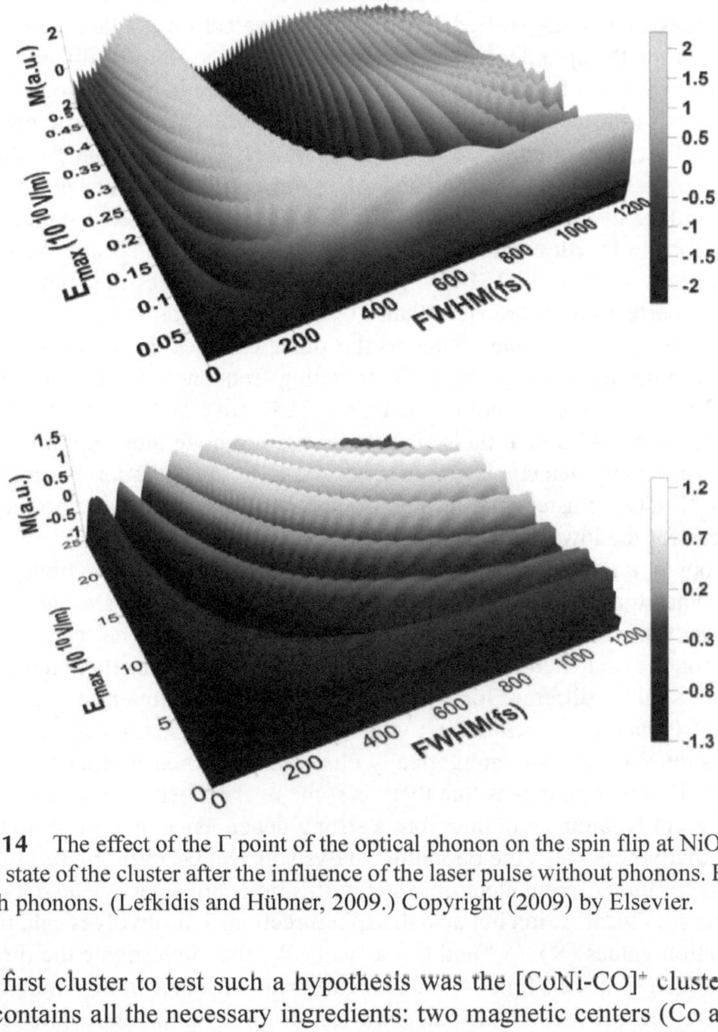

FIGURE 14 The effect of the Γ point of the optical phonon on the spin flip at NiO_5^{-8}. Up: magnetic state of the cluster after the influence of the laser pulse without phonons. Bottom: idem with phonons. (Lefkidis and Hübner, 2009.) Copyright (2009) by Elsevier.

The first cluster to test such a hypothesis was the $[CoNi\text{-}CO]^+$ cluster. This cluster contains all the necessary ingredients: two magnetic centers (Co and Ni) and a CO attached to it. The purpose of the latter is that it is very clearly detectable in infrared spectroscopy (stretch frequency, roughly in the region 1700–1900 cm^{-1}), while all the metal-metal and CO-metal frequencies lies well below. Additionally the CO stretch frequency is very sensitive to its immediate chemical environment. We calculate again the vibrational frequencies of the cluster using the force-matrix method (as above, that is we manually dislocate all atoms and recalculate all electronic energies). We find that not only the electronic states attain their energy minimum at slightly different geometries (different CO-cluster and C-O distances) but also that the curvature of the energy hypersurfaces is different. This means that within the harmonic approximation for the phonons the frequencies are slightly shifted depending on the electronic state of the system. At the same time the spin density distribution as obtained from Mulliken population analysis shows that for different electronic states the spin may be localized at different atoms (similarly to most metallic clusters with two magnetic centers, like the metallic chains mentioned above). For $[CoNi\text{-}CO]^+$ the spin for the ground electronic state is localized on Ni and the optimized C-O distance is 1.1518 Å, for the second excited state it is localized on Co and the optimized distance is 1.1509 Å, and for the fourth excited state it is again localized on Ni and the optimized distance now becomes 1.1514 Å. For higher excited states the spin localization is not so perfect any more (less than 80% is on one atom only). We notice that moving the spin from one center to the other changes the optimized distance, which, in turn, also affects the C-O stretching frequency: for the ground state it is 2252 cm^{-1}, for the second excited state 2257 cm^{-1} and for the fourth excited state 2208 cm^{-1}. Although their differences are not huge they certainly lie within an experimentally detectably frame (5 to 20 cm^{-1}). The situation is similar for the $[CoMg_2Ni\text{-}CO]^+$ cluster as well. Fig. 15 shows schematically the optimized C-O distances for the lowest electronic states.

In our opinion the most important effect of the lattice vibrations, however, occurs when the interactions between the magnetic centers are stronger, and therefore more sensitive to the geometry of the cluster. This effect has been extensively studied on the branched Ni_3Na_2 chain. The molecule contains three magnetic centers with slightly different local geometries. These are enough to break the symmetry to induce spin-localization while the electronic states with different spin localizations still remain energetically close enough to each other to allow spin-transfer. This feature means that there is some overlap between atomic orbitals of the different Ni atoms and therefore a strong dependence on their spatial separation. Exactly here kicks the phononic analysis in. We perform the full vibrational analysis for the different electronic states, this time, however, we do not only consider the spin localization but also the spin direction. This involves calculating the expectation values $\langle S_x \rangle$, $\langle S_y \rangle$ and $\langle S_z \rangle$ as well. We then investigate the direction of the localized spins as a function of the equilibrium distance of the interatomic distances (remember that this distance may differ for different electronic states) and find that there is very strong dependence. Moreover this dependence is not linear (nor gradual in any other way) but quite abrupt at some critical distances, thus creating "magnetic phases." Within each of the altogether five magnetic phases the spin direction remains always the same. Fig. 16 shows the structure and together with the five magnetic phases (Xiang et al., 2012).

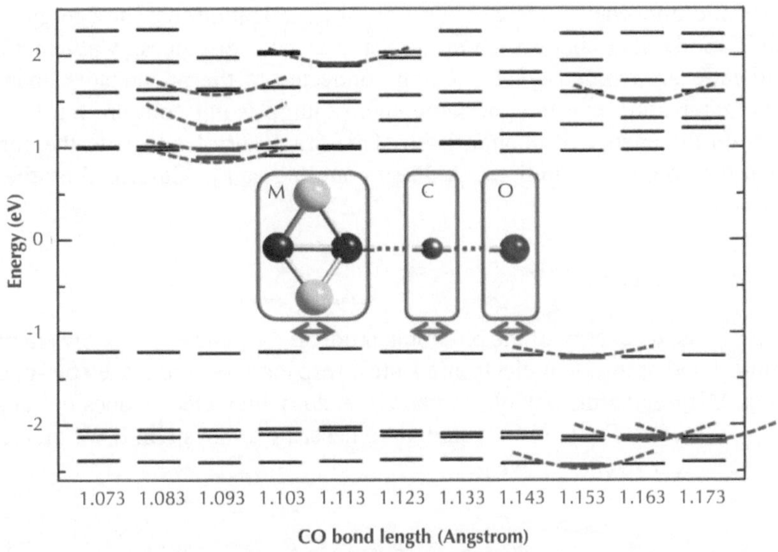

FIGURE 15 Calculated low-lying magnetic (triplet) states with respect to different CO bond lengths. The thick solid levels indicate the minimum energy of a state with respect to the CO bond lengths (the dashed curves near these levels show the minimum trends). The inset shows the separation of the cluster into three blocks for the frequency calculations. (Li et al., 2009.) Copyright (2009) by American Physical Society.

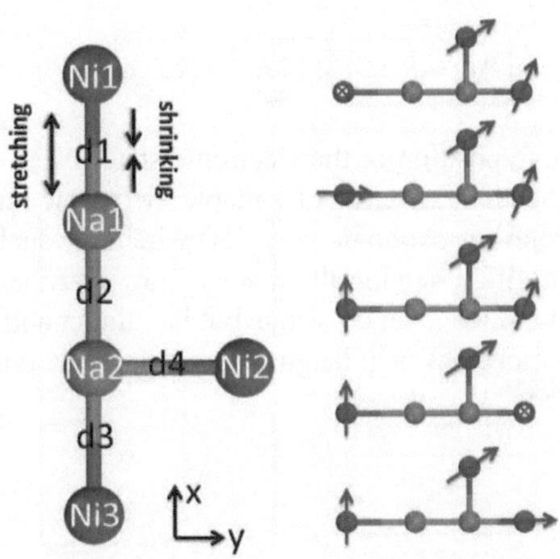

FIGURE 16 (a) The geometry structure of Ni_3Na_2 with the atoms and the interatomic distances. Stretching and shrinking is understood with respect to the original value of 3.6 Å. (b) The five magnetic phases due to the different interatomic distances (see text). The arrows indicate the spin direction when the spin density is localized on the respective atoms. (Xiang et al., 2012.) Copyright (2012) by American Physical Society.

A second important finding stems from the realization that the energy of the phononic mode also changes abruptly at the critical distances. Mathematically this indicates an avoided level-crossing connected to the phononspin coupling. It is also interesting that the phononic energy jump is not constant but strongly depends on the energy of phonon itself. This practically leads us to the conclusion that the coupling is indirect. A direct coupling can be described by the 2×2 Hamiltonian

$$\hat{H}_{\text{direct}} = \begin{pmatrix} E_{\text{phonon}} + E_{\text{spin-up}} & J \\ J & E_{\text{phonon}} + E_{\text{spin-down}} \end{pmatrix} \tag{7}$$

where E_{phonon} is the energy of the phononic mode, $E_{\text{spin-up}}$ and $E_{\text{spin-down}}$ the energy of the spin-up and spin-down electronic states, respectively, and J the spin-phonon coupling. Diagonalizing $\hat{H}$ yields a gap $E_{\text{gap}} = 2J$, which clearly does not depend on E_{phonon}, therefore Eq. (7) is not enough to describe our system. If we include an electronic state as well then we get:

$$\hat{H}_{\text{indirect}} = \begin{pmatrix} E_{\text{phonon}} + E_{\text{spin-up}} & 0 & J_1 \\ 0 & E_{\text{phonon}} + E_{\text{spin-down}} & J_2 \\ J_1 & J_2 & E_{\text{elec}} \end{pmatrix} \tag{8}$$

where E_{elec} is the energy of the mediating electronic state, and J_1 and J_2 the couplings of this state to the spin-up and spin-down states, respectively. Now the energy gap becomes

$$E_{gap} = \frac{1}{2}\left[\Delta E - \sqrt{(\Delta E)^2 + (2J_1)^2 + (2J_2)^2}\right] \tag{9}$$

and depends on the relative position of the electronic state (ΔE). This is further corroborated by investigating the existence of suitable electronic states (energetically only slightly above the relevant phononic energies) whenever such an avoided level crossing occurs. In a simplified version the results show that the spin-phonon coupling is strong enough to induce level crossings but is indirect and mediated through electronic states, or, in a more physical language, mediated via phonon-electron and electron-spin scattering (Fig. 17).

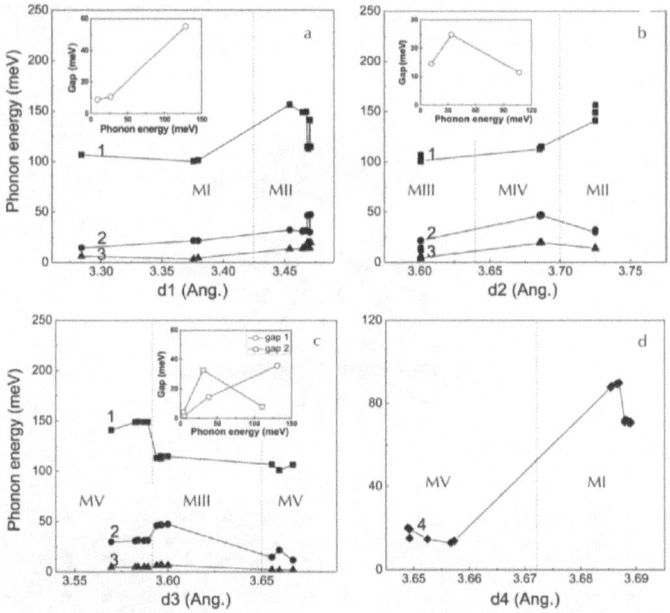

FIGURE 17 Phonon energies as a function of the equilibrium interatomic distances d1(a), d2(b), d3(c), and d4(d). For the definitions of d1, d2, d3, and d4 see Fig. 16. 1, 2, and 3 stand for the three normal modes along the x axis and 4 stands for one normal mode along the z axis. The red, dashed lines indicate the critical distances at which the magnetic phase transitions occur (see text). The insets in (a), (b), and (c) show the energy gap at the critical distances as a function of the phonon energy (center of the gap). (Xiang et al., 2012.) Copyright (2012) by American Physical Society.

4.8 SUBSTRATE

Of course it is not only the structure itself and its phononic modes that can alter the optical selection rules of our clusters. As mentioned earlier one long-term vision is to deposit functional clusters on a surface, which we would like to consider inert but, for most of the cases, are not. As an example, we refer to the deposition of Pt dimers on a Cu(001) surface (Pal et al., 2009). To model the Cu surface we use a layered cluster of 74 Cu atoms and calculated the 20 lowest many-body states with SAC-CI and a window of 534 molecular orbitals. In Fig. 18, one clearly sees that some of the molecular orbitals have a large contribution from the surface. To determine the exact adsorption geometry we also reoptimize the upper layers of the Cu cluster (surface relaxation). The energetically most favorable landing is at the aligned-hollow position at 1.8 Å. At this distance the Pt-Pt distance becomes somehow larger than the free Pt dimer (2.556 Å vs. 2.34 Å).

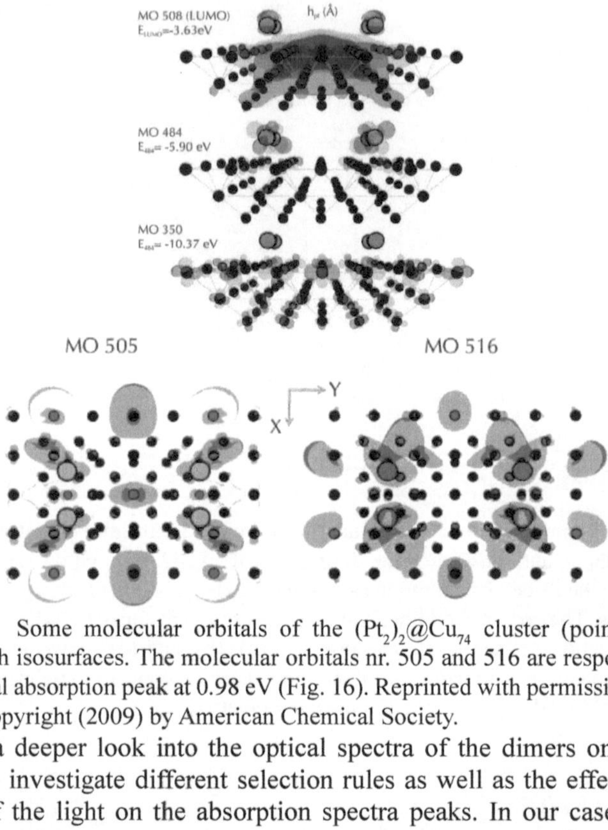

FIGURE 18 Some molecular orbitals of the $(Pt_2)_2@Cu_{74}$ cluster (point group C_{2v}), presented with isosurfaces. The molecular orbitals nr. 505 and 516 are responsible for the highest optical absorption peak at 0.98 eV (Fig. 16). Reprinted with permission from Pal et al. (2009). Copyright (2009) by American Chemical Society.

Taking a deeper look into the optical spectra of the dimers on the surface allows us to investigate different selection rules as well as the effect of the polarization of the light on the absorption spectra peaks. In our case, the transitions $A_1 \leftrightarrow A_1$ and $A_2 \leftrightarrow A_2$ require polarization of the incident light perpendicular to the surface, while for the transitions $A_1 \leftrightarrow B_1$, $A_1 \leftrightarrow B_2$, $A_2 \leftrightarrow B_1$, and $A_2 \leftrightarrow B_2$, field polarization parallel to the surface is needed. These selection rules are obviously dictated by the presence of the surface and the two dimers. If the dimers were in the gas phase and isolated then the rotational symmetry and the inversion symmetry would impose considerably stricter selection rules (*gerade* to *ungerade*) and higher-level degeneracy. Fig. 19 shows the absorption spectra starting from the ground states for different dimer-surface distances.

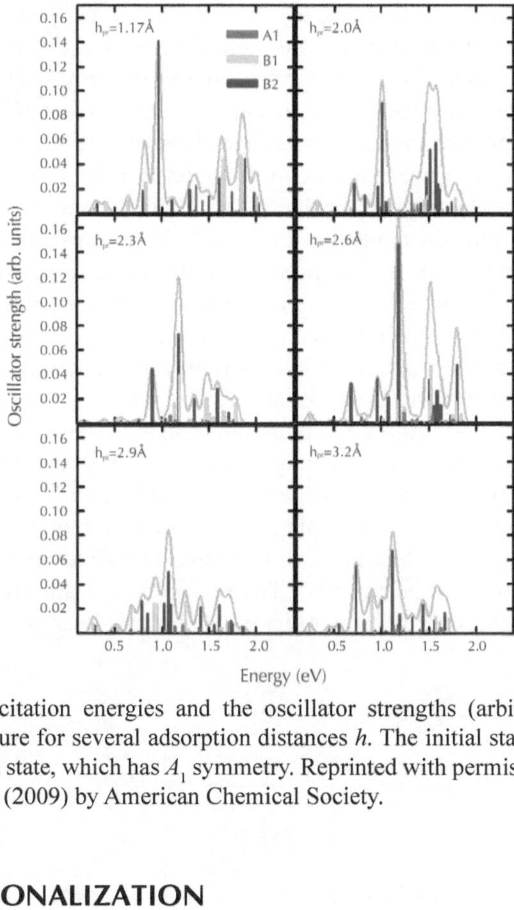

FIGURE 19 Excitation energies and the oscillator strengths (arbitrary units) of the $(Pt_2)_2$@Cu_{74} structure for several adsorption distances *h*. The initial state of the excitation is the SAC ground state, which has A_1 symmetry. Reprinted with permission from Pal et al. (2009). Copyright (2009) by American Chemical Society.

4.9 FUNCTIONALIZATION

As already mentioned in the beginning of this chapter one of the main visions of coherent, ultrafast spin manipulation is to derive spintronics devices, and if possible with logic functionalization. In other words one would like to have a structure capable of performing logic operations using spin as the information carrier instead of charge, as is the case with conventional electronic circuits. This would, among others, significantly reduce the energy involved in the operations since spins are much lighter and hence easier to move around than electrons. Of course this necessitates spin-charge separation, which is possible when dealing with higher multiplicities. Another prerequisite is the existence of at least spin-flip and spin-transfer possibilities in the same structure, as well as technical unification of these processes with similar laser pulses. Since a minimal logic gate needs three poles (two input bits and one output bit) it stands to reason to use structures with three magnetic centers. According to what we said already, it turns out that the best choice is when the centers are the same (i.e., the same atom-type, Ni, Co, or Fe) with a well-defined, low-symmetry geometry (to only slightly lift degeneracies). For the purpose of proving the concept, we chose a branched chain, which is hypothetically deposited on an inert surface. This surface serves only to stabilize the cluster and so takes over the role of the ligands in the $[Ni^{II}_2(L\text{-}N_4Me_2)(emb)]$ cluster. We choose Na atoms to bridge the magnetic centers for three reasons: (i) they are metallic but not magnetic thus facilitating spin- and charge-dynamics separation, (ii) the provide an odd number of electrons, so by varying their number we can not only adjust the length of the chain, but also ensure that the whole structure has an even number of electrons and we can so work with triplet states instead of doublets, and last but not least (iii) they are computationally cheap. Out of several similar structures the most successful proved to be the Ni_3Na_2 cluster. First of all it has a very high degree of spin-density localization (at least for the low-lying electronic states) on the magnetic centers. For two of the Ni atoms the direction of the spin is in-plane, while for the third one it is mainly out-of-plane (Fig. 20). This fact has a strong implication on the spin transfer: though it is possible to transfer the spin while keeping its direction, this has not been the case for the transfer to and from the third Ni. The reason is that this would imply a spin rotation by 90 degrees at the same time, which has not been possible.

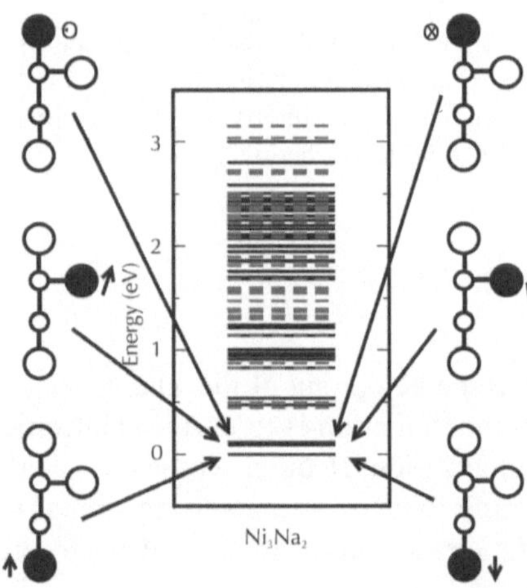

FIGURE 20 Level scheme of Ni_3Na_2 structure (including fine structure). The solid-black lines are spin triplets and the dashed red ones spin singlets. The six structures next to the level scheme show the spin localization. Large circles represent Ni atoms and small circles magnetically inert Na atoms. Solid circles indicate the spin localization of each state (arrows next to the sphere show its easy-axis direction). Note that the upper two states have the spin perpendicular to the molecular plane (*xy* plane). Hübner et al., (2009). Copyright (2009) by American Physical Society.

Searching through all possible Λ processes we find that we can realize all spin-flips (that is for every spin localization). Interestingly we were able to find two combinations, which were governed by the same laser pulse but two different orientations of the external magnetic field (Fig. 21). Thus we can use the *B* field as one of the input bits and some spin information as the second. Then we irradiate with the laser pulse (that is we perform the logic operation) and read the outcome which, depending on the new specific spin information, can be mapped to a Boolean truth table.

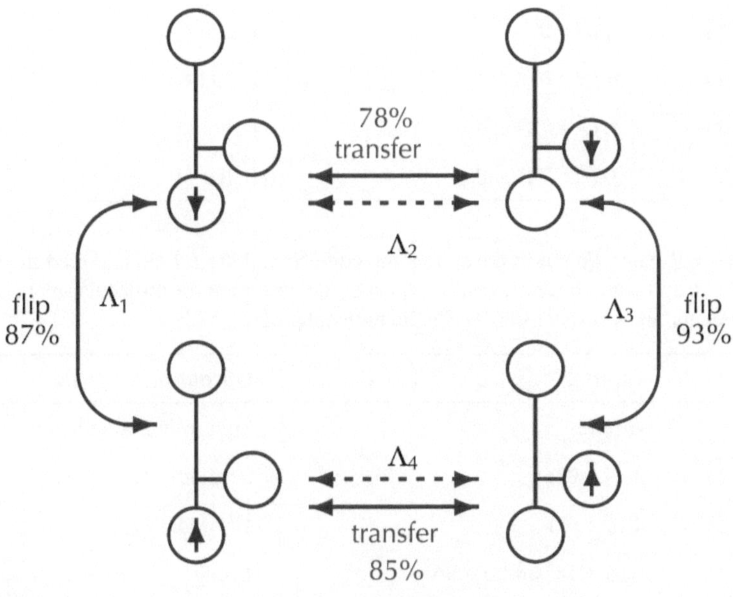

FIGURE 21 Spin flips and transfers in Ni_3Na_2. Spheres indicate the magnetic centers and arrows the localization and direction of the spin. The numbers show the fidelity of the Λ processes. All four mechanisms are possible if the *B* field has $\theta = 155°$ and $\varphi = 270°$ or $\theta = 78°$ and $\varphi = 96°$ (solid arrows). If the *B* field has $\theta = 0°$, i.e., if it is perpendicular to the cluster plane, then only transfer is possible and no switch. Thus the orientation of the static *B* field opens and closes the spin-switch channel. A *B* field along the molecule axis allows for a spin flips at the edge Ni (process Λ_1) only with a much longer laser pulse (approximately 450 fs) while spin flips at the middle Ni (process Λ_3) and transfer can be achieved with shorter pulses (<100 fs). Hübner et al., (2009). Copyright (2009) by American Physical Society.

Using the fact that some processes are possible for some experimental parameters and others not, and with a proper mapping of the spin state onto bit information we can construct several logic gates. This way we are able to construct two AND, an OR and a XOR gate (Tables 2 and 3).

TABLE 2 AND gate. We put in the spin at the edge Ni and the *B* field and read the middle Ni "up" state. θ and φ are the angles with respect to the normal of the molecular plane and the short molecular axis, respectively (*xy* is the molecular plane).

Input 1	Input 2	Output
spin	*B* field	Spin (orientation and position)
1 (edge ↑)	1 ($\theta = 0°$)	1 (middle ↑)
0 (edge ↓)	1 ($\theta = 0°$)	0 (middle ↓)
1 (edge ↑)	0 ($\theta = 78°$ and $\varphi = 96°$)	0 (edge ↑)
0 (edge ↓)	0 ($\theta = 78°$ and $\varphi = 96°$)	0 (edge ↓)

TABLE 3 OR gate. We put in the spin at the edge Ni and the *B* field and read the middle Ni up state. θ and φ are the angles with respect to the normal of the molecular plane and the short molecular axis, respectively (*xy* is the molecular plane).

Input 1	Input 2	Output
spin	*B* field	Spin (orientation and position)
1 (edge ↑)	1 ($\theta = 0°$)	0 (middle ↑)
0 (edge ↓)	1 ($\theta = 0°$)	1 (middle ↓)
1 (edge ↑)	0 ($\theta = 78°$ and $\varphi = 96°$)	1 (edge ↑)
0 (edge ↓)	0 ($\theta = 78°$ and $\varphi = 96°$)	1 (edge ↓)

Obviously the functionalization does not include only logic operators. For a hypothetical computer to work we need the logic circuits (presented above), the propagation of information (spin transfer in magnetic chains), the storing of information (spin flip), but also the ability to prepare our system in a desired state. In other words we need the ERASE functionality as well, which will drive the system from state $|\alpha\rangle$ to state $|\beta\rangle$, but leave it unaffected if it is already in state $|\beta\rangle$. All the scenarios presented till here are reversible, which means the *same* pulse always induces the opposite scenario as well, i.e., $|\alpha\rangle \rightarrow |\beta\rangle$ and $|\beta\rangle \rightarrow |\alpha\rangle$. In order to create an irreversible scenario we introduce the time-invariance-breaking element in the laser pulse itself by chirping it. More specifically we linearly vary the frequency of the laser pulse around the central one, which is the optimized value obtained with the genetic algorithm. Then we define a detuning parameter Δ_r, which gives the percentage of the detuning of the sech^2-shaped laser pulse at time when the envelope attains half its maximum. The rest of the laser parameters are left unchanged. Fig. 22 shows the effect of the detuning for chirped pulses on spin-flipping scenarios in several linear chains. The most pronounced effect is for Fe, for which a slight positive or negative detuning clearly moves the optimal frequency to one direction for the $|\uparrow\uparrow\rangle \rightarrow |\downarrow\downarrow\rangle$ process and in the opposite direction for the $|\downarrow\downarrow\rangle \rightarrow |\uparrow\uparrow\rangle$ process. The effect becomes almost negligible for Ni, while for the Co it lies somehow in-between: the curve exhibits a plateau-like behavior. After the first partial fidelity loss a stable region appears, in which Co can withstand much detuning. This seems to be in line with the stability order of the spin-flip processes established by unchirped pulses (Co>Fe>Ni) and can be attributed to the cascade-like behavior of Co (several imperfect Rabi oscillations are needed to completely transfer the population from the one spin-direction-state to the opposite one). Other structures result in the same findings as well.

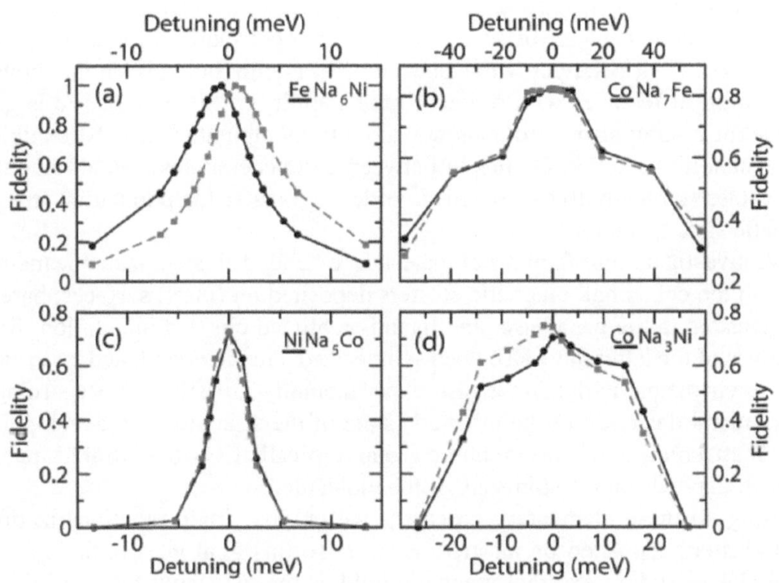

FIGURE 22 Fidelity vs. detuning (in meV) for spin-flipping Λ processes in four different magnetic molecules: (a) $FeNa_6Ni$, (b) $CoNa_7Fe$, (c) $NiNa_5Co$, and (d) $CoNa_3Ni$. The underlined elements are the atoms on which the spin density is localized and changes direction after the laser pulse. The solid black lines correspond to flipping from spin-up to spin-down, and the dashed red lines from spin-down to spin-up. Zhang et al., (2012). Copyright (2012) by American Institute of Physics.

4.10 CONCLUSIONS

In this chapter, we show that it is possible to coherently manipulate the spin degree of freedom using a subpicosecond laser pulse. In this way one can drive the magnetic state of a small cluster or an extended system to a different magnetic state through an intermediate excited state. We call the whole procedure a Λ process. Depending of the spin-density localization and spin-direction of the initial and the final (target) state, one can differentiate between spin-flip, spin-transfer and mixed spin-flip-and-transfer processes. Especially during spin-flip the light plays an important role as an angular momentum reservoir, which can be seen from the analysis of the light-induced polarization in the material. These processes can be thought of as the elementary constituents of spintronic devices. We also extend our studies to finite temperatures by including phonons/vibrons. We find that these changes the optical selection rules and hence affect the Λ processes. Furthermore, by quantizing the phonons, we find that in some cases there is a strong, phonon-spin coupling, which creates magnetic phases. The coupling is mediated through energetically higher-lying electronic states.

The magnetic centers we deal with are highly correlated $3d$ transition metals (Fe, Co, Ni). Their behavior with respect to laser manipulation is similar, however they exhibit differences with respect to the stability of the process. Fe is generally the most straightforward magnetic atom to manipulate, while Ni exhibits an almost chaotic behavior. Co lies in between and necessitates a small amount of Rabi cycles (typically five to seven) in order to transfer the population from one magnetic state to another.

We investigate four families of materials: extended clusters with one magnetic center in the cell, small magnetic clusters deposited on (inert) surfaces, bare metallic clusters in the gas phase, and ligand-stabilized clusters in solution. Representatives of the latter have also been synthesized, characterized, and, as an added value, their magnetic dynamics also experimentally investigated. Regarding the experimental detection of the magnetic state of the structures we develop a scenario of attaching some chromophore group (typically CO), the infrared spectrum of which depends on the spin state of the molecule.

Using all these elementary processes we propose logic functionalization of small clusters deposited on an inert surface. To this goal we use the spin state of the cluster and an external magnetic field as the two input bits, and the spin state of the cluster after the irradiation with an optimized laser pulse as the output bit. This way we build several Boolean-operator truth tables. We also realize the ERASE functionality by chirping the laser pulses in order to break the time inversion symmetry of the processes.

Concluding we believe that exploiting the ultrafast, laser-spin interaction opens exciting possibilities of translating new, fascinating physics to down-to-earth computational applications. These can pave the way to move magnetism from saving information to actively processing it as well.

ACKNOWLEDGMENTS

I am in debt to W. Hübner for offering me the possibility to work and investigate the fascinating field of ultrafast magnetism. I would also like to thank C. Li, G. Pal, H. P. Xiang, G. P. Zhang, and W. Jin for their long-lasting collaboration, as well as G. Katsoulos, C. A. Tsipis, J.-Y. Bigot, R. Broer-Braam, A. Kirilyuk and H. C. Schneider for many inspiring discussions over the past years (this is by far not an exhaustive list!).

Finally, I would like to acknowledge direct and indirect financial support from the German Research Foundation (among others through the Collaborative Research Center SFB/TRR 88 "3 MET" and several Individual Grants), the German Physical Society, the European Union for offering Initial Training Networks, the National Natural Science Foundation of China, as well as the Carl-Zeiss Foundation.

KEYWORDS

- **cascade-like behavior**
- **Faraday effect**
- ***gerade***
- **magnetic phases**
- **plateau-like behavior**
- ***ungerade***

REFERENCES

Allwood, D. A.; Xiong, G.; Faulkner, C. C.; Atkinson, D.; Petit, D.; Cowburn, R. P. *Science* **2005,** *309,* 1688.

Aryasetiawan F.; Gunnarsson, O. *Phys. Rev. Lett.* **1995,** *74,* 3221.

Baletto, F.; Ferrando, R. *Rev. Mod. Phy.* **2005,** *77,* 371.

Beaurepaire, E.; Merle, J.-C.; Daunois, A.; Bigot, J.-Y. *Phys. Rev. Lett.* **1996,** *76,* 4250.

Bialach, P. M.; Braun, M.; Luchow, A.; Gerhards, M. *Phys. Chem. Chem. Phys.* **2009,** *11,* 10403.

Bialach, P. M.; Funk, A.; Weiler, M.; Gerhards, M. J. *Chem. Phys.* **2010,** *133,* 194304.

Bigot, J.-Y.; Vomir, M.; Beaurepaire, E. *Nature Phys.* **2009,** *5,* 515.

Bovensiepen U. *Nature Phys.* **2009,** *5,* 461.

Cash, R. R.; Karp, A. H. *ACM Trans. Math. Software* **1990,** *16,* 201.

Chovan, J.; Kavousanaki, E. G.; Perakis, I. E. *Phys. Rev. Lett.* **2006,** *96,* 057402.

Church, G. M.; Gao, Y.; Kosuri, S. *Science* **2012,** *337,* 1628.

de Graaf, C.; Broer, R.; Nieuwport, W. C. *Chem. Phys.* **1996,** *208,* 35.

de Silva, A. P.; Uchiyama, S. *Nature Nanotechn.* **2007,** *2,* 399.

Faleev, S. V.; van Schilfgaarde, M.; Kotani, T. *Phys. Rev. Lett.* **2004,** *93,* 126406.

Fiebig, M.; Fröhlich, D.; Lottermoser, T.; Pavlov, V. V.; Pisarev, R. V.; Weber, H.-J. *Phys. Rev. Lett.* **2001,** *87,* 137292.

Geleijns, M.; de Graaf, C.; Broer, R.; Nieuwpoort, W. *Surf. Sci.* **1999,** *421,* 106.

Gómez-Abal, R.; Ney, O.; Satitkovitchai, K.; Hübner, W. *Phys. Rev. Lett.* **2004,** *92,* 227402.

Gorschlüter, A.; Merz, H. *Phys. Rev. B* **1994,** *49,* 17293.

Hartenstein, T.; Li, C.; Lefkidis, G.; Hübner, W. J. Phys. D: *Appl. Phys.* **2008,** *41,* 164006.

Hübner, W.; Kersten, S.; Lefkidis, G. *Phys. Rev. B* **2009,** *79,* 184431.

Imre, A.; Csaba, G.; Ji, L.; Orlov, A.; Bernstein, G. H.; Porod, W. *Science* **2006,** *311,* 205.

Jin, W.; Rupp, F.; Chevalier, K.; Wolf, M. M. N.; Lefkidis, G.; Krüger, H.-J.; Diller, R.; Hübner, W. *Phys. Rev. Lett.* **2012,** *109,* 267209.

Kimel, A.; Kirilyuk, A.; Usachev, P. A.; Pisarev, R. V.; Balbashov, A. M.; Rasing, Th. Nature (London) **2005,** *435,* 655.

Kirilyuk. A.; Kimel, A. V.; Rasing, T. *Rev. Mod. Phys.* **2010,** *82,* 2731.

Koopmans, B.; Ruigrok, J. J. M.; Dalla Longa, F.; de Jonge, W. J. M. *Phys. Rev. Lett.* **2005,** *95,* 267207.

Koseki, S.; Schmidt, M. W.; Gordon, M. S. *J. Phys. Chem A* **1998,** *102,* 10430.

Lefkidis, G; Zhang, G. P.; Hübner, W. *Phys. Rev. Lett.* **2009,** *103,* 217401.

Lefkidis, G.; Hübner, W. J. *Magn. Magn. Mat.* **2009,** *321,* 979.

Lefkidis, G.; Hübner, W. *Phys. Rev. B* **2006,** *74,* 155106.

Lefkidis, G.; Hübner, W. *Phys. Rev. B* **2007,** *76,* 014418.

Lefkidis, G.; Hübner, W. *Phys. Rev. Lett.* **2005,** *95,* 077401.

Lefkidis, G.; Li, C.; Pal, G.; Blug, M.; Kelm, H.; Krüger, H.-J.; Hübner, W. J. *Phys. Chem. A* **2011,** *115,* 1774.

Li, C.; Hartenstein, T.; Lefkidis, G; Hübner, W. *Phys. Rev. B* **2009,** *79,* 180413(R).

Li, C.; Jin, W.; Xiang, H. P.; Lefkidis, G.; Hübner, W. *Phys. Rev. B* **2011,** *84,* 054415.

Li, C.; Zhang, S. B.; Jin, W.; Xiang, H. P.; Lefkidis, G.; Hübner, W. J. *Magn. Magn. Mat.* **2012,** *324,* 4024.

Lichtenstein, A. I.; Katsnelson, M. I. *Phys. Rev. B* **1998,** *57,* 6884.

Meier, D; Maringer, M.; Lottermoser, Th.; Becker, P.; Bohatý, L.; Fiebig, M. *Phys. Rev. Lett.* **2009,** *102,* 107202.

Pal, G.; Lefkidis, G.; Hübner, W. J. *Phys. Chem. A*, **2009,** *113,* 12071.

Pal, G.; Lefkidis, G.; Hübner, W. *Phys. Status Solidi B*, **2010,** *247,* 1109.

Radu, I.; Woltersdorf, G.; Kiessling, M; Melnikov, A.; Bovensiepen, U.; Thiele, J.-U.; Back, C. H. *Phys. Rev. Lett.* **2009,** *102,* 117201.

Rubano, A.; Satoh, T.; Kimel, A.; Kirilyuk A.; Rasing, T.; Fiebig, M. *Phys. Rev. B* **2010,** *82,* 174431.

Satitkovitchai, K.; Pavlyukh, Y.; Hübner, W. *Phys. Rev. B* **2003,** *67,* 165413.

Sawatzky, G. A.; Allen, J. W. *Phys. Rev. Lett.* **1984,** *53,* 2339.

Schwefel, H. P. Numerical Optimization of Computer Models; John Wiley & Sons Ltd., 1981.

Sebetci, A. *Chem. Phys.* **2008,** *354,* 196.

Zhang, G. P.; Hübner, W.; Lefkidis, G.; Bai, Y.; George, T. F. *Nature Phys.* **2009,** *5,* 499.

Zhang, G. P.; Lefkidis, G.; Hübner, W.; Bai, Y. J. *Appl. Phys.* **2012,** *111,* 07C508.

CHAPTER 5

ALTERNANT CONJUGATED ORGANIC OLIGOMERS AS TEMPLATES FOR SUSTAINABLE CARBON NANOTUBE-BASED MOLECULAR NANOWIRE TECHNOLOGIES

SERGIO MANZETTI

CONTENTS

Abstract 166
5.1 Introduction 166
5.2 A Brief history of Carbon-based Conducing Polymers 167
5.3 Encapsulation of Molecules in Carbon Nanotubes 169
5.4 Organic and Sustainable Encapsulated Oligomers/Polymers as Nanowires 172
5.5 Computational Quantum Chemistry Study of Five 4*N*/4*N*+2 Oligomers 175
5.5.1 Methods 175
5.5.2 Preliminary Structure Optimization 176
5.5.3 Structure Preparation 176
5.5.4 Structure Optimization 176
5.5.5 Statistical Analysis 176
5.5.6 Wavefunction Analysis 177
5.6 Results 177
5.6.1 Wavefunction Analysis 181
5.6.2 A Lumo Symmetry Rule for Conductance 184
5.7 Conclusions 186
Acknowledgments 187
Keywords 187
References 188

ABSTRACT

Green chemistry and green nanotechnology are two exciting fields in material sciences and represent state-of-the-art disciplines in the development of advanced third and fourth-generation energy technologies (solar power, energy harvesting and advanced microelectronics). The nanomaterials and exotic materials, which are explored in this quest, provide insulating and conducting properties in nanosensors, nanounits and nanocircuits as well as advanced energy transfer and storage properties. Many solutions involve the use of metal alloys including rare and expensive metals (Gold, Platinum, Silver, Cerium and Europium) and evolve around the special energetic properties of crystals and their conductive characters. Many of these metals, such as Cadmium and Gallium-Arsenide, imply though toxicity issues and severe environmental hazards. Carbon clusters and carbon chains are on the other hand inexpensive, poorly – if not, nontoxic and have electronic properties with direct relevance to develop green nanocircuits and nanomaterials. Arranged in a specific and polymerized manner, carbon atoms can satisfy the *4n/4n+2 π-electron localization/delocalization rule,* which has the potential to be the basis for *semiconducting or conducting nanowires.* With their hybrid electronic potential, carbon-based organic polymers with *4n/4n+2* resonance can thus give unique tunable electronic properties. At the same time, these molecular ensembles have the potential to be both recyclable and biodegradable, and when encapsulated in carbon nanotubes, be physically elastic compared to metal-based nanowires and potentially provide appealing band-gap effects. This study reviews the literature on specific carbon-based conductive polymers, as well as the approaches for encapsulation of oligomers and molecules. The study also includes a brief quantum chemical analysis of five exemplary 4n/4n+2 oligomers for future development of encapsulated alternant conjugated organic structures as tunable carbon-based conductive nanowires, and shows important correlations between the oligomer chemistry, LUMO orbitals, SOMO and HOMO energies and their electronic excitability. The chapter illustrates the intriguing combination of properties of nanomaterials, the principles of organic oligomer chemistry and the method of computational quantum chemistry to provide and stimulate the knowledge of new conductive and green materials.

5.1 INTRODUCTION

Green and sustainable nanoarchitectures with conductive- and energy-storing properties are future potential components for alternative energy solutions and next generation microelectronics. The physical and energetic properties of novel conductive materials are central subjects of research and are prioritized in order to extract and increasingly transfer energy from unique atomic and electronic characteristics for microengineering and microelectronics application (Lindner et al., 2011; Wunderlich et al., 2010). Recent developments in this interesting field within nanotechnology include the regulation and reversal of photonic helicity of photocurrents with crystal insulators, Quantum Hall effects in single nanowires, and application of photoactive compounds for energy production among others (Herrero et al., 2012; Storm et al., 2012; Wunderlich et al., 2010). Material design within Nanotechnology includes also conductive materials and nanowires, which have a promizing and have a future role in third and fourth generation telecommunication and microelectronic devices (Tang et al., 2011). However, in order to stimulate this innovation further for the emphasis on green chemistry and sustainability, this chapter focuses on conductive carbon-based structures, and reviews recent techniques of encapsulation for exploiting the metallic properties of carbon, that results when carbon is organized in oligomers and polymers (Alcarazo, 2011).

5.2 A BRIEF HISTORY OF CARBON-BASED CONDUCING POLYMERS

Carbon-based conducting polymers were first explored and synthesized in 1977 by MacDiarmid, Shirakawa and co-workers (Chiang et al., 1977; Shirakawa et al., 1977). In their approach, the first electricity-conducing polymer was constructed based on polyacetylene composed of sp2 hybridized carbons with σ-bonded hydrogens, giving the possibility of π-electrons to delocalize into a band and create a metallic behavior of a carbonic polymer (Chiang et al., 1977). The structures synthesized by Shirakawa et al. (1977) were however unstable, given that the electronic gaps were too small, and became fully saturated at the π–π* transition during electron-transfer. The bond alternation in the polymers gave however semiconducting properties and various polymer-variants were tested to tune different conductivities (Chiang et al., 1977). The work by Shirakawa et al. (1977) and Chiang et al. (1977) aimed to modulate conductivity based on changing the *cis-/* and *trans*content in the polymers. The *trans*isomers showed to be more thermodynamically stable than *cis*-isomers, and the possibility of modulating conductance based on alternating *cis/trans*content was experimentally achieved by applying an earlier approach of thermal isomerization (Ito et al., 1975). These studies showed that the *trans* form of polyacetylene $(CH)_x$ had a higher conductivity (σ/Ωcm) than the *cis* conformation (Chiang et al., 1977). The studies by Chiang et al. (1977) further indicated that the polymers of both *cis* and *trans* were also highly sensitive to impurities in the dopants. For instance, large drops in conductivity were noted when polyacetylene was saturated (doped) with ammonia.

The rationale for this observation was that addition of ammonia depleted electron holes and saturated the exciton sites. The level of concentration of the dopant was also already known at that time to destroy the semiconducting properties of organic semiconductors upon full saturation (Mott, 1961), which thus inhibits the possibility of fine-tuning organic nanowires when fully engulfed in dopant.

The work on carbon-based conductive polymers resulted in wires that were produced as films of polymers, which were connected to electrical contacts using platinium wires and electroday (Ito et al., 1975). The films had the capacity to conduct enough electrical current to light a flashlight bulb when connected to two 1.5 V batteries (Chiang et al., 1977). The electrical conductivity was sustained by the use of metallic dopants such as sodium, AsF_5 and gaseous Hg, however a decrease in the conductive properties was noted after several days when exposed to air (Chiang et al., 1978). This led to the realization that effects from the surroundings can play a significant role for the properties of conducting polymers, and that protecting techniques were needed to be developed.

The role of the effects from the chemical surroundings during polymer and biopolymer synthesis and operation is extensively discussed by Yang et al. (2003). Yang and colleagues (2003) found that when they attempted to dissolve polymers in solutions in order to increase the stability of their conductive polymers towards external chemical stress, the increased stability induced twists and bends in the film resulting in breaks in the π-conjugation, and thus loss in conductance. The resulting physical properties of these polymers displayed an inhomogeneous distribution of adsorption energy, resulting in defects that slowed movement of charge carriers (electrons or Fermi holes) along the polymer relative to intact amorphous film-polymers (Yang et al., 2003). It was also found that the excitation energy was directly related to the polymer length. The longer the polymer, the higher the energy required to promote the formation of excitons. These difficulties in controlling the polymers length, orientation, uniformity and electronic qualities are thus prime challenges in the synthesis of carbon-based conducting polymers, which is especially relevant with the synthesis of carbon-based nanocomposite polymers/films, which do not have the same contiguous electronic arrangement as metals (Swol et al., 2003). Other challenges, which are difficult to solve for carbon-based polymers/oligomers, are the process of polymer infiltration in a solution phase for protective purposes, given that the most of these stable polymers are insoluble and strongly hydrophobic (Yang et al., 2003). This has however been partly resolved by applying silicon dioxide layers around the polymer/composite, which reduces permeability and leads to the oxidation of the nanowires (Yang et al., 2003). These discussed approaches and challenges have however potential solutions in *encapsulation* approaches, where the film and composite structure is replaced by single tube conducting structures, as nanowires for nano- and microelectronic circuits.

5.3 ENCAPSULATION OF MOLECULES IN CARBON NANOTUBES

Encapsulation of polymers is a method that can reduce the degree of physical and chemical stress exerted on conductive polymers/oligomers, protect their conductive properties and also give the final encapsulated carbon-molecular nanowires a higher physical elasticity and flexibility compared to that of heavy metals, such as Cadmium-based wires. Encapsulation in carbon nanotubes has been discussed in theoretical papers (Kim et al., 2005; McIntosh et al., 2003) however, it was achieved empirically several years prior to these first hypotheses on encapsulation of molecules in CNTs. The first carbon-based molecules successfully inserted (cozynthesized) in carbon nanotubes were short carbon chains, formed spontaneously as deposits from a double-anode arc-discharge process (Wang et al., 2000). These deposits were analyzed using electron microscopy, showing a majority of nanostructures to be arranged in a closed fashion, or arranged as three-dimensional graphitic sheets and as multiwalled carbon nanotubes, as also achieved by others (Warner et al., 2010). Other spontaneous-insertion/synthesis studies in CNTs had also been simultaneously published around the time of Wang et al. (2000), however, regarding crystals of potassium iodide (Meyer et al., 2000) and later followed by insertion of metal halides in CNTs (Bendall et al., 2006).

In the carbon-chain encapsulation approach, which is directly relevant to this chapter, Wang and co-workers observed that the nanowires formed spontaneously inside CNTs and organized as needle-like structures. These unique structures had a C_{n-}chain within the nanotube at a distance of 3.4Å from the carbon nanotube inner wall (Wang et al., 2000). The mechanism for growth of carbon needles was summarized by Wang et al. (2000) as applying the graphite arc-discharge in order to produce C_2 moieties, which were then linked to active sites of soot particles. During the assemblage into carbon nanotubes, a key intermediate, a half-C_{60} cage was formed, where the center would spontaneously attract carbon atoms from the discharge process. This additional atom could interact with new single carbons, thus growing the atomic chain. This mechanism assumes that the newly added atom is stabilized by electrostatic interactions with the half-C_{60} cage (Wang et al., 2000). Warner et al. (2010) used another approach and synthesized encapsulated carbon chains, linear benzene-chains, alkane-chains and polyyne chains (also accomplished by Nishide et al., 2006) inside in CNTs. The generated encapsulated chains were classified as "aggregates" encapsulated within CNTs of diameters of 2–3 nm at lengths of 1–6 nm. A similar approach was described by Zhao et al. (2003), who reported the formation of 20 nm-length carbon nanowires with approximately 100 carbon atoms in a 1D orientation inside multiwalled carbon nanotubes. Also, Sheng et al. (2012) proposed a similar method very recently for the full growth of carbon chains within multiwalled carbon nanotubes. Their method employed the synthesis of a 1D sp-hybridized carbon chain, followed by growing carbon nanotubes using a Raman scattering technique. During this study, Sheng et al. (2012) observed that carbon structures formed on the outside of the nanotube and eventually migrated into the nanotube under high temperatures, as also observed in similar experiments with structures (Koshino et al., 2007). The thermochemical effect on proper growth of polymers/oligomers in CNTs is thus pivotal, and suggested to be ideal at 1500°C (Sheng et al., 2012).

Growth of polymers, ribbons and larger molecular assemblies has also been performed very recently, with the impressive achievement of Chamberlain et al. (2012) who explored the formation of large molecular groups inside CNTs. The growth of molecular systems inside CNT was performed using tetrathiafulvalene, which yielded the formation of helix-twisted sheets encapsulated inside the CNTs. As in the study by Sheng et al. (2012), this group also applied very high temperatures during the treatment, where the thermal treatment under argon or e-beam irradiaition in vacuum purged the TTF molecules to decompose and transform into CNT-encapsulated sulfur-terminated nanoribbons.

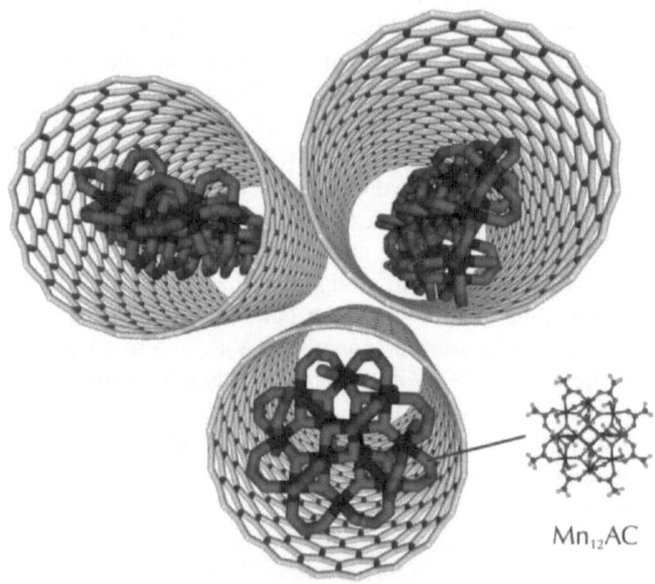

FIGURE 1 Encapsulation of small molecular magnets of $Mn_{12}Ac$ into carbon nanotubes. (Courtesy of Dr Maria del Carmen Gimenez-Lopez at the School of Chemistry, University of Nottingham.)

The insertion of smaller molecules is also of particular interest for the development of encapsulated carbon-based molecular wires or oligomers. Gimenez-Lopez et al. (2012) inserted $Mn_{12}Ac$ single molecular magnets in single or multiwalled carbon nanotubes (Fig. 1). This experimental accomplishment depended on the use of supercritical CO_2, which exerts low viscosity, high diffusivity and zero surface tension, and facilitates an easy and unrestricted insertion of the molecular magnets inside the carbon nanotube. Insertion of small carbon-based molecules in CNTs was also reported earlier by Takenobu et al. (2003). Takenobu and co. inserted organic molecules in carbon nanotubes in order to expand their performance in conductive experiments, and showed that *p*- and *n*-doping was achieved through the electrostatic interaction by the CNT to the inserted molecules with large electron affinity and small ionization potential. The results of this accomplishment facilitated field-effect nanotransistors with a novel potential of fine-tuning the electronic properties, by changing the chemical character of the inserted molecule.

The insertion process suggested by Takenobu and colleagues (2003) included encapsulation of the organic material as powder in the carbon nanotube loaded in a sealed glass tube, which was heated just above the sublimation temperature of the inserted organic molecules (anthracene, tetracene, pentacene among several tested molecules). The heating leads the molecules to enter the SWCNTs spontaneously. Once inserted in the nanotubes, Takenobu and colleagues were able to observe distinct charge-transfers between the inserted molecules and the carbon nanotube walls. The group also observed hole doping of SWCNT from one of the inserted molecules (Tetracyano-p-quinodimethane). This charge-transfer analyzes show interestingly that the devices with pristine SWCNT exhibited *p*-type operation, while SWCNT doped with tetrathiafulvalene and tetramethyltetraselenafulvalene molecules exhibited n-type operation (Takenobu et al., 2003). The charge-transfer between SWNTs and inserted organic molecules was furthermore found to be dependent on the relative ionization energy and/or the electron affinities of the inserted organic compounds.

Meunier and Sumpter (2005) showed in a separate study that the *n*-type nanotubes used in the study by Takenobu et al. (2003) experienced zero de-doping effects with the inserted organic molecule, which suggests an increased stability of the molecule@SWCNT complex. The increased stability is explained by the electrostatic forces between the inserted molecule and the inner wall of the CNT, effectively raizing the oxidation potential of the system (Meunier and Sumpter, 2005). This pivotal effect taking place between the encapsulating CNT and the encapsulated molecule/oligomer was also reported by Dinadayalane et al. (2007) who showed that the interaction energy increases with the length of the linear carbon chain encapsulated in the CNT.

The synthesis of the encapsulating materials, either CNTs or MWCNTs has also experienced dramatic changes in recent years. There are now many methods for synthesizing nanotubes in a variety of networks and arrangements. Branched and cross-linked CNTs (Liu et al., 2012) have demonstrated phononscattering effects upon thermal contact. This has the potential for making branched and cross-linked CNTs as excellent components in electronic devices. Therefore, the application of CNTs in this format can have many functions and applications for micro and nanoelectronics, and be synthesized for encapsulation purposes.

The effects of CNTs as encapsulators and protectors has also been explored extensively by theoretical means (Dinadayalane et al., 2007; Ilie et al., 2011; Kuwahara et al., 2011; Lee et al., 2011; Meunier and Sumpter 2005; Shi et al., 2007; Tran-Duc and Thamwattana, 2011). Focusing particularly on the studies working with carbon-based structures, encapsulation can occur spontaneously given the potential energy-vacuum in the CNTs, making the encapsulation energetically favorable (Kuwahara et al., 2011). Encapsulation promotes also increased stability of the π-electrons on the surface of the nanotubes from oxidative attacks and can act as nonreacting, permanent *n*- and *p*-type dopants during operation (Takenobu et al., 2003; Tran-Duc and Thamwattana 2011). A combination of hollow CNTs with metal dopants is also feasible (Zhou et al., 2010), however, metals are known to be rather difficult to use when making low carrier-density materials and the doping renders tuning and adjustment of the electronic and conductive properties more difficult, if one aims to synthesize a sensitive and tunable molecular nanowire.

The application of encapsulated metal wires instead of carbon has also been suggested (Ilie et al., 2011; Lee and Yang, 2011); however, this removes the potential of fine-tuning between conductive and semiconductive content within the molecular wire, and sustain operations with fixed conductivities. Instead, organic molecules, oligomers or polymers have the advantage of tunability at low doping levels as shown by Takenobu et al. (2003). This was also shown through computational modeling studies, where the interplay of the van der Waals forces between the encapsulated molecule and the inner wall of the CNT would create electrostatic tension (Tran-Duc and Thamwattana, 2011). In order to synthesize and encapsulated metal-like semiconductive carbon-based molecular wires, this chapter describes and suggests linear polycyclic conjugated compounds with $4n/4n+2$ alternant chemistry as candidates (Randic, 2003).

5.4 ORGANIC AND SUSTAINABLE ENCAPSULATED OLIGOMERS/POLYMERS AS NANOWIRES

The relationship between conductivity and encapsulation as discussed above requires knowledge on the (quantum) chemistry of the inserted molecular nanowire. Nanowires are conventionally regarded as nano-sized wires, which are applied as electricity-carrying components in ultramodern solar-cell technologies, nanoenergetic applications and other novel energy-harvesting technologies. Their versatility of metal-based nanowires lies in their high charge-collection and charge-transport as well as light-absorption properties through light-trapping including conductive and semiconductive properties (Beck et al., 2011; Guo et al., 2009; Tian et al., 2007). Such nanowires are conventionally applied as solid and rigid structures, sized 40–200 nm in diameter and are made by the use of Cadmium sulfide, Copper sulfide, Gallium arsenide and other metal-alloys, connected in series or parallel configurations (Tang et al., 2011).

Nanowires are therefore important components in future computing and electronic technologies, and can be engineered in complex with silicon units in nanotransistors, energy storage units and batteries, and have already been applied in present third generation energy technologies for energy harvesting (Law et al., 2005).

As mentioned above, nanowires built with Cadmium, Gallium-Arsenide and other heavy metals present environmental threats and health-risks (Tanaka, 2004) and have slow transformation degree in their life cycle and involve also health-risks during their fabrication process (Miles et al., 2005). Carbon is instead a far more environmental friendly solution than heavy-metal based nanowires, and is known from recent and older studies discussed (*vide supra)* to have a conductive nature if arranged in a proper fashion (Chiang et al., 1978; Curl and Smalley, 1988; Ijima and Ichihashi, 1993; Lee et al., 2011; Varma et al., 1991; Yang et al., 2012). Large carbon molecules and oligomers, and also many polymers are additionally biodegradable (Zhao et al., 2011; Zhang et al., 2006), and the use of carbon-based polymers with conductive properties (as shown by Chiang et al., 1977, 1978; Shirakawa et al., 1978) is therefore more environmental and health-friendly option than most metal-based structures. The use of polyacetylene strings with a *cis* or *trans* geometry (as synthesized by Chaing et al., 1977, 1978; Shirakawa et al., 1978) in an encapsulated format, induces steric clashing with the nanotube, given their nonlinear geometries.

Linear *benzenoid* molecules or polymers on the other hand, are expected to be structurally compatible with nanotubes (Fig. 2), and based on the discussed techniques mentioned above, favorable conducting effects can be achieved by encapsulating linear $4n/4n+2$ alternant benzenoid oligomer/polymers.

In this context, certain sustainable materials and polymers have already been explored for conducting purposes in context with plastic optoelectronics (Sun et al., 2006) and hydrogen storage purposes (Wang et al., 2009). However, the metallic and semiconductive nature of carbon (Alcarazo, 2011) is still not applied in a linear tunable and encapsulated format in CNTs. Benzenoid compounds with an antiaromatic/aromatic ($4n/4n+2$) chemistry, have a significantly different electron distribution than those systems with only aromatic or antiaromatic bonds individually (Clar, 1964, 1972). As it has been noted in an extensive review by Randic (Randic, 2003) and references therein, these hybrid molecules of antiaromatic/aromatic chemistry result in a change in π-electron delocalization, yielding electron localization interchanged with delocalization. The alternation of the electronic properties of aromatic/antiaromatic oligomeric or polymeric structures is thus crucial to give a tunability of the conductive properties of candidate carbon-based molecular oligomers in an encapsulated format (Fig. 2). Interestingly, reviews of the literature (Manzetti, 2012), show that the electron density and the electronic orbitals change considerably during a linear elongation of aromatic molecules (from benzene to naphthalene, further to anthracene and towards tetracene and pentacene) and also for antiaromatic/aromatic molecules, such as oligomers with $4n/4n+2$ chemistry (Fig. 3). These changes imply important variations in the electronic configuration that have to be endeavored quantum mechanically (QM).

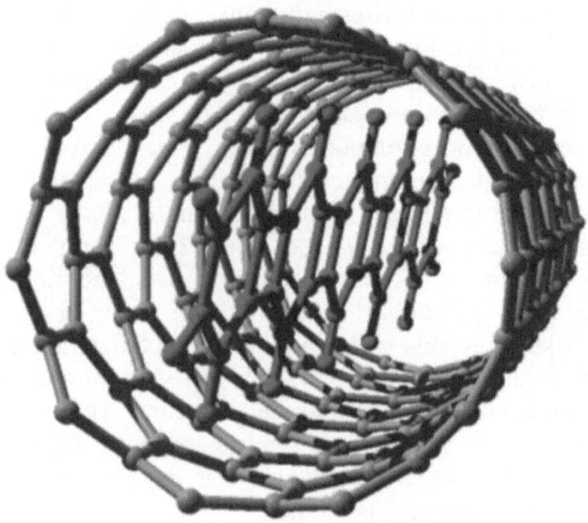

FIGURE 2 A cyclobutadiene/benzene oligomer (red) encapsulated in a carbon nanotube (gray). The model is not computationally optimized and is used for illustration of the concept of encapsulating $4n/4n+2$ chemistry in CNTs.

The electronic properties of such systems are therefore important to be revealed in context with research on encapsulated carbon-based molecular wires, and is studied here by starting with a preliminary analysis of the electronic structure of five short $4n/4n+2$ oligomers (Fig. 3). Previous studies on relevant $4n/4n+2$ systems were performed on odd membered-ring nonaltemant polycyclic hydrocarbons using the MMP2-method in quantum chemistry (McDougall et al., 1988; McWeeny et al., 1968; Pople et al., 1977), and showed a nontrivial relationship between the HOMO/LUMO energies and the orbital coefficients (Aoki et al., 1990). Aoki et al. (1990) further reported on the high electron-donating alternant chemistries where the extremely high HOMO energies and extremely low LUMO orbital energies are directly related to empirical cyclic voltammetry values (Sugihara et al., 1987). This set of quantum chemical studies show that alternating benzenoid systems (as in Fig. 3) have a valuable potential in electronic conductance. A complete electronic analysis of the selected five $4n/4n+2$ systems (Fig. 3) is therefore included in this chapter to highlight the important QM characteristics of carbon-based oligomers for encapsulated nanowire architectures.

A

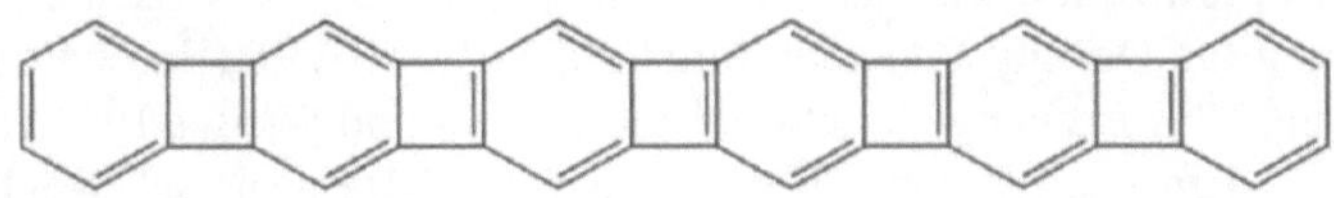

B

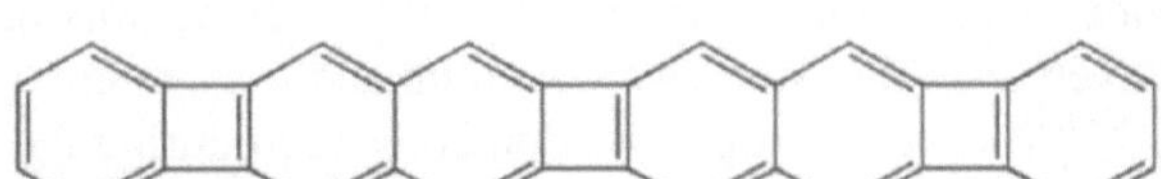

C

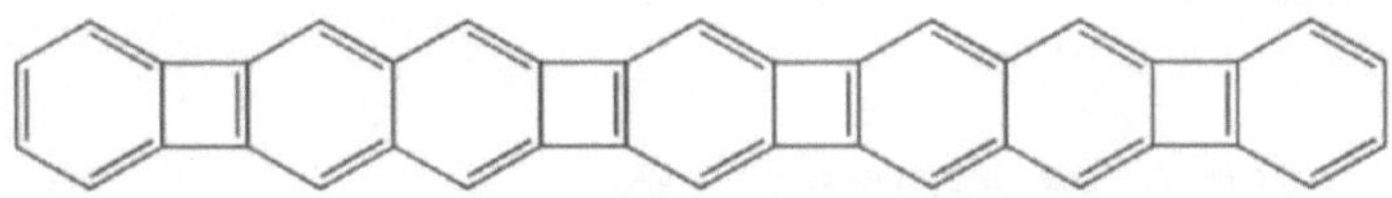

D

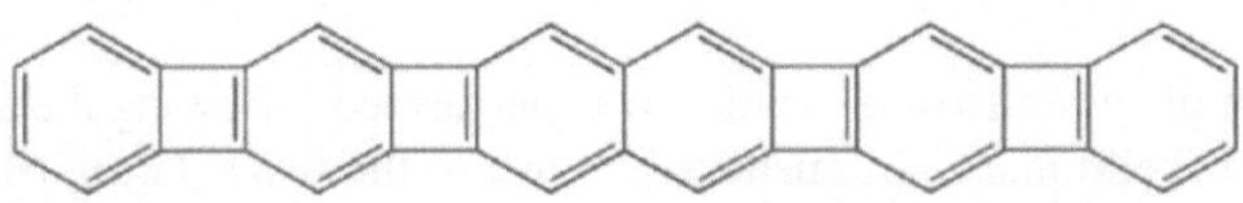

E

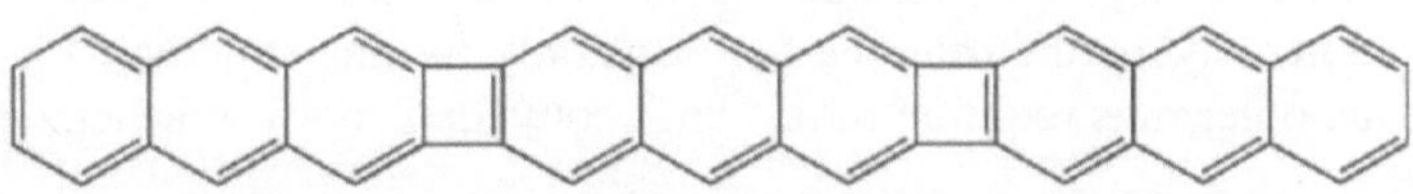

FIGURE 3 Five short $4n/4n+2$ oligomers of alternating conjugated resonance. From top: A: Repetitions of an sp2-unit with a one aromatic unit in between (1/1 ratio $4n/4n+2$); B: Repetitions of two aromatic units with one cyclobutadiene units in between (aromaticity interweaved with antiaromaticity at a 2/1 ratio); C: A repetition of 2–1-2 and 1–1-1 sequence of benzene/cyclobutadienes (7/4 ratio of π/sp2); D: 6/4 ratio π/sp2; E: 9/2 ratio of π/sp2.

5.5 COMPUTATIONAL QUANTUM CHEMISTRY STUDY OF FIVE 4N/4N+2 OLIGOMERS

5.5.1 METHODS

Five oligomer-molecules of alternating conjugation (Fig. 3) were drawn and prepared for analysis using GabEdit and Gaussview (Allouche, 2011; Frisch et al., 2003). Calculations were subdivided into three steps, (a) preliminary structure optimization, (b) a structure optimization and (c) a geometrical refining of the electronic configuration for the five neutral molecules (Fig. 3).

5.5.2 PRELIMINARY STRUCTURE OPTIMIZATION

Preliminary optimization was performed using Gaussian 03 (Frisch et al., 2003) package using the Density Functional method MPW1PW91 (Barone's Modified Perdew-Wang 1991 exchange functional and Perdew and Wang's 1991 correlation functional) (Adamo and Barone, 1991) with the 3–21G basis set (Raghavachari et al., 1980; Rassolov et al., 1998). The MPW1PW91 method was used given optimal results achieved on earlier studies on medium and large aromatic systems (Klein and Zottola, 2006). Structure optimization converged during the first runs for most of the five structures; however, a few were subjected to reruns until convergence was reached.

5.5.3 STRUCTURE PREPARATION

The second step of structure optimization was performed following the same procedure as in the preliminary optimization the use of the 6–31+G(2d, p) basis set (Raghavachari et al., 1980; Rassolov et al., 1998; Pople, et al., 1998). Structure optimization converged during the first runs only for one structure (Fig. 3a), and most of the other runs required reruns until convergence was reached.

5.5.4 STRUCTURE OPTIMIZATION

The third and last step of structure optimization was performed as above with the use of the 6–311+G(2d, p) basis set (Raghavachari et al., 1980; Pople, et al., 1998). Structure convergence was reached after calling for reruns from the CHK file with the geom=allcheck command.

5.5.5 STATISTICAL ANALYSIS

Statistical analysis performed with the software XPlot (Xplot.org), using a confidence interval of 95%.

5.5.6 WAVEFUNCTION ANALYSIS

Wavefunction analysis was performed with the program Multiwfn (Lu and Chen, 2011). The Coulomb Integral pr oligomer was also calculated using Gabedit (Allouche, 2011), which was solved by numerical method with a Schwartz cut-off of 1×10^{-8} in GabEdit in resolution, with the Poisson equation. The Coulomb Integral is calculated accordingly to the integral function:

$$C = \int \Phi(r_1)^2 \Phi(r_2)^2 (\frac{1}{r_{12}}) dr_1 dr_2 \quad (1)$$

describing the electrostatic potential integrated over the position-space wave functions (Φ) for the electrons (r_n). Although weak in such molecules, the electrostatic potential (ESP) was calculated as mentioned above, to identify potential statistical patterns of relationship between chemistry and ESP.

5.6 RESULTS

The five oligomers (Fig. 3) where primarily analyzed for their HOMO/LUMO gaps in order to elucidate their electronic excitability as also done for carbon-materials by others (Yu et al., 2011). The results show a very interesting and direct relationship between $4n/4n+2$ content and HOMO/LUMO gap width (Table 1, Fig. 4). The oligomers with the reduced aromatic content and thus increased antiaromatic content (A, D, C) have also the narrowest HOMO/LUMO gaps, suggesting the A oligomer to be the most excitable and conductive, with a HOMO/LUMO gap of only 0.6966 eV. The other oligomers increase linearly their HOMO/LUMO gaps with their aromatic content (see Fig. 4).

The regression analysis shows a weak correlation between the five oligomers, however without the 5th oligomer with the anthracene islands (oligomer E), the regression reaches a R^2=0.9932, indicating a statistically significant correlation (CI: 95%) between the chemistry of the A, B, C and D oligomers and their excitability. Still including the E oligomer, the statistical tests show all five oligomers to be within a 95% confidence band (Fig. 4). The common feature among the most statistical correlated oligomers (A–D) is that they do not have larger aromatic units than naphthalene islands, and have at least three $4n$-units (antiaromatic units). The oligomer with the lowest number of $4n$-units (cyclobutadiene) is the B oligomer, and it results as the oligomer with the largest HOMO-LUMO gap among A–D (Fig. 4, right). The presence of $4n$-units has therefore a role in decreasing the HOMO-LUMO gap, when interarranged with $4n+2$ units in such alternant conjugated linear oligomers. The elusive loss of statistical relevance of the E oligomer from the set of five oligomers is however putatively caused by its weaker and distorted 4*n*+2 chemistry by the large anthracene islands, causing a different electron localization and changing the morphology of the LUMO orbitals of the 4*n* units (when compared to the other four oligomers, Figs. 6, and 7). This is in tune with the findings by Yang et al. (2004) who noted the number of aromatic rings in an oligomer/polymer to be determinant for the electron localization *outside* the ring, affecting the HOMO-LUMO gap.

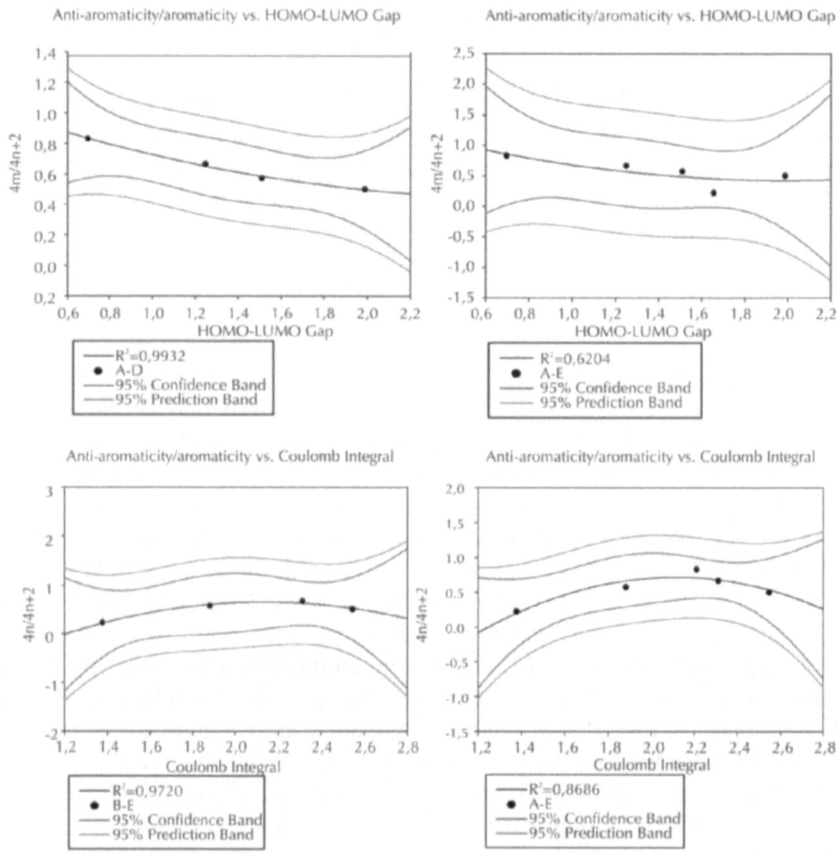

FIGURE 4 A statistical analysis of the correlation between chemistry and oligomer HOMO-LUMO gap for two sets of oligomers (at ground state). Top: Left: A–E oligomers. Right: A–D oligomers. The statistical analysis shows a quadratic correlation coefficient of 0.9932 for oligomers A–D, for 4*n*/4*n*+2 ratio and HOMO-LUMO gap, which however decreases to R^2=0.625 when oligomer E is included (right). The E oligomer is the only

oligomer with anthracene islands. Bottom: Left: Oligomer B-E; Right: Oligomers A-E. The quadratic regression line shows a weak but significant correlation (R^2=0.8562) of the chemistry of all oligomers with their Coulomb Integral, which increases with the removal of the A oligomer to 0.9720. The A oligomer has no naphthalene nor anthracene islands. The lowest Coulomb value is assigned to the E oligomer which as the lowest number of 4*n* units, and largest number of 4*n*+2 systems (anthracene).

TABLE 1 Electronic properties of A–E oligomers (Ranked by excitability). States: G–S; Ground State, (+): Cation; (–): Anion. $\Delta\gamma$ = |[SOMO-LUMO Gap(+)]| – |[SOMO-LUMO Gap(—)]|. 4*n*/4*n*+2: ratio between the counted number of cyclobutadiene and benzene rings. All units in eV, except 4*n*/4*n*+2. Red highlights the most excitable oligomer. SOMO: Single Occupied Molecular Orbital. HOMO: High Occupied Molecular Orbital. LUMO: Low Unoccupied Molecular Orbital.

Polymer	HOMO-LUMO (G–S)	SOMO-LUMO (+)	SOMO-LUMO (—)	$\lvert\Delta\gamma\rvert$	4n/4n+2
A	0.6966	0.4163	0.4761	0.0598	0.8333
D	1.2490	1.1047	0.3864	0.7183	0.6667
C	1.5102	1.2571	0.4217	0.8353	0.5714
E	1.6553	1.6272	0.4952	1.1319	0.2222
B	1.9891	1.7932	0.1278	1.6653	0.5000

Interestingly, the electronic properties at ground state reflect also a quadratic statistical relationship with the difference between the occupied and unoccupied orbitals of the cation and anions (Table 1, Fig. 5). The statistical analysis of the ionic forms of the oligomers shows that there is a significant statistical relationship within a confidence interval of 95% of R^2=0.9841 between $\lvert\Delta\gamma\rvert$ (the difference between the respective SOMO-LUMO values of the cation and anions of the oligomers) and their HOMO-LUMO gaps at ground state. Interestingly, this adds an additional factor to the properties of conducing oligomers of 4*n*/4*n*+2 chemistry, which is the narrow difference between the gap-energies of their cation and anion states. During a conductive state, the traveling electron generates anion and cation species of the oligomers, and the energy potential between these two states during a conductive operation has thus to be as minimal as possible, in order to facilitate electron transfer across the oligomer (see Fig. 5).

The observed pattern of a relationship between 4*n* and 4*n*+2 chemistry in the five oligomers shows thus a reduced conductivity (excitability) with increasing aromatic content, and vice-versa: an induced conductivity by increasing antiaromatic content (Fig. 4). The relationship between HOMO/LUMO gap and electric conductivity is known (Sun et al., 2010), and a better conductive effect is expected to be exerted by the oligomers with the lowest HOMO-LUMO gaps in particular, compared to larger HOMO-LUMO gaps. This is expected to rely therefore on the presence of the 4*n* units, in the listed alternant conjugated systems.

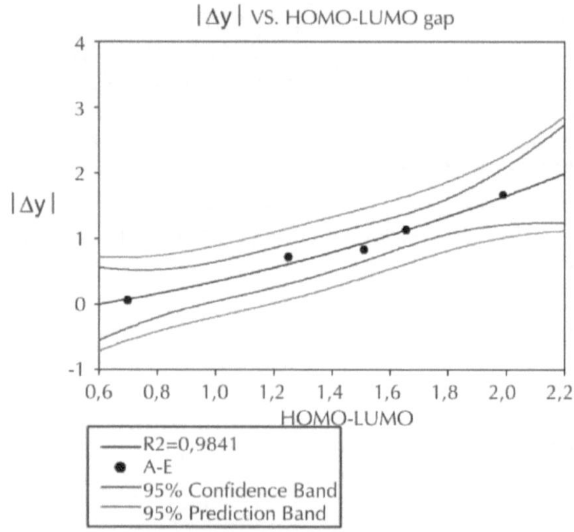

FIGURE 5 The correlation between anion and cation states and ground state. The statistical correlation plot of R^2=0.9841 depicts the crucial relationship of the absolute value of the difference in the SOMO-LUMO gaps between the cation and the anion forms of the oligomers (*y*-axis) with the HOMO-LUMO gaps of their respective ground state configuration (see Table 1 caption for further clarification on $|\Delta\gamma|$). The statistical analysis shows that the most conductive oligomers have also the lowest energetic expense (for the traveling electron) between a cation and anion state. The energetic conversion from anion and cation state during a single electron transfer requires thus less energy for the most conductive oligomers.

To further investigate the electronic chemistry of these alternant systems for encapsulating and conductive purposes by QM, a Coulomb integral analysis was performed in order to identify additional effects from the 4*n*/4*n*+2 ratios in the A–E oligomers on their Coulomb properties based on their partial charges. The relationship between the Coulomb integral potential and the oligomers is reported in Table 2 and Fig. 4. The regression analysis of the 4*n*/4*n*+2 content and the Coulomb integral showed a slightly reduced statistical correlation compared with the HOMO/LUMO gap regression (Fig. 4), with however a strong quadratic statistical regression of R^2=0.9720 for oligomers B–E, reduced to 0.8686 when the A-oligomer is included.

This difference between the correlation of the partial charges and $4n/4n+2$ chemistry (Fig. 4, bottom) and the equivalent correlation of the HOMO-LUMO gaps (Fig. 4, top) suggests that the morphology of the wavefunction (thus the arrangement of electron densities and Fermi holes) can be considerably different among the A–E oligomers and affect the subdivision of the Coulomb integral over space more than it affects the localized energy of the electronic configurations (HOMO-LUMO properties). This can therefore be the cause of the reduced correlation between a crude chemistry ratio ($4n/4n+2$) and the Coulomb integration method, which still shows a predominance of Coulomb Integrals > 2 for the most electronic excitable oligomers (Table 2).

Accounting for the Coulomb analysis (Fig. 4), the HOMO-LUMO gap correlation to chemistry (Fig. 4) and the particularly intriguing SOMO-LUMO differential analysis between anion and cation states (Fig. 5), the most conductive oligomers can thus be summarized in decreasing sequence as A>D>C>E>B (Tables 1, 2; Figs. 4, 5).

TABLE 2 Coulomb Integral (CI) of A-E oligomers. **G–S**: Ground State; **CI(+)**: Cation; **CI(–)**: Anion; **$4n/4n+2$**: ratio between the counted number of cyclobutadiene and benzene rings. All units in eV, except $4n/4n+2$, which is unit-less.

Polymer	CI (G–S)	CI (+)	CI (–)	$4n/4n+2$
A	2.2104	2.2212	1.9296	0.8333
B	2.5452	2.6028	2.3112	0.5000
C	1.8816	0.1806	1.6800	0.5714
D	2.3112	2.3292	2.3232	0.6667
E	1.3770	1.2420	1.2420	0.2222

In summary, the statistical analyzes show that the method of relating $4n/4n+2$ content to either HOMO-LUMO gap, SOMO-LUMO differences between cations and anions, or Coulomb properties yield a predictive power (within 95%) for developing algorithms for tuning $4n$ and $4n+2$ content with desired HOMO-LUMO gaps, *at the oligomer level.*

5.6.1 WAVEFUNCTION ANALYSIS

For a full understanding of the wavefunction properties of A–E oligomers, a topology analysis of their electronic structures is necessary. In a topology analysis, the critical points (CP) give an interpretation of the approximated position of where the gradient of the electron localization function is zero (Bader, 1990; Silvi and Savin, 1994). In respect to the structures A–E, one group of critical points will be considered here: (3, +1) points. (3, +1) points represent regions where two of three eigenvalues are positive, and appear often at the center of ring-systems, displaying steric- and ring-effects such as aromaticity and antiaromaticity (Fuentealba et al., 2007). (3, +1) points are known as ring-critical points (RCP) (Matta and Boyd, 2007). The properties of ring-critical points are particularly useful when studying intrinsic relationships in a molecule (McDougall et al., 1989).

The ring critical points (RCP) are analyzed and reported with particular emphasis on the total electron density (ED) and the electron localization function (ELF), which are pivotal properties to study the electronic subdivision among and within antiaromatic and aromatic units (Bader, 1990; Bader et al., 1996; Fuentealba et al., 2007; MacDougall et al., 1989; Poater et al., 2005). The relationship between HOMO/LUMO gaps of the oligomers and the *band* gaps of their polymerized states has been suggested by others to be highly interdependent, and to be tunable through changes in chemistry and the relationship to the monomer units (Chattopadhyaya et al., 2012; Yang et al., 2004). With focus on the alternant conjugated benzenoid systems (Fig. 3), an ELF/ED analysis of their wavefunction topology *with particular emphasis on their central and peri-central sections*, and not their absolute extremities, can therefore give a useful picture for further conductive modeling and analyzes. The ELF/ED analysis (Fig. 6) shows that the sp2-units (4*n* systems – cyclobutadiene) are gathering electrons and have considerable higher electron density and lower localization functions (ELF) than aromatic units.

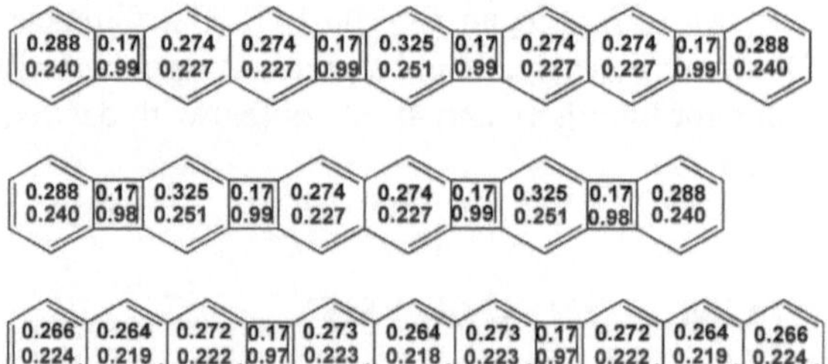

FIGURE 6 The relationship between 4*n*/4*n*+2 chemistry, ELF and Electron density (ED). ELF: Black color (top); ED: Blue color (bottom) for each ring. The smaller electron density pr. aromatic ring for larger aromatic islands is visible from the relationship: benzene-islands $_{\text{ED pr. ring}}$ (A, C, D) > naphthalene islands $_{\text{ED pr. ring}}$ (B, C, D) > anthracene islands $_{\text{ED pr. ring}}$ (E). A possible pattern of decreasing ED for the cyclobutadiene units with increasing size of aromatic islands occurs (compare E, C and A).

With a lower ELF value, the probability of confining the electrons into shells is reduced (Fuentealba et al., 2007). In a more extensive description, ELF is associated with the probability density of finding a second same-spin electron at or near the reference point (Steinmann et al., 2011). Thus, the ELF values of the $4n$ systems compared to the $4n+2$ systems show that the $4n$ systems have fewer same-spin electrons in their ensembles and that these are less confined to shells.

In combination with the HOMO/LUMO analysis and correlation with the $4n/4n+2$ content, it is therefore feasible to propose that a high number of $4n$ systems give a higher excitability by generating a higher number of electron-dense units with a higher abundance of opposite-spin electrons that are not confined to shells. Accounting for the direct relationship between oligomers and polymers (Chattopadhyaya et al., 2012), a conductive effect is expected to be represented principally by the $4n$ units in these linear alternant conjugated systems (Figs. 4, and 6). Thus, carbon-based molecular nanowires that are extrapolated from such oligomers should nevertheless be based on units that have high-localized electron density ($4n$ systems) and low localization of these electrons into shells (low ELF values) in order to give better conductive properties. The $4n+2$ (aromatic) units have lower electron densities pr ring, and higher ELF functions. The $4n+2$ units are therefore expected to provide weaker conductive properties than 4n units, by simple considerations of the ELF function and lower electron density properties. *Semi*-conductance and insulating effects can thus be included when tuning using the $4n+2$ chemistries, which give a higher number of shells and occurrence of same-spin electrons in the $4n+2$ units.

Interestingly, the $4n+2$ units are found to lose electron density (ED) by their monomer-multiplication across the oligomers (benzene islands ED pr. ring > naphthalene islands ED pr. ring > anthracene islands ED pr. ring). This emphasizes anticonductive properties of increasing 4n+2 systems, by their fewer and less dense electron holes, and can also explain the regression line disruption by oligomer E (Fig. 6). Also, the ELF functions of each ring system decrease upon multiplication of the aromatic monomers: Anthracene $_{\text{ELF pr ring}}$ < Napthalene $_{\text{ELF pr ring}}$ < Benzene $_{\text{ELF}}$. This implies that the multiplication of aromaticity over contiguous regions (long $4n+2$ moieties) induces a lower probability of finding same-spin electrons into shells, which can have unexpected effects at the polymerized level. In a CNT-encapsulated state, this property can putatively create regions with lower electrostatic interchange between the molecular wire and the intraface of the carbon nanotube, considering the conclusions of Dinadayalane et al. (2007), and thus furthermore affect conductivity.

Interestingly, the electron density of $4n$ systems is reduced (0.97) when found in between anthracene islands (E) compared to in between naphthalene islands (0.99) (see A and D). There is also an observed reduction in electron density in $4n$ units found 1 or 2 rings from the extremities (see oligomer A). The behavior of the wavefunction of cyclobutadiene units can therefore require further studies, however, at the current level of theory, it suggests that the elongated aromaticity in the oligomeric ensembles reduces the electron-attracting properties of $4n$-units, putatively creating a more homomorphic conductive configuration when $4n+2$ units are larger than naphtalene units ($n>2$).

5.6.2 A LUMO SYMMETRY RULE FOR CONDUCTANCE

During a conductive operation, the LUMO orbitals are expected to play a central role, by hosting the traveling electrons across the oligomer or the polymerized state. A series of interesting recent studies have emerged from Yoshizakawa, Tada and colleagues (Yoshizakawa et al., 2008; Yoshizakawa, 2012) where they have rationalized the path of travel of electrons in between orbital and energy levels. Their research has led to the development of an orbital rule for electron transport, which states that there are favorable and forbidden paths for an electron in order to travel through a molecule. In this context, an interesting relationship between the morphology of the LUMO orbitals of the five oligomers and their excitability was found which is further elucidated in a subsequent study (Manzetti and Lu, 2013) (Fig. 7).

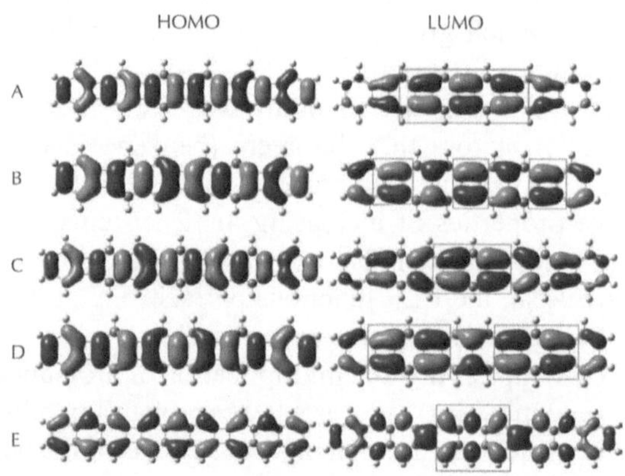

FIGURE 7 The HOMO and LUMO orbitals of the five oligomers. Accounting for the excitability values reported in Fig. 4, it is interesting how the LUMO orbitals are shaped. By the sequence of excitability from high to low, A, D, C, B and E have the LUMO elements (red squared around) with similar symmetries adjacent or separated from one another. In decreasing sequence of the oligomers excitability and their LUMO components: Excitability A (3 joined coherent pairs of homomorphic elements) > Excitability D (2 separated pairs of homomorphic elements) > Excitability C (1 pair of homomorphic elements) > Excitability E (1 triple small joined homomorphic elements) > Excitability B (3 separated single pairs of homomorphic elements).

The visual analysis shows an intriguing relationship between the excitability levels and the number of homologue/heterologue symmetries within the LUMO profile and their joined or separated arrangement (Fig. 7). The profiling of LUMO orbitals indicates that the most conductive LUMO elements are expected to be aligned and arranged in a symmetric fashion (Ψ+ / Ψ–), where the center of the A oligomer illustrates this in particular. In this symmetric order, the pairs of LUMO orbitals are arranged in an improper symmetry hence to one another (1-fold rotation over their central x-axis, followed by a vertical inversion). This repeated symmetric pattern is less conserved in the other oligomers, and is rather replaced by increasingly disjoined or heteromorphic symmetries (Fig. 7). The interesting feature of this putative correlation is that it can aid in tailoring molecular conductive properties by the analysis of the improper arrangement of LUMO entities (as an additional computational method to *band*-gap calculations). Although conductance is not only allocated to LUMO, but involves also charge-hopping transport, where the electron travels either from LUMO to LUMO, or from or hole to hole (HOMO to HOMO), as well as between hole and LUMO during transport, the putative correlation of improper symmetric arrangements of the LUMO spatial distributions in oligomers may add to the methodology of estimating conductive properties of single molecules, oligomers and potentially to single-chain polymers. This rule suggests that the following relationship among LUMO entities respects the given criteria for higher conductance:

$$(\text{LUMO}_i, \text{LUMO}_j) \perp (\text{LUMO}_k, \text{LUMO}_l)$$

The symmetric definition shown in this rule defines that each side of the symmetry operation must include LUMOs of inverse sign to one another. For instance, the wavefunction sign of LUMO_i is inverse to the wavefunction sign of $\text{LUMO}_{j.}$, and the pair LUMO_i, LUMO_j is homomorphic to the pair LUMO_k and $\text{LUMO}_{l,}$ thus as similar as possible (if not identical) when a symmetry operation of vertical inversion is performed (Fig. 8). The reduction in conductance of the given oligomers shows both a loss of improper symmetry, homomorphism and contiguous pattern (without interruptions by alternative shapes and wavefunction signs).

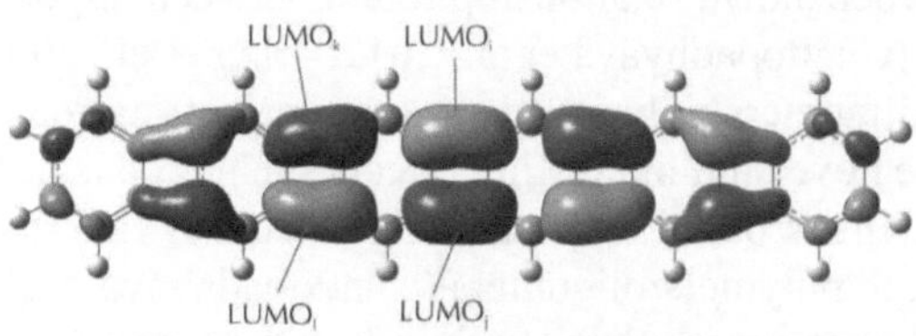

FIGURE 8 Symmetry rule for conductive oligomers of $4n/4n+2$ chemistry. The LUMO pairs are arranged in an improper symmetric fashion to one another, where both wavefunction signs and shape/size are relevant. By observing the application of this rule to the LUMOs in Fig. 7, a violation of this rule is increasingly observed in the less conductive oligomers: either interrupted by alternative shapes and sizes, or alternative wavefunction sign symmetries (see oligomer E wavefunction symmetry).

In order to estimate conductance of the molecular and polymeric extensions of such systems a mapping of decoherence and relaxation as well as spin-orbits and spin-polarization is required (Hammar et al., 1999; Potok et al., 2002) (which are all factors that will not be elucidated here). Methods for further studying such systems with computational algorithms have also been recently suggested by Yang and colleagues (2004) in particular however this points to the following subjects of further research:

- Mapping the relationship between the geometries and symmetries and eventual degeneracies of the LUMO orbitals of $4n/4n+2$ systems and their conductivity properties.
- How these properties can be used to develop molecular electronics approaches to estimate the relationship between the electronic properties of $4n/4n+2$ oligomers and the conductivity of their respective polymers, for future algorithms and computational applications.
- How encapsulation of $4n/4n+2$ oligomers/polymers in CNTs can improve their conductive properties for green and carbon based conductive molecular nanowires.

5.7 CONCLUSIONS

This review reports on the development of encapsulation and molecular insertion procedures in CNTs, as well as background from the early conductive carbon-based polymers. This chapter also includes an initial quantum chemical study on the electronic properties of five $4n/4n+2$ alternant conjugated oligomers, which are candidate templates for further molecular electronics studies (Manzetti and Lu, 2013). The reported quantum chemical analysis suggest that the $4n$ content in the $4n/4n+2$ alternant conjugated oligomers is the responsible part that gives narrow HOMO/LUMO gaps and is thus expected to be the main component to give conductive properties in their respective *polymers*. This statement is supported by the interdependent relationship found between oligomers and polymers in similar studies (Chattopadhyaya et al., 2012; Yang et al., 2004). Encapsulation of such $4n/4n+2$ oligomers/polymers is also an interesting route of research, and is expected to give novel and interesting effects such as considerable modulations of HOMO/LUMO gaps based on π-stacking effects as reported from studies of stacked single-chain polymers/oligomers (Chattopadhyaya et al., 2012) and from π–π effects from larger encapsulated carbon-based structures (Lee et al., 2002).

The initial analysis presented herein shows that aromatic units in the oligomeric structures make a base for expanding HOMO/LUMO gaps, and can tune the semiconductance and resistivity in final full-scale polymers. Alternating the

$4n/4n+2$ chemistries at different ratios can therefore give various semiconductance properties and become a base to develop novel carbon-based conductive and semiconductive materials for microelectronic technologies.

Extensive studies are further required to understand and construct semiconductive polymers for novel microengineering and nanotechnology applications that are entirely based on carbon. The use of $4n/4n+2$ chemistry can prove valuable in replacing metal-dependent nanowire technologies, through their protective encapsulation in CNTs, and also through the higher elasticity compared to rigid metal-nanowires.

The combination of $4n/4n+2$ nanowires@CNTs is therefore a potentially promizing direction for future microelectronics, processor technologies and a potential paradigm-shift in the field of microelectronics and engineering.

ACKNOWLEDGMENTS

The author would like to thank Dr M. Zottola for his participation on the $4n/4n+2$ structures. The author would also like to thank Prof. Johan Liu and Dr. Murali Murugesan at BioNano Systems Laboratory, Department of Microtechnology and Nanoscience, Chalmers University of Technology, Sweden, and Prof David van der Spoel at Computational Systems and Biology, Biomedical Centre, Uppsala University, Sweden for the guest-researcher position at the institute. Finally, the author would like to thank Dr Maria del Carmen Gimenez-Lopez at the School of Chemistry, University of Nottingham for kindly supplying illustration of insertion of small molecular magnets in CNTs.

KEYWORDS

- **4n/4n+2 structures**
- **carbon nanotubes**
- **carbon-based polymers**
- **conducting nanowires**
- **delocalization rule**
- **encapsulation**
- **HOMO/LUMO gaps**
- **linear benzenoid molecules**

REFERENCES

Adamo, C.; Barone, V. Exchange functionals with improved long-range behavior and adiabatic connection methods without adjustable parameters: The mPW and mPW1PW models. *J. Chem. Phys.* **1998,** *108,* 664–675.

Alcarazo, M. On the metallic nature of carbon in allenes and heterocumulenes. *Dalton Trans.* **2011,** *40,* 1839–1845.

Allouche, A. R. Gabedit – A graphical user interface for computational chemistry software. *J. Comp Chem,* **2011,** *32,* 174–182.

Aoki, Y.; Imamura, A.; Murata, I. A molecular orbital study on the electron donating and accepting abilities of alternant polycyclic conjugated hydrocarbons. *Tetrahedron* **1990,** *46,* 6659–6672.

Bader, R. F. W. Atoms in Molecules. In: *A Quantum Theory*, Clarendon Press, Oxford, 1990.

Bader, R. F. W.; Johnson, S.; Tang, T. H.; Popelier, P. L. A. The electron pair. *J. Phys Chem,* **1996,** *100,* 15398.

Bendall, J. S.; Ilie, A.; Welland, M. E.; Sloan, J.; Green, M. L. Thermal stability and reactivity of metal halide filled single-walled carbon nanotubes. *J. Phys. Chem. B.* **2006,** *110,* 6569–6573.

Chamberlain, T. W.; Biskupek, J.; Rance, G. A.; Chuvilin, A.; Alexander, T. J.; Bichoutskaia, E.; Kaiser, U.; Khlobystov, A. N. Size, structure, and helical twist of graphene nanoribbons controlled by confinement in carbon nanotubes. *ACS Nano* **2012,** *6,* 3943–3953.

Chattopadhyaya, M.; Sen, S.; Alam M, and Chakrabarti, S. The role of relativity and dispersion controlled interchain interaction on the band gap of thiophene, selenophene, and tellurophene oligomers. *J. Chem. Phys.* **2012,** *136,* 094904.

Chiang, C. K.; Druy, M. A.; Gau, S. C.; Heeger, A. J.; Louis, E. J.; MacDiarmid, A. G.; Park, Y. Q.; Shirakawa, H. Synthesis of Highly conducting films of derivatives of polyacetylene, $(CH)_x$. *JACS,* **1978,** *100,* 1013–1015.

Chiang, C. K.; Fincher, C. B.; Park, Y. W.; Heeger, A. J.; Shirakawa, H.; Louis, E. J.; Gau, S. C.; MacDiarmid, A. G. Electrical Conductivity in Doped Polyacetylene. *Phys Rev Lett.* **1977,** *39,* 1098–1101.

Clar, E. Polycyclic aromatic hydrocarbons. Academic Press: London, 1964.

Clar, E. The Aromatic Sextet; J. Wiley & Sons: London, 1972.

Curl, R. F.; Smalley, R. E. Probing c60. *Science.* **1988,** *242,* 1017–1022.

Dinadayalane, T. C.; Gorb, L.; Simeon, T.; Dodziuk, H. Cumulative – Interaction Triggers Unusually High Stabilization of Linear Hydrocarbon Inside the Single-Walled Carbon Nanotube. *Int. J. Quantum. Chem.* **2007,** *107,* 2204–2210.

Frisch, J.; G. W. Trucks, H. B. Schlegel, G. E. Scuseria, M. A. Rob, J. R. Cheeseman, J. A. Montgomery Jr., T. Vreven, K. N. Kudin, J. C. Burant, J. M. Millam, S. S. Iyengar, J. Tomasi, V. Barone, B. Mennucci, M. Cossi, G. Scalmani, N. Rega, G. A. Petersson, H.

Nakatsuji, M. Hada, M. Ehara, K. Toyota, R. Fukuda, J. Hasegawa, M. Ishida, T. Nakajima, Y. Honda, O. Kitao, H. Nakai, M. Klene, X. Li, J. E. Knox, H. P. Hratchian, J. B. Cross, V. Bakken, C. Adamo, J. Jaramillo, R. Gomperts, R. E. Stratmann, O. Yazyev, A. J. Austin, R. Cammi, C. Pomelli, J. W. Ochterski, P. Y. Ayala, K. Morokuma, G. A. Voth, P. Salvador, J. J. Dannenberg, V. G. Zakrzewski, S. Dapprich, A. D. Daniels, M. C. Strain, O. Farkas, D. K. Malick, A. D. Rabuck, K. Raghavachari, J. B. Foresman, J. V. Ortiz, Q. Cui, A. G. Baboul, S. Clifford, J. Cioslowski, B. B. Stefanov, G. Liu, A. Liashenko, P. Piskorz, I. Komaromi, R. L. Martin, D. J. Fox, T. Keith, M. A. Al-Laham, C. Y. Peng, A. Nanayakkara, M. Challacombe, P. M. W. Gill, B. Johnson, W. Chen, M. W. Wong, C. Gonzalez, and, J. A. Pople, Gaussian 03 (Gaussian, Inc., Wallingford, CT, 2003).

Fu, Y.; Carlberg, B.; Lindahl, N.; Lindvall, N.; Bielecki, J.; Matic, A.; Song, Y.; Hu, Z.; Lai, Z.; Ye, L.; Sun, J.; Zhang, Y.; Zhang, Y.; Liu, J. Templated growth of covalently bonded three-dimensional carbon nanotube networks originated from graphene. *Adv Mater*. **2012,** *24,* 1576–1581.

Fuentealba, P.; Chamorro E, and Santos, J. C. Understanding and using the electron localization function. Chapter 5. A. Toro-Labbé (Editor). Theoretical Aspects of Chemical Reactivity. **2007.** Published by Elsevier B.V.

Giménez-López del Carmen, M.; Moro, F.; La Torre, A.; Gómez-García, C. J.; Brown PD, van Slageren, J.; Khlobystov, A. N. Encapsulation of single-molecule magnets in carbon nanotubes. *Nat Commun*. **2011,** *2,* 407.

Hammar, P. R.; Bennett, B. R.; Yang, M. J.; Johnson, M. Observation of Spin Injection at a Ferromagnet-Semiconductor Interface. *Phys. Rev. Lett.* **1999,** *83,* 203.

Herrero, C.; Costentin, C.; Aukauloo, A. Converting Photons to Electron and Proton Shifts from Water for Fuel Production. In: *Molecular Solar Fuels, RSC Energy and Environment Series* No. *5,* Eds. Wydrzynski, T. J.; Hillier, W. Royal Society of Chemistry 2012.

Ilie, A.; Bendall, J. S.; Nagaoka, K.; Egger, S.; Nakayama, T.; Crampin, S. Encapsulated inorganic nanostructures: a route to sizable modulated, noncovalent, on-tube potentials in carbon nanotubes. *ACS Nano* **2011,** *5,* 2559–2569.

Ito, T.; Shirakawa, H.; Ikeda, S. Thermal cis–trans isomerization and decomposition of polyacetylene. *Journal of Polymer Science: Polymer Chemistry Edition* **1975,** *12,* 1943–1950.

Kim, G.; Kim, Y.; Ihm, J. Encapsulation and polymerization of acetylene molecules inside a carbon nanotube. *Chem. Phys. Lett.* **2005,** *415,* 279.

Klein, R. A.; Zottola, MA. Pople versus Dunning basis-sets for group IA metal hydrides and some other second row hydrides: The case against a De Facto standard. *Chem. Phys. Lett.* **2006,** *419,* 254–258.

Koshino, M.; Tanaka, T.; Solin, N.; Suenaga, K.; Isobe, H.; Nakamura, E. Imaging of single organic molecules in motion. *Science*. **2007,** *316,* 853.

Kuwahara, R.; Kudo, Y.; Morisato, T.; Ohno, K. Encapsulation of carbon chain molecules in single-walled carbon nanotubes. *J Phys Chem A*. **2011,** *115,* 5147–5156.

Law, M.; Greene, L. E.; Johnson, J. C.; Saykally, R.; Yang, P. Nanowire dye-sensitized solar cells. *Nat. Mat.* **2005,** *4,* 455–459.

Lee, C. H.; Yang, C. K. Structural and Electronic Properties of Bismuth and Lead Nanowires Inside Carbon Nanotubes. *J. Phys. Chem. C* **2011,** *115,* 10524.

Lee, J.; Kim, H.; Kahng, S. J.; Kim, G.; Son, Y. W.; Ihm, J.; Kato, H.; Wang, Z. W.; Okazaki, T.; Shinohara, H.; Kuk, Y. Bandgap modulation of carbon nanotubes by encapsulated metallofullerenes. *Nature* **2002,** *415,* 1005–108.

Lee, S. H.; Teng, C. C.; Ma, C. C.; Wang, I. Highly transparent and conductive thin films fabricated with nano-silver/double-walled carbon nanotube composites. *J Colloid Interface Sci.* **2011,** *364,* 1–9.

Lee, S. U.; Belosludov, R. V.; Mizuseki, H.; Kawazoe, Y. Electron transport characteristics of organic molecule encapsulated carbon nanotubes. *Nanoscale.* **2011,** *3,* 1773–9.

Lindner, N. H.; Refael, G.; Galitski, V. Floquet topological insulator in semiconductor quantum wells. *Nature Phys.* **2011,** *7,* 490–495.

Lu, T.; Chen, F. Multiwfn: A multifunctional wavefunction analyzer. *J Comp Chem* **2012,** *33,* 580–592.

Manzetti, S. Quantum chemical properties of Polycyclic Aromatic Hydrocarbons: a review. pp 1–34. In: *Polycyclic Aromatic Hydrocarbons: Chemistry, Occurrence and Health Issues*. Eds. Bandeira, G. C.; Meneses, H. E. Novascience Publishers 2012–2013.

Manzetti S and Lu T. Alternant conjugated oligomers with tunable and narrow HOMO-LUMO gap as sustainable nanowires. RSC Advances, **2013**, in press. 10.1039/C3RA41572D.

Matta, C. F. Hydrogen-Hydrogen Bonding: the Non-Electrostatic Limit of Closed-Shell Interaction Between two Hydrogen Atoms (A Review). In: *Hydrogen Bonding – New Insights. Grabowski, Slawomir (Ed.).* Series: Challenges and Advances in Computational Chemistry and Physics **2006,** *3,* 337–376.

McDougall, J. J. W.; Peasley, K.; Robb, M. A. A Simple MC-SCF Perturbation Theory: Orthogonal Valence Bond Møller-Plesset 2 (OVB-MP2). *Chem. Phys. Lett.* **1988,** *148,* 183–89.

McDougall, P. J.; Hall, M. B.; Bader RFW and Cheeseman, J. R. Extending the VSEPR Model Through the Properties of the Laplacian of the Charge Density. *Can. J. Chem.* **1989,** *67,* 1842.

McIntosh, G. C.; Tomanek, D.; Park, Y. W. Energetics and electronic structure of a polyacetylene chain contained in a carbon nanotube. *Phys. Rev. B* **2003,** *67,* 125419.

McWeeny, R.; Dierksen, G. Self-consistent perturbation theory. 2. Extension to open shells. *J. Chem. Phys.* **1968,** *49,* 4852.

Meunier, V, Sumpter, B. G. Amphoteric doping of carbon nanotubes by encapsulation of organic molecules: electronic properties and quantum conductance. *J Chem Phys.* **2005,** *123,* 24705.

Meyer, R. R.; Sloan, J.; Dunin-Borkowski, R. E.; Kirkland, A. I.; Novotny, M. C.; Bailey, S. R.; Hutchison, J. L.; Green, M. L. Discrete atom imaging of one-dimensional crystals formed within single-walled carbon nanotubes. *Science.* **2000,** *289,* 1324–1327.

Mott, N. F.; Twose, W. D. The theory of impurity conduction. *Adv. Phys.* **1961,** *10,* 107–163.

Nishide, D.; Dohi, H.; Wakabayashi, T.; Nishibori, E.; Aoyagi, S.; Ishida, M.; Kikuchi, S.; Kitaura, R.; Sugai, T.; Sakata, M. Single-wall carbon nanotubes encaging linear chain C10H2 polyyne molecules inside. *Chem. Phys. Lett.* **2006,** *428,* 356–360.

Poater, J.; Duran, M.; Solà, M.; Silvi, B. Theoretical evaluation of electron delocalization in aromatic molecules by means of AIM and ELF topological approaches. *Chem Rev.* **2005,** *105,* 3911–3947.

Pople, J. A.; Seeger, R.; Krishnan, R. Variational Configuration Interaction Methods and Comparison with Perturbation Theory. *Int. J. Quant. Chem.* **1977,** Suppl. Y-*11,* 149–63.

Potok, R. M.; Folk, J. A.; Marcus, C. M.; Umansky V. Detecting Spin-Polarized Currents in Ballistic Nanostructures. *Phys. Rev. Lett.* **2002,** *89,* 266602.

Raghavachari K., Binkley JS., Seeger R., Pople, J. A. Self-Consistent Molecular Orbital Methods. 20. Basis set for correlated wave-functions. *J. Chem. Phys.* **1980,** *72,* 650–654.

Randic, M. Aromaticity of Polycyclic Conjugated Hydrocarbons. *Chem. Rev.* **2003,** *103,* 3449–3605.

Rassolov, V. A.; Pople, J. A.; Ratner, M. A.; Windus, T. L. 6–31G* basis set for atoms K through Zn. *J. Chem. Phys.* ***1998,*** *109,* 1223–1229.

Sheng, L.; Jin, A.; Yu, L.; An, K.; Ando, Y.; Zhao, X. A simple and universal method for fabricating linear carbonchains in multiwalled carbon nanotubes. *Mat. Lett.* **2012,** *81,* 222–224.

Shi, X. Q.; Dai, Z. X.; Zhong, G. H.; Zheng XH, and Zeng, Z. Spin-Polarized Transport in Carbon Nanowires Inside Semiconducting Carbon Nanotubes. *J. Phys. Chem.* C **2007,** *111,* 10130–10134.

Shirakawa, H.; Louis, E. J.; MacDiarmid, A. G.; Chiang, C. K.; Heeger, A. J. Synthesis of electrically conducting organic polymers: halogen derivatives of polyacetylene, (CH)*x*. *JACS. Chem. Comm.* **1977,** 578.

Silvi, B.; Savin, A. Classification of chemical bonds based on topological analysis of electron localization functions. *Nature* **1994,** *371,* 683–686.

Steinmann, S. N.; Mo, Y.; Corminboeuf, C. How do electron localization functions describe π-electron delocalization? *Phys. Chem. Chem. Phys.* **2011,** *13,* 20584–20592.

Storm, K.; Halvardsson, F.; Heurlin, M.; Lindgren, D.; Gustafsson, A.; Wu, P. M.; Monemar, B.; Samuelson, L. Spatially resolved Hall effect measurement in a single semiconductor nanowire. *Nat Nano.* doi:10.1038/nnano.2012.190

Sugihara, Y.; Yamamoto, H.; Mizoue, K.; Murata, I. Cyclohepta[a]phenalene: A Highly Electron-Donating Nonalternant Hydrocarbon. *Angew. Chem. Int. Ed. Eng.* **1987,** *26,* 1247–1249.

Sun, P.; Oeschler, N.; Johnsen, S.; Iversen, B. B.; Steglich, F. Narrow band gap and enhanced thermoelectricity in FeSb2. *Dalton Trans.* **2010,** *39,* 1012–1019.

Sun, Y.; Giebink, N. C.; Kanno, H.; Ma, B.; Thompson, M. E.; Forrest, S. R. Management of singlet and triplet excitons for efficient white organic light-emitting devices. *Nature* **2006,** *440,* 908–912.

Swol, F.; Lopez, G. P.; Burns, A. R.; Brinker, C. J. Functional nanocomposites prepared by self-assembly and polymerization of diacetylene surfactants and silicic acid. *JACS* **2003,** *125,* 1269–1277.

Takenobu, T.; Takano, T.; Shiraishi, M.; Murakami, Y.; Ata, M.; Kataura, H.; Achiba, Y.; Iwasa, Y. Stable and controlled amphoteric doping by encapsulation of organic molecules inside carbon nanotubes. *Nat Mater.* **2003,** *2,* 683–688.

Tang, J.; Huo, Z.; Brittman, S.; Gao, H.; Yang, P. Solution-processed core–shell nanowires for efficient photovoltaic cells. *Nat. Nanotech.* **2011,** *6,* 568–572.

Tran-Duc, T.; Thamwattana, N. Modeling encapsulation of acetylene molecules into carbon nanotubes. *J Phys Condens Matter*. **2011,** *23,* 225–302.

Varma, C. M.; Zaanen, J.; Raghavachari, K. Superconductivity in the fullerenes. *Science* **1991,** *254,* 989–992.

Wang, X.; Maeda, K.; Chen, X.; Takanabe, K.; Domen, K.; Hou, Y.; Fu, X.; Antonietti, M. Polymer semiconductors for artificial photosynthesis: hydrogen evolution by mesoporous graphitic carbon nitride with visible light. *JACS* **2009,** *131,* 1680–1681.

Wang, Zhenxia; Ke, Xuezhi; Zhu, Zhiyuan; Zhang, Fengshou; Ruan, Meiling; Yang, Jinqing. Carbon-atom chain formation in the core of nanotubes. *Phys. Rev. B* **2000,** *61,* R2472–R2474.

Warner, J.; Rümmeli, M. H.; Bachmatiuk, A.; Büchner, B. Structural transformations of carbon chains inside nanotubes. *Phys Rev* B **2010,** *81,* 155419.

Wunderlich, J.; Park, B. G.; Irvine, A. C.; Zârbo, L. P.; Rozkotová, E.; Nemec, P.; Novák V, Sinova, J.; Jungwirth, T. Spin Hall effect transistor. *Science* **2010,** *330,* 1801–1804.

Yang, L.; Kong, J.; Yee, W. A.; Liu, W.; Phua, S. L.; Toh, C. L.; Huang, S.; Lu, X. Highly conductive graphene by low-temperature thermal reduction and in situ preparation of conductive polymer nanocomposites. *Nanoscale* **2012,** *4,* 4968–4971.

Yang, S.; Olishevski, P.; Kertesz, M. Bandgap calculations for conjugated polymers. *Synth. Met.* **2004,** *141,* 171–177.

Yang, Y.; Lu, Y.; Lu, M.; Huang, J.; Haddad, R.; Xomeritakis, G.; Liu, N.; Malanoski, A. P.; Sturmayr, D.; Fan, H.; Sasaki, D. Y.; Assink, R. A.; Shelnutt JA, van Swol, F.; Lopez, G. P.; Burns, A. R.; Brinker, C. J. Functional nanocomposites prepared by self-assembly and polymerization of diacetylene surfactants and silicic acid. *JACS* **2003,** *125,* 1269–1277.

Yoshizakawa, K. An Orbital Rule for Electron Transport in Molecules. *Acc. Chem. Res.,* **2012,** *45,* 1612–1621.

Yoshizakawa, K.; Tada, T.; Staykov, A. Orbital Views of the Electron Transport in Molecular Devices. *JACS.* **2008,** *130,* 9406–9413.

Yu, W. J.; Liao, L.; Chae, S. H.; Lee, Y. H.; Duan, X. Toward tunable band gap and tunable dirac point in bilayer graphene with molecular doping. *Nano Lett.* **2011,** *11,* 4759–4763.

Zhao, X.; Ando, Y.; Liu, Y.; Jinno, M; Suzuki, T. Carbon nanowire made of a long linear carbon chain inserted inside a multiwalled carbon nanotube. *Phys. Rev. Lett.* **2003,** *90,* 187401–187404.

Zhao, Y.; Allen, B.; Star, A. Enzymatic Degradation of Multiwalled Carbon Nanotubes. *J Phys Chem A.* **2011,** *115,* 9536–9544.

Zhou, J.; Song, H.; Chen, X.; Huo, J. Diffusion of metal in a confined nanospace of carbon nanotubes induced by air oxidation. *JACS* **2010,** *132,* 11402–11405.

CHAPTER 6

DESIGNING OF SOME NOVEL MOLECULAR TEMPLATES SUITABLE FOR HYDROGEN STORAGE APPLICATIONS: A THEORETICAL APPROACH

SUKANTA MONDAL, ARINDAM CHAKRABORTY, SUDIP PAN and PRATIM K. CHATTARAJ

CONTENTS

Abstract 197
6.1 Introduction 197
6.2 Application of Nanomaterials for Hydrogen Storage 199
6.2.1 Chemically Hydrogen Storing Materials 199
6.2.2 Physically Hydrogen Storing Materials 204
6.3 A Theoretical Approach to Model an Effective Hydrogen Storage Medium 212
6.3.1 Theoretical Background 212
6.3.2 Computational Details 215
6.3.3 Exohedral and Endohedral Hydrogen Trapping by Cage-Like Structures 216
6.4 Metal Clusters Interacting with Hydrogen 224
6.4.1 Interaction Between Transition Metal (TM) and H_2 in TM Doped Ethylene Complex and Hydrogen Storage 224

6.4.2 M_3^+ (M=Li, Na) Cluster 226
6.4.3 Be_3-M_2, Mg_3-M_2 and Al_4-M_2 (M=Li, Na, K) Clusters 227
6.5 Hydrogen Storage of Different Lithium Doped Molecular Clusters 228
6.5.1 Li_3Al_4- Cluster 228
6.5.2 2-D Stars 228
6.5.3 3-D Stars 230
6.5.4 Boron-Lithium Clusters 231
6.5.5 Li-Doped Boron Hydride 232
6.5.6 Li-Doped Annular Systems 233
6.5.7 Li-Doped Borazine Derivatives 234
6.5.9 $(N_4C_3H)_6Li_6$ and an Associated 3-D Functional Material 236
6.6 Hydrogen Storage Capacity of Super-Alkali Clusters 237
6.7 Conclusions 239
Acknowledgments 240
Keywords 240
References 240

ABSTRACT

Modeling of new molecular networks and aggregates – one of the most "sought after" topics in current chemical research is investigated on the basis of the theoretical paradigm of conceptual density functional theory and its various reactivity variants. The utility of these molecular materials as plausible storage templates for hydrogen gas is also investigated. The stability of these molecules and their hydrogen-loaded analogs is assessed through the dual perspectives of a charge analysis on the active atomic centers of the given systems as well as a comparison of the nucleus-independent chemical shift (NICS) values. Effects of the application of an external electric field and construction of relevant T–P phase diagrams reveal a thermodynamically spontaneous hydrogen binding process for many template moieties with a conspicuous increase in loading potential with an increase in the field gradient. *Ab initio* as well as classical molecular dynamics simulations are also carried out for few systems to assess their bulk properties as well as hydrogen trapping potentials.

6.1 INTRODUCTION

The branch of science, which deals with the objects having one or more dimensions of the order of 1–100 nm, is called nanoscience. Rigorous research on nanoscience and nanotechnology has been carried out since the early 1970s (Website, 2012). The word "nano" means dwarf or extremely small, which is derived from Greek word "nanos" (Website, 2012). Nanoscience has become of great importance as it serves well to explain the bridge between bulk material and atomic or molecular systems. Bulk materials consist of uniform physical properties whereas the same material at the nano-scale may exhibit different properties.

Size-dependent properties are observed such as quantum confinement in semiconductor particles, surface plasmon resonance in metal particles and superparamagnetism in magnetic materials. Such nanoscale particles are used in biomedical applications as drug carriers, imaging agents etc. Nanoscience is a vibrant topic and is the branch of science and engineering that exploits the unique behavior of materials at this scale as well as the fabrication of devices on this same length scale. It is a multidisciplinary grouping of physical, chemical, biological, engineering, and electronic processes, in which the defining characteristic originates from its size. It is obviously a revolution in science and can be defined in its simplest term as "engineering at a very small scale." When we talk about the study at micro or nano level then theoretical approach becomes indispensible. Theoretical study deals with the proper combination of mathematical expressions and fundamental laws of physics as well as chemistry. Computation through the proper combination of mathematical algorithms and basic laws of physics and chemistry exploring the different properties of matter at macro level as well as in micro level is named as follow:

Computational physics, when it deals with the physical properties of matter.

Computational chemistry, when it deals with the processes of chemical significance of matter.

We can say, computational chemistry is a branch of chemistry, which uses equations encapsulating the behavior of matter on an atomistic scale and uses computers to work out these expressions to determine structures and properties of molecules, gases, liquids and solids. Computational chemistry includes electron dynamics, quantum mechanical modeling, time independent *ab initio* calculations, semiempirical calculations and classical molecular dynamics. Almost one and a half decade before, it was mentioned by Davidson (2000) that chemists were mostly concerned with synthesis and characterization of materials whereas computational modeling of chemistry needed to account for reaction mechanism, structure, and spectra. It was also mentioned in the same review that we were far from the time when artificial intelligence methods can be used to conceive of new materials with predesigned properties and synthetic route. But now we have a handful of computational tools and superfast computers to do this job with different theoretical approaches. In this review chapter we have analyzed the results obtained through experimental and computational methods on different nanomaterials with endo or exo type hydrogen trapping potential.

It is estimated that fossil fuels cause a huge worldwide damage in addition to human discomfort and market price, which counts \$2360 billion per year or \$460 per capita per year (Barbir et al., 1990). It is reported that in 2005 feasting of fossil fuels surpassed 83 million barrel which in turn replies to our environment 11 billion metric tons of CO_2 with other greenhouse gases (Website). An alternative fuel, which will not destroy our environment and will generate only environment friendly byproducts, is of actual concern. Hydrogen is a potential fuel not only due to its ecological property and competence but also its store preservation, low cost of transportation and decrease in inflation (Awad and Veziroğlu, 1984). It is a good energy carrier as it is carbon free, highly abundant and the mass energy density is also exceptional (Berg and Arean, 2008). Comparison of hydrogen with other fuels leads to the fact that hydrogen has the superlative features with many environments friendly and appropriate properties (Veziroglu, 1992). Thus it is a potential and sought after topic to research on hydrogen storage capacity of different materials, because among all probable considered fuels hydrogen stances out as the best.

There are mainly three kinds of hydrogen storage techniques, namely, (a) chemical storage of hydrogen, (b) physical storage of hydrogen and (c) new emerging methods. Chemical storage of hydrogen includes sodium borohydrides, ammonia, methanol, and alkali metal hydrides. In physical storage of H_2 there are four different ways: compressing H_2, liquefying hydrogen cryogenically, in sponge metal hydrides, and in carbon nanofibers. The new emerging methods of hydrogen storage are amminex tablets, DADB (predicted), solar zinc production and alkali metal hydride slurry (Dillon and Heben, 2001; Murray et al., 2009; Yang et al., 2010).

6.2 APPLICATION OF NANOMATERIALS FOR HYDROGEN STORAGE

In order to search for effective hydrogen storing materials, different types of nanostructured molecular assemblies like metal hydrides, sheet-like frameworks, cage-like structures, tube-like cylindrical media are scrutinized. Both the chemically and physically hydrogen storing systems are investigated in this aspect.

6.2.1 CHEMICALLY HYDROGEN STORING MATERIALS

6.2.1.1 METAL HYDRIDES

1. Metal hydrides are a category of compounds comprizing mainly two components, one is the metal center and another is the anionic part containing rich amount of hydrogen. The main advantages of metal hydrides to use as hydrogen storage materials are: (i) they possess good energy density by volume, therefore the target set by the United States DOE (Department of Energy) can be easily achieved by these, (ii) the urge of security imposes priority of it over the storage of hydrogen in the form of gas or liquid, cryogenic condition is required for hydrogen storage in liquid form (Weast et al., 1983). Therefore, the metal hydrides are advantageous in terms of both safety and volume efficiency. The probable applicability of alanates and other metal hydrides as effective hydrogen storage media is well documented in the literature (Sakintuna et al., 2007; Schüth et al., 2004). Moreover, the same holds for the $NaAlH_4$ and $NaBH_4$, which are tested and thoroughly elaborated elsewhere (Bogdanović et al., 2009). In addition, the associated kinetics for the hydrogenation and dehydrogenation of $NaAlH_4$ system can be modified by applying various nanomaterials (Berseth et al., 2009; Xing et al., 2005). A couple of papers represent the direct synthesis of nanocrystalline complex metal hydrides, $LiAlH_4$ and $NaAlH_4$ for the sake of their applicability as hydrogen binding materials (Wang et al., 2007; Xiao et al., 2009). Vajo et al. (2004) have

successfully employed LiH and MgH_2 as hydrogen storage material by introducing destabilization into the hydride through incorporation of Si, which forms alloy. Very recently Setijadi et al. have shown that the nano-crystalline MgH_2 derived from the hydrogenolysis of Grignard (2012) reagents has very remarkable hydrogen storage capacity (6.8 mass%) and kinetics compared to that of nanoconfined magnesium. The hydrogenation and dehydrogenation kinetics and different thermodynamical properties of MgH_2 can be further modified by incorporating TiH_2 into it generating MgH_2–0.1TiH_2 material system (Lu et al., 2009). The significant improvement of thermodynamics for hydrogen cycling in a coupled $LiBH_4/MgH_2$ system compared to the individual isolated one is also reported (Pinkerton et al., 2007). The phenomenon of hydrogen sorption by intermetallic systems is also quite impressive and the hydrogen storage capacity of some selected systems is provided in Table 1 (Sandrock and Thomas, 2001; Schlapbach and Züttel, 2001).

TABLE 1 Intermetallic compounds and their hydrogen-storage properties (This table is reproduced from the work of Sandrock et al. with permission from Springer-Verlag, Berlin.)

Type	Metal	Hydride	Structure	mass %	p_{eq}, T
Elemental	Pd	$PdH_{0.6}$	Fm3m	0.56	0.020 bar, 298 K
AB_5	$LaNi_5$	$LaNi_5H_6$	Pblmmm	1.37	2 bar, 298 K
AB_2	ZrV_2	$ZrV_2H_{5.5}$	Fd3m	3.01	10^{-8} bar, 323 K
AB	FeTi	$FeTiH_2$	Pm3m	1.89	5 bar, 303 K
A_2B	Mg_2Ni	Mg_2NiH_4	*Pb*222	3.59	1 bar, 555 K
Body-centred cubic	TiV_2	TiV_2H_4	b.c.c.	2.6	10 bar, 313 K

The $LaNi_5$ based alloys have some astonishing features like fast sorption kinetics, reversibility, and quite large cycling lifetime. But the main disadvantage of these systems is low gravimetric wt% since both La and Ni are heavy elements e.g., $LaNi_5H_{6.5}$ has below 2 wt% of hydrogen (Latroche et al., 1999; Sakai et al., 1995; Schlapbach et al., 1994). However, there are some examples of hydride form of intermetallic compounds such as $Li_3Be_2H_7$ and $BaReH_9$ in which the hydrogen content reaches to higher extent (Yvon, 1998; Zaluska et al., 2001). An article comparing the hydrogen storage capacity of $NaBH_4$ and NH_3BH_3 is also well appreciated in this aspect (Demirci and Miele, 2009).

A pictorial presentation (Fig. 1) regarding the percentage of hydrogen per mass or per volume of these hydrides may be very helpful in understanding which hydride is more efficient in this regard.

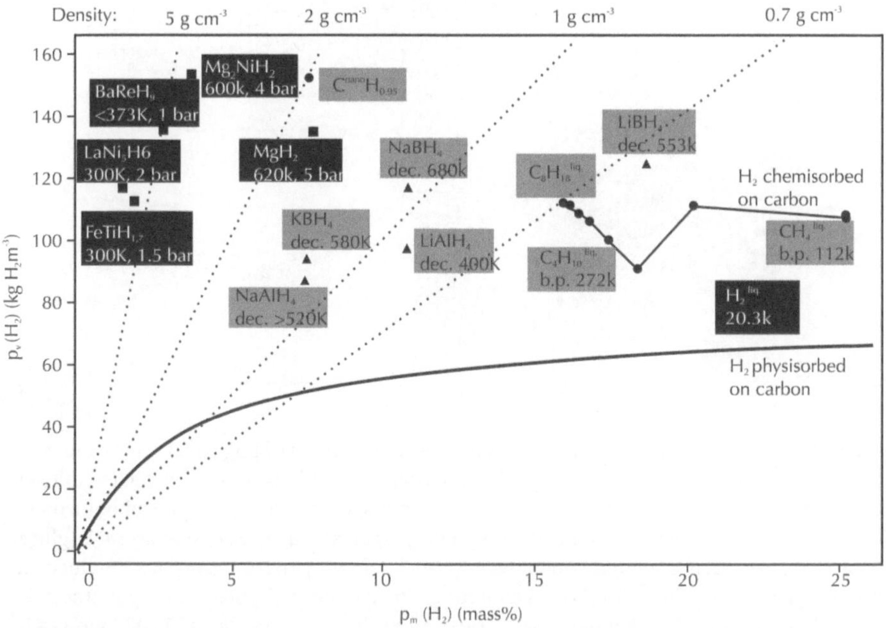

FIGURE 1 Stored hydrogen per mass and per volume. Comparison of metal hydrides, carbon nanotubes, petrol and other hydrocarbons. (This figure is reproduced from the work of Schlapbach et al. (1994) with permission from Nature.)

The applicability of the hydrogenated magnesium nickel boride ($MgNi_{2.5}B_2$) as promizing hydrogen storage material is also mentioned in the literature (Li et al., 2010).

6.2.1.2 CARBOHYDRATES

Zhang et al. have demonstrated a synthetic biological pathway to produce hydrogen from a mixture of starch, 13 enzymes and water. The stoichiometric process can be represented as $C_6H_{10}O_5$ (l) + $7H_2O$ (l) → $12H_2$ (g) + $6CO_2$ (g), which appears to be spontaneous and unidirectional (Zhang et al., 2007; Zhang, 2009). A systematic scheme depicting the process to produce hydrogen from starch is provided in Fig. 2. Mild reaction condition and high gravimetric wt% of hydrogen (14.8 wt% of hydrogen) make it superior to the other systems.

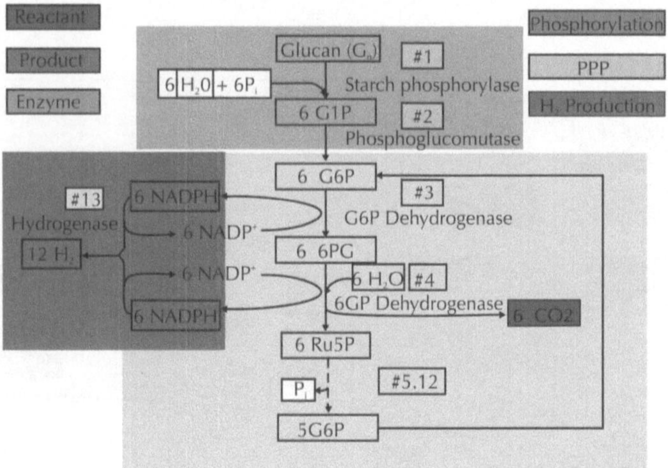

FIGURE 2 The synthetic metabolic pathway for conversion of polysaccharides and water to hydrogen and carbon dioxide. The abbreviations are: PPP, pentose phosphate pathway; G1P, glucose-1-phosphate; G6P, glucose-6-phosphate; 6PG, 6-phosphogluconate; Ru5P, ribulose-5-phosphate; and Pi, inorganic phosphate. The enzymes are: #1, glucan phosphorylase; #2, phosphoglucomutase; #3, G-6-P dehydrogenase; #4, 6-phosphogluconate dehydrogenase, #5 Phosphoribose isomerase; #6, Ribulose 5-phosphate epimerase; #7, Transaldolase; #8, Transketolase, #9, Triose phosphate isomerase; #10, Aldolase, #11, Phosphoglucose isomerase: #12, Fructose-1, 6-bisphosphatase; and #13, Hydrogenase. (This figure is reproduced from the work of Zhang et al. (2007) with permission from *PLoS ONE*.)

A review elaborating the hydrogen storage property of carbohydrates is here worth mentioning (Zhang, 2011).

6.2.1.3 SYNTHESIZED HYDROCARBONS AND LIQUID ORGANIC HYDROGEN CARRIERS

The regular hydrocarbons like propane and methane can also be nice substitutes of hydrides as hydrogen carriers (Wikipedia, 2012). But the problems for real application of these fuel cells are: (i) requirement of high temperatures (ii) slower start up time. The synthetic hydrocarbon fuel known as "synfuel" is also employed for hydrogen carrying purposes. The Fischer Tropsch (F-T) process is in frontier line for producing this (Website, 2012). The synthetic gas is formed according the reaction $2C + ½ O_2 + H_2O = 2CO + H_2$ for coal whereas the water gas shift reaction produces more hydrogen by the reaction $CO + H_2O = H_2 + CO_2$. The unsaturated hydrocarbons are also used for hydrogen storing purpose. The main feature of liquid organic hydrogen carriers (LOHC) is the unsaturation associated with the structures and hence, they can be hydrogenated and dehydrogenated in a regular basis giving recyclable hydrogen storage dias (Teichmann et al., 2012). Presently, the most exciting LOHC is N-ethylcarbazole ($C_{14}H_{13}N$), which upon hydrogenation gives perhydro N-ethylcarbazole ($C_{14}H_{19}N$). So this excess hydrogen can fulfill the on-board hydrogen demand (Teichmann et al., 2011).

6.2.1.4 AMMONIA AND AMMONIA-BORANE COMPLEX

Effective efforts to use ammonia as hydrogen releasing material are also made. Klerke et al. in a featured article have nicely elaborated the advantages of ammonia over other hydrogen storage materials (2008). The advantages of ammonia can be listed as: (i) it has a very high hydrogen density, (ii) it can be easily stored in the liquid form at general temperature and pressure, (iii) it is produced world-wide in large quantities and hence easily available, (iv) it produces hydrogen without any harmful side-components. But the only disadvantage is that liquid ammonia is toxic in nature and exists in trace amounts in the hydrogen produced from its decomposition. Ammonia-borane (H_3NBH_3) complex has also been widely used as a hydrogen storage material. Different chemical approaches to free hydrogen from ammonia-borane are reported. Keaton et al. (2007) have used first row transition metals as catalysts for dehydrogenation of ammonia-borane whereas Stephens et al. (2007) have shown that the dehydrogenation of ammonia-borane can be initiated by an acid. The improvement of dehydrogenation kinetics of ammonia-borane in presence of ionic liquid is also reported (Heldebrant et al., 2008). New modified synthetic processes to make ammonia-borane for hydrogen storage material should also be mentioned (Heldebrant et al., 2008).

6.2.1.5 FORMIC ACID AND IMIDAZOLIUM IONIC LIQUID

A lucky breakthrough appears in the field of hydrogen storage with the invention of formic acid as hydrogen storage material through the production of hydrogen and carbon dioxide by the catalytic reaction of formic acid (Laurenczy et al., 2008). The application of ruthenium water-soluble complexes (Ru/TPPTS; TPPTS=*meta*-trisulfonated triphenylphosphine) as catalyst to break formic acid into hydrogen and carbon dioxide is also widely appreciated (Fellay et al., 2008). Another related article showing the hydrogen storage property of formic acid and elaborating the significance of this invention should also be cited (Joo, 2008). A very recent discovery of catalyst, $RuCl_3$ dissolved in 1-ethyl-2,3-dimethylimidazolium acetate ($RuCl_3$/(EMMIM)(OAc)) for $HCOOH \rightarrow H_2 + CO_2$ process is also reported (Berger et al., 2011). Recently ionic liquid based hydrogen storage materials are also turned out as effective in serving the hydrogen storing purpose. Stracke et al. (2007) have explored the hydrogen storage property of 1-alkyl (aryl)-3-methyl imidazolium N-bis(trifluoro methanesulfonyl) imidate salts. Methyl guanidinium Borohydride is also reported as ionic liquid based hydrogen carrying material (Doroodian et al., 2010).

6.2.2 PHYSICALLY HYDROGEN STORING MATERIALS

6.2.2.1 METAL ORGANIC FRAMEWORKS

Metal organic frameworks (MOFs) are special type of 3D-frameworks, which include metal centers as coordination sites with organic parts, therefore, it can be divided into two parts, the part consisting of metal ions and another composed of organic linkers. The successful application of these extended porous 3D-MOFs towards hydrogen storage creates craze in this field recently. The interaction of hydrogen with MOFs in a physisorption manner makes the kinetics of adsorption – desorption phenomena reversible. This weak interacting force is of nonbonded type resembling London dispersion type of interaction (LDI) (Kuc et al., 2008). MOFs often show some geometrical features like structural rigidity, chemical functionality and chirality. Generally high experimental yielding of MOFs is another reason of getting attention for their practical utility. The molecular bridging is also important since one can tune the void region by changing the length of linkers (Rowsell andYaghi, 2005). A depiction of increment of void region with carefully tuning the linker is given in Fig. 3.

The astonishing property of MOFs having huge surface area has accelerated their application as effective hydrogen storing candidates. The reports are available of many MOFs having surface areas larger than 1000 $m^2 g^{-1}$ (higher than zeolites). Rosi et al. have shown that an improvement of hydrogen storage up to 2.0 wt % of hydrogen in IRMOF-8 at 293 K and 10 bar is possible by MOFs modification of the connecting ligand in isoreticular regions (2003). A relevant report elaborating the fairly good hydrogen storage property of MOFs comprizing Zn centers linked with aromatic di-carboxylic acids is also available in the literature (Kesanli et al., 2005). The MOFs made of $(Zn_4O)^{6+}$ unit connected with an aromatic carboxylate group like 1, 4 benzene dicarboxylate (BDC) show potentiality towards storage of hydrogen (Kaye et al., 2007). However, it has been found very often that in many MOFs, the gaps between the corresponding walls are too large to adsorb hydrogen effectively. MOF-5 is a classic example showing this defect in which the distance between two phenylene faces is 15 Å, much greater than kinetic diameter of hydrogen (2.89 Å) (Breck, 1974). An approach to overcome this pore size problem is the insertion of another adsorbate material within MOFs having large pore. Technically it can be done in two ways: (i) the impregnation with a nonvolatile component and (ii) the catenation by another similar system. It has been reported that C_{60} and Reichardt"s dye can be successfully encapsulated within MOF-177 (Chae et al., 2004). Few relevant reviews describing the novel techniques to model and synthesize MOFs and their further implication in storing hydrogen are here worth to mention (Eddaoudi et al., 2001; James, 2003). The potency of MOFs to adsorb different adsorbates in addition to store hydrogen is investigated thoroughly (Britt et al., 2008; Collins and Zhou, 2007; Han et al., 2007; Hu and Zhang, 2010; Ma and Zhou, 2010; Meek et al., 2011; Murray et al., 2009; Rosi et al., 2003; Rowsell and Yaghi, 2005; Rowsell et al., 2005; Zhao et al., 2004, 2008). The strategy of metal doping on MOF is also been adopted to improve its hydrogen storage capability (Dincă and Long, 2008; Dincă et al., 2006; Peterson et al., 2006).

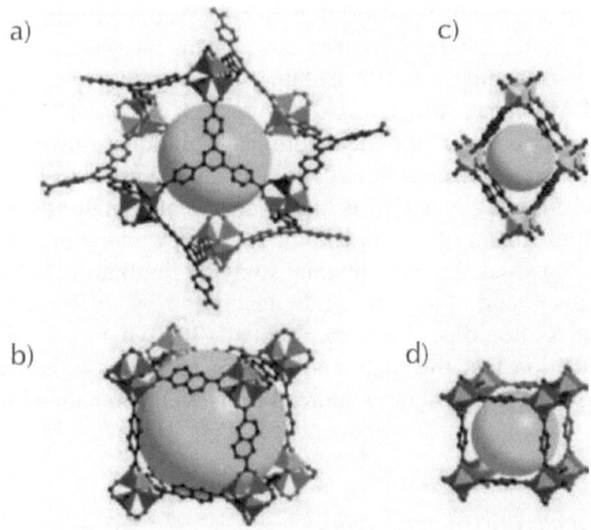

FIGURE 3 Examples of metal–organic frameworks (MOFs) studied for hydrogen adsorption include a) MOF-177, $Zn_4O(btb)_2$ (btb=benzene-1,3,5-tribenzoate), b) IRMOF-8, $Zn_4O(ndc)_3$ (ndc=naphthalene-2,6-dicarboxylate), c) MIL-53, M(OH)(bdc) (M=Al_3^+ or Cr_3^+), and d) $Zn_{2-}(bdc)_2(dabco)$ (dabco=1,4-diazabicyclo(2.2.2)octane). The propertiesof these compounds are listed in Table 1. Pores in the evacuated crystalline frameworks are illustrated by yellow spheres that contact thevan der Waals radii of the framework atoms (C: black, N: green, O: red, Zn: blue polyhedra, M: green octahedra). (This figure is reproduced from the work of Rowsell et al. (2004) with permission from John Wiley.)

6.2.2.2 COVALENT ORGANIC FRAMEWORKS

Covalent organic frameworks (COFs) are porous, extended structures connected by covalent bonds in which the constituent elements are mainly H, B, C, O, and N. Some crucial points with regard to the hydrogen storage ability via physisorption are: (i) its surface area, (ii) its porous volume, and (iii) hydrogen adsorption enthalpy. Yaghi et al. have synthesized a brand new class of systems, called them as three-dimensional covalent-organic frameworks (3D-COFs), which are enable to satisfy most of the norms and conditions predicted to be an effective hydrogen storage material (2007). The light composing elements of COFs make it advantageous over MOFs in spite of both maintaining the same structural features. A multiscale theoretical investigation on 3D COFs have shown that COF-108 can store up to 21 wt % of hydrogen at 77 K temperature and 100 bar pressure (Klontzas et al., 2008). A related grand canonical Monte Carlo simulation study exploring hydrogen, argon and methane adsorption on a series of COFs is here worth mentioning (Garberoglio, 2007). Another grand canonical Monte Carlo simulation shows that COF-105 and COF-108 are superior for hydrogen storage (Han et al., 2008), each of which can store hydrogen up to 10.0 wt % at 77 K. Moreover, the successful implication of multiscale theoretical techniques towards designing 3D COFs is also done (Klontzas et al., 2010). The much improved hydrogen storage capacity upon Li-doping on COFs is highly appreciated in this aspect (Cao et al., 2009). The improvement of hydrogen binding energy upon doping of Li and Mg on COFs is also studied. An experimental study by Furukawa et al. (2009) on the hydrogen, methane and carbon di oxide adsorption by different COFs at 1–85 bar and 77–298 K should be cited here. Li et al. (2010) have shown via their first principle calculation that the hydrogen binding energy is effectively enhanced upon substitution of the bridging nonmetal rings with some metal included rings in COFs.

6.2.2.3 SHEET-LIKE NETWORKS

Graphene, a very special contender of a class comprizing 2-dimensional material is made of hexagonal display of sp^2-hybridized carbon atoms (Geim and Novoselov, 2007). Some useful applications of graphene can be listed as: (i) hydrogen storage material, (ii) transistors, (iii) gas sensing agents, (iv) supporting membranes in transmission electron microscopy (TEM), and (v) inert coating agents. However, here we discuss its applicability as plausible hydrogen storage material. Its low cost, nontoxic nature and easily synthesizable property in addition to its capability to interact with hydrogen draw considerable attention to use it as a hydrogen storage material. The unfortunate thing is that the single layered graphene sheet does not show that much of proficiency towards capturing hydrogen but one can adopt the technique of doping metal on to graphene sheets to improve its hydrogen binding capability significantly. A well-known fact about the metal site is that it can bind with molecular hydrogen by inducing a dipole moment within hydrogen molecule and finally interacting through an ion-induced dipole type of interaction. Moreover, hydrogen has a high quadrupole moment, so an ion-quadrupole moment type of interaction also plays some contribution there. Srinivasu et al. (2008) have nicely demonstrated by taking the six-membered carbon ring, mimicking a unit of carbon nanotube that it as such is not a preferable candidate to bind hydrogen but the creation of charged surface by doping with metal ions (here it is alkali metal ions) and also the curvature significantly modify its hydrogen storage capacity. Another first principle wave calculation by Ataca et al. (2008) have shown the ability of graphene to adsorb Li via an electron transfer from 2 s orbital to π^* orbital of graphene, which further serves as effective hydrogen storage medium possessing 12.8 wt% of hydrogen. The study exploring the hydrogen storage capacity of boron-substituted graphenes, doped with alkaline earth metals particularly Mg and Ca and the basic mechanism of interaction therein, is a valuable database (Kim et al., 2009). A further relevant study has shown that a technique of boron-substitution on to graphene is enabled to overcome the Li-clustering problem by enhancing the binding strength between Li and grapheme (Liu and Zeng, 2010). This is presumably due to a shift from *s-p* hybridization in case of Li-graphene complex to *p-p* hybridization in case of Li-dispersed boron-doped graphene. The hydrogen storage on Li-doped B_2C graphene leading to 7.54 wt% of hydrogen is also investigated by first principles density functional theory (An et al., 2011). A couple of reports provide ample justification to the efficacy of porous graphene sheet towards hydrogen binding capability (Du et al., 2010; Lin et al., 2008). Ab initio calculations by Okamoto et al. (2001) further shown that the interaction energy of hydrogen molecule physisorbed on to a graphene sheet is almost independent of binding sites and have also demonstrated that the planar graphene in close-packed trigonal lattice may serve as a better physisorption template than that in a square lattice. Although graphene is widely considered as a physisorption system of hydrogen storage, a study by Subrahmanyam et al. (2011) have shown its applicability as a chemical hydrogen storage material. By means of Birch reduction of few-layer graphenes, a 5% hydrogen containing sample can be reached, which can be released as free hydrogen by exposing it to a temperature of 500°C or to an irradiation with UV or laser light.

Other modified network structures are multilayered graphene-oxide frameworks (GOFs), which are formed by oxidizing graphene sheets. Significant researches are done to demonstrate its utility as a plausible hydrogen storage material. In this regard, the efforts of Yildirim et al. (2010) exploring the hydrogen storage capacity of GOFs deserve to be specially mentioned. Wang et al. (2009) have successfully overcome the problem of Ti-clustering by using GOFs as a dais for Ti-doping since O centers of hydroxyl groups present in GOFs bind Ti with sufficiently large binding energy to avoid clustering. Moreover, they have shown that each Ti center can bind multiple hydrogen molecules providing overall 4.9 wt% of hydrogen. Recently another relevant inspection made by Chan et al. (2011) testing the hydrogen binding ability inside GOFs should be cited. In addition, the same for modified graphene oxide frameworks namely graphene oxide derived carbons (GODC) is also checked and found to be much efficient in storing hydrogen (Website, 2012). The performance of three-dimensional pillared graphene frameworks is also tested via some multiscale theoretical studies and it is found that multidecker GOFs can store at least hundred times larger hydrogen than general graphene oxide frameworks. A couple of papers by Dimitrakakis et al. (2008) and Han et al. (2009) show the efficacy of these 3D-GOFs towards storing hydrogen. The probable utility of 3D- covalently bonded graphenes (CBGs) as hydrogen storage material is also investigated by Park et al. (2007) and found to be much improved hydrogen capturing power than an isolated grapheme. They have also doped transitional metals on to CBGs to make it better hydrogen-binding templates.

In addition to graphene, graphyne and graphdiyne are also successfully employed to serve the hydrogen storing purpose. Li et al. (2011) have shown that Ca dispersed graphyne sheet has higher hydrogen storage capability than Ca-doped fullerene both in terms of binding energy and gravimetric wt%. Another theoretical calculation regarding the hydrogen storing ability of Li-doped graphyne and graphdiyne is made by Srinivasu et al. (2012).

6.2.2.4 TUBULAR NANOSTRUCTURES (NANOTUBES) AND CAGE-LIKE MOLECULES

Nanomaterials, in different structural forms are tested as suitable storage templates for hydrogen. Relevant studies by Kroto et al. (1985) and Krätschmer et al. (1990) show the existence and synthesis of the cage-like fullerene (C_{60}) molecule. Further study by Iijima (1991) hints at the existence of "needle-like" carbon-based structures, which eventually lead to the formation of the carbon nanotubes (Bethune, 1993; Iijima and Ichihashi, 1993). The plausibility of using these cage-like and tubular nanostructures as hydrogen binding templates, both exohedrally as well as endohedrally is also tested. A host species, preferably a charged moiety (like a metal ion) sitting on the surface of the nano-cage or tube influences the rate of hydrogen adsorption on to the same. It is also shown earlier that a metal-ion doped graphene-like sheet is more apt towards hydrogen adsorption compared to its bare counterpart (Srinivasu et al., 2008). Modeling of H_2O-decorated cage-like structures, called clathrate hydrates have also drawn considerable interest among the researchers. The viability of hydrogen storage on the clathrate hydrate surface in its bare form or in the presence of a guest species is also studied.

The utility of the carbon nanotubes (CNTs) as effective reaction templates and storage media for hydrogen in both the exohedral and endohedral fashions is reviewed. In this regard, Dillon and co-workers (1997) are supposed to be the first proponents who provided experimental evidences of hydrogen binding onto the CNT surfaces. Subsequent reviews endow the reader with a series of works proving the efficacy of these carbon-based nanomaterials as effective hydrogen storage media (Dillon and Heben, 2001; Ding et al., 2001; Hirscher, 1997; Hynek et al., 2003). Jiménez et al. (2010) have recently investigated the aptitude of adsorbing hydrogen on to a variety of carbon structures containing activated carbon (AC) and carbon nanofibers (CNFs). Upon suitable variation of the pressure and temperature, a linear relationship between the capacity of hydrogen-loading (through physisorption) and specific micropore surface area is established. High level theoretical calculations also elucidate the physisorption of hydrogen on a single-walled carbon nanotube (SWCNT). Ferre Vilaplana (2005) in a benchmark study demonstrated the physisorption of hydrogen on to a single-walled carbon nanotube (SWCNT) and concluded that the physisorption of molecular hydrogen on to SWCNTs (both for inside and outside adsorption) have got a direct bearing on the tube diameter which is almost independent of the chirality. Combined *ab initio* and grand canonical Monte Carlo simulation studies on the effect of curvature and chirality in the SWCNTs upon hydrogen loading also provide similar results (Mpourmpakis et al., 2006, 2007). Porous carbon nanomaterials like activated charcoal owing to their larger surface areas are also found to be quite suitable towards hydrogen adsorption. Kowalczyk et al. (2007) have shown theoretically that the most effective structures along with appropriate metal dopants can satisfy the dual norms of DOE. Further *ab initio* and Monte Carlo simulation studies by Firlej et al. (2009) have shown that the hydrogen adsorption capacity in activated carbon (AC) can be increased upon doping or intercalation by a foreign moiety leading to an increment in its heat of adsorption. Substitution of about 5–10% of C-atoms by boron atoms in a nanoporous assembly leads to an increase in the adsorption energy by 10–13.5 kJ/mol. The substitution of the carbon centers by boron atoms seems to improve the binding of Li-atom on the wall-surface whereby the Li-center acts as the primal site for hydrogen binding and hence increases the power of adsorption (Wu et al., 2008). Analogous metal-boron combined nanostructures, called metal-diborides are also found to be quite promizing for the purpose (Meng et al., 2007). The modeling of Si-doped carbon nanotubes (SiCNTs) has been reported to possess an increased hydrogen binding aptitude owing to favorably improved net charges on the tubular walls (Mpourmpakis et al., 2006).

The utility of Ti-doped CNTs (Yildirim and Ciraci, 2005), Ti-nanoclusters and bi-metallic Ti-Al nanoclusters as suitable templates for the reversible storage of hydrogen is explored (Tarakeshwar et al., 2009). Han et al. investigated the ability of the single-walled BN-nanotubes (BNNTs) for their ability as hydrogen storage materials (2005). Similar, aluminum doped one-dimensional AlN nanowires as well as different types of AlN nanostructures (nanocages, nanocones, nanotubes, and nanowires) are checked for their capability to act as a hydrogen-binding surface (Li et al., 2009; Wang et al., 2009). High-level multiscale approaches by Mpourmpakis et al. (2007) have further demonstrated the utility of available carbon nanoscrolls as potential hydrogen storage materials. The utility of heteroatom substituted carbon scaffolds as effective hydrogen binding media is recently reported (Jin et al., 2010).

Fullerenes, and their various allied cage-like nanostructural analogs designed so far have also proved their worth as effective molecular templates for hydrogen storage. The water molecules can also be arranged in a similar fashion to create a cage-like cluster assembly called clathrate hydrate. These clathrate hydrate structures are being vastly studied towards their potency to act as fruitful storage materials. The C_{56} cage is capable of binding hydrogen in an endohedral fashion (Türker and Gümüş, 2004). Studies on the exohedral and endohedral hydrogen storage ability of different sized fullerene cages are also reported (Yue et al., 2007). However, at par with the CNTs, the utility of alkali-metal doping on to the cage-like clusters is also well pronounced as it visibly increases its hydrogen binding capacity (Chandrakumar and Ghosh, 2008; Loutfy et al., 2002; Sun et al., 2006). Among the different metal dopants, Li is mostly preferred as it is the lightest and is well compatible with the gravimetric and volumetric standards set by the DOE. Yoon et al. (2007) investigated the effect of charge on the hydrogen binding potential of fullerenes. The capability of alkali-metal functionalized fullerenol derivatives ($M^{+}_{12}(C_{60}O_{12})^{12-}$ (M=Li and Na)) as effective hydrogen binding templates is also studied (Peng et al., 2009). Recent studies also portray the utility of carbon-nitride heterofullerenes as suitable endohedral trapping templates for engulfing hydrogen (Javan et al., 2011). Experimental studies by Saha et al. (2011) have further demonstrated the improved hydrogen adsorption capability of Pd and Ru dispersed C_{60} fullerene compared to the corresponding nondispersed ones. Owing to the "extra" strength of a B – N bond in a boron-nitrogen system due to a favorable electron donation from $N_{(2p\pi)}$ to $B_{(2p\pi)}$ orbitals, a lot of research groups have aimed to model $(BN)_n$ cage clusters with the B – N bond as the repeating unit. These $(BN)_n$ cages are able to entrap molecular hydrogen in both endohedral and exohedral modes. Relevant first principles study on the $B_{36}N_{36}$ cage reveal that the given molecule is able to incorporate up to 18 hydrogen molecules endohedrally at zero temperature leading to a gravimetric density of 4% (Sun et al., 2005). The results however show a conflicting trend at room temperature (300K) upon employing high-level molecular dynamics using Nose algorithm. Thus, the hydrogen binding potential of a $B_{36}N_{36}$ cage can be enhanced upon incorporation of a suitable charged species like a metal ion which, in turn, will attract the incoming hydrogen molecules with an added ion-induced dipole type of interaction in addition to the already existing weaker van der Waals type of forces (Wen et al., 2008). The plausible usage of metal-doped borazine networks and boron buckyballs (B_{80}) for hydrogen storage purpose is also investigated (Shevlin and Guo, 2006; Wu et al., 2009). Another study shows that C-doped B – N cages can bind hydrogen with a gravimetric density of up to 7.43% hence the given $B_{11}N_{12}C$ cage behaves as a potential storage material capable of a thermodynamically favorable hydrogenation and dehydrogenation kinetics under ambient conditions (Wu et al., 2010). A scrutiny of the hydrogen binding on the $B_{16}N_{16}$ cage and its associated smaller analogs reveals that the interaction energy falls off with increasing size of the cage cavity (Cui et al., 2010). The nature of hydrogen interaction and its storage within the $B_{16}N_{16}$ molecule upon incorporation of a charged guest species inside the cage cavity is also reported (Cui et al., 2010).

The clathrate hydrates, introduced by Powell consist of water molecules stacked together in a cage-like fashion (Powell, 1948). The possible utility of the clathrates as novel templates for hydrogen storage purposes has recently become an active area of research for both the experimentalists and the theoreticians. The term clathrate derives its name from the latin word "*clathratus*" meaning "enclosed or protected by cross bars of a grating" are in fact H_2O-modeled solid crystalline structures mimicking the ice frameworks. So, alike ice, hydrogen bonding serves as the primary binding force towards the physical modeling of the clathrate cage structures. The cage cavity entraps small guest molecules, both polar and nonpolar having large hydrophobic chains. However, in the absence of any guest molecule, the cage-like clathrate hydrate framework ceases to exist and transforms itself into the conventional ice-lattice arrangement or even liquid water by a first order phase transition. Some relevant articles nicely delineates the usage of several clathrate hydrate molecules towards entrapping hydrogen and thus acting as an effective storage material (Lang et al., 2010; Sloan and Koh, 2007; Struzhkin et al., 2007; Strobel et al., 2007). In this regard the semiclathrate hydrates containing an ionic moiety as a guest species also deserve a special mention (Chapoy et al., 2007; Jeffrey, 1984; Shimada et al., 2005). Results show that the semiclathrates containing quaternary ammonium ion as the guest species can store hydrogen in considerable volumetric amounts at low pressures (Chapoy et al., 2007). Relevant Raman spectroscopic and phase equilibrium measurements by Hashimoto et al. (2010) further explore the thermodynamic stability and hydrogen tenancy of the hydrogen + tetra-n-butyl ammonium bromide semiclathrate hydrates. Other spectroscopic methods including Raman, IR, X-ray and neutron diffraction are also executed to explore the hydrogen storage efficacy in sII type clathrates under ambient pressure and low temperatures (Mao et al., 2002; Mao and Mao, 2004). The utility of binary clathrate hydrates with tetrahydrofuran

(THF) as useful templates for hydrogen storage is also studied in detail (Florusse et al., 2004; Lee et al., 2005). These binary THF-clathrates, however, capture hydrogen with a low gravimetric density followed by a slow desorption kinetics and hence pose a difficulty towards their widespread applications as a fruitful storage material (Talyzin, 2008). The use of polymer hydrogels (Su et al., 2009) or polymerized high internal phase emulsion (polyHIPE) are found to improve the hydrogen storage capacity and desorption kinetics of THF-H_2O clathrate hydrates (Su et al., 2008). Some studies reveal that *tert*-butylamine bound binary clathrate hydrates can also serve as potential storage stuffs for molecular hydrogen (Aladko et al., 2010; Prasad et al., 2009).

6.3 A THEORETICAL APPROACH TO MODEL AN EFFECTIVE HYDROGEN STORAGE MEDIUM

6.3.1 THEORETICAL BACKGROUND

In this section we delineate the computational methodology used in our laboratory. Among these computational tools mainly we are using density functional theory. During last two decades conceptual density functional theory (CDFT) has become indispensible as a theoretical tool not only to chemists but also to people in other fields of science (Chakraborty et al., 2012; Chattaraj, 2009; Chattaraj and Giri, 2009; Geerlings et al., 2003; Parr and Yang, 1989). Density functional theory has given us few parameters by which we can explain the underlying properties of matter; these are called conceptual DFT based reactivity descriptors. These reactivity descriptors give an idea about the bonding, stability and reactivity of molecular systems. Study of global reactivity descriptors like electronegativity (χ), hardness (η) and electrophilicity (ω) reveals the stability and reactivity of the molecular motif whereas local reactivity descriptors: atomic charges (Q_k) and Fukui functions (f_k), describe the site-selectivity of a chemical species (Chattaraj and Roy, 2007; Chattaraj, 1992; Chattaraj et al., 2006, 2011; Mulliken, 1955; Parr et al., 1978, 1999; Parr and Pearson, 1983; Parr and Yang, 1984; Pearson, 1997).

The expressions of electronegativity (χ), hardness (η) and electrophilicity (ω) for an N-electron system are:

$$\chi = -\left(\frac{\partial E}{\partial N}\right)_{v(\mathbf{r})} = -\mu \tag{1}$$

$$\eta = \left(\frac{\partial^2 E}{\partial N^2}\right)_{v(\mathbf{r})} \tag{2}$$

and $$\omega = \left(\frac{\chi^2}{2\eta}\right) = \left(\frac{\mu^2}{2\eta}\right) \tag{3}$$

where $v(r)$ and μ are the external potential and chemical potential respectively (Chattaraj, 1992; Chattaraj and Roy, 2007; Chattaraj et al., 2006, 2011; Parr et al., 1978, 1999; Parr and Pearson, 1983; Pearson, 1997).

Electronegativity and hardness can be expressed as follow with the help of a finite difference method:

$$\chi = \frac{(I + A)}{2} \text{ and } \eta = (I - A) \tag{4}$$

where I and A represent the ionization potential and electron affinity of the system respectively and are computed applying Koopmans' theorem (1933).

$$\chi = -\frac{(E_{HOMO} + E_{LUMO})}{2} \text{ and } \eta = (E_{LUMO} - E_{HOMO}) \tag{5}$$

where E_{HOMO} and E_{LUMO} are the energies of the highest occupied and the lowest unoccupied molecular orbitals. Again we can express them in terms of the energies of the N and $N \pm 1$ electron systems using ΔSCF technique. For an N-electron system with energy $E(N)$ they may be expressed as follow:

$$I \approx E(N-1) - E(N) \tag{6}$$

$$A \approx E(N) - E(N+1) \tag{7}$$

The local reactivity descriptor, Fukui function (FF) measures the change in electron density at a given point when an electron is added to or removed from a system at constant $v(\vec{r})$ (Parr and Yang, 1984). It may be written as:

$$f(\vec{r}) = \left(\frac{\partial \rho(\vec{r})}{\partial N}\right)_{v(\vec{r})} = \left(\frac{\delta \mu}{\delta v(\vec{r})}\right)_N \tag{8}$$

In terms of electron population p_k of an individual atomic site k in a molecule the following expressions for Fukui function, $f(\vec{r})$, may be written as (Yang and Mortier, 1986)

$$f_k^+ = p_k(N+1) - p_k(N) \text{ for nucleophilic attack} \tag{9a}$$

$$f_k^- = p_k(N) - p_k(N-1) \text{ for electrophilic attack} \tag{9b}$$

$$f_k^0 = [p_k(N+1) - p_k(N-1)]/2 \text{ for radical attack} \tag{9c}$$

Calculation of local philicity (ω_k^α) reveals the site selectivity for a particular atomic site in a molecule. The local philicity (ω_k^α) variants for the *kth* atomic site in a molecule are expressed as

$$\omega_k^\alpha = \omega . f_k^\alpha \tag{10}$$

where α = +, – and 0 represent nucleophilic, electrophilic and radical attacks respectively.

The maximum hardness principle (MHP), minimum polarizability principle (MPP) and minimum electrophilicity principle (MEP) serve well to assess the stability order of reactants and products (Ayers and Parr, 2000; Chamorro et al., 2003; Chattaraj and Sengupta, 1996; Fuentealba et al., 2000; Parr and Chattaraj, 1991; Parthasarathi et al., 2005; Pearson, 1987). In addition to these molecular electronic structure principles there are few thermodynamic parameters, which help to conclude the effectiveness of hydrogen binding. These parameters are: interaction energy (*IE*), interaction energy per H_2 molecule (ΔE), gain in energy (*GE*), reaction enthalpy (ΔH), dissociative chemisorption energy (ΔE_{CE}). The working formulas to get these parameters are following:

$$IE = E_{nH_2X} - [E_X + nE_{H_2}];\ n = \text{no. of molecular } H_2 \tag{11}$$

$$\Delta E = (1/n)[E_{X(H_2)_n} - \left(E_X + nE_{H_2}\right)];\ \text{n=no. of molecular } H_2 \tag{12}$$

$$GE = E_{(n-1)H_2X} + E_{H_2} - E_{nH_2X} \tag{13}$$

$$\Delta E_{CE} = \frac{2}{n}[E_X + \frac{n}{2}E_{H_2} - E_{XH_n}];\ n = \text{no. of H atoms} \tag{14}$$

where E_X is the energy of the parent moiety.

There are different aromaticity indices but among them the nucleus-independent chemical shift (NICS) has become one of the extensively used aromaticity probes (Schleyer, 2001; Schleyer et al., 1996). NICS reveals the extent of electron delocalization and it is equal to the negative of isotropic nuclear magnetic shielding tensor of the ghost atom, which has been used to measure the aromatic character of different compounds. The NICS values at the cage centers are computed by using the gauge-independent atomic orbital (GIAO) method at the B3LYP/6–31G

(d) level of theory with the B3LYP/6–31G (d) geometries. The NICS-rate at a given distance r from the ring center is calculated with the following mathematical formula given as (Jimenez-Halla et al., 2009; Li et al., 2008; Noorizadeh and Dardaba, 2010; Stanger, 2006):

$$\text{NICS-rate}(r) = \frac{d\text{NICS}}{dr} \cong \text{Lim}_{\Delta r \to 0} \frac{\text{NICS}(r + \Delta r) - \text{NICS}(r)}{\Delta r} \quad (15)$$

6.3.2 COMPUTATIONAL DETAILS

Most of the computations are carried out by using GAUSSIAN 03 and GAUSSIAN 09 program packages (Frisch et al., 2009; GAUSSIAN 03, 2003). Full geometry optimization as well as harmonic vibrational frequency calculation for all the systems (excluding sI hydrogen hydrate and $C_{12}N_{24}$ nano tube) are carried out at different levels of theory with various basis sets. Used levels of theories are: B3LYP, DFT-D-B3LYP, MP2, MPW1K, MO52X, MO6, PBE0; whereas the different basis sets are: 6–31G, 6–31+G(d), 6–311+G(d), 6–311+G(d, p), cc-pvdz. Harmonic vibrational frequency calculation confirms that all the reported structures correspond to minima on the potential energy surface (PES). All the global as well as local reactivity descriptors are calculated by the equations mentioned in the theoretical background section. Effect of electric field has been examined on different isomers with a single adsorbed hydrogen molecule. In presence of electric field the geometries are optimized with dispersion correction, at various levels of theory. The electric field has been applied in the *x* direction. Binding energies in presence of electric field are calculated by using the following formula:

$$\Delta E^F = -[E^F{}_{system} - E{}_{system}] \quad (16)$$

where ΔE^F and $E^F{}_{system}$ indicate the binding energy and total energy of the system respectively in the presence of the electric field F whereas E_{system} indicates the energy of the system in the absence of the same. We have used the graphical software GaussView 3.0 to model all the initial geometries as well as to understand the associated structures[193] (Frisch et al., 2009).

For all the elemental constituents projector augmented wave (PAW) potentials are employed and the used kinetic energy cutoff is 550 eV (Blöchl, 1994; Kresse and Joubert, 1999). The exchange–correlation energy density functional $E_{xc}(\rho)$ has been treated using the Generalized Gradient Approximation (GGA) of Perdew-Burke-Ernzerhof (PBE) (1996). All the electronic degrees of freedom were relaxd until the change in energy in the self-consistent field iteration is less than 1×10^{-6} eV. The ionic optimizations are carried out at constant cell volume until the Hellmann-Feynman force components on each atom are less than 0.01

eV $Å^{-1}$. The Brillouin zone is sampled using the automatically generated 1x1x6 Monkhorst-Pack set of *k*-points (1976). The reported Fig.s of the periodic systems are generated using the graphical software XCrySDen and the periodic calculations are carried out through the first principles periodic density functional methods using the Vienna *ab initio* Simulation Package (VASP) (Kresse and Furthmüller, 1996; Kresse and Hafner, 1993, 1994; Kokalj, 2003).

Extended SPC water model is used to model the structure of hydrates with different number of hydrogen molecules (Mizan, 1994; Teleman et al., 1987). An imaginary sphere without any atomic charge, which resembles hydrogen molecule, is incorporated instead of hydrogen molecule for the entire classical simulation (Klauda and Sandler, 2003). The potential energy equation between two atoms *i* and *j* can be written as:

$$V_{ij} = \sum_{m\epsilon i}\sum_{n\epsilon j}\frac{q_m q_n}{r_{mn}} + 4\epsilon_{ij}\left[\left(\frac{\sigma_{ij}}{r_{ij}}\right)^{12} - \left(\frac{\sigma_{ij}}{r_{ij}}\right)^{6}\right] \tag{17}$$

where q_m and q_n are the charges on atoms *i* and *j*, respectively, r_{mn} is the distance between the q_m and q_n charged *i*th and *j*th atoms, ε_{ij}, σ_{ij} are Lennard-Jones parameters and r_{ij} is the distance between the two interacting atoms *i* and *j*.

Lorentz-Berthelot rule is used to calculate the interaction potential of unlike atoms (Allen and Tildesley, 1996; Hirschfelder et al., 1954). Temperature and pressure scaling are done with the Nose-Hoover thermostat and barostat as employed in the LAMMPS package (Plimpton, 1995). The bond lengths and bond angle of water molecules are frozen by using the fix shake algorithm. A PPPM (*particle-particle particle-mesh*) solver is used with a precision of 1×10^{-4} to calculate the long-range Coulombic interactions. The simulation is carried out with a time step of 1.0 fs. During energy minimization and temperature ramping the neighbor lists are updated at every time step but for equilibration at every 10 steps. NPT and NVT ensembles are used during temperature and pressure scaling. The VMD (Visual Molecular Dynamics) program package is used to generate pictorial presentations (Humphrey et al., 1996). We have used LAMMPS package to perform all the molecular dynamics simulations.

6.3.3 EXOHEDRAL AND ENDOHEDRAL HYDROGEN TRAPPING BY CAGE-LIKE STRUCTURES

6.3.3.1 HYDROGEN HYDRATES

Chattaraj et al. (2011) have shown the potentiality of clathrate hydrates as a molecular hydrogen-trapping agent. An analysis of the results at B3LYP/6–31G(d) level reveals that only 5^{12}, $5^{12}6^2$ and $5^{12}6^8$ clathrate structures possess minima on PES among the other cage units of clathrate hydrates. The size and structure of the clathrate cage unit are the key factors while we talk about the capacity of storing molecular hydrogen. It has been observed that the interaction energy is negative up to encapsulation of one hydrogen molecule in case of 5^{12} and two hydrogen molecules in case of $5^{12}6^2$. In case of $5^{12}6^8$ interaction energy remains negative up to encapsulation of six hydrogen molecules which was as expected as the cage size of $5^{12}6^8$ is much more greater than that of 5^{12} and $5^{12}6^2$. Conceptual DFT based reactivity descriptors like hardness (η), electrophilicity (ω) are well known in rationalizing stability of a molecular or ionic system. The increasing trend of hardness and decreasing trend of electrophilicity in most of the cases with gradual encapsulation of molecular hydrogen indicate an increased stability with gradual hydrogen loading according to maximum hardness principle (MHP)[179–181] and minimum electrophilicity principle (MEP)[184,185] (Ayers and Parr, 2000; Chamorro et al., 2003; Parr and Chattaraj, 1991; Parthasarathi et al., 2005; Pearson, 1987). Therefore, the unit clathrate hydrate cages are found to be potential molecular hydrogen trapping agents. In Fig. 4, optimum hydrogen encapsulated 5^{12}, $5^{12}6^2$ and $5^{12}6^8$ cages are provided.

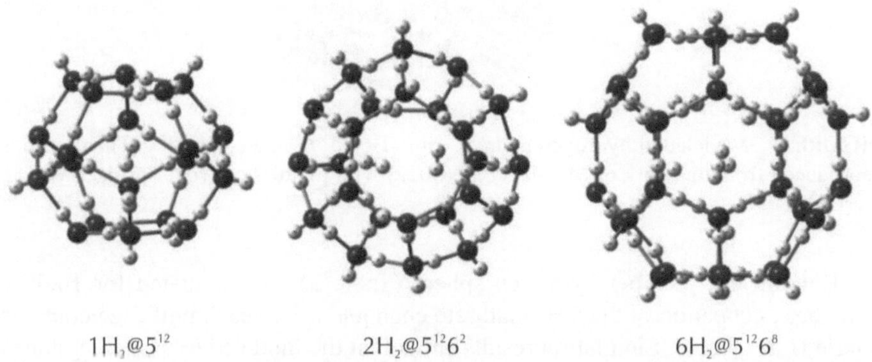

FIGURE 4 Hydrogen encapsulated 5^{12}, $5^{12}6^2$ and $5^{12}6^8$ cages at B3LYP/6–31G(d) level of theory. (This figure is reproduced from the work of Chattaraj et al. (2011) with permission from American Chemical Society.)

In a recent classical molecular dynamics simulation Mondal et al. (2013) have further investigated the nature of 5^{12} and $5^{12}6^2$ clathrate cages in SI hydrogen hydrate from the view point of molecular hydrogen holder. The SI hydrogen hydrate is modeled with different number of hydrogen molecules. Interaction potentials of the atoms are given in the Table 2.

TABLE 2 Lennard-Jones interaction parameters and atomic charges for SPC/E water and the spherical H_2 molecule (This table is reproduced from the work of Mondal et al. with permission from Springer-Verlag, Berlin.)

Atom	σ_{ij} (Å)	ε_{ij} (kcal/mol)	q (e)
O	3.166	0.1553	–0.8476
H (water)	0.0	0.0	0.4238
H (guest)	3.140	0.0190	0.0

Loaded hydrogen molecule goes into each $5^{12}6^2$ cage of each unit cell whereas 5^{12} cage remains vacant (Fig. 5).

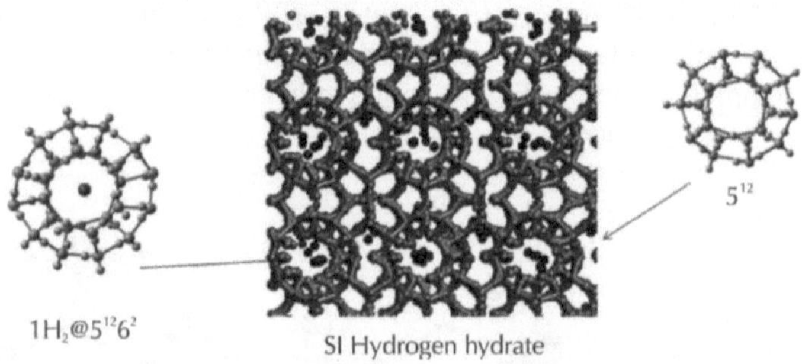

FIGURE 5 Modeled sI hydrogen hydrate with $1H_2@5^{12}6^2$ and empty 5^{12}. (This figure is reproduced from the work of Mondal et al. (2012) with permission from Springer-Verlag, Berlin.)

Binding energies per hydrogen sphere (Table 3) are calculated for 16.66% and 50% occupation of the $5^{12}6^2$ clathrate channels where each unit cage contains single H_2 molecule. Simulation results show that the modeled hydrogen hydrates are stable up to ~200 fs where as the SI hydrate is unstable.

TABLE 3 Binding energy per hydrogen sphere for the loading of n hydrogen spheres in the $5^{12}6^2$ channel of sI clathrate slab. (This table is reproduced from the work of Mondal et al. with permission from Springer-Verlag, Berlin.)

% Occupied	No. of H2 molecules (n)	ΔE/n (kcal/mol)
16.66	216	–3.82
50	648	–4.00

When the $5^{12}6^2$ clathrate channels of sI hydrogen hydrate is half occupied and each unit cage contains single H_2 molecule, the gravimetric weight percent of hydrogen storage becomes only 0.72. But when the 5^{12} and $5^{12}6^2$ cages of the clathrate slab would be occupied by one and two hydrogen spheres respectively then the gravimetric weight percent of hydrogen storage will increase substantially.

6.3.3.2 $B_{12}N_{12}$ CAGE

In a computational investigation Giri et al. studied the stability as well as aromaticity of $B_{12}N_{12}$ cage and their hydrogen-loaded analogs (Giri et al., 2011). These $B_{12}N_{12}$ clathrate is different from the clathrate hydrates in the manner of trapping molecular hydrogen. In case of $B_{12}N_{12}$ the hydrogen adsorption takes place exohedrally. The nitrogen centers of $B_{12}N_{12}$ possess slight negative charge (Fig. 6), which can bind H_2 molecules exohedrally thus N center acts as the hydrogen-binding site. The interaction energy per H_2 molecule is calculated for the adsorption of all the hydrogen. Obtained negative value of interaction energy concludes a favorable interaction for adsorption of molecular hydrogens with the N centers whereas positive gain in energy by the molecular system shows the greater stability of the nH_2-trapped (n = 1–12) $B_{12}N_{12}$ systems compared to the empty $B_{12}N_{12}$ cage.

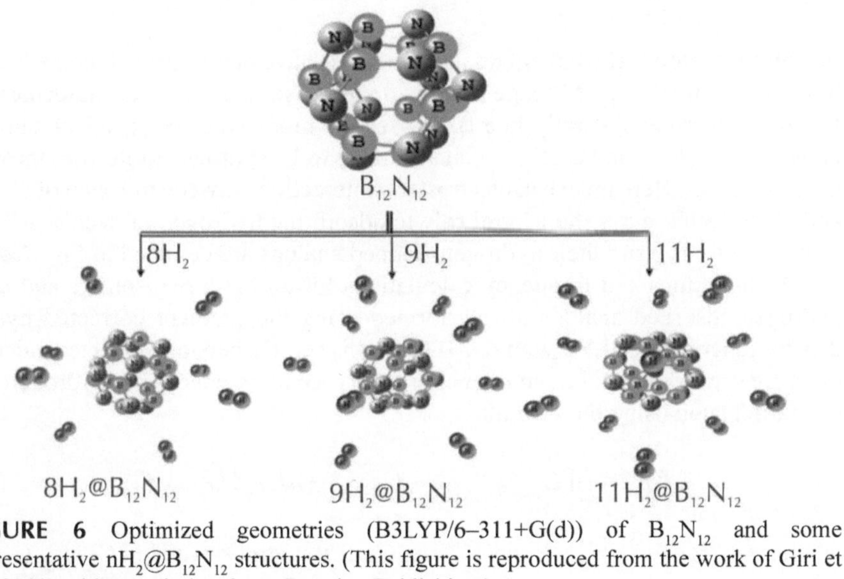

FIGURE 6 Optimized geometries (B3LYP/6–311+G(d)) of $B_{12}N_{12}$ and some representative $nH_2@B_{12}N_{12}$ structures. (This figure is reproduced from the work of Giri et al. (2011) with permission from Coaction Publishing.)

Stability and reactivity of hydrogen adsorbed $B_{12}N_{12}$ moieties are assessed by employing various conceptual DFT based reactivity descriptors. As the number of hydrogen molecule adsorbed on to the $B_{12}N_{12}$ cage increases, the electronegativity (χ) of the resulting molecular system are found to decrease, which indicates further entry of electrons into the system, is not favorable. Hardness (η) value shows an increasing trend whereas electrophilicity (ω) values show mostly a decreasing trend. The molecular electronic structure principles like maximum hardness principle (MHP) and minimum electrophilicity principle (MEP) corroborates well with the increasing number of adsorbed hydrogen molecule (Ayers and Parr, 2000; Chamorro et al., 2003; Pearson, 1987; Parr and Chattaraj, 1991; Parthasarathi et al., 2005). The reaction electrophilicity values are calculated for the adsorption of H_2 on to the $B_{12}N_{12}$ cage and the resulting negative values indicate that the processes would be favorable. The cage aromaticity in terms of nucleus-independent chemical shift (NICS) is calculated at the cage center for the bare as well as hydrogen adsorbed $B_{12}N_{12}$ systems. The calculated negative NICS values for the bare as well as H_2 loaded analogs indicate that the systems are aromatic in nature. From the justification of negative interaction energy per H_2 molecule, positive gain in energy by the system, negative reaction electrophilicity and the presence of aromaticity it is clear that the $B_{12}N_{12}$ cage is an effective hydrogen-trapping agent.

6.3.3.3 $C_{12}N_{12}$ CAGE

In a recent theoretical work Mondal et al. (2012) have demonstrated the hydrogen trapping potential of $C_{12}N_{12}$ cage isomers. In this case also hydrogen molecules are found to adsorb exohedrally like $B_{12}N_{12}$. All the modeled $C_{12}N_{12}$ isomers namely, $C_{12}N_{12}$-A, $C_{12}N_{12}$-B and $C_{12}N_{12}$-C can adsorb up to 12 hydrogen molecules through their N centers. Here the weak electrostatic interaction between nitrogen of $C_{12}N_{12}$ and H_2 molecule plays the central role to adsorb the hydrogen molecule. All the isomers of $C_{12}N_{12}$ and their hydrogen-trapped analogs are depicted in Fig. 7.

Optimizations and frequency calculations for all the three isomers and their hydrogen-adsorbed analogs are performed using the gradient corrected hybrid density functional B3LYP with 6–31G(d) basis set. Dispersion corrected calculations are carried out at the same level of theory. Binding energy per hydrogen has been calculated using the formula,

$$\Delta E = -[E_{nH_2@C_{12}N_{12}} - E_{C_{12}N_{12}} - nE_{H_2}]/n \tag{18}$$

where, $E_{nH_2@C_{12}N_{12}}$, $E_{C_{12}N_{12}}$ and E_{H_2} denote the energies of $nH_2@C_{12}N_{12}$, $C_{12}N_{12}$ and H_2 respectively.

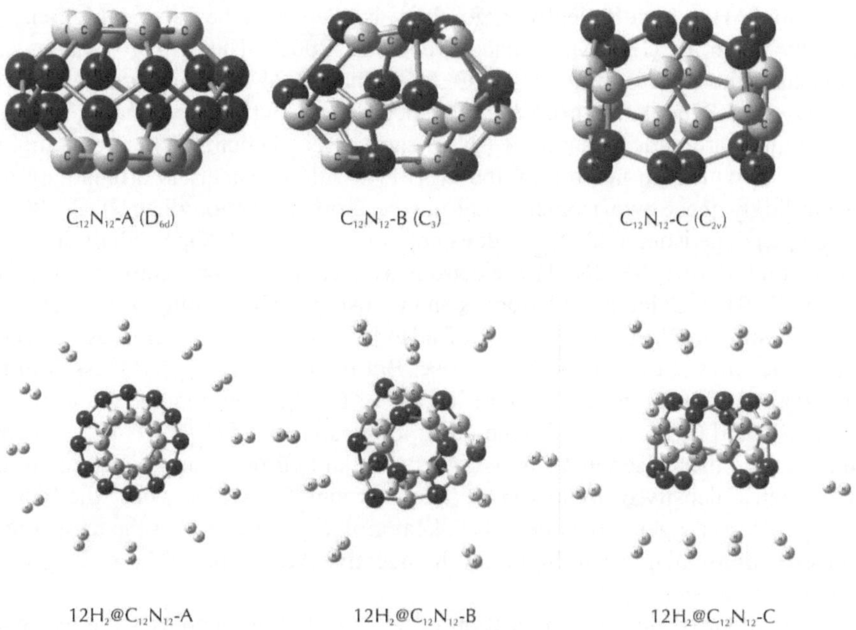

FIGURE 7 Optimized geometries of three $C_{12}N_{12}$ isomers and their hydrogen-adsorbed analogs at DFT-D-B3LYP/6–31G(d) level of theory.

The stability orders of three isomers are, $C_{12}N_{12}$-A <$C_{12}N_{12}$-B <$C_{12}N_{12}$-C where $C_{12}N_{12}$-C is 55.41 kcal/mol more stable than $C_{12}N_{12}$-B and 66.11 kcal/mol more stable than $C_{12}N_{12}$-A. The extra stability of $C_{12}N_{12}$-C isomer originates from its open cage skeleton and as it contains only five membered rings so it has less strain (Douglas, 2006; Karleta et al., 2007).

Gas phase heat of formation (ΔH°_f) values is quite high for the synthesis of $C_{12}N_{12}$. Gas phase heats of formation of these $C_{12}N_{12}$ isomers are calculated by using the following isodesmic reaction:

$$C_{12}N_{12} + 4\ H_2C{=}CH_2 + 3NH_3 \rightarrow 5\ C_4H_4NH + 5\ H_2N\text{-}NH_2 \qquad (19)$$

To calculate the ΔH°_f of the above reaction experimental ΔH°_f values of ethylene, ammonia, pyrrole, and hydrazine are used. Obtained ΔH°_f (eV) values are 77.470, 208.574, and 210.918 for $C_{12}N_{12}$-A, $C_{12}N_{12}$-B and $C_{12}N_{12}$-C, respectively (Chas, 1998; Scott et al., 1967). From these ΔH°_f values we can say that $C_{12}N_{12}$ could be synthesized and the isomer $C_{12}N_{12}$-C is the most stable one. Since all the values of ΔH°_{f} are sufficiently large, these isomers may be used as high-energy density materials (HEDMs) (Liang-Wei et al., 2008). The NICS(0) values calculated at the centers of $C_{12}N_{12}$-A, $C_{12}N_{12}$-B, and $C_{12}N_{12}$-C cages are found to be −2.6006, −5.8764, and −4.6387 ppm respectively which indicate that these cages are slightly aromatic in nature. $C_{12}N_{12}$-A and $C_{12}N_{12}$-B both of them contain 24π electrons, which do not follow the $2(N+1)^2 \pi$ rule of spherical aromaticity and accordingly these two isomers may not be aromatic (Zhongfang et al., 2001). The open cage isomer, $C_{12}N_{12}$-C does not obey the rule of "Open-Shell Spherical Aromaticity" of $(2N^2+2N+1)\ \pi$ electrons as it contains 24π electrons (Jordi and Miquel, 2011). Calculating hardness and by using MHP stability of a system can also be supported. As the number of adsorbed H_2 molecule increases, hardness values for all the three isomers increase. But the increment of hardness with the adsorption of H_2 molecule is more in case of $C_{12}N_{12}$-A in comparison to that of $C_{12}N_{12}$-B and $C_{12}N_{12}$-C. Hydrogen adsorption capacity of $C_{12}N_{12}$-A is better than that of the others. Isomer $C_{12}N_{12}$-A is a potential hydrogen tapping agent ant the gravimetric density of hydrogen storage is around 7.2 wt %, where the binding energy per hydrogen molecule is 1.22 kcal/mol. Fig. 8 represents the temperature pressure phase diagram to highlight the negative ΔG region of the hydrogen adsorption process.

It is reported that the application of an electric field increases the hydrogen storage capacity of the adsorbent without destroying any molecular integrity (Zhoua et al., 2010). Electric field is applied along the *x* direction from 0.001 au to 0.005 au with an interval of 0.001 au. With an increase in the applied electric field there is a change in orientation of the adsorbed H_2 molecule of $H_2@C_{12}N_{12}$-B at field strength of 0.003 a.u. The trend of increasing binding energy is good in case of $H_2@C_{12}N_{12}$-A and $H_2@C_{12}N_{12}$-C for the adsorption of the single H_2 molecule. Electric field of strength 0.005 a.u. increases the binding energy by 0.46 kcal/mol in the case of $C_{12}N_{12}$-A. Removal of electric field decreases binding energy which can facilitate the desorption of hydrogen. Binding energy per hydrogen molecule calculated at DFT-D-B3LYP/6–31G(d) method (1.22 kcal/mol) is less than the one which is obtained by ab initio calculation, using PBE method (1.44 kcal/mol). Modeled $C_{12}N_{24}$ nanotube (Fig. 9) can adsorb more hydrogen molecule than that of $C_{12}N_{12}$-A, with a gravimetric hydrogen density of 9.1 wt%. A detailed study on the intermolecular interactions of these $C_{12}N_{12}$ isomers result a one-dimensional cluster assembled material (Fig. 9).

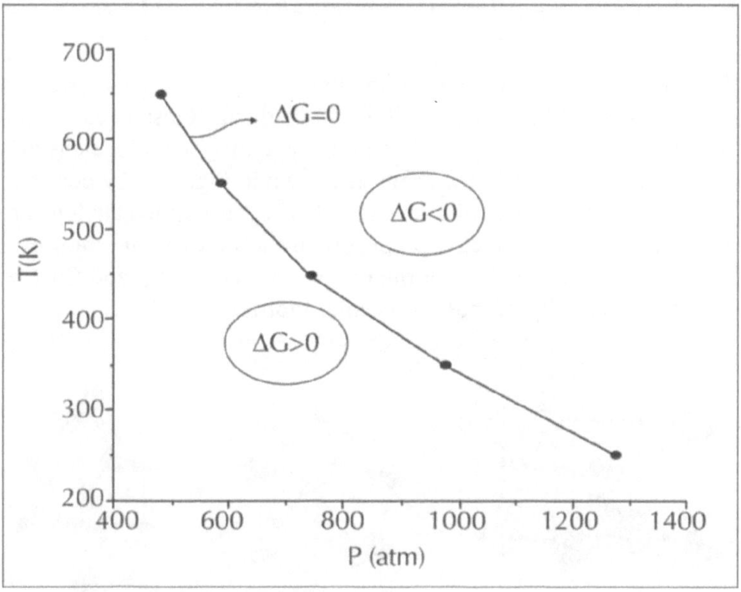

FIGURE 8 T-P phase diagram depicting negative ΔG region for the adsorption process ($12H_2 + C_{12}N_{12}$-A) to $12H_2@C_{12}N_{12}$-A.

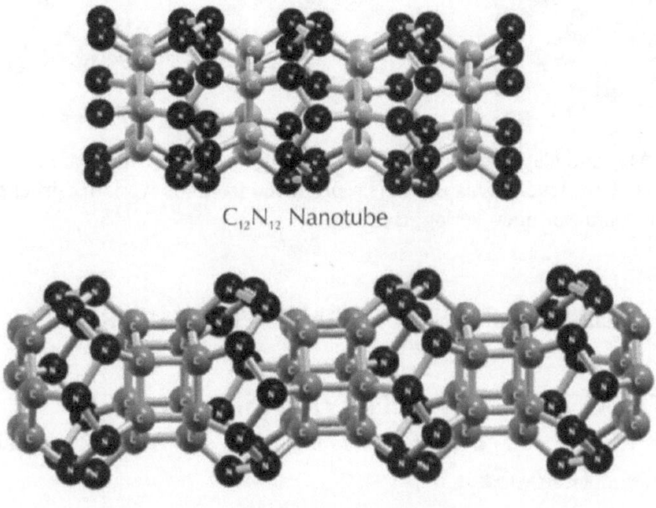

FIGURE 9 The optimized geometries of the unit cell (a) and its 2 × 1 × 1 super cell (b) of the one-dimensional cluster assembled material based on $C_{12}N_{12}$-A.

6.3.3.4 Mg_n AND Ca_n (n=8–10) CAGE-LIKE CLUSTERS

Giri et al. (2011) did a study on the endohedral hydrogen storage capacity of Mg_n and Ca_n (n=8–10) cage-like clusters. All the studied cage clusters can adsorb molecular hydrogen except Ca_{10} cage. It is found that hydrogen molecule is dissociated into corresponding atoms in Ca_{10} cage as shown in Fig. 10. The corresponding values of hardness and electrophilicity of the hydrogen-trapped analogs are quite high and low respectively, thereby, support their thermodynamic stability. Negative NICS(0) and NICS(1) values of the hydrogen-loaded Mg_n and Ca_n (n=8–10) clusters justify the stability of these molecular motifs.

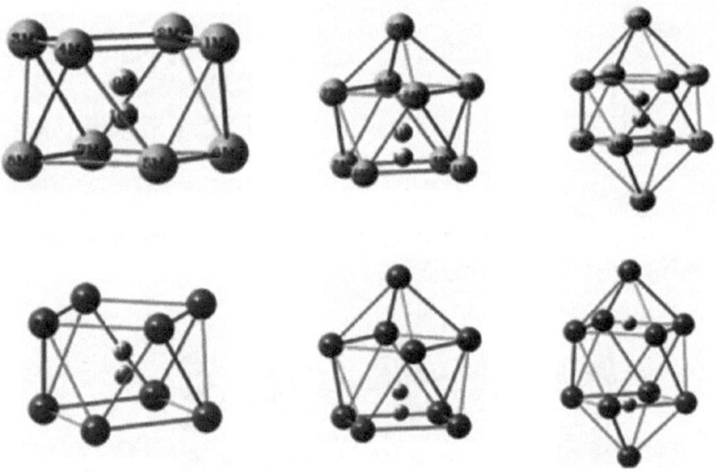

FIGURE 10 Mg_n and Ca_n (n=8–10) clusters trapping hydrogen endohedrally; optimized at B3LYP/6–311+G(d) level. (This figure is reproduced from the work of Giri et al. (2011) with permission from Springer-Verlag, Berlin.)

6.4 METAL CLUSTERS INTERACTING WITH HYDROGEN

6.4.1 INTERACTION BETWEEN TRANSITION METAL (TM) AND H_2 IN TM DOPED ETHYLENE COMPLEX AND HYDROGEN STORAGE

The mechanism and molecular interaction between hydrogen and the transition metal-ethylene complexes are studied theoretically (Chakraborty et al., 2011;

Durgun et al., 2006; Zhou et al., 2007). Doping of ethylene molecule has been done with four different transition metals (TM) namely Sc, Ti, Fe and Ni. The general formula of the doped mother moiety is ($M_n-(C_2H_4)$) where M = Sc, Ti, Fe, Ni and n = 1, 2. Hydrogen molecules are adsorbed on the TM centers (Fig. 11).

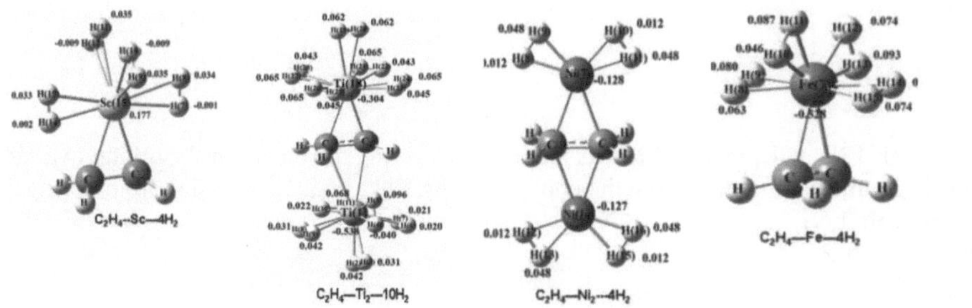

FIGURE 11 M-ethylene (M=Sc, Ti, Fe, Ni) complex and its corresponding H_2-trapped analogs studied at B3LYP/6–311+G(d, p) level. (This figure is reproduced from the work of Chakraborty et al. (2011) with permission from Springer-Verlag, Berlin.)

As the number of adsorbed hydrogen molecules on the complex ($M_n-(C_2H_4)$) increases the stability of the resulting hydrogen adsorbed complex increases, which justifies the hydrogen trapping potential. Hardness and electrophilicity values calculated at B3LYP and MP2 levels for ($Sc_n-(C_2H_4)$) and ($Ti_n-(C_2H_4)$) complexes do not corroborate well with each other with gradual hydrogen loading; where as in case of Fe and Ni-doped complexes, they correlate with each other at B3LYP level but fluctuate at MP2 level. Fe and Ni-doped ethylene complexes at B3LYP level of theory satisfy both maximum hardness principle (MHP) and minimum polarizability principle (MPP). Calculated values of reaction electrophilicity ($\Delta\omega$) and the reaction enthalpy (ΔH) support the feasibility of hydrogen adsorption process on ($M_n-(C_2H_4)$) complexes. The interactions through which the molecular hydrogen adsorbs on the metal site in ($M_n-(C_2H_4)$) complexes are electrostatic and of Kubas (2001) type in nature. The adsorbed H_2 having dihydrogen character interact through diplole-induced dipole and dipole-quadrupole type of interactions with metal. Kubas interaction explains the dihydrido character of hydrogen in which σ electrons of molecular hydrogen come to the metal d-orbital and back-donation of electrons occur from metal d-orbital to σ* orbital of H_2 molecule (Fig. 12).

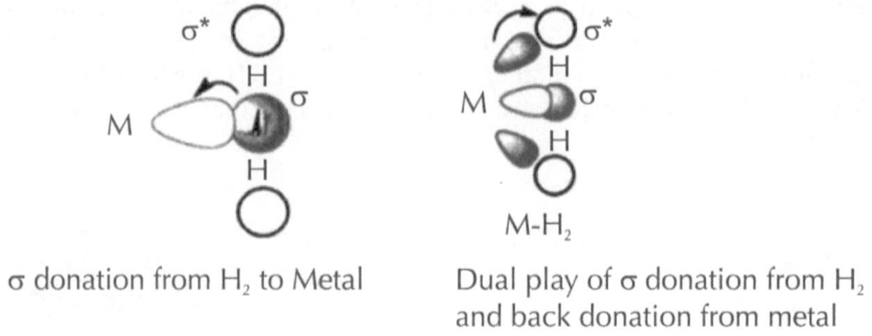

FIGURE 12 A depiction of Kubas interaction showing the nature of the orbital overlap. (This figure is reproduced from the book edited by Kubas with permission from Springer-Verlag, Berlin.)

6.4.2 M_3^+ (M=Li, Na) CLUSTER

Giri et al. (2011) in a recent article discover the possibility of holding hydrogen molecules on small all-metal aromatic clusters like Li_3^+ and Na_3^+. In Fig. 13, optimized geometries of H_2-trapped Li_3^+ and Na_3^+ are provided. It is found that when the vertices of the trigonal all-metal rings adsorb hydrogen, the latter retains its molecular nature but it dissociates upon trapping on the trigonal plane.

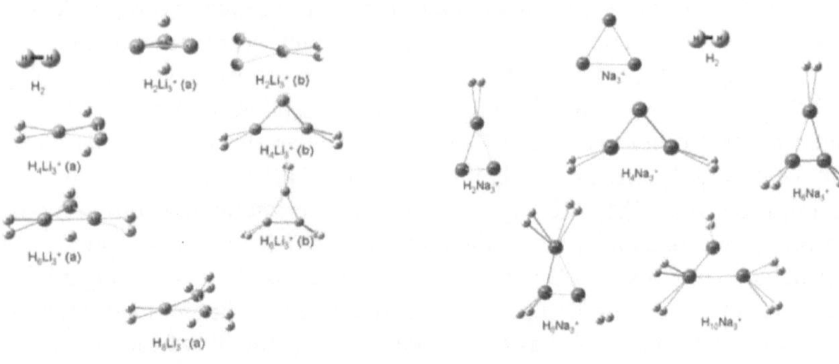

FIGURE 13 Optimized geometries of hydrogen trapped Li_3^+ and Na_3^+ clusters at B3LYP/6–311+G(d) level. (This figure is reproduced from the work of Giri et al. (2011) with permission from Springer-Verlag, Berlin.)

Calculated negative interaction energy of various Li_3^+ and Na_3^+ clusters with steady hydrogen trapping reveals the potential of these clusters as effective hydrogen trapping systems. Calculated negative interaction energy supports well the hardness (η), electrophilicity (ω) and $NICS_{zz}$ values.

6.4.3 BE_3-M_2, MG_3-M_2 AND AL_4-M_2 (M=Li, Na, K) CLUSTERS

Srinivasu et al. (2012) studied the potential of binary all-metal aromatic clusters as hydrogen trapping agents. They got these binary all-metal aromatic clusters by doping them with counter ions like Li^+, Na^+ and K^+. As these doped systems contain alkali atoms so there is a slight positive charge on each of them and that is the possible reason for adsorption of molecular hydrogen through dipole-induced dipole type of interaction. Fig. 14 shows the hydrogen adsorbed moieties of Be_3M_2.

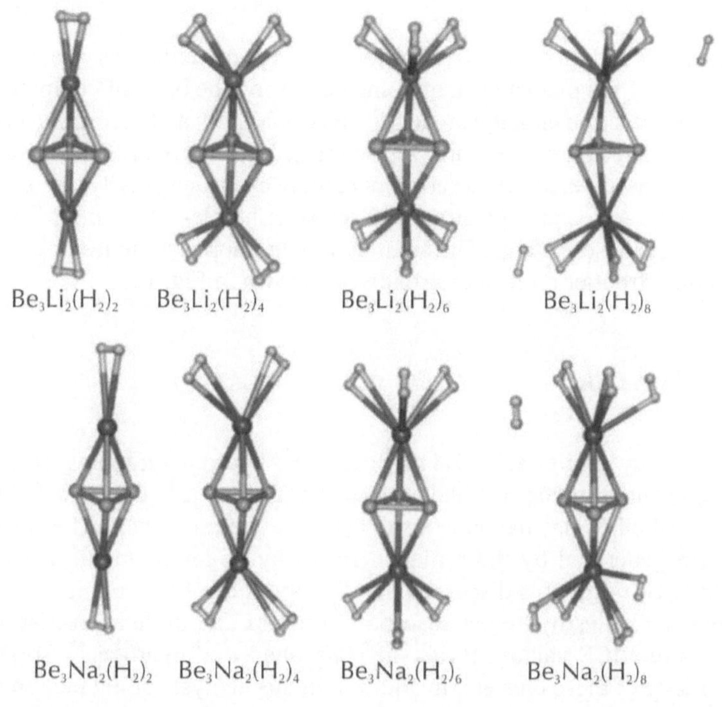

FIGURE 14 Optimized moieties of hydrogen-trapped Be_3M_2 (M=Li, Na) analogs at MP2/6–31++G(2d,2p) level. (This figure is reproduced from the work of Srinivasu et al. (2012) with permission from Royal Society of Chemistry)

Compared to Be_3M_2 and Al_4M_2 the net positive charge on M of Mg_3M_2 is less and due to this the hydrogen trapping potential is also degraded. Calculated interaction energy for the hydrogen adsorption of Be_3K_2 and Al_4K_2 is considerably less, which was not expected due to their greater positive charge. Reaction enthalpy (ΔH) and reaction electrophilicity (Δω) values for the hydrogen adsorption processes of Be_3M_2 and Al_4M_2 (M=Li, Na) are corroborated well with their energy and enthalpy of hydrogen adsorption.

6.5 HYDROGEN STORAGE OF DIFFERENT LITHIUM DOPED MOLECULAR CLUSTERS

6.5.1 $Li_3Al_4^-$ CLUSTER

In a different study Pan et al. (2012) rationalized the effectiveness of bond-stretch isomers of $Li_3Al_4^-$ as plausible H_2 trapping agents, on the basis of computed interaction energy, reaction enthalpy, reaction electrophilicity and desorption energies. All these calculated parameters are well supported by the conceptual density functional theory based reactivity descriptors namely electronegativity, hardness, and electrophilicity. Associated electronic structure principles MHP and MEP served well to analyze these systems. Different hydrogen trapped moieties of $Li_3Al_4^-$ and corresponding frontier molecular orbitals are shown in Fig. 15.

6.5.2 2-D STARS

In a recent study Giri et al. (2011) reported the hydrogen trapping potential of aromatic and antiaromatic star-like molecular clusters C_4Li_4, $C_5Li_5^-$ and C_6Li_6 and that of several other star-like clusters (Chakraborty et al., 2012). These star-like clusters are generated by the replacement of hydrogen atoms of planar C_4H_4, $C_5H_5^-$ and C_6H_6 rings. It is discussed that the presence of Li atom in a molecular framework facilitates hydrogen adsorption process due to their positive charge. An analysis of NICS values reveals the consequence of hydrogen adsorption on the aromaticity of these clusters. In addition to this analysis of interaction energy, reaction enthalpy and reaction electrophilicity explain the potential of theses star-like clusters as hydrogen trapping agents. In Fig. 16, the hydrogen-adsorbed moieties of C_4Li_4, $C_5Li_5^-$ and C_6Li_6 are shown.

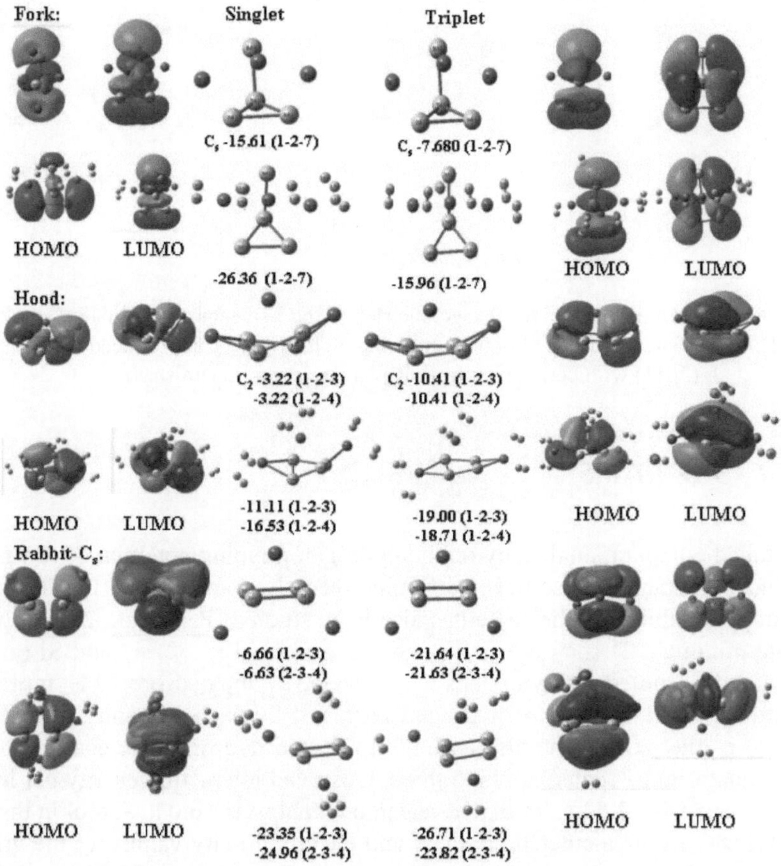

FIGURE 15 Optimized geometries of different isomeric $Li_3Al_4^-$ clusters, their hydrogen trapped analogs and their corresponding frontier molecular orbitals. (NICS values are given below the structures.) (This figure is reproduced from the work of Pan et al. (2012) with permission from Mexican Chemical Society.)

FIGURE 16 Optimized molecular geometries of the Li-decorated "star-like" molecules and their associated hydrogen bound complexes. (This figure is reproduced from the work of Giri et al. (2011) with permission from Royal Society of Chemistry.)

6.5.3 3-D STARS

A detailed computational study on the hydrogen trapping potential of few global minima 3-D stars has also been performed recently (Pan et al., 2012). The noble gas trapping ability of the same has also been studied (Pan et al., 2013).Studied global minima 3-D stars are $C_5Li_7^+$, $Si_5Li_7^+$ and $Ge_5Li_7^+$. $C_5Li_7^+$ and $Si_5Li_7^+$ can bind up to 21 molecular hydrogens whereas $Ge_5Li_7^+$ may adsorb 19 H_2 molecules according to the number of hydrogen adsorbed by each Li center. In the case of $Ge_5Li_7^+$ the axial Li atoms can bind up to two hydrogen molecules whereas each equatorial Li center can bind three. Observed interaction energy per hydrogen molecule is −2.8 kcal/mol and reaction enthalpy is −6.0 kcal/mol in the case of $21H_2@C_5Li_7^+$. Calculated hardness and electrophilicity values for the gradual hydrogen adsorption lend additional support to the stability of hydrogen-loaded clusters. In Fig. 17, optimized geometries of star like clusters and their hydrogen-trapped analogs are displayed.

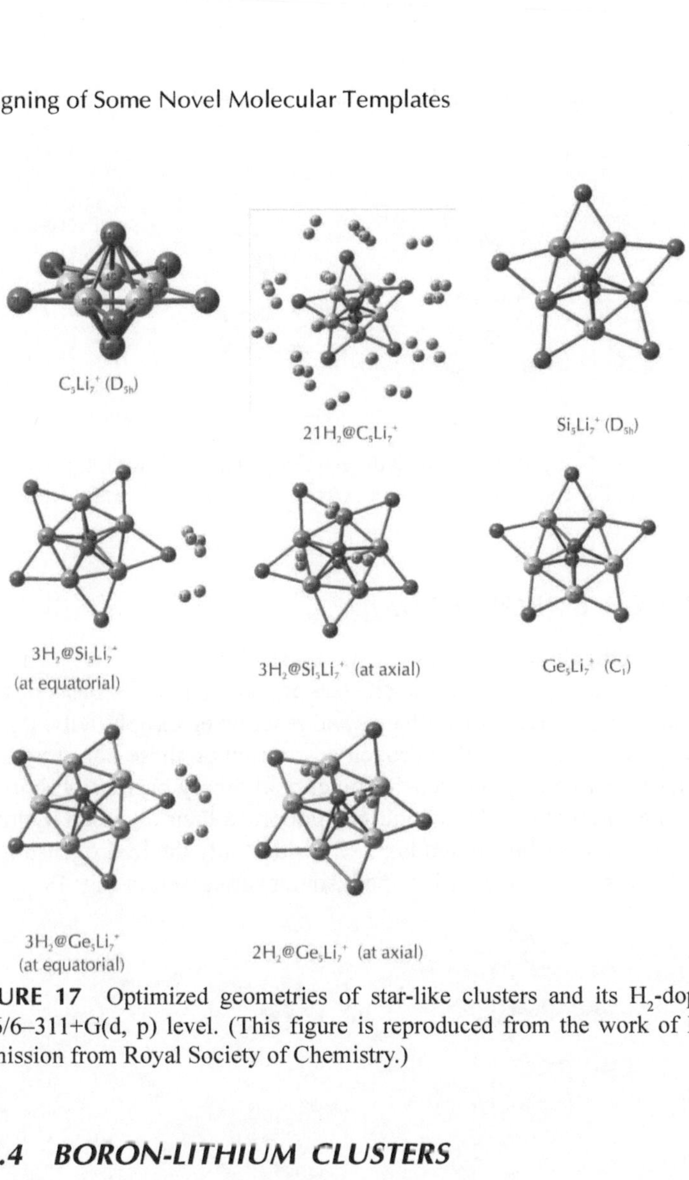

FIGURE 17 Optimized geometries of star-like clusters and its H_2-doped analogs at M06/6–311+G(d, p) level. (This figure is reproduced from the work of Pan et al. with permission from Royal Society of Chemistry.)

6.5.4 BORON-LITHIUM CLUSTERS

Bandaru et al. (2012) studied the stability as well as hydrogen trapping potential of Li- doped boron-lithium (B_xLi_y, $x = 2–6$; $y = 1, 2$) neutral and charged clusters. Fig. 18 depicts the geometries of hydrogen loaded neutral and charged boron-lithium (B_xLi_y, $x = 3–6$; $y = 1, 2$) clusters. Computed interaction energy (IE), reaction enthalpy (ΔH) and reaction electrophilicity ($\Delta\omega$) for the gradual hydrogen loading on this boron-lithium clusters reveal the fact that charged clusters are better hydrogen holder than that of neutral clusters.

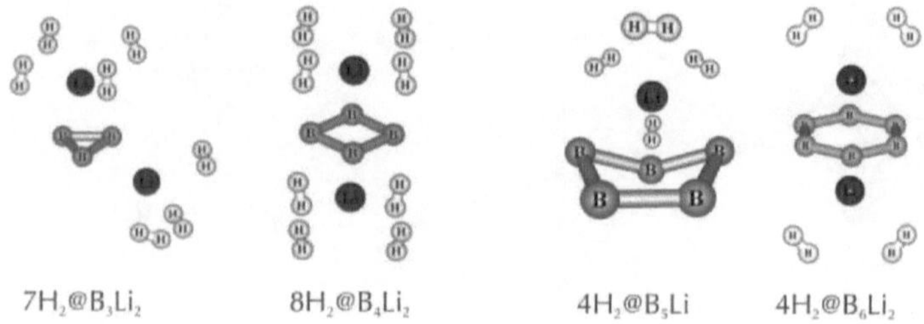

FIGURE 18 B_xLi_y (x=3–6; y=1,2) and its hydrogen trapped analogs at MP2/6–311+G(d, p) level of theory. (This figure is reproduced from the work of Bandaru et al. with permission from John Wiley.)

6.5.5 *Li-DOPED BORON HYDRIDE*

Pan et al. (2012) studied a few boron hydrides doped by Li atoms/ions. The possibility of hydrogen adsorption on these clusters are studied on the basis of computed interaction energies, reaction enthalpy and reaction electrophilicity. It is observed that the gravimetric wt% of hydrogen adsorption on these boron hydrides is quite high. Different conceptual density functional theory based reactivity descriptors as well as aromaticity lend addition support to their assessed hydrogen trapping potential. Among the studied boron hydrides only the hydrogen-trapped motifs of $B_3H_3^{2-}$ cluster and its Li/Li$^+$ doped isomers are given in Fig. 19.

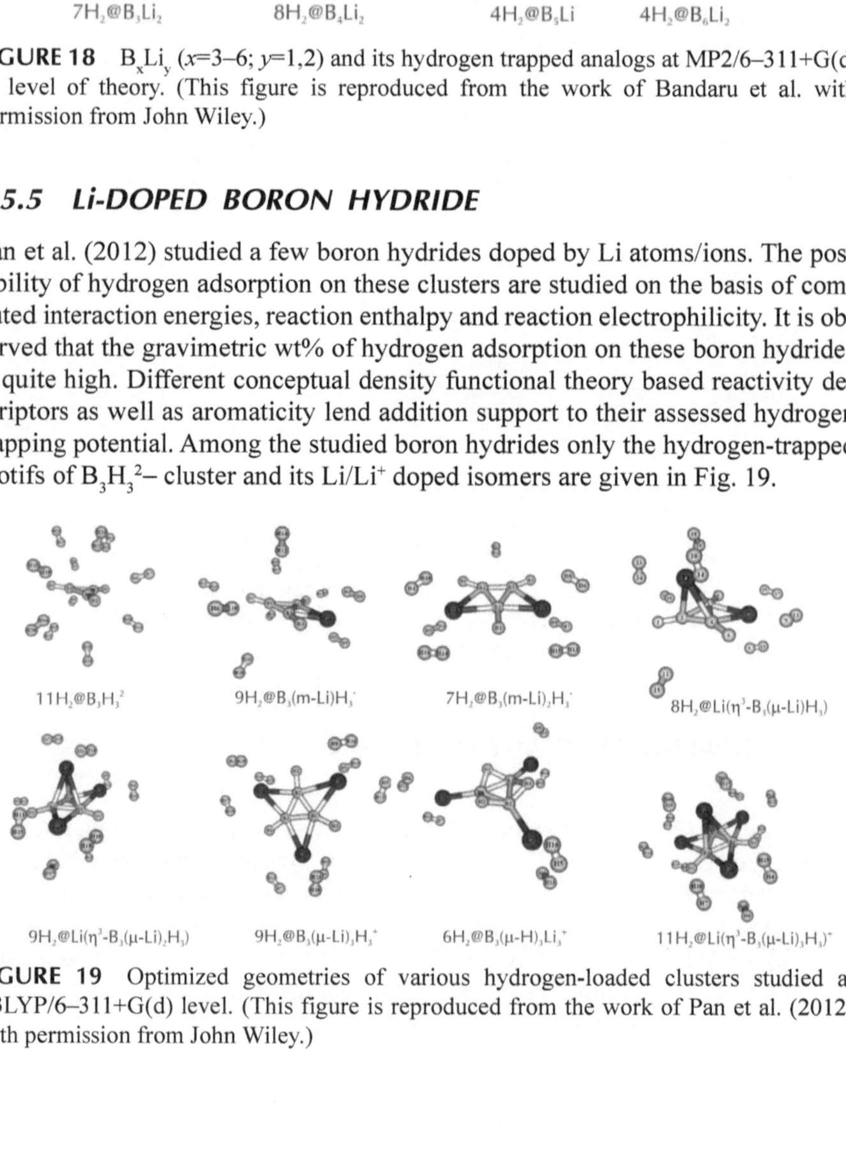

FIGURE 19 Optimized geometries of various hydrogen-loaded clusters studied at B3LYP/6–311+G(d) level. (This figure is reproduced from the work of Pan et al. (2012) with permission from John Wiley.)

The variation of ΔG versus temperature and pressure for the adsorption of H_2 molecules on $B_3(\mu\text{-Li})_3H_3^+$ cluster is presented in Fig. 20. The ΔG=0 region lies in between the two dotted lines, where the adsorbed and desorbed systems are in equilibrium. The region with black dots (ΔG<0) corresponds to the temperature pressure zone for favorable adsorption process whereas that with the red dots corresponds to the favorable desorption process.

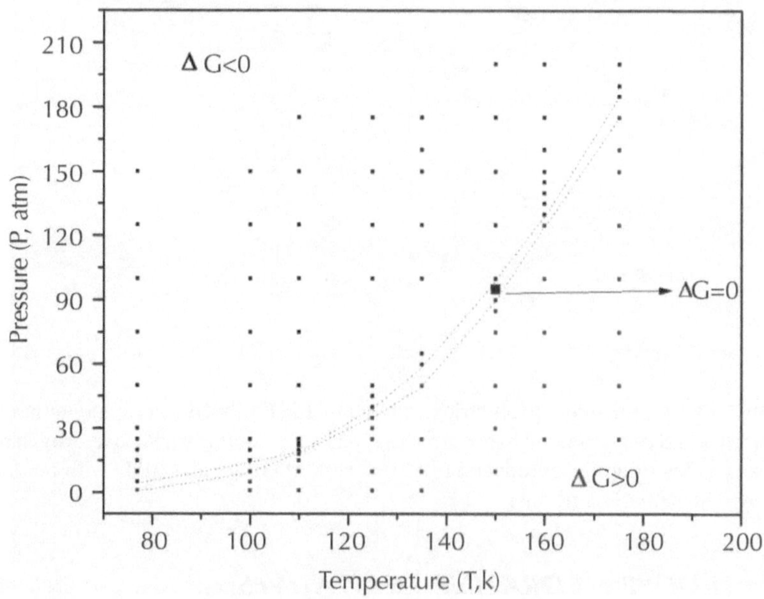

FIGURE 20 Temperature–pressure phase diagram showing the variation of ΔG for the $6H_2$ adsorption process on $B_3(\mu\text{-Li})_3H_3^+$ cluster. (This figure is reproduced from the work of Pan et al. (2012) with permission from John Wiley.)

6.5.6 Li-DOPED ANNULAR SYSTEMS

In another study all the hydrocarbons, starting from the three membered carbon ring to polyaromatic hydrocarbon phenanthrene, are doped by Li^+ and F^- (Giri et al., 2011). The negative charge on fluorine can bind molecular hydrogens due to the dipole-induced dipole type of interaction as we have seen in the case of doping Li. Interactions between the doped centers (Li/F) and adsorbed molecular hydrogens are assessed by calculated interaction energy, reaction enthalpy and reaction electrophilicity values. It is observed that all the doped systems can serve as potential hydrogen storage materials. A few representatives H_2-bound mononuclear as well as polynuclear rings are shown in Fig. 21.

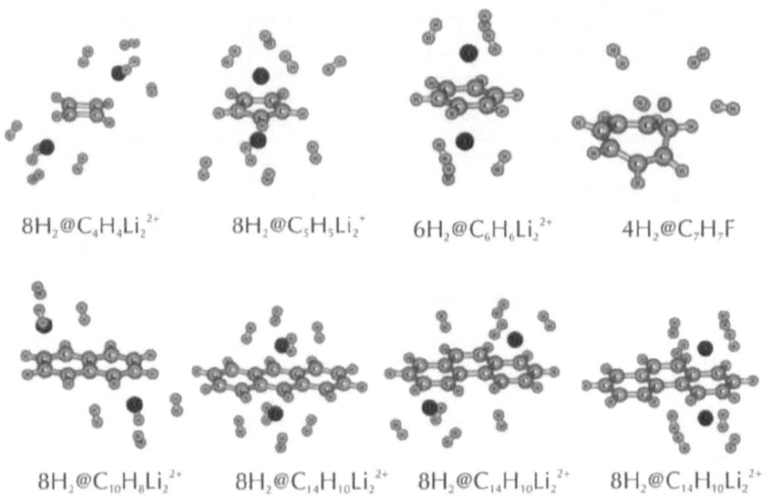

FIGURE 21 Optimized geometries of the Li^+/F^--doped aromatic/antiaromatic hydrocarbons and polyaromatic hydrocarbons (PAHs) and their associated hydrogen bound complexes. (This figure is reproduced from the work of Giri et al. (2011) with permission from Royal Society of hemistry.)

6.5.7 Li-DOPED BORAZINE DERIVATIVES

Recently, fruitful designing of some Li-decorated borazine derivatives and further application of them towards hydrogen trapping are also checked (Pan et al., 2012). The corresponding interaction energy and gravimetric wt% of hydrogen are found to be better on Li^+ doping. Fig. 22 shows the optimized geometries of various Li-decorated borazine derivatives and their highest hydrogen-captured analogs. The gradual increment in stability can be explained by an average increasing and decreasing trends of hardness and electrophilicity values, respectively, upon gradual hydrogen loading.

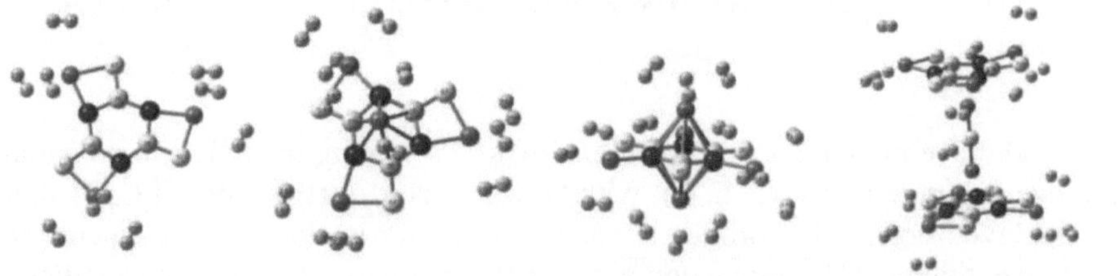

$9H_2@B_3N_3F_3Li_3$ $12H_2@B_3N_3F_3Li_4^{+}$ $14H_2@B_3N_3F_3Li_5^{2+}$ $17H_2@[(B_3N_3F_3Li_4)_2F]^{+}$

FIGURE 22 Hydrogen-trapped analogs of various Li-decorated borazine derivatives optimized at B3LYP/6–311+G(d) level. (This figure is reproduced from the work of Pan et al. (2012) with permission from Mexican Chemical Society).

6.5.8 N_4Li_2 AND N_6Ca_2 CLUSTERS

The planar nitrogen rings in analogy with cyclobutadiene and benzene are modeled by Duley et al. (2011). The calculated NICS values of these annular N_4^{2-} and N_6^{4-} rings are comparable with cyclobutadiene and benzene. N_4^{2-} and N_6^{4-} rings are doped by keeping Li^+ and Ca^{2+} cations, respectively, at each face of the rings. Resulting N_4Li_2 and N_6Ca_2 systems are stabilized by cation-π interaction. Calculated binding energy per hydrogen molecule of aromatic N_6Ca_2 and N_4Li_2 systems demonstrate the potential of them to be used as hydrogen storage material. The optimized geometries of the N_6Ca_2 and N_4Li_2 clusters and their corresponding hydrogen-trapped analogs are depicted in Fig. 23.

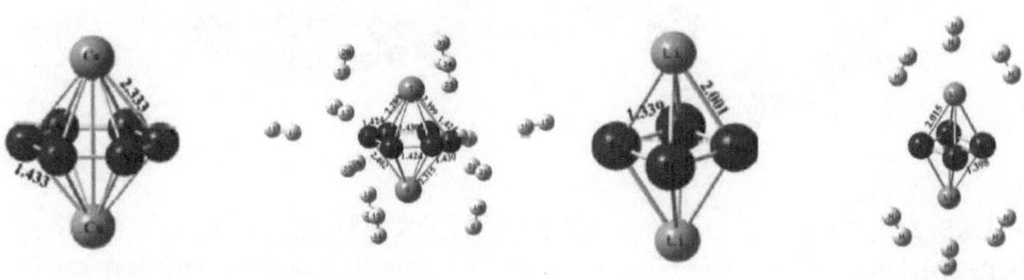

N_6Ca_2 $N_6Ca_2(H_2)_{12}$ N_4Li_2 $N_4Li_2(H_2)_6$

FIGURE 23 Optimized molecular geometries of N_6Ca_2 and N_4Li_2 and their hydrogen trapped complexes. (This figure is reproduced from the work of Duley et al. (2011) with permission from Elsevier.)

6.5.9 $(N_4C_3H)_6Li_6$ AND AN ASSOCIATED 3-D FUNCTIONAL MATERIAL

Das et al. have modeled a new 3-D framework by joining the unit of a potential hydrogen storage motif, $(N_4C_3H)_6$ with "C≡C" linker (2012). The $(N_4C_3H)_6$ unit has achieved potential hydrogen storage ability after Li doping. The calculated reaction enthalpy (ΔH), interaction energy (ΔE), hardness (η) and electrophilicity (ω) for adsorption of up to three molecular hydrogens by each Li atom is optimal. A representative diagram of the Li doped $(N_4C_3H)_6$ unit and the modeled 3-D framework with their hydrogen adsorbed analogs are given in Fig. 24.

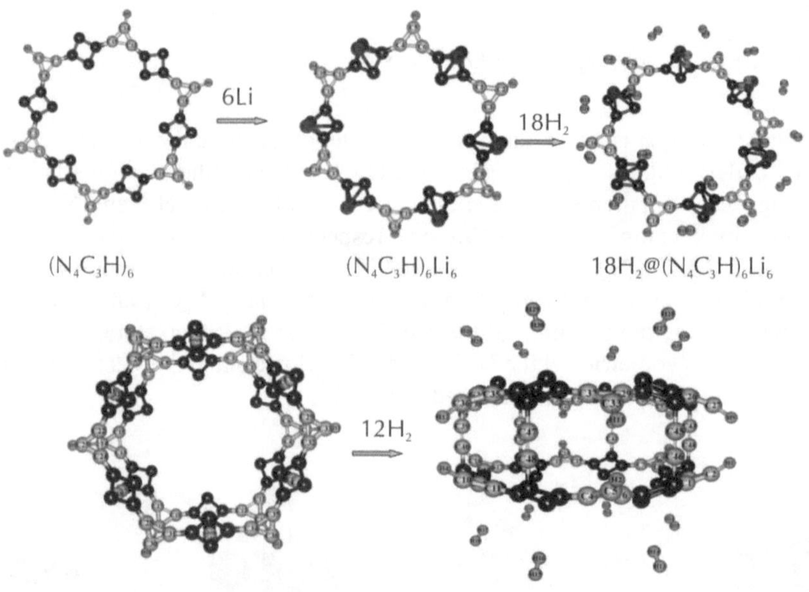

FIGURE 24 Optimized geometries of $(N_4C_3H)_6$, $(N_4C_3H)_6Li_6$, its highest hydrogen-trapped analog at B3LYP/6–31+G(d) level and 3-D molecular material (both bare and H_2-loaded) at B3LYP/6–31G(d) level. (This figure is reproduced from the work of Das et al. with permission from American Chemical Society.)

The variation of ΔG vs. T and P for the process of hydrogen adsorption is depicted in Fig. 25. The dotted line indicates the temperature and pressure where ΔG is equal to zero. The blue dots show the region where ΔG<0, hence, hydrogen adsorption process is thermodynamically feasible whereas the red dots indicate the T–P zone where the hydrogen desorption process is feasible.

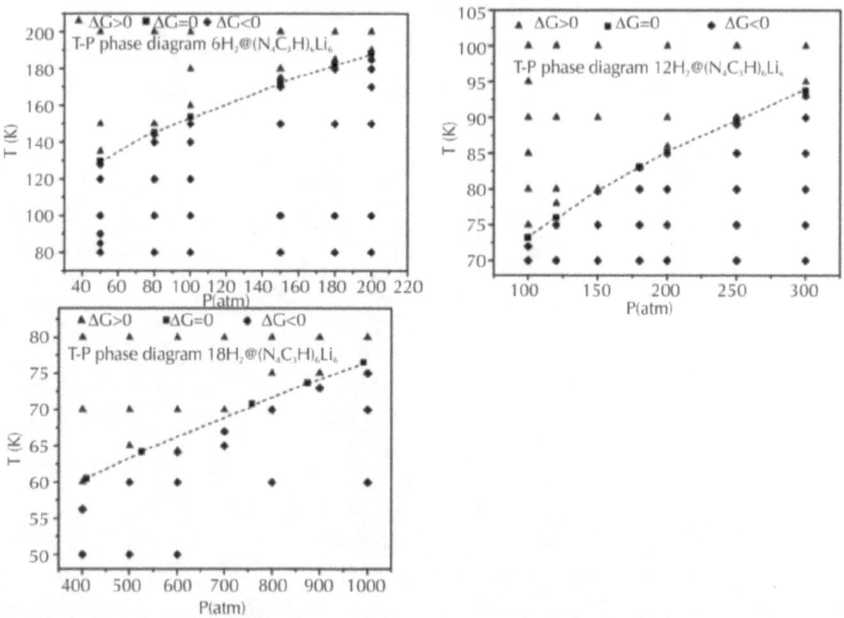

FIGURE 25 Temperature–pressure phase diagrams showing the variation of ΔG. (This figure is reproduced from the work of Das et al. with permission from American Chemical Society.)

6.6 HYDROGEN STORAGE CAPACITY OF SUPER-ALKALI CLUSTERS

Very recently the demonstration regarding the hydrogen storage capability of various global minima super-alkali ions has also been reported (Pan et al., 2012). For this purpose, both mononuclear as well as binuclear super-alkali ions like BLi_6^+, FLi_2^+, OLi_3^+ and $M_2Li_{2k+1}^+$ (k=1–5) are employed. Few examples of super-alkali clusters and corresponding hydrogen-bound analogs are displayed in Fig. 26. The theoretical investigation shows that they are apt in capturing molecular hydrogen maintaining acceptable interaction energy and reaction enthalpy values. The gravimetric wt% of hydrogen also satisfies the demand of DOE.

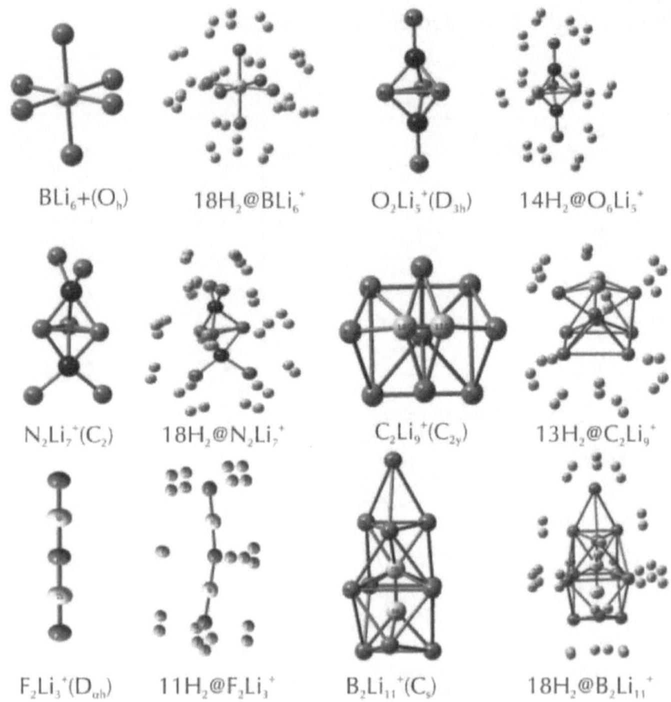

FIGURE 26 Various super-alkali ions and their hydrogen-trapped analogs optimized at M052X/6–311+G(d) level (This figure is reproduced from the work of Pan et al. (2012) with permission from Royal Society of Chemistry.)

An efficient tuning of interaction energy per hydrogen molecule can also be done by altering the external electric field in a suitable direction. A depiction about the nature of changing of interaction energy with the variation of external electric field from zero to 0.006 au is provided in Fig. 27 taking one H_2 adsorption process as reference. It is obvious from the Fig. that the interaction energy is gradually improved with increasing field strength. So, only by changing the applied electric field one can shift the equilibrium towards adsorption or desorption of hydrogen.

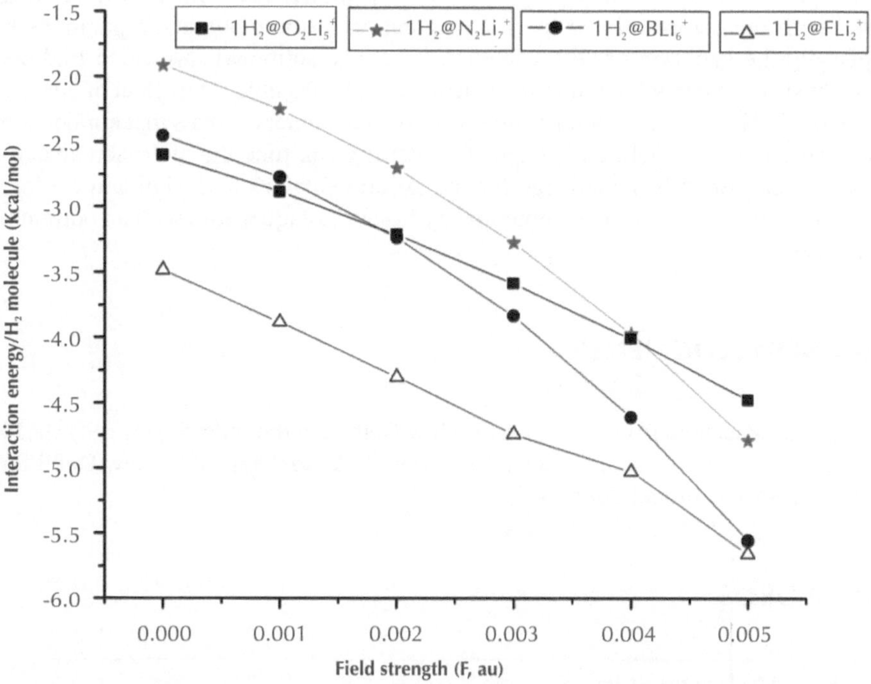

FIGURE 27 Plots of variation of interaction energy per hydrogen molecule (ΔE, kcal/mol) with the applied electric field strength (F, au) studied at the M052X/6–311+G(d) level. (This figure is reproduced from the work of Pan et al. (2012) with permission from Royal Society of Chemistry.)

6.7 CONCLUSIONS

The molecular frameworks, metal organic frameworks, cage networks and their different doped analogs have been modeled theoretically followed by further experimental verifications. In this era computation is very much indispensible, not only for the modeling but also to carry out the theoretical work to explain the different aspects of these materials. Hydrogen trapping potential of all the mentioned nano-materials are assessed on the basis of interaction energy, reaction enthalpy, reaction electrophilicity, aromaticity, etc. with gradual loading of molecular hydrogen. Conceptual density functional theory based global and local reactivity descriptors are already proven as good computational tools for the assessment of stability as well as reactivity of molecular systems. Various electronic structure principles like maximum hardness principle (MHP), minimum electrophilicity principle (MEP) and minimum polarizability principle (MPP) are corroborating well with the computational facts. Computation of thermodynamic parameters, specially the variation of Gibbs free energy lends additional support to find out the physical state in which the experiment would be feasible. The effect of electric field on $C_{12}N_{12}$ isomers as well as on few Li doped isomers reveals that application of external electric field with a specific intensity can tune the hydrogen adsorption process. So it is a challenge for the experimentalists to design and explore these materials and to provide potential hydrogen storage media with an optimum expense.

ACKNOWLEDGMENTS

Financial assistance from UGC, CSIR, New Delhi and the Indo-EU (HYPOMAP) project is gratefully acknowledged. PKC would like to thank DST, New Delhi for the J. C. Bose National Fellowship.

KEYWORDS

- **Hydrogen Storage**
- **Conceptual DFT**
- ***ab initio* Simulation**
- **Nucleus-Independent Chemical Shift**
- **T- P Phase Diagram**
- **External Electric Field**
- **Classical Molecular Dynamics**

REFERENCES

Aladko, E. Y.; Larionov, E. G.; Rodionova, T. V.; Aladko, L. S.; Manakov, A. Y. *J. Incl. Phenom. Macrocycl. Chem.*, **2010**, *68*, 381.

Allen, M. P.; Tildesley, D. J. *Computer Simulation of Liquids*. Oxford University Press, Oxford, **1996**.

An, H.; Liu, C.-S.; Zeng, Z.; Fan, C.; Ju, X. *Appl. Phys. Lett.*, **2011**, *98*, 173101.

Ataca, C.; Aktürk, E.; Ciraci, S.; Ustunel, H. *Appl. Phys. Lett.*, **2008**, *93*, 043123.

Awad, A .H.; Veziroğlu, T. N. *Int. J. Hydrogen Energy*, **1984**, *9*, 355.

Ayers, P. W.; Parr, R. G. *J. Am. Chem. Soc.*, **2000**, *122*, 2010.

Bandaru, S.; Chakraborty, A.; Giri, S.; Chattaraj, P. K. *Int. J. Quantum Chem.*, **2012**, *112*, 695.

Barbir, F.; Veziroğlu, T. N.; Plass Jr, H. J. *Int. J. Hydrogen Energy*, **1990**, *15*, 739.

Berger, M. E. M.; Assenbaum, D.; Taccardi, N.; Spiecker, E.; Wasserscheid, P. *Green Chem.*, **2011**, *13*, 1411.

Berseth, P. A.; Harter, A. G.; Zidan, R.; Blomqvist, A.; Araüjo, C. M.; Scheicher, R. H.; Ahuja, R.; Jena, P. *Nano Lett.*, **2009**, *9*, 1501.

Bethune, D. S.; Kiang, C. H.; de Vries, M. S.; Gorman, G.; Savoy, R.; Vazquez, J.; Beyers, R. *Nature*, **1993**, *363*, 605.

Blöchl, P. E. *Phys. Rev. B*, **1994**, *50*, 17953.

Bluhm, M. E.; Bradley, M. G.; Butterick III, R.; Kusari, U.; Sneddon, L. G. *J. Am. Chem. Soc.*, **2006**, *128*, 7748.

Bogdanović, B.; Felderhoff, M.; Streukens, G.; Serb, J. *Chem. Soc.*, **2009**, *74*, 183.

Breck, D. W. *Zeolite Molecular Sieves*, Wiley, New York, **1974**.

Britt, D.; Tranchemontagne, D.; Yaghi, O. M. *Proc. Natl. Acad. Sci.*, **2008**, *105*, 11623.

Burress, J.; Simmons, J.; Ford, J.; Yildirim, T. "Gas Adsorption Properties of Graphene-Oxide-Frameworks and Nanoporous Benzene-Boronic Bcid Polymers" (Presented at the March meeting of the American Physical Society (APS) in Portland, Ore), March 18, **2010**.

Cao, D.; Lan, J.; Wang, W.; Smit, B. *Angew. Chem. Int. Ed.* **2009**, *48*, 4730.

Chae, H. K.; Siberio-PSrez, D. Y.; Kim, J.; Go, Y.; Eddaoudi, M.; Matzger, A. J.; O'Keeffe, M.; Yaghi, O. M. *Nature*, **2004**, *427*, 523.

Chakraborty, A.; Bandaru, S.; Das, R.; Duley, S.; Giri, S.; Goswami, K.; Mondal, S.; Pan, S.; Sen, S.; Chattaraj, P. K. *Phys. Chem. Chem. Phys.*, **2012**, *14*, 14784.

Chakraborty, A.; Duley, S.; Giri, S.; Chattaraj, P. K. In: "*A Matter of Density: Exploring the Electron Density Concept in the Chemical, Biological, and Materials Sciences*," edited by N. Sukumar, (John Wiley and Sons) **2012**.

Chakraborty, A.; Giri, S.; Chattaraj, P. K. *Struct. Chem.*, **2011**, *22*, 823.

Chamorro, E.; Chattaraj, P. K.; Fuentealba, P. *J. Phys. Chem. A*, **2003**, *107*, 7068.

Chan, Y.; Hill, J. M. *Nanotechnology*, **2011**, *22*, 305403.

Chandrakumar, K. R. S.; Ghosh, S. K. *Nano Lett.*, **2008**, *8*, 13.

Chapoy, A.; Anderson, R.; Tohidi, B. *J. Am. Chem. Soc.*, **2007**, *129*, 746.

Chas, E. M. NIST-JANAF Thermochemical tables, 4th Edition. JPCRD. Monograph 9, **1998**; 1–1951.

Chattaraj, P. K. *Chemical Reactivity Theory: A Density Functional View*; (Taylor & Francis/CRC Press, Florida) **2009**.

Chattaraj, P. K. *J. Indian Chem. Soc.*, **1992**, *69*, 173.

Chattaraj, P. K.; Bandaru, S.; Mondal, S. *J. Phys. Chem. A*, **2011**, *115*, 187.

Chattaraj, P. K.; Giri, S. *Ann. Rep. Prog. Chem., Sect. C: Phys. Chem.*, **2009**, *105*, 13.

Chattaraj, P. K.; Giri, S.; Duley, S. *Chem. Rev.*, **2011**, *111*, PR43.

Chattaraj, P. K.; Roy, D. R. *Chem. Rev.*, **2007**, *107*, PR46.

Chattaraj, P. K.; Sarkar, U.; Roy, D. R. *Chem. Rev.*, **2006**, *106*, 2065.

Chattaraj, P. K.; Sengupta, S. *J. Phys. Chem.*, **1996**, *100*, 16126.

Collins, D. J.; Zhou, H. C. *J. Mater. Chem.* **2007**, *17*, 3154 and references therein.

Cui, X.-Y.; Jia, J.-F.; Yang, B.-S.; Yang, P.; Wub, H.-S. *J. Mol. Struct. (THEOCHEM)*, **2010**, *953*, 1.

Cui, X.-Y.; Yang, B.-S.; Wub, H.-S. *J. Mol. Struct. (THEOCHEM)*, **2010**, *941*, 144.

Das, R.; Chattaraj, P. K. *J. Phys. Chem. A*, **2012**, *16*, 3259.

Davidson, E. R. *Chem. Rev*. **2000**, *100*, 351.

Demirci, U. B.; Miele, P. *Energy Environ. Sci*., **2009**, *2*, 627.

Dillon, A. C.; Jones, K. M.; Bekkedahl, T. A.; Klang, C. H.; Bethune, D. S.; Heben, M. J. *Nature (London)*, **1997**, *386*, 377.

Dillon, A.C.; Heben, M. J. *Appl. Phys. A* **2001**, *72*, 133.

Dimitrakakis, G. K.; Tylianakis, E.; Froudakis, G. E. *Nano Lett*., **2008**, *8*, 3166.

Dincă, M.; Dailly, A.; Liu, Y.; Brown, C. M.; Neumann, D. A.; Long, J. R. *J. Am. Chem. Soc*., **2006**, *128*, 16876.

Dincă, M.; Long, J. R. *Angew. Chem. Int. Ed*, **2008**, *47*, 6766.

Ding, R. G.; Lu, G. Q.; Yan, Z. F.; Wilson, M. A. *J. Nanosci. Nanotechnol*, **2001**, *1*, 7.

Doroodian, A.; Dengler, J. E.; Genest, A.; Rösch, N.; Rieger, B. *Angew. Chem. Int. Ed*., **2010**, *49*, 1871.

Douglas, L. S. *J. Phys. Chem. A*, **2006**, *110*, 7228.

Du, A.; Zhu, Z.; Smith, S. C. *J. Am. Chem. Soc*., **2010**, *132*, 2876.

Duley, S.; Giri, S.; Sathymurthy, N.; Islas, R.; Merino, G.; Chattaraj, P. K. *Chem. Phys. Lett*. **2011**, *506*, 315.

Durgun, E.; Ciraci, S.; Zhou, W.; Yildirim, T. *Phys. Rev. Lett*., **2006**, *97*, 226102.

Eddaoudi, M.; Moler, D. B.; Li, H.; Chen, B.; Reineke, T. M.; O'Keeffe, M.; Yaghi, O. M. *Acc. Chem. Res*., **2001**, *34*, 319.

El-Kaderi, H. M.; Hunt, J. R.; Mendoza-Cortes, J. L.; Cote, A. P.;Taylor, R. E.; O'Keeffe, M.; Yaghi, O. M. *Science*, **2007**, *316*, 268.

Energy Information Administration, Official Energy Statistics from the U.S. Government, International Energy Annual 2005, available at: http://www.eia.doe.gov/emeu/iea/ (accessed on 17/09/12).

Fellay, C.; Dyson, P. J.; Laurenczy, G. *Angew. Chem. Int. Ed*., **2008**, *47*, 3966.

Ferre Vilaplana, A. *J. Chem. Phys*., **2005**, *122*, 214724.

Firlej, L.; Roszak, S.; Kuchta, B.; Pfeifer, P.; Wexler, C. *J. Chem. Phys*., **2009**, *131*, 164702.

Florusse, L. J.; Peters, C. J.; Schoonman, J.; Hester, K. C.; Koh, C. A.; Dec, S. F.; Marsh, K. N.; Sloan, E. D. *Science*, **2004**, *306*, 469.

For an exhaustive review of the Fischer-Tropsch process, Syntroleum Inc. sponsors a website: http://www.fischer-tropsch.org/.

Frisch, M.J., et al. Gaussian 09, Revision A.1.

Fuentealba, P.; Simon-Manso, Y.; Chattaraj, P. K. *J. Phys. Chem. A*, **2000**, *104*, 3185.

Furukawa, H.; Yaghi, O. M. *J. Am. Chem. Soc.*, **2009**, *131*, 8875.

Garberoglio, G. *Langmuir*, **2007**, *23*, 12154.

GAUSSIAN 03, Revision B.03; Gaussian, Inc.: Pittsburgh, PA, **2003**.

Geerlings, P.; De Proft, F.; Langenaeker, W. *Chem. Rev.*, **2003**, *103*, 1793.

Geim, A. K.; Novoselov, K. S. *Nature Materials*, **2007**, *6,* 183.

Giri, S.; Bandaru, S.; Chakraborty, A.; Chattaraj, P. K. *Phys. Chem. Chem. Phys*., **2011**, *13*, 20602.

Giri, S.; Chakraborty, A.; Chattaraj, P. K. *J. Mol. Model*, **2011**, *17*, 777.

Giri, S.; Chakraborty, A.; Chattaraj, P. K. *Nano Reviews*, **2011**, *2*, 5767.

Graphene Oxide Derived Carbons (GODC); High-Surface Area NanoPorous Materials for Hydrogen Storage and Carbon Capture Author: T Yildirim – Bulletin of the American Physical Society, 2012 – APS (http://meetings.aps.org/Meeting/MAR12/Event/162605).

Han, S. S.; Deng, W.-Q.; Goddard III, W. A. *Angew. Chem. Int. Ed*, **2007**, *46*, 6289.

Han, S. S.; Furukawa, H.; Yaghi, O. M.; Goddard III, W. A. *J. Am. Chem. Soc*., **2008**, *130*, 11580.

Han, S. S.; Jang, S. S. *Chem. Commun. (Camb)*, **2009**, *28*, 5427.

Han, S. S.; Kang, J. K.; Lee, H. M.; van Duin, A. C. T.; Goddard III, W. A. *J. Chem. Phys.*, **2005**, *123*, 114704.

Hashimoto, S.; Tsuda, T.; Ogata, K.; Sugahara, T.; Inoue, Y.; Ohgaki, K. *J. Thermodynamics*, **2010**, Article ID 170819, doi:10.1155/2010/170819.

Heldebrant, D. J.; Karkamkar, A.; Linehan, J. C.; Autrey, T. *Energy Environ. Sci*., **2008**, *1*, 156.

Hirscher, M.; Becher, M. *J. Nanosci. Nanotechnol*, **2003**, *3*, 3.

Hirschfelder, J. O.; Curtiss, C. F.; Bird, R. B. *Molecular Theory of Gases and Liquids* Wiley, New York, **1954**.

http://en.wikipedia.org/wiki/Hydrogen_storage#Carbohydrates (accessed on 17/09/12).

http://nanogloss.com/nanotechnology/the-history-of-nanotechnology/#axzz24T5jHSBK (accessed on 17/09/12).

http://wiki.answers.com/Q/What_is_the_meaning_of_nano (accessed on 17/09/12).

Hu, Y. H.; Zhang, L. *Adv. Mater*, **2010**, *22*, 117.

Humphrey, W.; Dalke, A.; Schulten, K. *J. Molec. Graphics*, **1996**, *14*, 33.

Hynek, S.; Fuller, W.; Bentley, J. *Int. J. Hydrogen Energy*, **1997**, *22*, 601.

Iijima, S. *Nature*, **1991**, *354*, 56.

Iijima, S.; Ichihashi, T. *Nature,* **1993**, *363*, 603.

James, S. L. *Chem. Soc. Rev*., **2003**, *32*, 276.

Javan, M. B.; Ganji, M. D.; Sabet, M.; Danesh, N. *J. Comp. Theo. Nanosci*., **2011**, *8*, 803.

Jeffrey, G. A. In "Inclusion Compounds" edited by Atwood, J. L.; Davies, J. E. D.; MacNicol, D. D. Vol. 1, (Academic Press) **1984**, pp. 135.

Jimenez-Halla, J. C.; Matito, E.; Blancafort, L.; Robles, J.; Sola, M. *J. Comp. Chem*., **2009**, *30*, 2764.

Jiménez, V.; Sánchez, P.; Díaz, J. A.; Valverde, J. L.; Romero, A. *Chem. Phys. Lett*., **2010**, *485*, 152.

Jin, Z.; Sun, Z.; Simpson, L. J.; O"Neill, K. J.; Parilla, P. A.; Li, Y.; Stadie, N. P.; Ahn, C. C.; Kittrell, C.; Tour, J. M. *J. Am. Chem. Soc*., **2010**, *132*, 15246.

Joó, F. *ChemSusChem*, **2008**, *1*, 805.

Jordi, P.; Miquel, S. *Chem. Commun*., **2011**, *47*, 11647.

Karleta, D. C.; Roshawnda, C.; Douglas, L. S. *J. Chem. Theory Comput*., **2007**, *3*, 2176.

Kaye, S. S.; Dailly, A.; Yaghi, O. M.; Long, J. R. J. *Am. Chem. Soc*., **2007**, *129*, 14176.

Keaton, R. J.; Blacquiere, J. M.; Baker, R. T. *J. Am. Chem. Soc*., **2007**, *129*, 1844.

Kesanli, B.; Cui, Y.; Smith, M. R.; Bittner, E. W.; Bockrath, B. C.; Lin, W. *Angew Chem Int Ed*., **2005**, *44*, 72.

Kim, G.; Jhi, S.-H.; Lim, S.; Park, N. *Phys. Rev. B*, **2009**, *79*, 155437.

Klauda, J. B.; Sandler, S. I. *Chem. Eng. Sci*. **2003**, *58*, 27.

Klerke, A.; Christensen, C. H.; Nørskov, J. K.; Vegge, T. *J. Mater. Chem*., **2008**, *18*, 2304.

Klontzas, E.; Tylianakis, E.; Froudakis, G. E. *J. Phys. Chem. C*, **2008**, *112*, 9095.

Klontzas, E.; Tylianakis, E.; Froudakis, G. E. *Nano Lett*., **2010**, *10*, 452.

Kokalj, A. *Comp. Mater. Sci*., **2003**, *28*, 155.

Koopmans, T. A. *Physica*, **1933**, *1*, 104.

Kowalczyk, P.; Hołyst, R.; Terrones, M.; Terrones, H. *Phys. Chem. Chem. Phys*., **2007**, *9*, 1786.

Krätschmer, W.; Lamb, L.; Fostiropoulos, K.; Huffman, D. R. *Nature*, **1990**, *347*, 354.

Kresse, G.; Furthmüller, *J. Comput. Mat. Sci*., **1996**, *6*, 15.

Kresse, G.; Furthmüller, J. *Phys. Rev. B*, **1996**, *54*, 11169.

Kresse, G.; Hafner, J. *Phys. Rev. B*, **1993**, *47*, 558.

Kresse, G.; Hafner, J. *Phys. Rev. B*, **1994**, *49*, 14251.

Kresse, G.; Joubert, D. *Phys. Rev. B*, **1999**, *59*, 1758.

Kroto, H. W.; Heath, J. R.; O'Brien, S. C.; Curl, R. F.; Smalley, R. E. *Nature*, **1985**, *318*, 162.

Kuc, A.; Heine, T.; Seifert, G.; Duarte, H. A. *Chem. Eur. J*., **2008**, *14*, 6597.

Lang, X.; Fan, S.; Wang, Y. *J. Natural Gas Chem*., **2010**, *19*, 203.

Latroche, M.; Percheron-Guegan, A.; Chabre, Y. *J. Alloys Compounds*, **1999**, *295*, 637.

Laurenczy, G.; Fellay, C.; Dyson, P. J. *PCT Int. Appl.* (2008), CODEN: PIXXD2 WO 2008047312 A1–20080424 AN 2008:502691.

Lee, H.; Lee, J.-w.; Kim, D. Y.; Park, J.; Seo, Y.-T.; Zeng, H.; Moudrakovski, I. L.; Ratcliffe, C. I.; Ripmeester, J. A. *Nature*, **2005**, *434*, 743.

Li, C.; Li, J.; Wu, F.; Li, S.-S.; Xia, J.-B.; Wang, L.-W. *J. Phys. Chem. C*, **2011**, *115*, 23221.

Li, F.; Zhao, J.; Johansson, B.; Sunhave, L. *Int. J. Hydrogen Energy*, **2010**, *35*, 266.

Li, W.; Vajo, J. J.; Cumberland, R. W.; Liu, P.; Hwang, S.-J.; Kim, C.; Bowman, Jr. R. C. *J. Phys. Chem. Lett.*, **2010**, *1*, 69.

Li, Y.; Zhou, Z.; Shen, P.; Zhang, S. B.; Chen, Z. *Nanotechnology*, **2009**, *20*, 215701.

Li, Z.; Zhao, C.; Chen, L. *J. Mol. Struct. (THEOCHEM)*, **2008**, *854*, 46.

Liang-Wei, S.; Bin, C.; Jun-Hong, Z.; Tao, Z.; Qiang, K.; Min-Bo, C. *J. Phys. Chem. A*, **2008**, *112*, 11724.

Lin, Y.; Ding, F.; Yakobson, B. I. *Phys. Rev. B*, **2008**, *78*, 041402(R).

Liu, C.-S.; Zeng, Z. *Appl. Phys. Lett.*, **2010**, *96*, 123101.

Loutfy, R. O.; Wexler, E. M. In "*Perspectives of Fullerene Nanotechnology*," edited by E Õsawa, (Kluwer Academic Publishers), **2002**, pp 289.

Lu, J.; Choi, Y. J.; Fang, Z. Z.; Sohn, H. Y.; Rönnebro, E. *J. Am. Chem. Soc.*, **2009**, *131*, 15843.

Ma, S.; Zhou, H.-C. *Chem. Commun.*, **2010**, *46*, 44.

Mao, W. L.; Mao, H. *Proc. Natl. Acad. Sci. U.S.A.*, **2004**, *101*, 708.

Mao, W. L.; Mao, H.; Goncharov, A. F.; Struzhkin, V. V.; Guo, Q.; Hu, J.; Shu, J.; Hemley, R. J.; Somayazulu, M.; Zhao, Y. *Science*, **2002**, *297*, 2247.

Meek, S. T.; Greathouse, J. A.; Allendorf, M. D. *Adv. Mater*, **2011**, *23*, 249.

Meng, S.; Kaxiras, E.; Zhang, Z. *Nano Lett.*, **2007**, *7*, 663.

Metal Dihydrogen and Bond Complexes-Structure, Theory and Reactivity, edited by Kubas, G. J. (Kluwer Academic/Plenum Publishing, New York) **2001**.

Mizan, T. I.; Savage, P. E.; Ziff, R. M. *J. Phys. Chem.*, **1994**, *98*, 13067.

Mondal, S.; Ghosh, S.; Chattaraj, P. K. *J. Mol. Model*, **2013**, *19*, 2785.

Mondal, S.; Srinivasu, K.; Ghosh, S.; Chattaraj, P. K. *RSC Advances*, **2013**, *3*, 6991.

Monkhorst, H. J.; Pack, J. D. *Phys. Rev. B*, **1976**, *13*, 5188.

Mpourmpakis, G.; Froudakis, G. E.; Lithoxoos, G. P.; Samios, J. *J. Chem. Phys.*, **2007**, *126*, 144704.

Mpourmpakis, G.; Froudakis, G. E.; Lithoxoos, G. P.; Samios, J. *Nano Lett.*, **2006**, *6*, 1581.

Mpourmpakis, G.; Tylianakis, E.; Froudakis, G. E. *Nano Lett.*, **2007**, *7*, 1893.

Mpourmpakis, G.; Tylianakis, E.; Froudakis, G. *J. Nanosci. Nanotechnol*, **2006**, *6*, 87.

Mulliken, R. S. J. *Chem. Phys.*, **1955**, *23*, 1833.

Murray, L. J.; Dincă, M.; Long *J. R. Chem. Soc. Rev.*, **2009**, *38*, 1294.

Noorizadeh, S.; Dardaba, M. *Chem. Phys. Lett.*, **2010**, *493*, 376.

Okamoto, Y.; Miyamoto, Y. *J. Phys. Chem. B*, **2001**, *105*, 3470.

Pan, S.; Banerjee, S.; Chattaraj, P. K. *J. Mex. Chem. Soc.*, **2012**, *56*, 229.

Pan, S.; Contreras, M.; Romero, J.; Reyes A.; Merino, G.; Chattaraj, P. K. *Chem. Euro. J.*, **2013**, *19*, 2322.

Pan, S.; Giri, S.; Chattaraj, P. K. *J. Comp. Chem.*, **2012**, *33*, 425.

Pan, S.; Merino, G.; Chattaraj, P. K. *Phys. Chem. Chem. Phys.*, **2012**, *14*, 10345.

Park, N.; Hong, S.; Kim, G.; Jhi, S. H. *J. Am. Chem. Soc.*, **2007**, *129*, 8999.

Parr, R. G.; Chattaraj, P. K. *J. Am. Chem. Soc.*, **1991**, *113*, 1854.

Parr, R. G.; Donnelly, R. A.; Levy, M.; Palke, W. E. *J. Chem. Phys.*, **1978**, *68*, 3801.

Parr, R. G.; Pearson, R. G. *J. Am. Chem. Soc.*, **1983**, *105*, 7512.

Parr, R. G.; Szentpaly, L. v.; Liu, S. *J. Am. Chem. Soc.*, **1999**, *121*, 1922.

Parr, R. G.; Yang, W. *Density Functional Theory of Atoms and Molecules*, (Oxford University Press, New York) **1989**.

Parr, R. G.; Yang, W. *J. Am. Chem. Soc.*, **1984**, *106*, 4049.

Parthasarathi, R.; Elango, M.; Subramanian, V.; Chattaraj, P. K. *Theor. Chem. Acc.*, **2005**, *113*, 257.

Pearson, R. G. *Chemical Hardness: Applications from Molecules to Solids*, (Wiley-VCH: Weinheim) **1997**.

Pearson, R. G. *J. Chem. Edu.*, **1987**, *64*, 561.

Peng, Q.; Chen, G.; Mizuseki, H.; Kawazoe, Y. *J. Chem. Phys.*, **2009**, *131*, 214505.

Perdew, J. P.; Burke, K.; Ernzerhof, M. *Phys. Rev. Lett.*, **1996**, *77*, 3865.

Peterson, V. K.; Liu, Y.; Brown, C. M.; Kepert, C. J. *J. Am. Chem. Soc.*, **2006**, *128*, 15578.

Pinkerton, F. E.; Meyer, M. S.; Meisner, G. P.; Balogh, M. P.; Vajo, J. J. *J. Phys. Chem. C*, **2007**, *111*, 12881.

Plimpton, S. J. *J. Comp. Phys.*, **1995**, *117*, 1.

Powell, H. M. *J. Chem. Soc.*, **1948**, 61.

Prasad, P. S. R.; Sugahara, T.; Sum, A. K.; Sloan, E. D.; Koh, C. A. *J. Phys. Chem. A*, **2009**, *113*, 540.

Rosi, N. L.; Eckert, J.; Eddaoudi, M.; Vodak, D. T.; Kim, J.; O'Keeffe, M.; Yaghi, O. M. *Science*, **2003**, *300*, 1127.

Rowsell, J. L. C.; Yaghi, O. M. *Angew. Chem. Int. Ed.*, **2005**, *44*, 4670.

Rowsell, J. L.; Millward, A. R.; Park, K. S.; Yaghi, O. M. *J. Am. Chem. Soc.*, **2004**, *126*, 5666.

Saha, D.; Deng, S. *Langmuir*, **2011**, *27*, 6780.

Sakai, T., Natsuoka, M.; Iwakura, C. *Handb. Phys. Chem. Rare Earths*, **1995**, *21*, 135.

Sakintuna, B.; L-Darkrimb, F.; Hirscherc, M. *Int. J. Hydrogen Energy,* **2007**, *32,* 1121.

Sandrock, G.; Thomas, G. *Appl. Phys. A,* **2001**, *72*, 153.

Schlapbach, L.; Felix Meli, F.; Züttel, A.; Westbrook, J. H.; Fleischer, R. L. (eds.) in Intermetallic Compounds: Principles and Practice Vol. 2, Ch. 22, Wiley, **1994**.

Schlapbach, L.; Züttel, A. *Nature*, **2001**, *414*, 353.

Schleyer, P. V. R. *Chem. Rev*., **2001**, *101*, 1115.

Schleyer, P. V. R.; Maerker, C.; Dransfeld, A.; Jiao, H.; Hommes, N. J. R. V. E. *J. Am. Chem. Soc*., **1996**, *118*, 6317.

Schüth, F.; Bogdanović, B.; Felderhoff, M. *Chem. Commun.,* **2004**, 2249.

Scott, D. W.; Berg, W. T.; Hossenlopp, I. A.; Hubbard, W. N.; Messerly, J. F.; Todd, S. S. *J. Phys. Chem*., **1967**, *71*, 2263.

Setijadi, E. J.; Boyer, C.; Aguey-Zinsou, K.-F. *Phys. Chem. Chem. Phys.,* **2012**, *14*, 11386.

Shevlin, S. A.; Guo, Z. X. *Appl. Phys. Lett*., **2006**, *89*, 153104.

Shimada, W.; Shiro, M.; Kondo, H.; Takeya, S.; Oyama, H.; Ebinuma, T.; Narita, H. *Acta Crystallogr C*, **2005**, *61*, 65.

Sloan, E. D.; Koh, C. A. *Clathrate Hydrates of Natural Gases*, 3rd Edn, (Taylor & Francis – CRC Press, London), **2007**.

Srinivasu, K.; Chandrakumar, K. R. S.; Ghosh, S. K. *Phys. Chem. Chem. Phys*., **2008**, *10*, 5832.

Srinivasu, K.; Ghosh, S. K. *J. Phys. Chem. C*, **2012**, *116*, 5951.

Srinivasu, K.; Ghosh, S. K.; Das, R.; Giri, S.; Chattaraj, P. K. *RSC Adv*., **2012**, *2*, 2914.

Stanger, A. *J. Org. Chem*., **2006**, *71*, 883.

Stephens, F. H.; Baker, R. T.; Matus, M. H.; Grant, D. J.; Dixon, D. A. *Angew. Chem. Int. Ed.*, **2007**, *46*, 746.

Stracke, M. P.; Ebeling, G.; Cataluña, R.; Dupont, J. *Energy and Fuels*, **2007**, *21*, 1695.

Strobel, T. A.; Koh, C. A.; Sloan, E. D. *Fluid Phase Equilibria*, **2007**, *261*, 382.

Struzhkin, V. V.; Militzer, B.; Mao, W. L.; Mao, H.-k.; Hemley, R. J. *Chem. Rev*., **2007**, *107*, 4133.

Su, F.; Bray, C. L.; Carter, B. O.; Overend, G.; Cropper, C.; Iggo, J. A.; Khimyak, Y. Z.; Fogg, A. M.; Cooper, A. I. *Adv. Mater*, **2009**, *21*, 2382.

Su, F.; Bray, C. L.; Tan, B.; Cooper, A. I. *Adv. Mater*, **2008**, *20*, 2663.

Subrahmanyam, K. S.; Kumar, P.; Maitra, U.; Govindaraj, A.; Hembram, K. P. S. S.; Waghmare, U. V.; Rao, C. N. R. *Proc. Nat. Acad. Sci. (USA)*, **2011**, *108*, 2674.

Sun, Q.; Jena, P.; Wang, Q.; Marquez, M. *J. Am. Chem. Soc*., **2006**, *128*, 9741.

Sun, Q.; Wang, Q.; Jena, P. *Nano Lett*., **2005**, *5*, 1273.

Talyzin, A. *Int. J. Hydrogen Energy*, **2008**, *33*, 111.

Tarakeshwar, P.; Dhilip Kumar, T. J.; Balakrishnan, N. *J. Chem. Phys*., **2009**, *130*, 114301.

Teichmann, D.; Arlt, W.; Wasserscheid, P.; Freymann, R. *Energy Environ. Sci*., **2011**, *4*, 2767.

Teichmann, D.; Stark, K.; Müller, K.; Zöttl, G.; Wasserscheid, P.; Arlt, W. *Energy Environ. Sci.*, **2012**, *5*, 9044.

Teleman, O.; Jonsson, B.; Engstrom, S. *Mol. Phys*., **1987**, *60*, 193.

Türker, L.; Gümüş, S. *J. Mol. Struct. (Theochem)*, **2004**, *681*, 21.

Vajo, J. J.; Mertens, F.; Ahn, C.C.; Bowman, R. C. Jr., Fultz, B. *J. Phys. Chem. B*, **2004**, *108*, 13977.

van den Berg, A. W. C.; Areán C.O. *Chem. Commun.*, **2008**, 668.

Veziroglu, T. N.; Barbir, F. *Int. J. Hydrogen Energy*, **1992**, *17*, 391.

Wang, J., Ebner, A. D.; Ritter, J. A. *J. Phys. Chem. C*, **2007**, *111*, 14917.

Wang, L.; Lee, K.; Sun, Y.-Y.; Lucking, M.; Chen, Z.; Zhao, J. J.; Zhang, S. B. *ACS Nano*, **2009**, *3*, 2995.

Wang, Q.; Sun, Q.; Jena, P.; Kawazoe, Y. *ACS Nano*, **2009**, *3*, 621.

Weast, R. C.; Astle, M. J.; Beyer, W. H. CRC Handbook of Chemistry and Physics. 64th Edn, (Boca Raton, FL, CRC Press) **1983**.

Wen, S.-H.; Deng, W.-Q.; Han, K.-L. *J. Phys. Chem. C*, **2008**, *112*, 12195.

Wu, G.; Wang, J.; Zhang, X.; Zhu, L. *J. Phys. Chem. C*, **2009**, *113*, 7052.

Wu, H. Y.; Fan, X. F.; Kuo, J.-L.; Deng, W.-Q. *Chem. Commun.*, **2010**, *46*, 883.

Wu, X.; Gao, Y.; Zeng, X. C. *J. Phys. Chem. C*, **2008**, *112*, 8458.

Xiao, X. Z.; Chen, L. X.; Fan, X. L.; Wang, X. H.; Chen, C. P.; Lei, Y. Q.; Wang, Q. D. *Appl. Phys. Lett.*, **2009**, *94*, 041907.

Xing, L. G.; Li, N. X.; Wei, Y. L.; Jun, C.; Qiang, X. *Mater Sci. Forum*, **2005**, *475,* 2437.

Yang, J.; Sudik, A.; Wolverton, C.; Siegel, D. *J. Chem. Soc. Rev.*, **2010**, *39*, 656.

Yang, W.; Mortier, W. J. *J. Am. Chem. Soc.*, **1986**, *108*, 5708.

Yildirim, T.; Ciraci, S. *Phys. Rev. Lett.*, **2005**, *94*, 175501.

Yoon, M.; Yang, S.; Wang, E.; Zhang, Z. *Nano Lett.*, **2007**, *7*, 2578.

Yue, X.; Zhao, J.; Qiu, J. *Computational Science – ICCS 2007, Lecture Notes in Computer Science*, **2007**, *4488*, 280.

Yvon, K. *Chimia*, **1998**, *52*, 613.

Zaluska, A.; Zaluski, L.; Stroem-Olsen, J. O. *Appl. Phys. A*, **2001**, *72*, 157.

Zhang, Y.-H. P. *Energy Environ. Sci.*, **2009**, *2*, 272.

Zhang, Y.-H. P.; Evans, B. R.; Mielenz, J. R.; Hopkins, R. C.; Adams, M. W. W. *PLoS ONE*, **2007**, *2(5)*, e456. doi: 10.1371/journal.pone.0000456.

Zhang, Y.-H. P.; Mielenz, J. R. *Energies*, **2011**, *4*, 25.

Zhao, D.; Yuan, D.; Zhou, H.-C. *Energy Environ. Sci.*, **2008**, *1*, 222.

Zhao, X.; Xiao, B.; Fletcher, A. J.; Thomas, K. M.; Bradshaw, D.; Rosseinsky, M. J. *Science*, **2004**, *306*, 1012.

Zhongfang, C.; Haijun, J.; Andreas, H.; Walter, T. *J. Mol. Model.*, **2001**, *7*, 161.

Zhou, W.; Yildirim, T.; Durgun, E.; Ciraci, S. *Phys. Rev. B*, **2007**, *76*, 085434.

Zhoua, J.; Wang, Q.; Sun, Q.; Jena, P.; Chen, X. S. *PNAS*, **2010**, *107*, 2801.

CHAPTER 7

COMPUTATIONAL INSIGHTS OF ADSORPTION AND CATALYSIS WITHIN NANOPOROUS ZEOLITES

GANG YANG

CONTENTS

Abstract 250
7.1 Introduction 250
7.2 Zwitterion Stablization by H-form Zeolite 252
7.3 Benzene Hydroxylation within Fe/ZSM-5 Zeolite 256
7.4 Hydrogen Adsorption and Storage 259
7.5 Conclusions 266
7.6 Acknowledgments 267
Keywords 267
References 267

ABSTRACT

Zeolites are an important type of nanoscale materials with an ever-increasing application in a variety of fields such as separation, adsorption and heterogenous catalysis. With the aid of computational tools, a series of complicated adsorption and reaction mechanisms within zeolites have been clarified at a molecular level. Three currently focusing research topics will be discussed in this chapter, as (1) H-ZSM-5 zeolite to stabilize the amino acid zwitterions and their transformation with the canonical isomers, (2) The exchange of the acidic proton in H-ZSM-5 zeolite with the iron species (i.e., Fe/ZSM-5 zeolite) catalyzes the direct benzene hydroxylation to produce phenol and (3) The alteration of exchanged metal ions (M) in M/ZSM-5 zeolites causes distinct adsorption behaviors for the hydrogen adsorption process, where the La^{III} ion is expected to have outstanding adsorption capacity and verified by the accompanying density functional calculations. The effects of zeolite framework are discussed. These computational findings greatly help to understand the adsorption and reactivity performances within nanoporous zeolites.

7.1 INTRODUCTION

Not a few researchers call zeolites "philosopher's stone" of modern chemistry (Arends et al., 2004). Zeolites can be classified into two types: natural and synthetic. The natural zeolites, mainly from the chemical reactions of volcanic rocks and ash layers with alkaline groundwater, are usually contaminated by other minerals, metals, quartz and other zeolites. Because of this reason, natural zeolites are not suitable for many commercial applications wherein the uniformity and purity are essentially required. In addition, the global production of natural zeolties is approximately 3 million tones per year, which seems far from sastifying the increasing demand of zeolites. As reported (Win, 2007), the fluid catalytic cracking (FCC) technology alone consumes 5 million tones annually. Thus, it becomes a necessity to synthesize and produce zeolites on an industrial scale, and the present chapter will be based on synthetic zeolites unless noted otherwise.

Since the discovery of ZSM-5 by Mobil Oil in the early 1970′s, zeolites have found applications in a wide range of fields (Čejka et al., 2010; Cariati et al., 2003; Sauer et al., 1994), with the concominant production of a series of novel types of zeolites, such as BEA, MCM-22, and SBA-15. Zeolites are generally a family of crystalline microporous aluminosicates with regular pore systems. Each framework aluminum (Al) site is balanced by one counterion nearby, usually proton (H^+). The catalytic activity of H-form zeolites is heavily dependent on the acidic property, which may vary significantly for the different types of zeolties or for the different T sites of one given zeolite. In addition, the acidic strength of H-form zeolites is closely related with the incorporated M^{3+} ions, which can be Al, Fe, B or Ga. Theoretical calculations at different levels show that the acidity increases as B- < Fe- < Ga- < Al-incorported zeolites (Chatterjee et al., 1998; Yuan et al., 2002; Wang et al., 2004). The extralattice acidic protons are ready to be exchanged with a wide range of ions such as Li^+, Na^+, Mg^{2+} and Fe^{3+}, which further causes the diversity of binding affinity and catalytic activity for adsorbents. Moreover, the chemical states of a given exchanged species are depenedent on the raw material, synthetic and pretreatment conditions; for example, the exchange with the iron species can lead to Fe^{2+}, Fe^{3+}, $(FeOH)^+$, $Fe(OH)_2^+$, $(FeO_2Fe)^+$, $(FeO(OH)Fe)^{2+}$ and/or other forms (Pirngruber et al., 2004; Ribera et al., 2000; Yang et al., 2010). The transformation among these various forms seems impossible to be observed experimentally, and it also remains a grand challenge for the experimental researchers to characterize the "short-life" intermediates and transition states within zeolite nanopores. Instead, theoretical calculations are capable of providing a clear description of these information as well as related reaction mechanisms. Kohn and Pople shared the 1998's Nobel Prize in chemistry owing to their outstanding contributions to the deveplement and application of computational methods, which have been extensively used in current reseach activities, including nanoporous zeolitic systems.

In this chapter, the application of computational methods to the nanoporous zeolitic systems will be presented with several of our recent works. The first is H-form zeolite to stabilize the amino acid zwitterion ($NH_3^+CHRCOO$-, *see* Scheme 1) and interconversion with the corresponding canonical isomer. Owing to the creation of strong electric fields around, zwitterions have been acknowledged as the driving forces that determine the structures, functions and catalysis of biomolecules such as amino acids, peptides and proteins. The stabilization within zeolite nanopores offers a good prototype to study the amino acid zwitterions, which play a significant function in a wide range of chemical and biological processes by generating strong electric fields around. Nonetheless, the zwitterions of amino acids cannot exist independently in gas phase (Gutowski et al., 2000; Kass, 2005; Kassab et al., 2000). The second is the direct hydroxylation of benzene to produce phenol catalyzed by Fe-exchanged zeolite. The catalysis by Fe/ZSM-5 zeolite has shown great potential to replace the industrial process via cumeme, which is relatively energy costing and seriously dependent on the "by-product" acetone (Panov et al., 1998). Owing to the large consumption of phenol, estimated to be 1.4 million tons in 2011, the current technology results in the aggravation of the already serious environmental crisis. The third is the hydrogen adsorption and storage over the various ion-exchanged zeolites. As compared to the depleting petroleum reserve, hydrogen represents a promizing renewable energy carrier and receives extensive attention from all around the world (Graetz, 2009; Kubas,

2007). These computational disussions are expected to help to understand the adsorption and reactivity performances within zeolite nanopores.

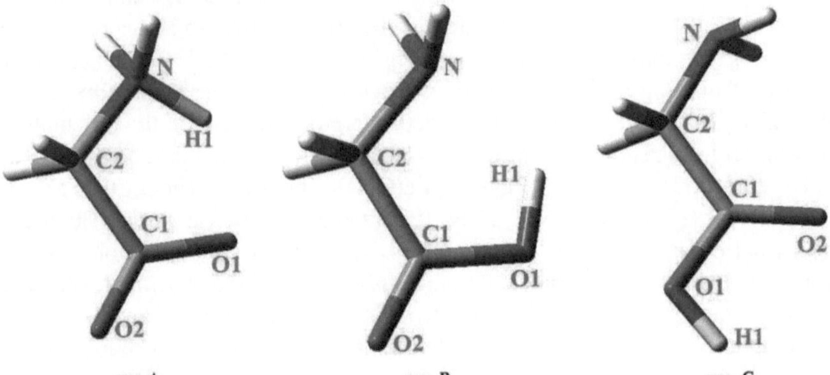

SCHEME 1 Several isomers of glycine. Gly^A is the zwitterion, Gly^B is the canonical isomer that can transform to Gly^A through intramolecular proton transfer, and Gly^C is the most stable isomer in gas phase.

7.2 ZWITTERION STABLIZATION BY H-FORM ZEOLITE

As indicated by many computational results, zwitterions of amino acids, including glycine, are generally not local energy minima in gas phase. With the use of density functional calculations, Kass (Kass, 2005) observes that the presence of the oxalic or malonic dianion renders the glycine zwitterion geometrically stable. Unfortunately, the oxalic and malonic dianions themselves are not stationary points on the potential energy surfaces (PES) just as the amino acid zwitterions; instead, they will decompose spontaneously by losing an electron. Later on, our group (Yang et al., 2008) screens out a number of dianions or monoanion pairs that can exist independently and stabilize the glycine zwitterion effectively, including SO_4^{2-}, HPO_3^{2-}, 2I- and other anions that are easily available in almost all chemistry laboratories. In addition, the stabilization effects are closely related with the proton affinities (PA) of these dianions and monoanion pairs as well as the proton-affinity differences ($\Delta|PA|$) of the comprizing two binding sites.

Zeolites are generally nanoporous aluminosilicates with negative framework, which can be potentially applied to the stabilization of amino acid zwitterions. The two-layer ONIOM(B3LYP/6–31+G**:B3LYP/3–21G) scheme is used to describe the local structures of H-form ZSM-5 zeolite. The glycine zwitterion (**Gly^A**, see Scheme 1), which by itself is geometrically unstable and transforms automatically to the canonical isomer, has been rendered to be a stationary point within the ZSM-5 nanopore (Fig. 1). The corresponding canonical isomer (**Gly^B**) and the lowest-energy isomer (**Gly^C**) have also been optimized at the same level of theory, *see* Fig. 1 (Yang et al., 2009a). The interaction energies with ZSM-5 zeolite at ONIOM(MP2/6–311++G**:B3LYP/3–21G)//ONIOM(B3LYP/6–31+G**:B3LYP/ 3–21G) are estimated to be −44.9, −30.0 and −20.3 kcal/mol for **Gly^A**, **Gly^B** and **Gly^C**, respectively. The presence of stronger electrostatic and H-bonding interactions causes larger interaction energy with the zwitterionic isomer **Gly^A**. As a result, **Gly^A** in ZSM-5 zeolite is 7.6 kcal/mol more stable than **Gly^C**, the isomer that predominates in gas phase. Accordingly, the nanoporous H-ZSM-5 zeolite successfully stabilizes glycine in the zwitterionic form.

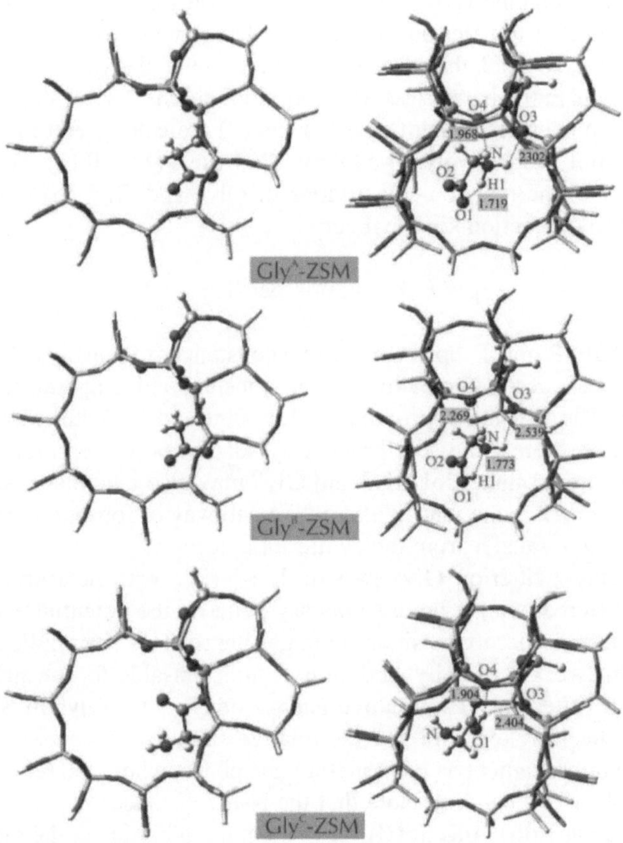

FIGURE 1 Optimized structures of **Gly^A** (top), **Gly^B** (middle) and **Gly^C** (bottom) within ZSM-5 zeolite.

The relative energy of Gly^A vs. Gly^B in ZSM-5 zeolite is rather small (2.7 kcal/mol); in addition, the intramolecular proton transfer is very facile, and the energy barrier is calculated to be 4.5 kcal/mol. In the transition state structure of proton transfer, the N-H1 and O1-H1 distances are optimized at 1.242 and 1.284 Å, respectively.

The direct adsorption of any of the three-glycine isomers (Scheme 1) on the H-form ZSM-5 zeolite will be protonated (Boekfa et al., 2008; Yang et al., 2009a). It is caused by the fact that H-ZSM-5 zeolite has a much lower proton affinity than glycine and will donate the acidic proton to glycine automatically when interacted. On the other hand, the deprotonated glycine should have a much larger proton affinity and is thus assumed to deprive the acid proton from H-form ZSM-5 zeolite. This is confirmed by the two-layer ONIOM calculations. The adsorption of deprotonated glycine ($NH_2CH_2COO^-$, Dep) via the amino group causes the spontaneous proton transfer and formation of the zwitterionic isomer, see Pathway (1) in Fig. 2. Instead, the adsorption of deprotonated glycine via the carboxylate anion results in the barrierless transformation to Gly^C, see Pathway (2) in Fig. 2. The reaction energies of Pathways (1) and (2) are calculated to be −56.03 and −48.46 kcal/mol, respectively. In addition, Pathways (1) and (2) are irreversible. In consequence, the reaction selectivities of Gly^A and Gly^C (ρ) can be derived using the parallel reaction kinetic theory.

$$\rho = k_1/k_2 \approx \exp(—\Delta E/RT) \quad (1)$$

where k_1, k_2, ΔE, R and T represent the rate constants of Pathways (1) and (2), the energy difference of the two pathways, gas constant and temperature (298.15 K), respectively. The ρ value approximates 350–000 and indicates that the glycine zwitterion is absolutely the sole product by adsorption of deprotonated glycine. Nonetheless, the formation of $\mathbf{Gly^B}$ and $\mathbf{Gly^C}$ may also take place, see Pathways (3) and (4) in Fig. 2. Note that an alternative pathway of forming the $\mathbf{Gly^B}$ isomer is through proton transfer from the zwitterionic form.

The glycine zwitterion $\mathbf{Gly^A}$ in Silicalite-1 (i.e., with no isomorphous substitution of heteroatoms) is not a stationary point on the potential energy surface (PES), which is similar to the situation in gas phase. This is an indication that the negative framework caused by Al doping is indispensable for the stabilization of amino acid zwittterions. The relative energy of $\mathbf{Gly^A}$ vs. $\mathbf{Gly^B}$ in Silicalite-1 is estimated to be 15.8 kcal/mol and the relative stability of the zwitterionic form shows some improvement as compared to gas phase, where the relative energy is approximately 18.0 kcal/mol. Note that the N-H1 distances of $\mathbf{Gly^A}$ in these two cases are fixed at 1.061 Å as in $\mathbf{Gly^{A\text{-}}ZSM}$. On the other hand, the relative stability of $\mathbf{Gly^B}$ vs. $\mathbf{Gly^C}$ equals 4.1 kcal/mol within Silicilate-1 and shows a dramatic increase when compared to gas phase (0.42 kcal/mol). Therefore, the inclusion of

Silicalite-1 framework reinforces the relative stability of the **Gly^C^** isomer whereas destabilizes the **Gly^B^** isomer.

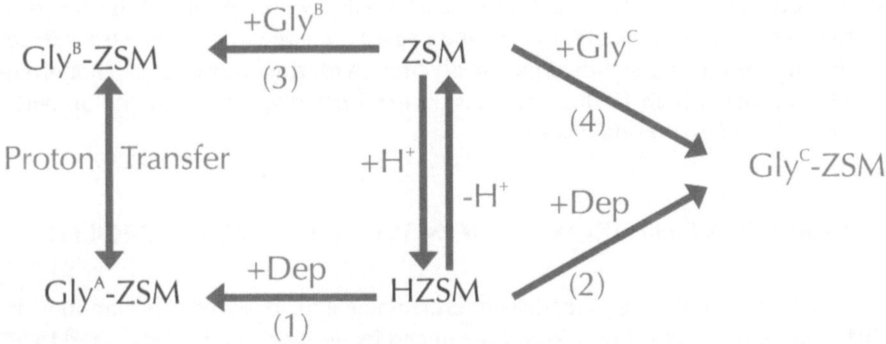

FIGURE 2 Formation and conversion pathways of the glycine isomers within H-ZSM-5 zeolite.

The **Gly^A** geometries in ZSM-5 zeolite and Silicalite-1 can be finely superposed with each other, which is also applicable to those of **Gly^B** but not to those of **Gly^C**. Accordingly, the respective roles of the framework (E_F^{ZSM}) and negative charge created by Al doping (E_N^{ZSM}) in H-ZSM-5 zeolite can be assessed by using the **Gly^B** isomer as the benchmark. As the **Gly^A** and **Gly^B** isomers in ZSM-5 and Silicalite-1 are in the respective proximal positions, the contribution from their framewok should be close to each other; that is,

$$E_F^{ZSM} \cong E_F^{SIL} \tag{2}$$

where ZSM and SIL represent ZSM-5 zeolite and Silicalite-1, respectively.

As there is no negative charge in Silicalite-1, its energy contribution to the zwitterion stabilization (E^{SIL}) comes solely from the framework; that is,

$$E^{SIL} = E_F^{SIL} \tag{3}$$

Thus, it leads to the following equation,

$$E_F^{ZSM} \cong E^{SIL} \tag{4}$$

Accordingly, the energy contribution of ZSM-5 zeolite during the zwitterion stabilization is decomposed into,

$$E^{ZSM} = E_N^{ZSM} + E_F^{ZSM} \cong E_N^{ZSM} + E^{SIL} \tag{5}$$

The E_N^{ZSM} and E_F^{ZSM} values calculated according to Eq. (5) are equal to −8.6 and −6.3 kcal/mol, respectively. The negative charge is indispensable and plays a larger role during the stabilization of glycine zwitterion; however, the role of ZSM-5 framework cannot be underestimated and accounts for 74% of that of the negative charge. That is, the collaboration of the negative charge and framework results in the formation and stabilization of glycine zwitterion within the nanoporous ZSM-5 zeolite, which further provides an ideal prototype to study the property and function of zwitterionic structures.

7.3 BENZENE HYDROXYLATION WITHIN FE/ZSM-5 ZEOLITE

Phenol is one of the top chemical intermediates and its global consumption in 2011 amounts to 1.4 million tons. Phenol and its derivatives are considered to be the key building blocks for a variety of chemicals such as polycarbonates, epoxies, nylon, pharmaceutical drugs and etc. Currently, the industrial production of phenol is based on the cumene process, see the equation below:

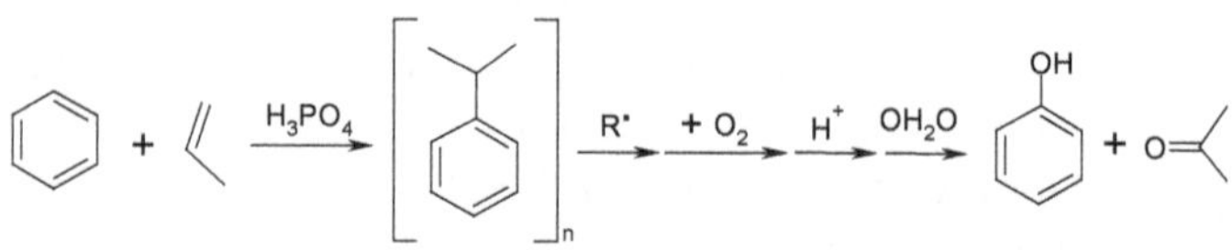

(6)

Obviously, this is a multistep and relatively enery-costing process; in addition, acetone will be coproduced in a 1:1 ratio with phenol, which causes the market price of phenol to be significantly dependent on that of acetone. In contrast, Fe-exchanged zeolite can hydrolyze benzene directly to phenol with high efficiency. The hydroxylation reaction contains only one step and the active sites can be generated by selective reduction of nitrous oxide (N_2O), thus providing an alternative way to use the harmful nitrous oxide. Accordingly, the direct hydroxylation of benzene to phenol catalyzed by Fe-exchanged zeolite has shown great potential to replace the industrial process via cumeme (Panov et al., 1998).

There are a number of experimental and computational studies on the active sites of Fe-exchange zeolites, which are assumed to be the mononuclear iron, binuclear iron (Fe, Fe), iron-peroxide ($Fe(O_2)^{2-}Fe$) or extralattice aluminum (Fe, Al) (Fellah et al., 2011; Hansen et al., 2007; Hensen et al., 2003; Ryder et al., 2002; Yang et al., 2006a, 2007). It has been testified that the chemical states of the active sites are dependent on many factors such as synthetic methods and conditions, pretreatments and so on. At low Fe/Al ratios, the mononuclear iron should at least be one of the competive active sites in Fe/ZSM-5 zeolite. The reactions of N_2O decomposition and benzene hydroxylation over the mononuclear iron in Fe/ZSM-5 zeolite can be written as,

$$N_2O + Fe^* \rightarrow Fe\text{–}O^* + N_2 \quad (7)$$

$$\text{benzene} + Fe\text{–}O^* \rightarrow \text{phenol} + Fe^* \quad (8)$$

Scheme 2 displays the reaction cylces of N_2O decomposition and benzene hydroxylation on Fe^{III}/ and Fe^{II}/ZSM-5 zeolite (Yang et al., 2013). For the trivalent Fe/ZSM-5 zeolite, the reaction can be divided into seven elementary steps, *see* Fig. 3. Step 1 is the adsorption of N_2O, wherein the N1-O4 bond is slightly elongated from 1.196 to 1.212 Å. Step 2 is the breaking of N1-O4 bond via the transition state **TS1**. Step 3 is the desorption of N_2 forming the α-O species. The geometries of the α-O species are different for the 5-T and 37-T cluster models (Yang et al., 2009b, 2013). In the larger cluster models (37-T), the FeO_3O_4 species falls nearly within the $O_1O_2Al_{12}$ plane, and the two extralattice O atoms (O_3 and O_4) of the α-O species should be oriented towards the zeolite framework rather than channel, which is obviously not beneficial for the subsequent chemical reactions such as benzene hydroxylation presently studied. This is different from the situation of 5-T cluster models where the two extralattice O atoms (O_3 and O_4) of the α-O species face the zeolite channels and are ready to initiate the reaction of benzene hydroxylation.

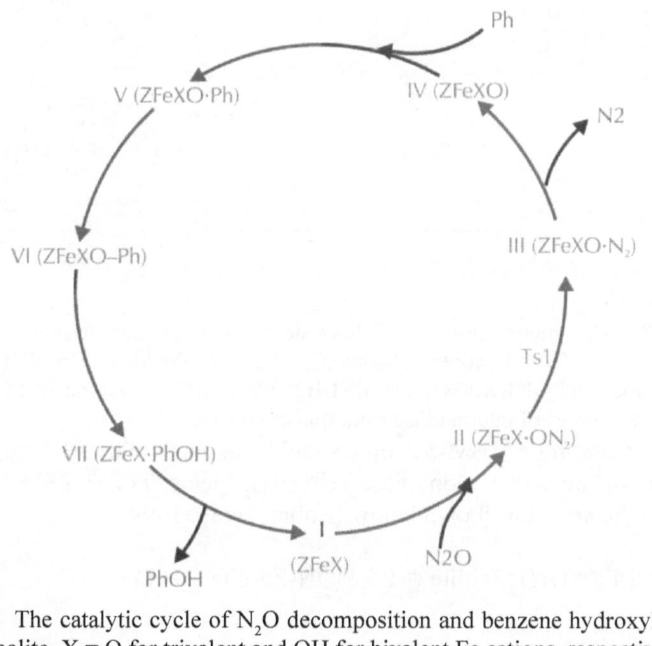

SCHEME 2 The catalytic cycle of N_2O decomposition and benzene hydroxylation over Fe-ZSM-5 zeolite. X = O for trivalent and OH for bivalent Fe cations, respectively.

Step 4 is the adsorption of benzene on the active site (i.e., α-O species), and in this adsorption structure (**v**), the two-extralattice O atoms have nearly equidistances with the C_1 atom of benzene. This adsorption structure is not reactive and requires the conformational transformation in order to become the active configuration (**V**), Following Step 5 (**v** ® V) is Step 6, where benzene approaches the extralattice O_4 atom and results in the formation of $C_{1-}O_4$ bond, see **VI** in Fig. 3. Step 7 is the formation of phenol through the proton transfer from C_1 to O_4. The final step (Step 8) is the desorption of phenol, which restores Fe/ZSM-5 zeolite to the original state that is ready for the next catalytic cycle.

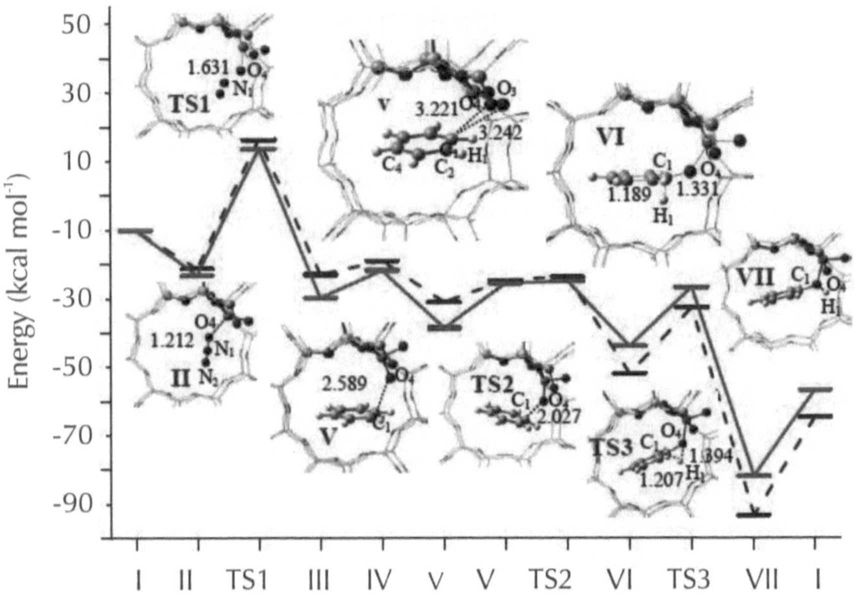

FIGURE 3 The energy profiles of N_2O decomposition and benzene hydroxylation reactions on Fe^{III}/ZSM-5 zeolite calculated at the ONIOM(M06L/BS1:B3LYP/6–31G*) (red solid line) and ONIOM(B3LYP/ BS1:B3LYP/3–21G*) (blue dashed line) levels as well as the structures of intermediates and transition states

The iron cations in Fe^{II}-ZSM-5 zeolite should exist in the $Fe^{II}OH$ form, which may come from the auto-reduction of the $Fe^{III}(OH)_2$ species in Fe^{III}-ZSM-5 zeolite, according to the equation shown below (Lobree et al., 1999):

$$2\ Fe^{III}(OH)_{2\text{-}}Zeolite \text{ ® } 2\ Fe^{II}OH\text{-}Zeolite + H_2O + {}^1/_2\ O_2 \qquad (9)$$

The vibrational frequency of the OH group in the $Fe^{II}OH$ form is calculated to be 3726 cm^{-1}, which is in good agreement with the experimental value of Fe-BEA zeolite at 3683 cm^{-1} (Nobukawa et al., 2004). We have tesitified that the Fe^{III} cations in Fe^{III}/ZSM-5 zeolite are facile to be reduced by H_2 treatment. The reaction steps of N_2O decomposition and benzene hydroxylation are exactly according to Scheme 2; that is, six elementary steps are included. The extralattice O atoms of the α-O species (**IV**) in Fe^{II}/ZSM-5 zeolite are oriented towards zeolite channel and can ignite the reaction of benzene hydroxylation directly. That is, there is no need for conformational transition as in the trivalent case.

As shown in Fig. 3, the M06L:B3LYP energy barriers of N_2O decomposition to form the active site (α-O species), benzene activation to form C-O bond and proton transfer to form phenol are equal to 37.0, 13.7 and 17.2 kcal/ mol for Fe^{III}/ ZSM-5 zeolite as well as 33.7, 3.0 and 19.1 kcal/mol for Fe^{II}-ZSM-5 zeolite, respectively. Note that in B1, all the elements are described by the 6–31+G** basis set, except Fe being treated with the LANL2DZ effective core potential (ECP) basis set. For the energy barriers of N_2O decomposition (Step 2), the activation barriers for the 5-T and 37 cluster models are consistent with each other; in addition, this step is easier to take place on the divalent state due to a lower energy barrier.

The reaction energies have been calculated using the various basis sets. It shows that the combination of the B1 and 6–31G* basis sets for the high- and low-level regions can obtain reliable reaction energies, respectively, while the 3–21G* basis set is not sufficient to describe the reaction energies of the low-level region. The energy differences of MP2:B3LYP and M06L/B1:B3LYP are mainly caused by those steps where the adsorbents move significantly within zeolites. For the intramolecular proton-transfer steps to form phenol, the adsorbents have no obvious migration within the zeolite nanopores and therefore the energy barriers at these two theoretical levels are rather close to each other. As revealed by the hybrid MP2:B3LYP energy calculations, the benzene adsorption is unfavorable, whether for the trivalent or divalent Fe/ZSM-5 zeolite. Especially, the adsorption in the divalent state seems to be thermodynamically prohibitive, with positive adsorption energy of 32.5 kcal/mol. There is less space left in the divalent state for the accommodation of the benzene molecule, which is further evidenced by the much smaller energy barrier of the following step of benzene activation in the divalent state.

7.4 HYDROGEN ADSORPTION AND STORAGE

The fossil fuels are expected to be depleted within the next few decades, and hydrogen represents a promizing alternative renewable energy carrier. More attractive, hydrogen energy is environmentally benign, because water is the only product of combustion. Hydrogen can be produced from steam reforming, partial oxidation of hydrocarbons, water photolysis or other methods. The utilization of hydrogen energy significantly relieves the increasingly serious global climate change. Unfortunately, there lacks effective means to store and transport the rather low-density hydrogen energy. Zeolites have been verified to be one of the materials with potential applications in this aspect (Areán et al., 2006; Assfour et al., 2011; Calleja et al., 2010; Georgiev et al., 2007; Nobukawa et al., 2004; Rorres et al., 2007).

Using 5-T cluster models, we study the adsorption behavior of H_2 on Fe^{III}/ZSM-5 zeolite, with the production of three different adsorption modes (Yang et al., 2006b). Two adsorption structures of close stability correspond to the high-spin Fe^{III} cations whereas one of lower relative energies corresponds to the low-spin Fe^{III} cations. The high-spin state predominates and agrees with the results of the inelastic neutron scattering (INS) experiments (Mojet et al., 2001). Later on, one adsorption structure of the high-spin state is also optimized with 25-T cluster models, see Fig. 4 (Yang et al., 2012). The adsorption of H_2 causes slight elongations of Fe-O_e and H_{a}-H_b bonds, from 1.668 and 0.744 Å to 1.676 and 0.754 Å, respectively. This is consistent with the results of 5-T cluster models. The H_2 adsorption structures on Al/, Co/, La/ and B/ZSM-5 zeolites are obtained as well, which are similar to that of Fe^{III}/ZSM-5 zeolite. The B/ZSM-5 zeolite is an exception, where the adsorbed H_2 molecule interacts with the extralattice O atom instead of the B site, for the B ion has already used up the *2 s* and *2p* orbitals by forming two single and one double B-O bonds. The analysis of M-H bonds indicates that the H_2 molecule forms two direct bonds with the metal ions in each case of Al, Co, Fe and La, which shows that the H_2 molecule tends to interact with the metal ions via the η^2-binding mode. The adsorption energies of H_2 on these ion-exchanged ZSM-5 zeolites increase as Al (9.3) < La (–2.4) < Co (–2.6) < Fe (–3.8), units in kcal/mol. It is surprizing to find that in the case of Fe^{III}/ZSM-5 zeolite, the adsorption energies of 5-T and 25-T cluster models are compable to each other, probably due to the canceling-out effects of the zeolite framework and dispersion energy. Note that the energy calculations are determined at the ONIOM(MP2/B1:B3LYP/6–31G*) level unless otherwise noted. For B1, the 6–31G(d) basis set is used for all except the heavy atoms and H_2-related species, which are treated with the LANL2DZ effective core potential (ECP) and 6–311++G(d, p) basis sets, respectively.

For 25-T cluster models, the H_2 reduction and restoring energy barriers are calculated to be 18.4 kcal/mol, which is rather small and suggests the facile reduction by hydrogen pretreatment. This is in agreement with the experimental results (Berlier et al., 2006; Jíša et al., 2009). Instead, the H_2 recombination will be hindered by the relatively much larger energy barrier (35.1 kcal/mol). The calculations of other metal ions indicate that the energy barriers of H_2 reduction are often smaller than those of H_2 recombination, especially for Al/ZSM-5 zeolite, where the energy barriers of H_2 reduction and recombination are equal to 7.7 and 62.5 kcal/mol, respectively. Accordingly, it seems not possible for the H atoms to recombine into the H_2 molecules. That is, zeolites with the $M^{III}O$ exchanged species are not suitable for hydrogen storage.

In the case of Al/ZSM-5 zeolite, the adsorption of a second H_2 molecule will not be adsorbed on the Al site; instead, it is pushed away and forms an H-bond with the O_e atom (Fig. 5a). The adsorption of a third H_2 molecule also forms H-bonding interaction with the O_e atom, but the H-bonding distances increase with the H_2 loading (n) indicative of a weakening interaction. When the number of adsorbed H_2 molecules (n) increase up to 5, the extralattice O_e atom has been covered up. Fig. 5b shows the adsorption structures with eight H_2 molecules. In this adsorption structure, the first adsorbed H_2 molecule (H_aH_b) forms direct bonds with the Al site via the η^2-binding mode, three subsequently adsorbed H_2 molecules (H_cH_d, H_eH_f and H_gH_h) form H-bonds with the extralattice O atom whereas the fourth incoming H_2 molecules surround the first-shell H_2 molecules without lateral interactions with the exchanged species. The $H_{a-}H_b$ bonds remain almost invariant (0.761~0.762 Å) during the increase of H_2 loading, except the slight elongation to 0.765 Å in the case of n = 3.

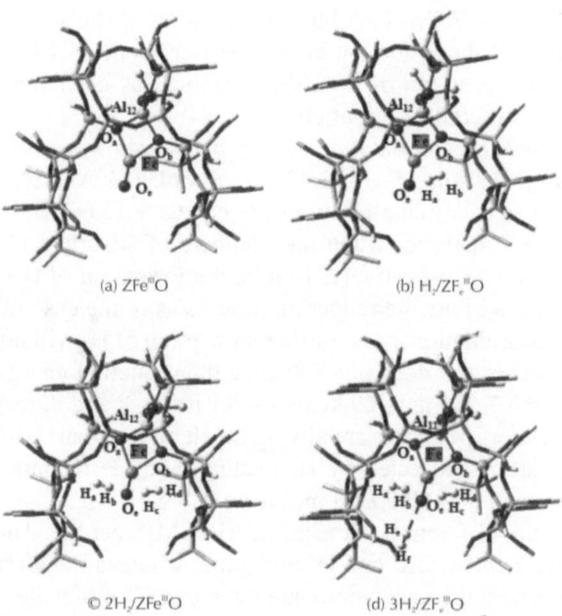

FIGURE 4 The adsorption structures of n-numbered H_2 molecules over the Fe^{III}-exchanged ZSM-5 zeolite (n = 0, 1, 2, 3).

In contrast to Al/ZSM-5 zeolite, the second H_2 molecule can be chemisorbed as the first one on the metal site in the case of Fe/, Co/ and La/ZSM-5 zeolites, see Fig. 4c. It can be seen that both H_2 molecules interact with the metal sites via the η^2-binding mode. The further adsorption will cause the formation of H-bonds with the extralattice O_e atom (Fig. 4d). The Mulliken charges of metal ions decrease with H_2 loading (n), indicating that the charge transfers take place from the $\sigma(H_2)$ to d(M) orbitals; meanwhile, the $d(M)\rightarrow\sigma^*(H_2)$ back-donations result in the negative charges on some H atoms. As compared to the adsorption of one H_2 molecule, the adsorbed H-H bonds are shorter and less activated, and the Fe-H and Co-H distances on the whole are larger, whereas the La-H distances are not apparently larger for La/ZSM-5 zeolite. The adsorption energies of the second H_2 molecule are calculated to be −1.5, −1.7 and −2.8 kcal/mol for Fe/, Co/ and La/ZSM-5 zeolites, respectively. It can be seen that the adsorption strengths of the second H_2 molecule are significantly reduced for Fe/ZSM-5 zeolite (by 60.5%) and less affected for Co/ZSM-5 zeolite with a reduction of adsorption strength of 34.6%. However, the adsorption strength of the second H_2 molecule seems to be comparable to that of the first one in the case of La/ZSM-5 zeolite, which has also been observed by other researchers during the H_2 adsorption study on K/FER zeolite (Areán et al., 2006).

As described above, two H_2 molecules can be chemisorbed simultaneously on the metal sites of Fe/ZSM-5 zeolite. The addition of the second H_2 molecule causes the increase of the reduction energy barrier, from 18.4 to 28.4 kcal/mol. It indicates that the reduction of Fe/ZSM-5 zeolite has been suppressed by the increased number of metal coordinations.

Meanwhile, the H_2 recombination energy barrier shows an increase with H_2 loading (n), changing from 35.1 and 43.5 kcal/mol. For Al/ZSM-5 zeolite, the reduction energy barrier by one H_2 molecule equals 3.5 kcal/mol, and increases to 4.0, 7.8, 8.2, 12.3 kcal/mol when the numbers of adsorbed H_2 molecule (n) amount to 2, 3, 4 and 5, respectively. That is, the formation of H-bonds with the extralattice O_e atom inhibits the reduction process as in the case of the increased number of metal coordinations. The further adsorption of H_2 will not interact with the exchanged species and does not influence the reduction energy barrier, with the values equal to 8.7, 8.6 and 8.9 kcal/mol for n = 6, 7 and 8, respectively. The H_2 recombination energy barriers are always much higher than those of reduction; in addition, they are not affected by H_2 loading, and the recombination energy barriers range from 58.3 to 60.9 kcal/mol for n = 1–8.

Fig. 5 shows the different Al species in Al/ZSM-5 zeolite. The above $Al^{III}O$ species can at most chemisorb one H_2 molecule, whereas the $Al^{III}(OH)_2$ species has no enough space left to accommodate even one H_2 molecule, where the adsorbed H_2 molecule will form an H-bond with the O_e atom. The Al^{III} species binds the H_2 molecule strongly, as observed from the large energy adsorption energy (–9.4 kcal/mol). As a result, the adsorbed H_2 molecule is much more activated on the Al^{III} rather than $Al^{III}O$ and $Al^{III}(OH)_2$ species. What is more important is that the Al^{III} species has a obviously larger adsorption capability of H_2 than the other two species, and three H_2 molecules can be chemisorbed at the same time. The adsorption energies of the second and third H_2 molecules are −65.8 and −10.5 kcal/mol, respectively. The sharp increase of adsorption energies due to the second H_2 molecule is due to the structural transformation, wherein the coordination number of the Al atom with lattice-O atoms is three instead of two as in the absence of H_2 molecules (Fig. 5d).

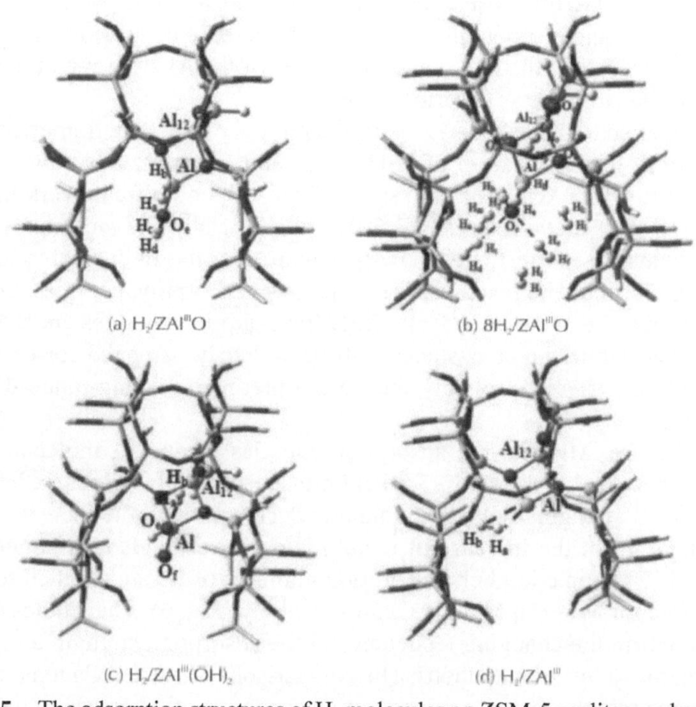

FIGURE 5 The adsorption structures of H_2 molecules on ZSM-5 zeolites exchanged with the various Al species.

A systematic study has been performed on the M^m species owing to the larger H_2 adsorption capacity than the other species. The Mg^{II} ion is the first to be discussed because it has often been used in hydrogen storage (Aguey-Zinsou et al., 2010; Torres et al., 2007). Four H_2 molecules can be simultaneously chemisorbed on the Mg site in Mg^{II}/ZSM-5 zeolite, whereas the fifth H_2 molecule is pushed away from the Mg site and interacts laterally with the previous H_2 molecules. For n = 1–4, all the H_2 molecules interact directly with the Mg sites via the η^2-

binding mode. The first four adsorbed H_2 molecules in the case of n = 5 also adopt the η^2-binding mode, even if the η^1-binding mode has been tried. In addition, the adsorption of the fifth H_2 molecule exerts slight influences on the $4H_2/ZMg^{II}$ geometries. The apparent adsorption energy averaged of the four H_2 molecules equals −8.2 kcal/mol, and the adsorption energy of the fifth H_2 molecule is negligible (0.6 kcal/mol). The studies of other metal ions ($M^m = Be^{II}, Ca^{II}, Li^I, Na^I$) also indicate that the H_2 molecules are preferred to be adsorbed through the η^2-binding mode, irrespective of the number of H_2 molecules (n). Among the above metal ions, only the Na ion forms three direct bonds to lattice-O atoms (Chu et al., 2011). Although the bicoordination charateristics of the other metal ions, one of the $M\text{-}O_{lattice}$ distances gradually shortens with increase of H_2 loading (n), and in the case of $2H_2/ZAl^{III}$, the coordination number of the Al atom with the lattice-O atoms increases to three as described before.

As compared to the naked Mg^{II} ions, the presence of zeolite framework causes a significantly less elongation of H-H bonds and a smaller adsorption energies from −53.4 to −16.1 kcal/mol. The same effects of zeolite framework have been detected for other metal ions ($M^m = Be^{II}, Ca^{II}, Li^I, Na^I$). The inclusion of zeolite framework increases the LUMO energies and accordingly decreases the Lewis acidity, which should be responsible for the lower adsorption energies. On the naked metal ions, the MP2:B3LYP and B3LYP adsorption energies are rather close to each other, while the consistency will be seriously damaged for calculations of the zeolitic systems, probably due to the presence of long-range dispersion interactions.

For M = Be, Mg, Ca, the adsorption energies of one H_2 molecule are calculated to be −16.1, −5.2 and 21.8 kcal/mol, respectively, which indicates that the adsorption strengths will have dramatic decreases with the increase of metal radius. In contrast, the increase of metal radius corresponds to a larger adsorption capacity. The maximal chemisorption numbers of H_2 are testified to be two, four and five for M = Be, Mg and Ca, respectively (Fig. 6). The studies of Li and Na ions confirm the changing tendencies of the adsorption strength and capacity along with the radius of metal ions. The comparisons of Na and Ca ions with close radii demonstrate the increase of adsorption capacity with increase of the valence state. The changes of adsorption capacity with metal radius and valence state can be explained with Lewis acidity, which increases with decrease of metal radius and with increase of valuence state. For each metal ion, the Lewis acidity gradually decreases with H_2 loading, as reflected by the LUMO energies. As Fig. 6 indicates, the apparent H_2 adsorption energies of the various ZM^m cluster models are equal to −1.7, −16.1, 0.8, −8.2, −28.6, 0.9 kcal/mol for Li (n = 2), Be (n = 2), Na (n = 3), Mg (n = 4), Al (n = 3) and Ca (n = 5), respectively.

Owing to the large radius and valence state, the La^{III} ion is expected to be a good candidate for hydrogen storage, which has been testified by the accompanying calculations. A maximum of six H_2 molecules can be simultaneously chemisorbed on the La site in La^{III}/ZSM-5 zeolite and the H_2 loading is higher than any of the above metal ions. This explains why the preference of using the La^{III} ion as an additive for the hydrogen storage materials (Gupta et al., 2008; Quyang et al., 2009). The La^{III} ion tends to bind the H_2 molecules via the end-on mode (η^1-binding), even in the case of adsorption of only one H_2 molecule. This is quite different from those of the other metal ions discussed above.

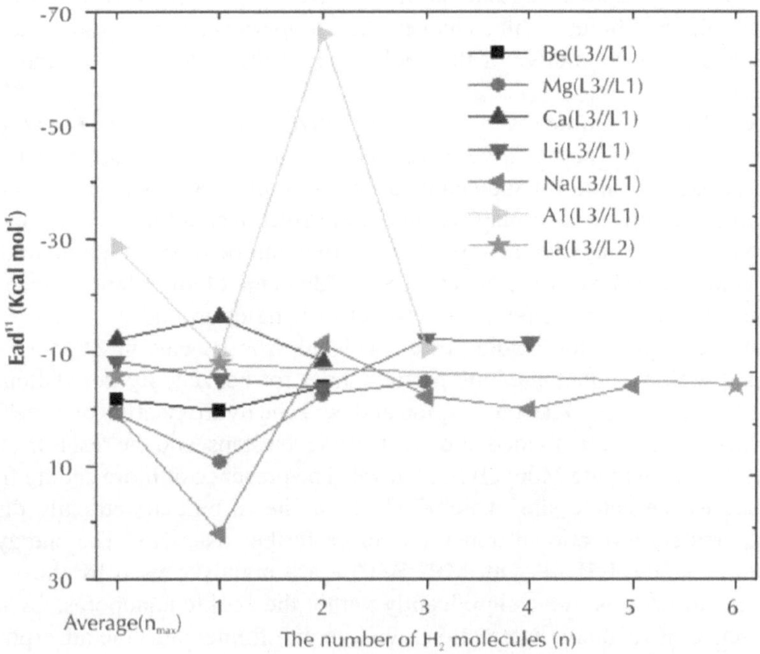

FIGURE 6 The H_2 adsorption energies on the ZM^m cluster models (M^m = Li^I, Be^{II}, Na^I, Mg^{II}, Ca^{II}, Al^{III}, La^{III}), where Average(n_{max}) corresponds to the average adsorption energy with the maximal H_2 molecules in the chemisorbed state (n_{max}).

Besides the hybrid MP2:B3LYP methodology that has been testified to reproduce reaction barriers with near chemical accuracy (de Moor et al., 2008; Svelle et al., 2009), the adsorption energies of the nH_2/ZMg^{II} cluster models are calculated at the MP2:HF and MP2 levels. The MP2:B3LYP calculated energies deviate within 1.0 kcal/mol as compared to the MP2 energies, and larger devations will be caused by using the MP2:HF method albeit the energies can still be regarded to be consistent with the MP2 method. Thus, it is demonstrated that long-range dispersion interactions from zeolite framework have no observable influences for the adsorption processes. As compared to the MP2 method, the hybrid MP2:DFT method costs significantly less computational resources and is therefore recommended for calculating the energies of similar zeolitic systems.

7.5 CONCLUSIONS

Currently, computational simulations are extensively used in a variey of scientific and technological fields. In this chapter, the nanoporous zeolitic systems are discussed as one representative of the application of the state-of-art computational methods.

The H-form zeolite is able to stabilize the amino acid zwitterion ($NH_3^+CHRCOO$-). The negative charge created by Al incorporation in H-form zeolite is indispensible for the stabilization of amino acid zwitterions, whereas the zeolite framework also plays an important role, accounting 74% of the contribution from the negative charge. It thus offers an ideal prototype to study the amino acid zwitterions, which function in a wide range of important chemical and biological processes by generating strong electric fields around.

The exchange of the acidic proton with the iron species in ZSM-5 zeolite endows it with excellent catalytic performances for benzene hydroxylation. The overall reactions of N_2O decomposition and benzene hydroxylation over Fe^{III}/ and Fe^{II}/ZSM-5 zeolites are divided into eight and seven steps, and the reaction energy barriers are given at the M06L:B3LYP level. The presence of more zeolite framework causes the active site of Fe^{III}/ZSM-5 zeolite to be geometrically distinct and requires conformational transition before further reactions. The energy differences of M06L:B3LYP with MP2:B3LYP are mainly caused by those steps where the adsorbents move significantly within the zeolite nanopores. Owing to the presence of repulsive interactions with zeolite framework, the adsorption of benzene becomes unfavorable in Fe/ZSM-5 zeolite, especially in the case of the divalent state.

The extensive studies of various exchanged metal ions show that they have quite distinct adsorption strengths and capacities for H_2. For the $M^{III}O$ species, the H_2 reduction energy barriers increase with more chemisorbed H_2 molecules or by formation of H-bonds with the extralattice O atom until covered up. The metal ions with larger radius and higher valence state have a larger capacity for H_2 adsorption, and as expected, the La^{III} ion is a good candidate with a maximal of six H_2 molecules being chemisorbed at the same time. The inclusion of zeolite framework decreases the Lewis acidity and thus causes a reduction of adsorption strength, as compared to the corresponding naked metal ions. The dispersion interactions among H_2 molecules and local exchanged sites instead of from zeolite lattices play a significant role. The hybrid MP2:DFT method is recommended for calculating the energies of zeolitic systems.

7.6 ACKNOWLEDGMENTS

The author would like to thank National Natural Science Foundation (No. 20903019) and Special Fund for Basic Scientific Research of Central Colleges (SWU113049) for financial supports.

KEYWORDS

- **amino acid zwitterion**
- **benzene hydroxylation**
- **computational simulations**
- **hydrogen adsorption**
- **zeolite framework**

REFERENCES

Aguey-Zinsou, K. F.; Ares-Fernandez, J. R. *Energy Environ. Sci.* **2010**, *3*, 526.

Areán, C. O.; Palomino, G. T.; Garrone, E.; Nachtigallová, D.; Nachtigall, P. *J. Phys. Chem. B* **2006**, *110*, 395.

Arends, I. W. C. E.; Sheldon, R. A.; Wallau, M.; Schuchardt, U. *Angew. Chem. Int. Ed.* **1997**, *36*, 1144.

Assfour, B. Leoni, S. Yurchenko, S. Seifert, G. *Int. J. Hydrogen Energy* **2011**, *36*, 6005.

Berlier, C.; Gribov, E.; Cocina, D.; Spoto, G.; Zecchina, A. *J. Catal.* **2006**, *238*, 243.

Boekfa, B.; Pantu, P.; Limtrakul, J. *J. Mol. Struct.* **2008**, *889*, 81.

Calleja, G.; Botas, J. A.; Sánchez-Sánchez, M.; Orcajo, M. G. *Int. J. Hydrogen Energy* **2010**, *35*, 9916.

Cariati, E.; Roberto, D.; Ugo, R.; Lucenti, E. *Chem. Rev.* **2003**, *103*, 3707.

Čejka, J.; Corma, A.; Zones, S., Eds. *Zeolites and Catalysis: Synthesis, Reactions and Applications*, Wiley, **2010**.

Chatterjee, A.; Iwasaki, T.; Ebina, T.; Miyamoto, A. *Micropor. Mesopor. Mat.* **1998**, *21*, 421.

Chu, Z. K.; Fu, G.; Guo, W. P.; Xu, X. *J. Phys. Chem. C* **2011**, *115*, 14754.

de Moor, B. A.; Reyniers, M. F.; Sierka, M.; Sauer, J.; *J. Phys. Chem. C* **2008**, *112*, 11796.

Fellah, M. F.; Pidko, E. A.; van Santen, R. A.; Onal, I. *J. Phys. Chem. C* **2011**, *115*, 9668.

Georgiev, P. A.; Albinati, A.; Mojet, B. L.; Ollivier, J.; Eckert, J. *J. Am. Chem. Soc.* **2007**, *129*, 8086.

Graetz, J. *Chem. Soc. Rev.* **2009**, *38*, 73.

Gupta, R.; Agresti, F.; Russo, S. L.; Maddalena, A.; Palade, P.; Principi, G. *J. Alloys Comp.* **2008**, *450*, 310.

Gutowski, M.; Skurski, P.; Simons, J. *J. Am. Chem. Soc.* **2000**, *122*, 10159.

Hansen, N.; Heyden, A.; Bell, A. T.; Keil, F. J. *J. Phys. Chem. C* **2007**, *111*, 2092.

Hensen, E. J. M.; Zhu, Q.; van Santen, R. A. *J. Catal.* **2003**, *220*, 260.

Jíša, K.; Nováková, J.; Schwarze, M.; Vondrová, A.; Sklenák, S.; Sobalik, Z. *J. Catal.* **2009**, *262*, 27.

Kass, S. R. *J. Am. Chem. Soc.* **2005**, *127*, 13098.

Kassab, E.; Langlet, J.; Evleth, E.; Akacem, Y. *J. Mol. Struct.* **2000**, *531*, 267.

Kubas, G. J. *Chem. Rev.* **2007**, *107*, 4152.

Lobree, L. J.; Hwang, I. C.; Reimer, J. A.; Bell, A. T. *J. Catal.* **1999**, *186*, 242.

Mojet, B. L.; Eckert, J.; van Santen, R. A.; Albinati, A.; Lechner, R. E. *J. Am. Chem. Soc.* **2001**, *123*, 8147.

Nobukawa, T.; Yoshida, M.; Kameoka, S.; Ito, S.; Tomishige, K. Kunimori, *Catal. Today* **2004**, *93–95*, 791.

Panov, G. I.; Uriarte, A. K.; Rodkin, M. A.; Sobolev, V. I. *Catal. Today* **1998**, *41*, 365.

Pirngruber, G. D.; Luechinger, M.; Roy, P. J.; Cecchetto, A.; Smirniotis, P. *J. Catal.* **2004**, *224*, 429.

Quyang, L. Z.; Xu, Y. J.; Dong, H. W.; Sun, L. X.; Zhu, M. *Int. J. Hydrogen Energy* **2009**, *34*, 9671.

Ribera, A.; Arends, I. W. C. E.; de Vries, S.; Pérez-Ramírez, J.; Sheldon, R. A. *J. Catal.* **2000**, *195*, 287.

Rorres, F. J.; Vitillo, J. G.; Civalleri, B.; Ricchiardi, G.; Zecchina, A. *J. Phys. Chem. C* **2007**, *111*, 2505.

Ryder, J. A.; Chakraborty, A. K.; Bell, A. T. *J. Phys. Chem. B* **2002**, *106*, 7059.

Sauer, J.; Ugliengo, P.; Garrone, E.; Saunders, V. R. *Chem. Rev.* **1994**, *94*, 2095.

Svelle, S.; Tuma, C.; Rozanska, X.; Kerber, T.; Sauer, J. *J. Am. Chem. Soc.* **2009**, *131*, 816.

Torres, F. J.; Civalleri, B.; Terentyev, A.; Ugliengo, P.; Pisani, C. *J. Phys. Chem. C* **2007**, *111*, 1871.

Wang, Y.; Zhou, D. H.; Yang, G.; Miao, S. J.; Liu, X. C.; Bao, X. H. *J. Phys. Chem. A* **2004**, *108*, 6730.

Win, D. T. *AU. J. T.* **2007**, *11*, 36.

Yang, G.; Guan, J.; Zhou, L. J.; Han, X. W.; Bao, X. H. *Catal. Surv. Asia* **2010**, *14*, 85.

Yang, G.; Zhou, D. H.; Liu, X. C.; Han, X. W.; Bao, X. H. *J. Mol. Struct.* **2006a**, *797*, 131.

Yang, G.; Zhou, L. J.; Liu, C. B. *J. Phys. Chem. B* **2009a**, *113*, 10399.

Yang, G.; Zhou, L. J.; Liu, X. C.; Han, X. W.; Bao, X. H. *Catal. Commum.* **2007**, *8*, 1981.

Yang, G.; Zhou, L. J.; Liu, X. C.; Han, X. W.; Bao, X. H. *J. Phys. Chem. B* **2006b**, *110*, 22295.

Yang, G.; Zhou, L. J.; Liu, X. C.; Han, X. W.; Bao, X. H. *J. Phys. Chem. C* **2009b**, *113*, 18184.

Yang, G.; Zhou, L. J.; Liu, X. C.; Han, X. W.; Bao, X. H. *Micropor. Mesopor. Mat.* **2012**, *161*, 168.

Yang, G.; Zu, Y. G.; Liu, C. B.; Fu Y. J.; Zhou, L. J. *J. Phys. Chem. B* **2008**, *112*, 7104.

Yang, Z. W.; Yang, G.; Zhou, L. J.; Han, X. W. *Catal. Lett.* **2013**, DIO: 10.1007/s10562-012-09537.

Yuan, S. P.; Wang, J. G.; Li, Y. W.; Jiao, H. J. *J. Phys. Chem. A* **2002**, *106*, 8167.

CHAPTER 8

CLUSTER BUNDLET MODEL OF SINGLE-WALL C, BC_2N AND BN NANOTUBES, CONES AND HORNS

FRANCISCO TORRENS and GLORIA CASTELLANO

CONTENTS

Abstract 272
8.1 Introduction 272
8.2 Computational Method 275
8.3 Leak-in Kinetics of Preequilibrium Membrane Pores: Mono- and Multi-exponential Kinetics 278
8.4 Enumeration of Heterofullerenes: A Survey 279
8.5 Advanced Methods in Magnetic Force Microscopy (MFM): Graphite and Graphene 279
8.6 Photocatalytic Degradation of Organic Pollutants in Semiconductor Aqueous Dispersion 281
8.7 Heterogeneous Photocatalysis: An Emerging Technology for Water Treatment 283
8.8 Serum Proteins on Intracellular Uptake and Phenol Red Cytotoxicity in Assays for C-NPs 286
8.9 Quantitative Structure–Activity Relationship/QSPR Toxicity Prediction in Nanotechnology 286
8.10 Insights into Co-operative-Effects Origin in Spin Transition: Supramolecular Intereactions 287
8.11 Calculation, Results and Discussion 288
8.12 Conclusions 302
Acknowledgments 303
Keywords 303
References 303

ABSTRACT

It is discussed the existence of single-wall C-nanocones (SWNCs), especially nanohorns (SWNHs), and BC_2N/boron nitride (BN) analogs in organic solvents in cluster form; a theory is developed based on a *bundlet* model describing the distribution function by size. The present phenomena unified the explanation in the bundlet model in which free energy of (BC_2N/BN-)SWNCs, involved in the cluster, is combined from two components: the volume one proportional to number of molecules *n* in cluster and the surface one, to $n^{1/2}$. The bundlet model enables describing distribution function of (BC_2N/BN-)SWNC clusters by size. From the geometrical differences, bundlet [(BC_2N/BN-)SWNCs]/droplet (C_{60}/$B_{15}C_{30}N_{15}$/$B_{30}N_{30}$) models predict dissimilar behaviors. Various disclinations (BC_2N/BN-)SWNCs are studied *via* energetic/structural analyzes. Several (BC_2N/BN-)SWNC's ends are studied, which are different because of the closing structure/arrangement type. Packing efficiencies and interaction-energy parameters of (BC_2N/BN-)SWNCs/SWNHs are intermediate between C_{60}/$B_{15}C_{30}N_{15}$/$B_{30}N_{30}$ and (BC_2N/BN-) single-wall C-nanotube (SWNT) clusters: in-between behavior is expected; however, properties of (BC_2N/BN-)SWNCs, especially (BC_2N/BN-)SWNHs, are calculated closer to (BC_2N/BN-)SWNTs. The structural asymmetry in different (BC_2N/BN-)SWNCs, characterized by cone angle, distinguishes properties of types: P2. BC_2N/BN, especially species isoelectronic with C-analogs, may be stable.

8.1 INTRODUCTION

Interest in nanoparticles (NPs) arises from the shape-dependent physical properties of nanoscale materials (Faraday, 1857; Murphy et al., 2010). Single-wall C-nanocones (SWNCs) were used to study nucleation/growth of curved C-structures, suggesting pentagon role. When a pentagonal defect is introduced into a graphitic sheet (graphene) *via* extraction of a 60° sector from a piece, a cone leaf is formed. The pentagon presence in SWNC apex is analog of single-wall C-nanotube (SWNT) tip topology. Balaban et al. (1994), Klein (2002) and Klein and Balaban (2006) analyzed the eight classes of positive-curvature graphitic nanocones. Klein (1992), Misra et al. (2009ab), Balaban and Klein (2009) and Klein and Balaban (2011) examined Clar theory for conjugated C-nanostructures (NSs). The SWNT ends predicted electronic states related to topological defects in graphite lattice (Tamura and Tsukada, 1995). Resonant picks in density of states were observed in SWNTs (Kim et al., 1999)/multiple-wall C-nanotubes (MWNTs) (Carroll et al., 1997). Table 1 compares both.

TABLE 1 Comparison between SWNTs and MWNTs.

SWNT	MWNT
Single layer of graphene	Multiple layer of graphene
Catalyst is required for synthesis	Can be produced without catalyst
Bulk synthesis is difficult as it requires proper control over growth and atmospheric condition	Bulk synthesis is easy
Purity is poor	Purity is high
A chance of defect is higher during functionalization	A chance of defect is lower but once occurred it is difficult to improve
Less accumulation in body	More accumulation in body
Characterization and evaluation is easy	It has a complex structure
It can be easily twisted and are more pliable	It cannot be easily twisted

The SWNCs with discrete opening angles θ of *ca.* 19°, 39°, 60°, 85° and 113° were observed in C-sample generated by hydrocarbon pyrolysis (Krishnan et al., 1997). Observation was explained by cone wall model composed of wrapped graphene sheets, where geometrical requirement for seamless connection naturally accounted for semidiscrete character/absolute angles θ. Total disclinations are multiples of 60°, corresponding to number ($P \geq 0$) of pentagons in SWNC apices. Considering graphene-sheet symmetry and Euler theorem, five SWNC types are obtained from continue graphene sheet matching to P values in 1–5. Angle θ is given by $\sin(\theta/2) = 1 - P/6$, leading to flat discs and caped SWNTs corresponding to $P = 0, 6$, respectively; SWNC most abundant with $P = 5$ pentagons ($\theta \approx 19°$) is single-wall C-nanohorn (SWNH). Several configurations exist for given SWNC angle, depending on pentagon arrangement: $\theta \approx 113°$ SWNC contains one pentagon in tip center and one configuration; other structures show isomers. According to *isolated pentagon rule* (IPR), configurations containing isolated pentagons lead to isomers that are more stable than including grouped ones (Kroto, 1987); another rules were derived form *ab initio* calculations (Han and Jaffe, 1998). Covalent functionalization of SWNCs with NH_4^+ improved solubility (Tagmatarchis et al., 2006), which was achieved by skeleton (Cioffi et al., 2006, 2007; Pagona et al., 2007a)/cone-end (Pagona et al., 2006a) functionalization and supramolecular π–π stacking interactions (Pagona et al., 2006b, 2007b; Zhu et al., 2003) with pyrenes/porphyrins. An MNDO calculation of BN substitutions in C_{60} showed that analogous one gave $B_{30}N_{30}$ (Xia et al., 1992). C-atom substitution in diamond by alternating B/N atoms provided BN-cubic (Silaghi-Dumiterscu et al., 1993). BN-hexagonal (h) resembles graphite since it consists of fused planar six-membered B_3N_3

rings; however, interlayer B–N interactions exist. BN nanotubes were visualized (Hamilton et al., 1993, 1995; Loiseau et al., 1996). BN-h was proposed (Rubio et al., 1994). BN nanocones were observed (Bourgeois et al., 1999, 2000; Terauchi et al., 2000)/calculated (Mota et al., 2003; Machado et al., 2003abc, 2004, 2005); most abundant ones presented 240/300° disclinations. BN/AlN nanotube junction was computed (Thesing et al., 2006). Theoretical studies on BC_2N tubules (Miyamoto et al., 1994) and graphite-like onion/nanotube production using layered materials, e.g., WS_2 (Tenne et al., 1992), MoS_2 (Margulis et al., 1993), BC_2N, BC_3 (Weng-Sieh et al., 1995) and BN (Chopra et al., 1995), allowed structures with oxidation resistance and low thermal conductivity/electronic behavior (*cf.* Table 2). The NSs of pyrolytically grown $B_xC_yN_z$ were studied: concentration profiles along/across tubes revealed that B, C and N are separated into C/BN domains; compound provides materials that are useful as robust nanocomposites (NCs)/ semiconductor devices enhanced towards oxidation (Kohler-Redlich et al., 1999; Madden, 2009; Terrones et al., 1996). Dense periodic packings (Chen et al., 2010; Betke and Henk, 2000) of tetrahedra (Kallus et al., 2010) and Platonic solids (Baker and Kudrolli, 2010) were examined.

TABLE 2 Some chemical and physical properties of some materials discussed in this work.

Cmp.	Electrongt. difference[a]	Moleclr. ionicity (%)	Crst. ionic. (%)	Melting point (°C)	Crystallgr. structure	Space group	Interlyr. distance (Å)	Intralyr. distance (Å)
WS_2	0.22	1	34	1250 (decomposes)	2Hb·MoS_2	P63/mmc	6.180	3.171
MoS_2	0.42	4	36	1184 (decomposes)	2Hb·MoS_2	P63/mmc	6.138	3.156
PbI_2	0.33	2	68	410	2H·CdI_2	P3 m1	6.986	4.558
BiI_3	0.64	9	55	408	BiI_3	R3	6.906	7.516
SbI_3	0.61	9	54	168	BiI_3	R3	6.966	7.480
CdI_2	0.97	22	74	387	2H·CdI_2	P3 m1	6.840	4.240

[a] Pauling's scale is used for electronegativity values.

In earlier publications, SWNT (Torrens and Castellano, 2005, 2007abcd, 2011)/(BC_2N/BN-)SWNC [Torrens and Castellano, 2010, 2012, 2013] cluster the *bundlet* model was presented. The aim of the present report is to perform a comparative study of different structures where electrons (e^-) are globally delocalized. A wide class of phenomena accompanying solution behavior is analyzed from a unique point of view, taking into account cluster formation. Based on the droplet model, (BC_2N/BN-)SWNCs bundlet is analyzed. The following section describes the method. The next section describes membrane pores. Following that, advanced methods in magnetic force microscopy are analyzed. Next, two sections explain heterogeneous photocatalytic degradation. The next section discusses the results. The last section summarizes our conclusions.

8.2 COMPUTATIONAL METHOD

The solubility mechanism is based on SWNC cluster formation in solution. The aggregation process changes SWNC thermodynamic parameters, which displays phase equilibrium and changes solubility. The *bundlet* model is valid when the characteristic SWNC number in cluster $n >> 1$. In saturated SWNC solution, the chemical potentials per SWNC for dissolved substance and crystal match. Equality is valid for SWNC clusters. Cluster free energy is made up of two parts: volume one proportional to number of SWNCs n in cluster and surface one, to $n^{1/2}$ (Bezmel'nitsyn et al., 1998; Gasser et al., 2001; Notman et al., 2006; Haluska et al., 2006; Neu et al., 2002). The model assumes that clusters, consisting of $n >> 1$ particle, present bundlet shape and permits Gibbs energy G_n for cluster of size n to be:

$$G_n = G_1 n - G_2 n^{1/2} \tag{1}$$

where $G_{1/2}$ are responsible for contribution to Gibbs energy of molecules placed inside the volume and on the surface of cluster. Chemical potential μ_n of cluster of size n is:

$$\mu_n = G_n + T \ln C_n \tag{2}$$

where T is the absolute temperature. Combining Eqs. (2) and (1) the result is:

$$\mu_n = G_1 n - G_2 n^{1/2} + T \ln C_n \tag{3}$$

where $G_{1/2}$ are expressed in temperature units. In saturated SWNC solution, cluster-size distribution function is determined *via* equilibrium condition linking clusters of specified size with a solid phase, which corresponds to equality between chemical potentials for SWNCs incorporated into clusters of any size and a crystal, resulting in an expression for the distribution function in saturated solution:

$$f(n) = g_n \exp\left(\frac{-An + Bn^{1/2}}{T}\right) \tag{4}$$

where A is the equilibrium difference between SWNC interaction energies with its surroundings in solid phase and cluster volume, B, similar difference for SWNCs located on cluster surface, and g_n, statistical weight of cluster of size n. One neglects $g_n(n, T)$ dependences in comparison with exponential Eq. (4). Normalization of distribution function Eq. (4):

$$\sum_{n=1}^{\infty} f(n)n = C \tag{5}$$

requires $A > 0$, and C is the solubility in relative units. As $n >> 1$, the normalization Eq. (5) results:

$$C = \bar{g}_n \int_{n=1}^{\infty} n \exp\left(\frac{-An + Bn^{1/2}}{T}\right) dn = C_0 \int_{n=1}^{\infty} n \exp\left(\frac{-An + Bn^{1/2}}{T}\right) dn \tag{6}$$

where $\bar{g}_n$ is the statistical weight of cluster averaged over range of n that makes major contribution to integral (6), and C_0, SWNC molar fraction. The A, B and C_0 were taken equal to those for C_{60} in hexane, toluene and CS_2: $A = 320K$, $B = 970K$ and $C_0 = 5 \cdot 10^{-8}$ ($T > 260K$). For polymeric BN [poly(BN)], A and B were renormalized with regard to $B_{30}N_{30}/C_{60}$ energies: $A = 350K$ and $B = 1062K$. Correction takes into account different packing efficiencies of C_{60}/SWNTs/SWNCs:

$$A' = \frac{\eta_{cyl}}{\eta_{sph}} A \text{ and } B' = \frac{\eta_{cyl}}{\eta_{sph}} B \text{ (SWNTs) } A' = \frac{\eta_{con}}{\eta_{sph}} A \text{ and } B' = \frac{\eta_{con}}{\eta_{sph}} B \text{ (SWNCs)} \tag{7}$$

where $\eta_{cyl} = \pi/2(3)^{1/2}$ is the cylinder packing efficiency in space (equal to that of circles on plane), $\eta_{sph} = \pi/3(2)^{1/2}$, that of spheres (face-centered cubic, FCC) and η_{con}, that of cones. As $\eta_{sph} < \eta_{con} < \eta_{cyl}$, SWNC behavior is expected to be intermediate between spherical fullerenes/cylindrical SWNTs. Dependences of the cluster-size distribution function on concentration/temperature lead to the thermodynamic/kinetic parameters characterizing SWNT behavior. For unsaturated solutions, the distribution function is determined by equilibrium condition for clusters. From Eq. (3) one can obtain the distribution function *vs*. the concentration:

$$f_n(C) = \lambda^n \exp\left(\frac{-An + Bn^{1/2}}{k_B T}\right) \tag{8}$$

where λ depends on the concentration; which is determined by normalization conditions:

$$C = C_0 \int_{n=1}^{\infty} n\lambda^n \exp\left(\frac{-An + Bn^{1/2}}{k_B T} \right) dn \tag{9}$$

where C_0 defines the absolute concentration; $C_0 = 10^{-4}$ mol·l^{-1} is found by requiring saturation in Eq. (9). The formation energy of a cluster of n SWNTs is:

$$E_n = n\left(An - Bn^{1/2} \right) \tag{10}$$

Using the cluster-size distribution function, one obtains a formula governing the thermal effect of SWNT solution per mole of dissolved substance:

$$H = \frac{\sum_{n=1}^{\infty} E_n f_n(C)}{\sum_{n=1}^{\infty} n f_n(C)} N_a = \frac{\sum_{n=1}^{\infty} n\left(An - Bn^{1/2} \right) \lambda^n \exp\left[\left(-An + Bn^{1/2} \right) / k_B T \right]}{\sum_{n=1}^{\infty} n\lambda^n \exp\left[\left(-An + Bn^{1/2} \right) / k_B T \right]} N_a \tag{11}$$

where λ depends on the solution total concentration by normalization condition Eq. (9). Equations (1)–(11) are modeled in a home-built program available from authors. A droplet cluster model of C_{60} is proposed following modified Eqs. (1′)–(11′).

$$G_n = G_1 n - G_2 n^{2/3} \tag{1′}$$

$$\mu_n = G_1 n - G_2 n^{2/3} + T \ln C_n \tag{3′}$$

$$f(n) = g_n \exp\left(\frac{-An + Bn^{2/3}}{T} \right) \tag{4′}$$

$$C = \bar{g}_n \int_{n=1}^{\infty} n \exp\left(\frac{-An + Bn^{2/3}}{T} \right) dn = C_0 \int_{n=1}^{\infty} n \exp\left(\frac{-An + Bn^{2/3}}{T} \right) dn \tag{6′}$$

$$f_n(C) = \lambda^n \exp\left(\frac{-An + Bn^{2/3}}{k_B T} \right) \tag{8′}$$

$$C = C_0 \int_{n=1}^{\infty} n\lambda^n \exp\left(\frac{-An + Bn^{2/3}}{k_B T} \right) dn \tag{9'}$$

$$E_n = n\left(An - Bn^{2/3}\right) \tag{10'}$$

$$H = \frac{\sum_{n=1}^{\infty} E_n f_n(C)}{\sum_{n=1}^{\infty} n f_n(C)} N_a = \frac{\sum_{n=1}^{\infty} n\left(An - Bn^{2/3}\right)\lambda^n \exp\left[\left(-An + Bn^{2/3}\right)/k_B T\right]}{\sum_{n=1}^{\infty} n\lambda^n \exp\left[\left(-An + Bn^{2/3}\right)/k_B T\right]} N_a \tag{11'}$$

8.3 LEAK-IN KINETICS OF PREEQUILIBRIUM MEMBRANE PORES: MONO- AND MULTI-EXPONENTIAL KINETICS

Heimburg (2010) reviewed the lipid ion channels. The amphitropic proteins of family B-cell lymphoma type 2 (Bcl-2) regulate apoptosis, controlling the permeability of outer mitochondrial membrane, which is because of the Bcl-2-associated X (Bax) pores. Nature, mechanism and properties remain elusive because of lack of structural information about Bax/membrane complexes; however, Fuertes et al. (2010) showed that the main-α structure of Bax allowed the design of minimal active versions. Peptides encompassing sequences of amphipathic helices from central hairpin of Bax, Bcl-2 homology domain-3 interacting domain death (Bid) and Bcl-extralarge (xL) exhibited membrane binding/permeabilization, which resemble complete proteins. Peptide fragments of Bax cause membrane poration. An action-mechanism study, e.g., modeling/quantitative analysis of vesicle-leakage kinetics, permitted count/sizing pores and studying formation, equilibration and dynamics. The pores formed by proapoptotic Bax fragment α5 relaxd to a smaller size and were kept at equilibrium. Fuertes et al. (2011) published a lipocentric view of peptide-induced pores. The time constant results:

$$\tau_{\text{flux}} = \frac{V}{A_0 P} = \frac{V}{A_0 D/m} \tag{12}$$

and allows calculating the pore area A_0 corresponding to the *initial* (short-term) pores, where V is the vesicle volume, D, the *diffusion* coefficient, m, the effective pore length and $P = D/m$, the pore *permeability* coefficient. The poresize change was modeled as an exponential decay, from a state corresponding to an initially *large* area A_0 into one of a *smaller* area A_∞, with relaxation time τ_{relax}:

$$A(t) = A_\infty + (A_0 - A_\infty) e^{-t/\tau_{\mathrm{relax}}} \tag{13}$$

meaning that the characteristic flux time of the dye leak-in is time dependent and consequently Eq. (12) converts into:

$$\tau_{\mathrm{flux}}(t) = \frac{V}{A(t) D/m} \tag{14}$$

8.4 ENUMERATION OF HETEROFULLERENES: A SURVEY

The chemical-compounds enumeration was accomplished using Pólya-Redfield theorem as the standard for combinatorial enumerations of graphs, polyhedra, chemical compounds, etc. Heterofullerenes are enumerated as fullerenes in which one/more C-atoms are replaced by heteroatoms, e.g., B and N. *Via* Pólya's theorem, Ghorbani (2012) computed number of permutational isomers of fullerene graphs.

8.5 ADVANCED METHODS IN MAGNETIC FORCE MICROSCOPY (MFM): GRAPHITE AND GRAPHENE

Leite et al. (2012) reviewed theoretical models for surface forces/adhesion and measurement *via* atomic force microscopy (AFM). Horsell et al. (2011) reported graphene mechanical measurements by AFM. Scanning force microscopy (SFM) allowed detecting short/long-range interactions; e.g., magnetic force microscopy (MFM) characterized the domain configuration in ferromagnetic materials (thin films grown by physical techniques, ferromagnetic NSs). Topography/magnetic signals were separated scanning at lift distance of 25–50 nm, such that long-range tip–sample interactions dominated. The MFM permitted detecting magnetic fields (MFs) from low-dimensional complex systems, e.g., organic nanomagnets, superparamagnetic NPs, C-based materials, etc. Magnetic nanocomponents/supporting substrates presented different electronic behavior, i.e., exhibited surface-potential differences causing heterogeneous electrostatic tip–sample interaction, which could be interpreted as magnetic one; in order to distinguish tip–sample-forces origin, Jaafar et al. (2011) proposed Kelvin probe force microscopy (KPFM)/MFM combination. The KPFM allows compensating in real time (RT) electrostatic tip–sample forces, minimizing electrostatic contributions to the frequency shift signal, which is a challenge in samples with low magnetic moment. They studied Co-NSs array that exhibited an electrostatic interaction with MFM tip;

with KPFM/MFM, they separated electr/magnetic tip–sample interactions: it is challenging to understand ferromagnetic mechanism in C-based materials, which contain only s/*p* e^-, in contrast to traditional ferromagnets based on 3*d*/4*f* e^-. Cervenka et al. (2009) showed ferromagnetic order locally at defect structures in highly oriented pyrolytic graphite (HOPG), *via* MFM, and bulk magnetization measurements at room temperature. Magnetic impurities were excluded as the origin of the signal. The ferromagnetic effect originated from localized electron states at HOPG grain boundaries, forming two-dimensional arrays of point defects. The theoretical magnetic ordering temperature, based on weak interlayer coupling/magnetic anisotropy, was comparable to experimental values. Unusual chemical environment of defects bonded in graphitic networks revealed s/*p* e^- role, creating routes for spin transport in C-based materials. Martínez-Martín et al. (2010) studied possible ferromagnetic order on graphite surface by MFM; they showed that tip–sample interaction along steps is external-MF independent. Combining KPFM/MFM, they separated electrosta/magnetic interactions along steps, obtaining upper bound for magnetic force gradient: $16\mu N{\cdot}m^{-1}$; they showed ferromagnetic-signal absence in graphite at room temperature.

Jaafar et al. (2012) introduced drive-amplitude modulation (DAM) AFM as dynamic mode with performance in all environments. As with frequency modulation, the DAM followed the feedback scheme with nested loops: the first kept cantilever-oscillation amplitude constant, regulating the driving force, and the second used this as a topography feedback variable. A phase-locked loop was used as parallel feedback allowing non/conservative-interaction separation. They also described DAM basis and exemplified performance in different environments. The DAM is stable, intuitive and easy-to-use, which is free of feedback instability associated with noncontact-to-contact transition, which occurs in frequency modulation. Jaafar et al. (2009) introduced a variable-external-field MFM microscope, which allowed stable images under the variable external MF, which was applied in/out-of-plane directions; they illustrated the microscope performances for samples: HOPG, longitudinal magnetic storage media, FePt thin films with in-plane anisotropy and Ni nanowires (NWs) with axial easy axis embedded in ceramic matrix; they used variable-field microscope as magnetic writ/reading technique. Tiberj et al. (2011) performed microRaman/transmission imaging experiments on epitaxial graphene, grown on C/Si-faces of on-axis 6H-SiC substrates. On C-face, they showed that SiC sublimation resulted in growth of long/isolated graphene ribbons ($\leq 600\mu m$), which were strain-relaxd and *p*-type doped, in which combining results of microRaman spectroscopy with microtransmission, they ascertained that uniform monolayer ribbons were grown and found Bernal stacked/misoriented bilayer ribbons. On the other hand, full graphene coverage of SiC surface was achieved, but anisotropic growth occurred because of step-bunched surface reconstruction. While in the middle of reconstructed terraces thin graphene stacks (≤ 5 layers) were grown, thicker graphene stripes appeared at the step edges; in both, graphene layers–SiC substrate interaction induced compressive thermal strain/*n*-type doping.

Escrig et al. (2007) studied the effect of macroscopic size of Ni NW array system on their remanence state; they developed a phenomenological magnetic model to obtain remanence *vs.* magnetostatic interactions in an array; they observed that because of long-range dipolar interactions between wires, sample size affected array remanence; they studied NWs magnetic state by variable-field MFM for remanent states; they deduced NWs distribution, with magnetization in up/down directions, and subsequent remanent magnetization from magnetic images. Two short-range magnetic orderings with similar energies explain the typical labyrinth pattern observed in MFM images of NW arrays. Imaging of MFM is useful to study locally the NSs magnetic state. Asenjo et al. (2006) used MFM to characterize an ordered array of Ni NWs embedded in porous membrane. Because of the large aspect ratio of wires (30 nm diameter, 1000 nm length) these presented an easy axis. Considering NWs as nearly single-domain structures, and calculating the amount of wires pointing to each direction, they obtained average magnetization. They introduced a method to analyze MFM data considering distribution functions of magnetic contrast, with which they studied NW-array magnetization and compared the results with major/minor hysteresis loops, measured by supercomputing quantum interface device (SQUID) magnetometer. Knorr and Vinzelberg (2012) reported charge writing/detection by electrostatic force microscopy (EFM)/KPFM scanning probe techniques.

8.6 PHOTOCATALYTIC DEGRADATION OF ORGANIC POLLUTANTS IN SEMICONDUCTOR AQUEOUS DISPERSION

Oppenländer (2003) evaluated the water/air photochemical purification. Anpo and Kamat (2010) examined environmentally benign photocatalysts with applications of TiO_2-based materials. Legrini et al. (1993) reviewed photochemical processes for water treatment. Hoffmann et al. (1995) revised environmental applications of semiconductor photocatalysis. Kuznetsov and Serpone (2006) analyzed heat/photoinduced absorption spectra of TiO_2-Degussa-P25/polymer NCs; they described spectra as sum of overlapping absorption bands (ABs) with maxima at 2.90eV (427 nm, AB1), 2.55eV (486 nm, AB2) and 2.05eV (604 nm, AB3); spectra correlated with experiment after TiO_2 reduction. They studied visible (VIS) spectra of TiO_2 photocatalysts reported in the literature. The relative narrow spectra were similar and photocatalyst-preparation independent. Average spectrum was described by the AB1/2 sum; the activation by VIS of TiO_2 specimens (anion doped/otherwise) implicated defects associated with O-vacancies, which started color centers displaying ABs and not narrowing the original band gap of TiO_2-anatase (E_{bg} = 3.2eV) *via* dopant/O-states mixing as suggested in the bibliography. Second-generation $TiO_{2-x}D_x$ photocatalysts doped with cat/anions (N, C, S) showed absorption edge red-shifted towards lower energies, enhancing the photonic efficiencies of photoassited surface redox reactions. Some researchers proposed that the shift is caused by narrowing the band gap of pristine TiO_2 (e.g., anatase absorption edge *ca.* 387 nm), while others suggested intragap localized states of dopants; by contrast, they showed that commonality in all doped TiO_2 rested with O-vacancies formation/advent of color centers (e.g., F, F^+, F^{2+}, Ti^{3+}), which absorbed VIS. Serpone (2006) argued that absorption-edge shift was caused by color centers formation and, while band-gap narrowing is not unknown in semiconductor physics, it needs heavy doping of metal-oxide semiconductor, producing materials that may present different chemical compositions from TiO_2 with different band-gap electronic structures. Kuznetsov and Serpone (2007) examined the TiO_2/polymer-NCs photocoloration and the color-centers photobleaching, at selected irradiation wavelengths from ultraviolet (UV) to near-infrared regions; they analyzed centers photoactivation irradiating into AB1–3; they observed different types of photostimulated absorbance changes: (1) rises and (2) decays. The latter is a direct experimental manifestation of photobleaching of the colored TiO_2/polymer NCs, which showed presence and photoinduced disappearance/destruction of TiO_2 color centers. Results indicated that photobleaching of colored TiO_2/polymer NCs originated from *intrinsic* light absorption by TiO_2 (at $h\nu > 3.2$eV) and *extrinsic* light absorption, by color centers at wavelengths corresponding to absorption spectral bands (*i.e.*, $h\nu < 3.2$eV), which were active in photodestruction centers. They proposed a photobleaching photochemical mechanism involving O-assisted annihilation of O-vacancies; they showed that TiO_2 VIS absorption originated from *only* color centers and *not* band-gap narrowing of pristine TiO_2. Unlike color centers, valence/conduction bands, which some researchers suggested that are involved in observed red shifts of absorption edges in VIS of doped TiO_2 because of the apparent narrowing of TiO_2 band gap, cannot be photodestroyed. They modeled competitive photoinduced forma/destruction of color centers, which resulted in qualitative agreement with experiment.

D'Oliveira et al. (1993) studied photocatalytic degradation of six dichlorophenols (DCPs) and three trichlorophenols (TCPs) in $TiO_{2(aq)}$ suspensions; they correlated apparent first-order rate constants k_{app} of both/monochlorophenols (MCPs) disappearance with Hammett constants σ/1-octanol–water partition coefficients k_{ow}: k_{aw} (h^{-1}) = $-10\ \sigma + 5.2\ \log k_{ow} - 7.5$ (correlation coefficient $R = 0.987$, if 2,4,6-TCP was omitted). Aromatic intermediates, identified by UV (high-performance liquid chromatography (HPLC) separation) and mass spectra (gas chromatography (GC) one), corresponded to aromatic-nucleus hydroxylation with reactivity: $p > o > m$. They detected *p*-benzoquinones with 0–2 Cl/0–1 OH and intermediates with two aromatic/quinone rings; they determined intermediates of temporal variations. All the compounds in which the subsisted C6 ring were unstable, their maximum concentration was low compared to the initial chlorophenol (CP). Release of Cl^-, and CO_2 evolution followed apparent first-order kinetics within the degradation's first part. Effects of the Cl atoms number/position on kinetics were discussed. Synthetic wastewaters contained substituted phenols: Ksibi et al. (2003) photocatalytically degraded two hydroxyphenols (hydroquinone (hydro) and resorcinol), 4-nitrophenol (4-NP), 2,4-dinitrophenol (2,4-DNP) and 2,4,6-trinitrophenol (2,4,6-TNP) in $TiO_{2(aq)}$ suspension. The substituent-nature effect on photodegradation was studied, comparing the initial degradation rates (v_0 relative to phenol) to substituted phenols, which reactivity followed Hammett law with regard to group-nature effect on phenolic functionality. The plot of $v_0 = f(\sigma)$ showed poor correlation when considering all substituted phenols; when nitrophenols (NO_2) were evaluated separately and hydroxyphenol (OH) was rejected, good correlations were observed in both. A p*H* effect upon kinetics of chemical oxygen demand (COD) disappearance was observed. Acidic p*H* was preferred over the COD removal for phenolics. Photocatalysis transformed NO_2 in nitrophenols into NO_3^-/NH_4^+ *via* NO_2^--intermediate formation. The amount of NO_3^- depended on the NO_2 number in nitrophenol. Selectivity in NO_3^- resulted 80, 56 and 66% for 4-NP, 2,4-DNP and 2,4,6-TNP, respectively. They found that in all phototreated solutions, 5-day biological oxygen demand ratios (BOD_5)/COD showed values higher than the initial solutions, which indicated a positive phototreatment effect.

8.7 HETEROGENEOUS PHOTOCATALYSIS: AN EMERGING TECHNOLOGY FOR WATER TREATMENT

Herrmann et at. (1993) surveyed the principle and literature. Based on studies on substituted-aromatics degradation in UV-irradiated $TiO_{2(aq)}$ suspensions, they indicated: kinetic characteristics, intermediates nature, degradation pathways and p*H* effect, common ions, etc.; they described photocatalytic recovery of noble metals and detoxification of water containing inorganic ions; they discussed the commercialized-photocatalysis advantages and disadvantages. The photocatalysis was based on the photocatalyst aptitude to adsorb simultaneously reactants/efficient photons. Herrmann (1999) described fundamental principles and effects of kinetic parameters (catalyst mass, wavelength, initial concentration, temperature, radiant flux). Besides the selective mild oxidation of organics performed in gas/liquid organic phase, the UV-irradiated TiO_2 became total oxidation catalyst once in water, because of the $OH^\bullet$ photogeneration by neutralization of the OH^- surface groups by photo-holes (h^+). Many organics could be degraded/mineralized into CO_2/harmless inorganic anions. Improvement in TiO_2 photoactivity by noble-metal deposition/ion-doping was detrimental. Toxic heavy-metal ions (Hg^{2+}, Ag^+, noble metals) were removed from water by photodeposition on TiO_2. Several water-detoxification photocatalytic devices were commercialized. Solar platforms are working with large-scale pilot photoreactors, in which pollutants are degraded with quantum yields comparable to those determined with artificial light in a laboratory. Herrmann (2005) explained photocatalysis fundamentals with parameters that govern kinetics: (1) catalyst mass, (2) wavelength, (3) reactants partial pressure/concentrations, (4) temperature and (5) radiant flux. Photocatalytic-reactions types concern (1) selective mild oxidation of hydrocarbons, (2) H_2 production and (3) total oxidation reactions of organics in water presence. The last constitute ensemble of recent photocatalysis developments. Most organic contaminants, e.g., dangerous pesticides, can be easily completely degrad/mineralized. Dyes are not only decolorized but also mineralized in colored aqueous effluents. Most abundant (azo) dyes have their groups –N=N– decomposed into N_2(g), which represents an ideal decontamination. Photocatalytic engineering is under development, using deposited TiO_2 in fixed beds. Some (solar) photocatalytic pilot reactor/prototypes were described. Solar-energy used as a source of activating UV–A irradiation is called *helio-photocatalysis*. Catalyst TiO_2 adsorbs (1) reactants and (2) efficient photons, which create the e^-/h^+ responsible for redox reactions. Photocatalysis is a complex polyphasic system involving solid, gaseous and liquid phases plus UV-irradiation (*electromagnetic* phase). The action of basic physicochemical parameters that control kinetics was described. In dry media, photocatalysis is used for selective mild oxidation reactions; in water, $OH^\bullet$ generation results in the mineralization to $CO_2 + H_2O$ of pollutants, pesticides, dyes in water; and pollutants degradation in (humid) air. Herrmann (2006a) reported some examples. A trend is the use of fixed beds of TiO_2 photocatalysts for air/water purification. Photocatalytic engineering is developing using pilot plants, e.g., in solar photocatalysis. *Catalysis by metals* was transposed to *bifunctional photocatalysis* based on noble metals deposited on TiO_2 (Herrmann, 2006b). Participations of (1) photo-active oxidic support and (2) deposited metal were analyzed, delimited and associated. For H-involving reactions [cyclopentane-D isotope exchange (CDIE) and ROH dehydrogenation], he evidenced the e^-/h^+ respective roles. Photogenerated e^- were spontaneously transferred to metal NPs because of Fermi-levels alignment of both solid phases; they neutralized H^+/D^+ into H/D before H_2/HD recombina/evolution. The h^+ neutralized anions in oxidations responsible for organics activation. Anions were OD^- surface groups in CDIE and RO^- in ROH dehydrogenation; both reactions exhibited stoichiometric threshold, which was exceeded hundreds of times, defining catalysis. Noble metals deposited on TiO_2 appeared as auxiliaries/cocatalysts, working independently of UV-irradiation for making reaction run catalytically. Bifunctional-photocatalysis accounted for interphasic photocatalyzes, e.g., H/e^--transfer, photosensitization in VIS *via* CdS addition. Photocatalysis became a major discipline owning to: (1) intuition of twentieth century pioneers and (2) mutual enrichment of scientists arizing from different fields: photochemistry, electrochemistry, analytical chemistry, radiochemistry, materials chemistry, surface science, electronics and catalysis; however, heterogeneous photocatalysis belongs to catalysis, which means that the bases must be respected and it became imperative to refocus the frame to avoid misfits/conceptions: (1) reaction-rate proportionality to catalyst mass (below plateau because of photon full absorption); (2) implication of Langmuir–Hinshelwood mechanism of kinetics with initial rate proportional to coverages θ in reactants; (3) obtaining conversions beyond *stoichiometric threshold*: number of potential active sites initially present at surface. Photonics should be respected with photocatalytic activity being (1) parallel to photocatalyst absorbance and (2) proportional to radiant flux Φ enabling to determine quantum yield, defined as ratio of reaction rate r (in molecules converted per second) to efficient photonic flux (in photons per second) received by a solid. Photocatalytic normalized tests should be established to prove catalytic activity of irradiated solids, independent of noncatalytic side-reactions; e.g., dye decolorization is a misleading test, which provides dye *visible/* apparent *disappearance*, photochemical but not photocatalytic. Thermodynamics must be respected: photon energy decay to VIS may be thermodynamically detrimental for generating active species, e.g., $OH^{\bullet}$. In solid-state chemistry, it is admitted that cationic doping is harmful to photocatalysis; anionic doping must be rapidly clarifi/abandoned. Recommendations must be addressed and experiments, operated in suitable conditions before claiming that one deals with a true photocatalytic reaction, whose veracity can be proved following the correct protocol (Herrmann, 2010).

Pino and Encinas (2012) compared photodegradation kinetics UV/TiO_2-mediated of 4-CP/2,6-DCP mixtures, under the same experimental conditions, with individual CPs. Their aim was to get approach to critical processes, involved in photoinduced heterogeneous catalysis of systems containing two contaminants in competition, and evaluate particle-size effect using commercial TiO_2–325 mesh/ P25 as catalysts. Determination of equilibrium adsorption constants, in dark, showed that adsorption of 2,6-DCP is higher than 4-CP in both particles but relation decays in mixture. Parameters affecting reaction rate were studied: initial phenol concentration, catalyst loading, p*H* change and particle-size distribution. Phenols degradation kinetics at low conversion showed similar reactivity towards photogenerating $OH^{\bullet}$. In agreement with reduced adsorption in mixture, their degradation rate was lower in mix than individual. Photodegradation rate was higher on TiO_2–325 mesh than -P25, which is opposite to particles surface area, indicating that other surface properties, e.g., pore size, are important. Photoinduced CP degradation showed two intermediate products that correspond to each-CP degradation.

8.8 SERUM PROTEINS ON INTRACELLULAR UPTAKE AND PHENOL RED CYTOTOXICITY IN ASSAYS FOR C-NPS

In order to explore effects of C-NPs novel properties on cytotoxicity, Zhu et al. (2009) investigated serum-proteins adsorption in cell culture medium on MWNTs and three kinds of C-blacks (CBs); they measured quantitatively CNPs uptake by Henrietta Lacks (HeLa) cells, *via* ^{99m}Tc radionuclide labell/tracing techniques, and examined CNPs-uptake dependence on serum proteins; they assayed CNPs cytotoxicity in medium with and without serum by [3-(4,5-dimethylthiazol-2-yl)-2,5-diphenyltetrazolium bromide] (MTT) method. Cellular uptake was much higher in cells exposed to CNPs in serum-free culture medium than those in culture medium with serum. Serum proteins adsorbed on CNPs attenuated this inherent cytotoxicity, and decrease the rose with increasing serum proteins adsorbed on CNPs. They indicated possible reasons responsible for considerable influence of serum proteins on cytotoxicity. In order to explore CNPs novel properties in nanotoxicity assays, Zhu et al. (2012) studied adsorption of phenol red (PR, a p*H* indicator for culture medium) by MWNTs and three kinds of CBs with nanosize, and its effects on cytotoxicity. The PR adsorb/delivered into cells by CBs was responsible for the toxicity to HeLa in medium without serum. Cellular uptake of PR was verified *via* ^{125}I-labeling. Size-dependent cytotoxicity of CBs correlates closely to PR adsorption, cellular uptake of PR–CB complexes and amount of PR delivered into cells by CBs. Although CBs were nontoxic/slightly toxic as PR vehicles, they played an essential role in PR-induced cytotoxicity; however, MWNTs showed intrinsic cytotoxicity independent of PR.

8.9 QUANTITATIVE STRUCTURE–ACTIVITY RELATIONSHIP/ QSPR TOXICITY PREDICTION IN NANOTECHNOLOGY

Berhanu et al. (2012) reviewed applications of quantitative structure–activity/ property relationships (QSARs/QSPRs), in toxicity prediction in nanotechnology. Numerous products based on nanotechnology entered the market and high quantities of NPs are produced annually (Barnard, 2009; Chatterjee, 2008). Unintended exposure of humans/ecological receptors to nanomaterials results in adverse effects, which differ from the bulk (Dreher, 2004; Som et al., 2010). Environmental protection plans, associated with nanomaterial manufacture and use, require understanding nanobiointerface interactions that govern nanomaterial bioactivity and potential toxicity (Rallo et al., 2011). QSARs studies involving nanomaterials appeared. Evaluation of desired and undesired bioeffects caused by manufactured NPs (MNPs) is an important issue for nanotechnology. Toxicological studies are time consuming, costly and impractical, calling for development of efficient computational approaches capable of predicting MNP bioeffects. Puzyn et al. (2011) investigated the cytotoxicity of 17 metal-oxide NPs towards *Escherichia coli* bacteria; based on toxicity and computed structural descriptors, they built a model to predict nanomaterials cytotoxicity. Fourches et al. (2010) developed a quantitative NS–activity relationship (QNAR) *via* machine learning approaches, e.g., support vector machine-based classification and *k*-nearest neighbors-based regression. The external prediction power resulted 73% for classification, presenting $R^2 = 0.72$ for the modeling of regression. QNARs studies could be employed for predicting nanomaterials bioactivity profiles and prioritizing the design/manufacture, towards better/safer products. The measurement distinguished the changes in NP toxicity (Dean, 2012).

8.10 INSIGHTS INTO CO-OPERATIVE-EFFECTS ORIGIN IN SPIN TRANSITION: SUPRAMOLECULAR INTEREACTIONS

Dîrtu et al. (2010) tracked thermally induced hysteric spin transition (ST), which occurs in polymeric chain compound $[Fe(NH_2trz)_3](NO_3)_2$ above room temperature ($T_c^{\uparrow} = 347K$, $T_c^{\downarrow} = 314K$), by ^{57}Fe Mössbauer spectroscopy, SQUID, differential scanning calorimetry (DSC) and X-ray powder diffraction (XRPD) at variable temperatures. From XRPD pattern indexation, they observed orthorhombic primitive cell with parameters: $a = 11.83(2)Å$, $b = 9.72(1)Å$, $c = 6.361(9)Å$ at 298K [low-spin (LS)] and $a = 14.37(2)Å$, $b = 9.61(4)Å$, $c = 6.76(4)Å$ at 380K [high-spin (HS)]. They evaluated enthalpy/ropy variations associated to ST by DSC: $\Delta H = 23(1)kJ \cdot mol^{-1}$ and $\Delta S = 69.6(1)J \cdot mol^{-1} \cdot K^{-1}$; they used thermodynamic data within two-level Ising-like model for statistical analysis of first-order reversal curve (FORC) diagram, which was recorded in cooling mode; they witnessed intramolecular cooperative effects by derived interaction parameter $J = 496K$; they obtained $[Cu(NH_2trz)_3](NO_3)_2 \cdot H_2O$ crystal structure thanks to high-quality single crystals prepared by slow evaporation after hydrothermal pretreatment. Catena poly[μ-tris(4-amino-1,2,4-triazole-*N1, N2*)Cu^{II}]$(NO_3)_2 \cdot H_2O$ crystallizes in monoclinic space group *C2/c*, with $a = 16.635(6)Å$, $b = 13.223(4)Å$, $c = 7.805(3)Å$, $\beta = 102.56(3)°$, $Z = 4$. Complex is one-dimensional (1D) infinite chain containing triple *N1, N2*-1,2,4-triazole bridges with intrachain distance of Cu…Cu = 3.903(1) Å. They observed dense H-bonding net with NO_3^- counterion involved in intra/erchain interactions. Such a supramolecular net could be at the origin of unusually large hysteresis loop ($\Delta T \sim 33K$), result of efficient propagation of elastic interactions *via* net. The hypothesis is strengthened by the crystal structure and absence of crystallographic phase transition over whole temperature range as shown by XRPD. Some $3d^n$ ($4 \leq n \leq 7$) transition metal compounds exhibit LS/HS cooperative transition, which is abrupt and occurs with thermal hysteresis, which confers memory effect on the system (Kahn and Martinez, 1998). Intersite interactions/cooperativity are magnified in polymeric compounds, e.g., $[Fe(Rtrz)_3]A_2 \cdot nH_2O$ in which Fe^{2+} are triply bridged by 4-R-substituted-1,2,4-triazole molecules. Moreover, in these compounds, ST is accompanied by well-pronounced change of color violet/white (LS/HS). Materials transition temperatures were tuned *via* approach based on molecular alloy. In particular, it is possible to design compounds for which room temperature falls in the middle of the thermal hysteresis loop. These materials present potential applications, e.g., as temperature sensors, active elements of various displays types and in information storage/retrieval. Roubeau et al. (2001) synthesized family of polymeric 1D chains of Fe^{II} species showing ST using 4-*n*-alkyl-1,2,4-triazoles as bridging ligands; they studied the effect of alkyl-tails length on triazole ligands on ST features, showing steepness degrading with increasing length; they used set of four counterions to access wider range of transition temperatures; they detected large hysteresis loops with small tails mainly for m/ethyl-substituted products. Longer tails weaken the cooperativity and hysteresis gradually decreases to zero; however, they showed that with certain anions the hysteresis remains even with long tails on triazoles. The cooperativity weakening arises from the diminution of polymeric-chains length with increasing alkyl tails on triazole. The effect is anion dependent. Strong interactions along polymeric chains are confirmed.

8.11 CALCULATION, RESULTS AND DISCUSSION

Table 3 lists the number of pentagons P, disclination angles D_θ, cone apex angles θ, solid angles Ω, number of cones in a sphere and solid-angle/spherecovering efficiencies in a graphene hexagonal network. A given disclination, e.g., 300° ($P = 5$), is usually built by extraction of one segment generating one distinct cone type (horn). Cone angle decays as number of pentagons increases from flat discs ($P = 0$) to cones ($P = 1$–5, e.g., SWNHs $P = 5$) to tubes ($P = 6$). Solid angle results: $\Omega = 2\pi[1-\cos(\theta/2)]$; maximum corresponds to sphere ($P = 12$): $\Omega_{sph} = 4\pi$. Solid-angle-covering efficiency discards uncomplete SWNCs; spherecovering efficiency corrects it by packing efficiency of parallel cylinders η_{cyl}: both drop as number of pentagons increases from discs ($P = 0$) to cones ($P = 1$–5).

TABLE 3 Numbers of pentagons *P*, cones, angles and covering efficiencies in graphene-hexagonal.

P[a]	Disclination angle [°]	Cone angle [°]	Solid angle [sr]	No. of cones	Solid-angle-covering efficiency	Spherecovering efficiency
0	0	180.00	6.28319	2	1.00000	0.90690
1	60	112.89	2.81002	4	0.89446	0.81118
2	120	83.62	1.59998	7	0.89125	0.80828
3	180	60.00	0.84179	14	0.93782	0.85051
4	240	38.94	0.35934	34	0.97225	0.88173
5	300	19.19	0.08788	142	0.99306	0.90060
6	360	0.00	0.00000	∞	1.00000	0.90690
12	720	360.00	12.56637	1	1.00000	0.90690

[a]*P* = 0 (disc), 1–5 (cone), 5 (horn), 6 (tube), 12 (sphere).

Table 4 collects the packing efficiencies η, correction factors and parameters *A,'* *B'* and C_0 determining molecule interaction energy. As $\eta_{sph} < \eta_{con} < \eta_{cyl}$, cone parameters are intermediate between spheres (*P* = 12) and cylinders (*P* = 6); e.g., SWNH (*P* = 5) parameters are closest to SWNTs (*P* = 6).

TABLE 4 Packing efficiencies/parameters for interaction energy. $C_0 = 5{\cdot}10^{-8}$ (molar fraction).

Molecule	No. of pentagons[a]	Packing efficiency	ETA-correction factor	*A'* [K]	*B'* [K]
SWNC ETA-correction[b]	0	0.90690	1.22474	392	1188
	1	0.81118	1,09548	351	1063
	2	0.80828	1.09156	349	1059
	3	0.85051	1.14859	368	1114
	4	0.88173	1.19075	381	1155
	5	0.90060	1.21624	389	1180
SWNT ETA-correction[c]	6	0.90690	1.22474	392	1188
C_{60}-face-centerd cubic[d]	12	0.74048	1.00000	320	970

[a] *P* = 0 (disc), 1–5 (cone), 5 (horn), 6 (tube), 12 (sphere).
[b] SWNC: single-wall carbon nanocone.
[c] SWNT: single-wall carbon nanotube.
[d] For *T* > 260K.

Table 5 lists the packing parameters *closeness*, dimension D and efficiency η of equal objects for atom clusters with short-range interaction (Betke and Henk, 2000; Conway and Torquato, 2006; Jiao and Torquato, 2011).

TABLE 5 Objects, closeness, packing dimensions D and efficiencies η of equal objects.

Objects	Closeness	D	Packing efficiency
Low-density sphere (LDS) I	Extremely low	3	0.042
Low-density sphere (LDS) II	Extremely low	3	0.045
Low-density sphere (LDS) III	Extremely low	3	0.056
Tetrahedron I	Not closest	3	18/49 = 0.36735
Sphere simple cubic (SC)	Not closest	3	$\frac{\pi}{6} \approx 0.52360$
Sphere random *loose* (RL)	Not closest	3	0.601±0.005
Sphere random close (RC)	Not closest	3	0.6366±0.0005
Tetrahedron II	Not closest	3	2/3 = 0.66667
Sphere body-centerd cubic (BCC)	Not closest	3	$\frac{\pi\sqrt{3}}{8} \approx 0.68017$
Truncated tetrahedron I	Not closest	3	207/304 = 0.68092
Tetrahedron III	Not closest	3	17/24 = 0.70833
Tetrahedron IV	Not closest	3	$\frac{139+40\sqrt{10}}{369} \approx 0.71949$
Sphere (FCC alias cubic closest packing, CCP or hexagonal closest packing, HCP)	Closest	3	$\frac{\pi}{3\sqrt{2}} \approx 0.74048$
Tetrahedron V	–	3	0.7786
Tetrahedron VI	–	3	0.7820
Truncated icosahedron	–	3	0.78499
Snub cube	–	3	0.78770
Snub dodecahedron	–	3	0.78864
Rhombic icosidodecahedron	–	3	0.80471
Tetrahedron VII	–	3	0.8226

TABLE 5 *(Continued)*

Truncated icosidodecahedron	–	3	0.82721
Icosahedron	–	3	0.83636
Truncated cubeoctahedron	–	3	$\frac{99}{992}\sqrt{66}-\frac{231}{1984}\sqrt{33}+\frac{2835}{992}\sqrt{2}-\frac{6615}{1984}=0.84937$
Tetrahedron VIII	–	3	0.85027
Tetrahedron IX	–	3	100/117 = 0.85470
Tetrahedron X	–	3	4000/4671 = 0.85635
Icosidodecahedron	–	3	0.86472
Rhombic cubeoctahedron	–	3	$\frac{16\sqrt{2}-20}{3}=0.87581$
Truncated dodecahedron	–	3	0.89779
Dodecahedron	–	3	0.90451
Cubeoctahedron	–	3	45/49 = 0.91837
Octahedron	–	3	18/19 = 0.94737
Truncated tetrahedron II	–	3	23/24 = 0.95833
Truncated cube	–	3	$\frac{9}{5+3\sqrt{2}}=0.97375$
Truncated tetrahedron III	–	3	207/208 = 0.99519
Cube	Closest	3	1.0
Truncated octahedron	Closest	3	1.0
Cylinder in space as square packing (SP) of circles on a plane	Not closest	2	$\frac{\pi}{4}\approx 0.78540$
Cone ($P = 2$)	Not closest	2	0.80828
Cone ($P = 1$)	Not closest	2	0.81118
Cone ($P = 3$)	Not closest	2	0.85051
Cone ($P = 4$)	Not closest	2	0.88173
Cone ($P = 5$, horn)	Not closest	2	0.90060
Cylinder (as hexagonal packing of circles on a plane)	Closest	2	$\frac{\pi}{2\sqrt{3}}\approx 0.90690$

For closest, not closest and extremely low packings, packing efficiency η variations *vs*. packing dimension D (*cf.* Fig. 1) show many superimposed points. On going from $D = 2$ to 3, $\eta_{\text{extremely low}}$ decays quicker than $\eta_{\text{not closest/closest}}$. For all cases, packing objects with lower packing dimension show best fits. Regressions turn out to be:

$$\eta_{\text{closest}} = 1.00 + 0.0334D - 0.0400D^2 \tag{15}$$

$$\eta_{\text{not closest}} = 1.00 + 0.0125D - 0.0463D^2,\ n = 16\ \ r = 0.833\ \ s = 0.093\ \ F = 14.8 \tag{16}$$

where n is number of points, r, correlation coefficient, s, standard deviation, and F, Fischer ratio. Results are improved if the data for tetrahedra I–IV and truncated tetrahedron I are suppressed:

$$\eta_{\text{not closest}} = 1.00 + 0.0192D - 0.0497D^2,\ n = 11\ \ r = 0.942\ \ s = 0.054\ \ F = 31.3 \tag{17}$$

For extremely low packing:

$$\eta_{\text{extremely low}} = 1.00 - 0.317D \tag{18}$$

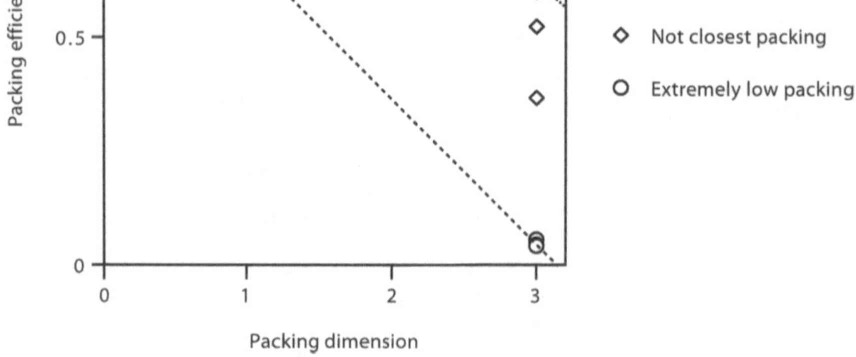

FIGURE 1 Packing efficiency η *vs*. packing dimension for closest/not closest/extremely low packings.

The parabolic nature of Eqs. (15)–(17) suggests that linearization would be achieved, if reciprocal packing dimension D^{-1} is used as abscissa instead of D. For closest, not closest and extremely low packings, packing efficiency η variations *vs*. D^{-1} (*cf*. Fig. 2) show many superimposed points. The $\eta_{\text{extremely low}}$ raises quicker than $\eta_{\text{not closest}}$ than η_{closest}.

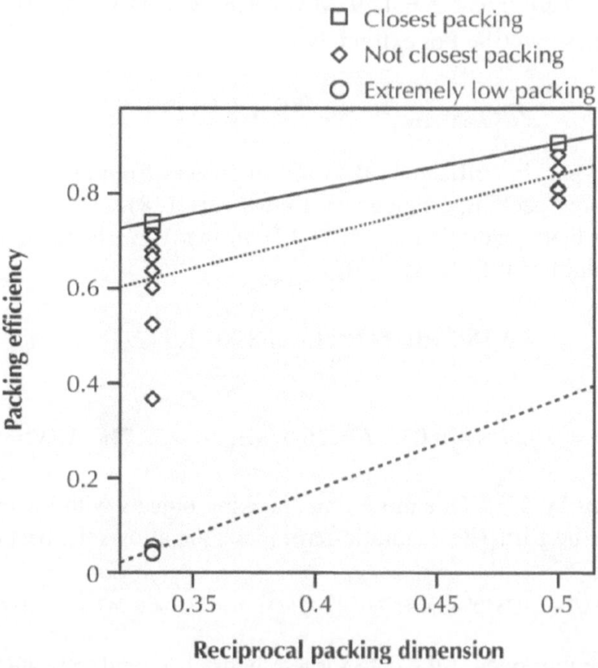

FIGURE 2 Packing efficiency *vs*. reciprocal dimension for closest/not closest/extremely low packings.

Again, packing objects with lower packing dimension D present the best fits, which result in:

$$\eta_{\text{closest}} = 0.408 + 0.999 D^{-1} \tag{19}$$

$$\eta_{\text{not closest}} = 0.182 + 1.31 D^{-1} \tag{20}$$

$n = 15 \;\; r = 0.780 \;\; s = 0.093 \;\; F = 20.2 \;\; \text{MAPE} = 9.10\% \;\; \text{AEV} = 0.3916$

where the mean absolute percentage error (MAPE) is 9.10%, and the approximation error variance (AEV), 0.3916. Results are improved if the data for tetrahedra I-IV and truncated tetrahedron I are suppressed:

$$\eta_{\text{not closest}} = 0.152 + 1.38D^{-1} \tag{21}$$

$n = 10$ $r = 0.918$ $s = 0.054$ $F = 42.9$ MAPE $= 6.38\%$ AEV $= 0.2366$
and AEV decays by 40%. For extremely low packing:

$$\eta_{\text{extremely low}} = -0.589 + 1.91D^{-1} \tag{22}$$

Rizing rate of packing efficiency η *vs.* D^{-1} increases from closest to not closest to extremely low packing efficiencies. Linear Eqs. (18)–(22) perform better for extrapolations than quadratic Eqs. (15)–(17). Property-*closeness* inclusion allows performing joint linear fit for $\eta_{\text{closest}}/\eta_{\text{not closest}}$:

$$\eta = 0.180 + 0.0976\,\text{closeness} + 1.31D^{-1} \tag{23}$$

$n = 12$ $r = 0.924$ $s = 0.053$ $F = 26.5$ MAPE $= 5.17\%$ AEV $= 0.1453$

and AEV drops by 63%. One more time, packing objects with lower packing dimension show best fit. The quadratic-term inclusion allows the best model:

$$\eta = 1.00 + 0.0102D + 0.0381\,\text{closeness}\cdot D - 0.0455D^2,\ n = 18\ r = 0.847\ s = 0.089\ F = 11.9 \tag{24}$$

The results are improved if the data for tetrahedra I-IV and truncated tetrahedron I are suppressed:

$$\eta = 1.00 + 0.0151D + 0.0402\,\text{closeness}\cdot D - 0.0481D^2 \tag{25}$$

$n = 13$ $r = 0.946$ $s = 0.051$ $F = 25.6$ MAPE $= 4.34\%$ AEV $= 0.1050$

and AEV decreases by 73%. Once more, packing objects with lower packing dimension present the best fit. Quadratic Eqs. (24)/(25) perform better than linear Eq. (20) for intrapolation. Predictions for packing objects with lower packing dimension show an improvement; e.g., for sphere (C_{60})/cylinder (SWNT), results are quite good.

Table 6 reports the disclination angles D_θ, numbers of 2-membered rings (2 MR), squares S and pentagons P and cone apex angles θ in a poly(BN) hexagonal network. A given disclination, e.g., 240°, can be built by the extraction of one segment generating one distinct cone type (2 MR = S = P = 0); however, the same disclination can be derived by the extraction of two separated segments of 120° each (S = 1, P = 2) or four unconnected segments of 60° each (S = 0, P = 4). The cone angles decay as the numbers of 2 MR, squares or pentagons increase from flat discs (D_θ = 0°, 2 MR = S = P = 0) to cones (D_θ = 60–300°, 2 MR = 0–1, S = 0–2, P = 0–5) to tubes (D_θ = 360°, 2 MR = 0, S = 0–3, P = 0–6). Structures observed in BN cones are attributed to lower energy of squares compared with pentagons; indeed, B–N present higher stability than B–B than N–N bonds, e.g., line defect D_θ = 300° and 2 MR = S = P = 0 would consist of B–B bonds.

TABLE 6 Angles, numbers of 2 MR, squares S and pentagons P in a poly(BN) hexagonal network.

Disclination angle [°]	2 MR	*S*	*P*	Cone angle [°]
0	0	0	0	180.00
60	0	0	1	112.89
120	0	0	2 in opposed ends of an edge	83.62
120	0	0	2 neighbors	83.62
120	0	0	2 isolated by a hexagon	83.62
120	0	1	0	83.62
180	0	0	3 in line	60.00
180	0	0	3 in an arrangement such that each ring has 2 pentagons as nearest neighbors	60.00
180	0	0	3 isolated by a hexagon	60.00
180	0	1	1	60.00
240	0	0	0; 2–2-coordinated atoms at the apex	38.94
240	0	0	4 isolated by 2 hexagons	38.94
240	0	0	4 neighbors sharing 2–3-coordinated atoms at the apex	38.94
240	0	1	2	38.94

TABLE 6 *(Continued)*

240	0	2	0	38.94
240	1	0	0	38.94
300	0	0	0; line defect consisting of like bonds	19.19
300	0	0	5	19.19
360	0	0	6	0.00
360	0	3	0	0.00
720	0	0	12	360.00

The equilibrium difference between the Gibbs free energies of interaction of an SWNC with its surroundings, in solid phase and the cluster volume/on surface (*cf.* Fig. 3), shows that results for $B_{15}C_{30}N_{15}/B_{30}N_{30}$ are superimposed on C_{60}, and (BC_2N/BN-)SWNC/SWNT on SWNT. On going from C_{60} (droplet) to SWNT (*bundlet*), the minimum is less marked (68% of C_{60}), which results in a lesser number of units in (BC_2N/BN-)SWNT/SWNCs ($n_{min} \approx 2$) than in $C_{60}/B_{15}C_{30}N_{15}/B_{30}N_{30}$ clusters (≈8). Moreover, the abscissa is longer in $C_{60}/B_{15}C_{30}N_{15}/B_{30}N_{30}$ ($n_{abs} \approx 28$) than in (BC_2N/BN-)SWNT/SWNCs (≈9). When going from C_{60} to $B_{15}C_{30}N_{15}$ to $B_{30}N_{30}$ (or from SWNT to BC_2N- to BN-SWNT or from SWNCs to BC_2N- to BN-SWNCs), the minimum is increasingly emphasized (4.6% and 9.5%, respectively) while contains the same number of units. In SWNCs/BC_2N-SWNCs/BN-SWNCs (bundlet), minima result 61–67% of $C_{60}/B_{15}C_{30}N_{15}/B_{30}N_{30}$, similar to those in ($BC_2N$/BN-)SWNT.

Temperature dependence of SWNC solubility, *cf.* Fig. 4, shows that the results for (BC_2N/BN-)SWNC/SWNT are superimposed on SWNT. Solubility decays with temperature because of cluster formation. At $T \approx 260$K, C_{60}-crystal presents an orientation disorder phase transition from FCC to simple cubic (SC). Solubility decays are less marked for (BC_2N/BN-)SWNT/SWNCs, in agreement with lesser numbers of units in the clusters (Fig. 3). In particular at $T = 260$K, on going from C_{60} to $B_{15}C_{30}N_{15}$ to $B_{30}N_{30}$ (droplet), the solubility rises by 22.8% and 52.5%, respectively. When going from C_{60} (droplet) to SWNT (bundlet), the solubility decays to 2.6% of C_{60}; SWNCs (bundlet) solubility drops to 2.0–2.5% of C_{60}. On going from $B_{15}C_{30}N_{15}$ (droplet) to BC_2N-SWNT (bundlet), the solubility decreases to 2.4% of $B_{15}C_{30}N_{15}$; from $B_{30}N_{30}$ to BN-SWNT (bundlet), the solubility diminishes to 2.2% of $B_{30}N_{30}$; BC_2N/BN-SWNCs solubilities decay to 1.8–2.3% of $B_{15}C_{30}N_{15}$ and 1.6–2.1% of $B_{30}N_{30}$.

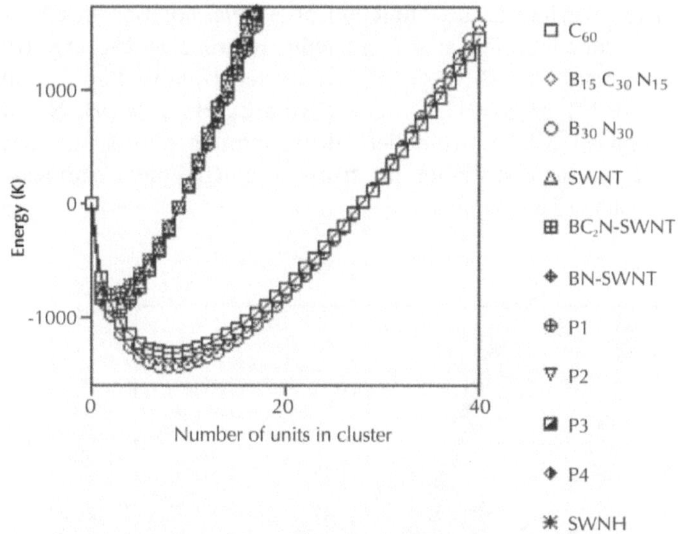

FIGURE 3 $C_{60}/B_{15}C_{30}N_{15}/B_{30}N_{30}$–($BC_2N$/BN-)SWNT–SWNH interaction energy with surroundings in cluster volume/surface.

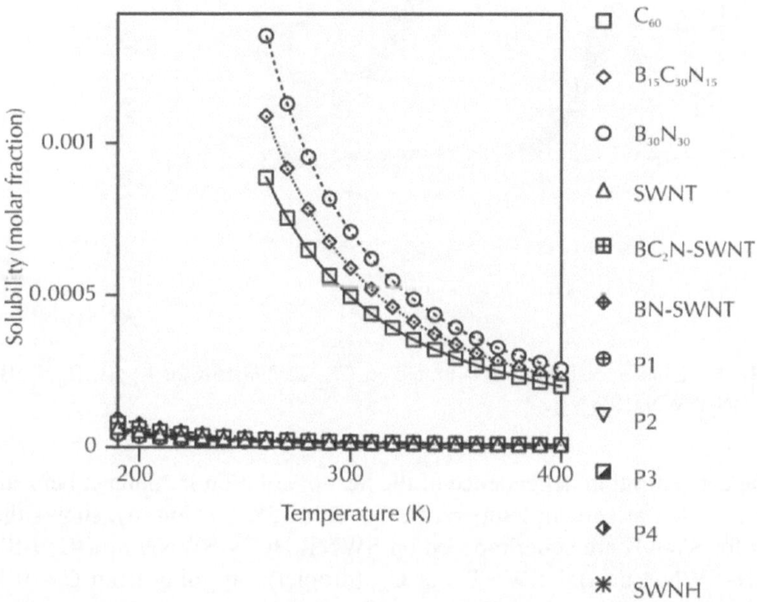

FIGURE 4 Temperature dependence of solubility of $C_{60}/B_{15}C_{30}N_{15}/B_{30}N_{30}$–($BC_2N$/BN-)SWNT/SWNH.

The cluster distribution size function of SWNC solution in CS_2, calculated the saturation concentration at solvent temperature T = 298.15K (*cf.* Fig. 5), shows that the results for $B_{15}C_{30}N_{15}/B_{30}N_{30}$ are superimposed on C_{60}, and (BC_2N/BN-)SWNC/SWNT on SWNT. On going from $C_{60}/B_{15}C_{30}N_{15}/B_{30}N_{30}$ (droplet) to (BC_2N/BN-)SWNT/SWNCs (bundlet), the maximum cluster size decays from $n_{max} \approx 8$ to ≈ 2 and the distribution is narrowed, in agreement with lesser number of units in clusters (Fig. 3).

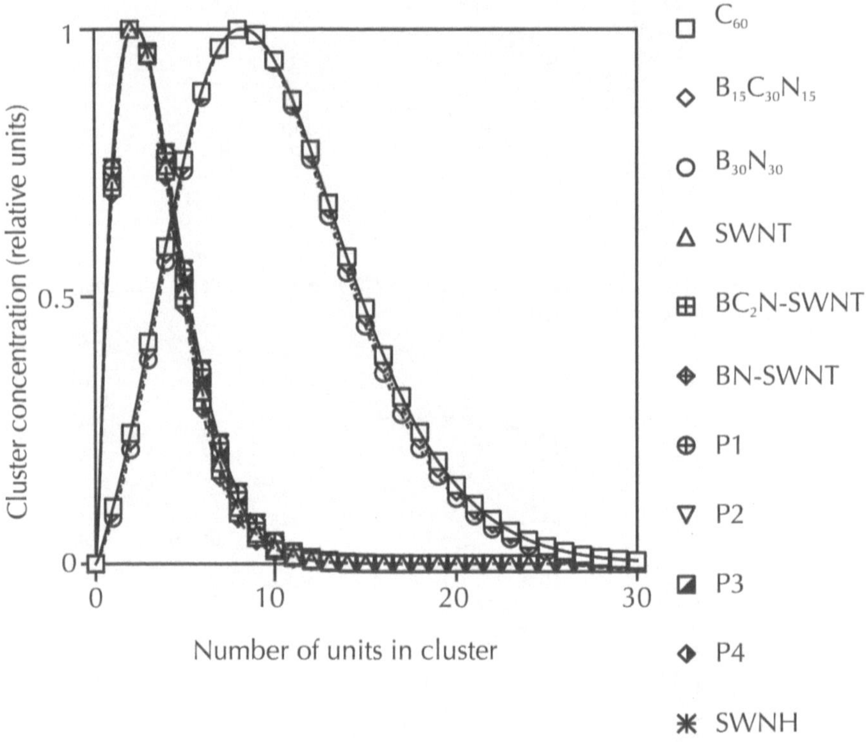

FIGURE 5 Cluster distribution saturated in CS_2 at 298.15K of $C_{60}/B_{15}C_{30}N_{15}/B_{30}N_{30}$–($BC_2N$/BN-)SWNT/SWNH.

The concentration dependence of the heat of solution in toluene, benzene and CS_2, calculated at solvent temperature T = 298.15K (*cf.* Fig. 6), shows that the results for SWNH are superimposed on SWNT, BC_2N-SWNH on BC_2N-SWNT, and BN–SWNH on BN–SWNT. For C_{60} (droplet), on going from $C < 0.1\%$ of saturated ($\langle n \rangle \approx 1$) to $C = 15\%$ ($\langle n \rangle \approx 7$), the heat of solution decays by 73%. In turn for SWNT (bundlet), the heat of solution increases by 54% in the same range, in agreement with lesser number of units in the clusters (Figs. 3 and 5). In SWNCs (bundlet), the heat of solution augments by 55–80% in accordance with smaller aggregations. In $B_{15}C_{30}N_{15}$ (droplet), the heat of solution drops by 74%; in turn for BC_2N-SWNT (bundlet), the heat of solution rises by 49% in agreement with smaller clusters. In BC_2N-SWNCs (bundlet), the heat of solution enlarges by 50–63% in accordance with smaller aggregations. In $B_{30}N_{30}$ (droplet), the heat of solution decays by 73%; in turn for BN-SWNT (bundlet), the heat of solution increases by 44% in agreement with smaller clusters. In BN-SWNCs (bundlet), the heat of solution enlarges by 45–57% in accordance with smaller aggregations. Discrepancies between various experimental data of heat of solution of fullerenes, poly(BC_2N/BN) and (BC_2N/BN-)SWNT/SWNCs may be ascribed to the sharp concentration dependence of the heat of solution. The effect of different number of pentagons *P* on the concentration dependence shows that the results for SWNC P2 are superimposed on P1, and SWNH on SWNT. The heat of solution varies with the following behavior: P2 ≈ P1 > P3 > P4 > SWNH ≈ SWNT >> C_{60}.

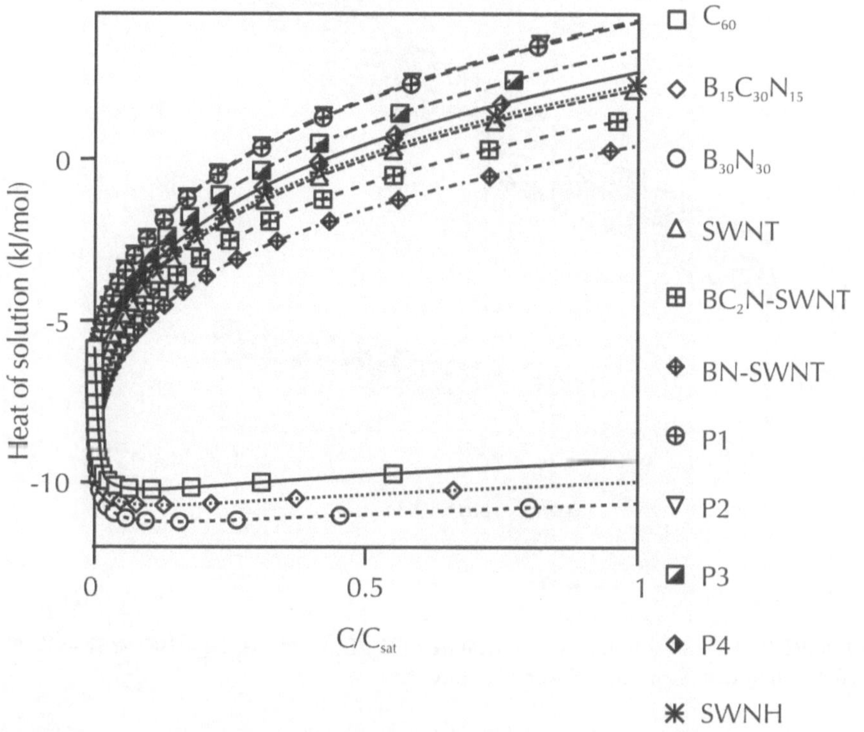

FIGURE 6 Heat of solution *vs.* concentration of C_{60}/$B_{15}C_{30}N_{15}$/$B_{30}N_{30}$–(BC_2N/BN-)SWNT/SWNC in toluene/benzene/CS_2 at 298.15K.

Figure 7 displays the temperature dependence of the heat of solution in toluene, benzene and CS_2 calculated for the saturation concentration. The results for SWNH are superimposed on SWNT, BC_2N-SWNH on BC_2N-SWNT, and BN-SWNH on BN-SWNT. The data of C_{60}, etc. are plotted for $T > 260$K after FCC/SC transition. For C_{60} (droplet) on going from $T = 260$K to $T = 400$K, heat of solution increases 2.7kJ·mol^{-1}. For SWNT and SWNCs (bundlet), the heat of solution augments 10.4 and 10.4–10.9kJ·mol^{-1}, respectively, in the same range. For $B_{15}C_{30}N_{15}$ (droplet), the heat of solution rises 2.5kJ·mol^{-1}. For BC_2N-SWNT and BC_2N-SWNCs (bundlet), the heat of solution augments 10.2 and 10.2–10.7kJ·mol^{-1}. For $B_{30}N_{30}$ (droplet), the heat of solution enlarges 2.3kJ·mol^{-1}. For BN-SWNT and BN-SWNCs (bundlet), the heat of solution rises 9.9 and 10.0–10.5kJ·mol^{-1}.

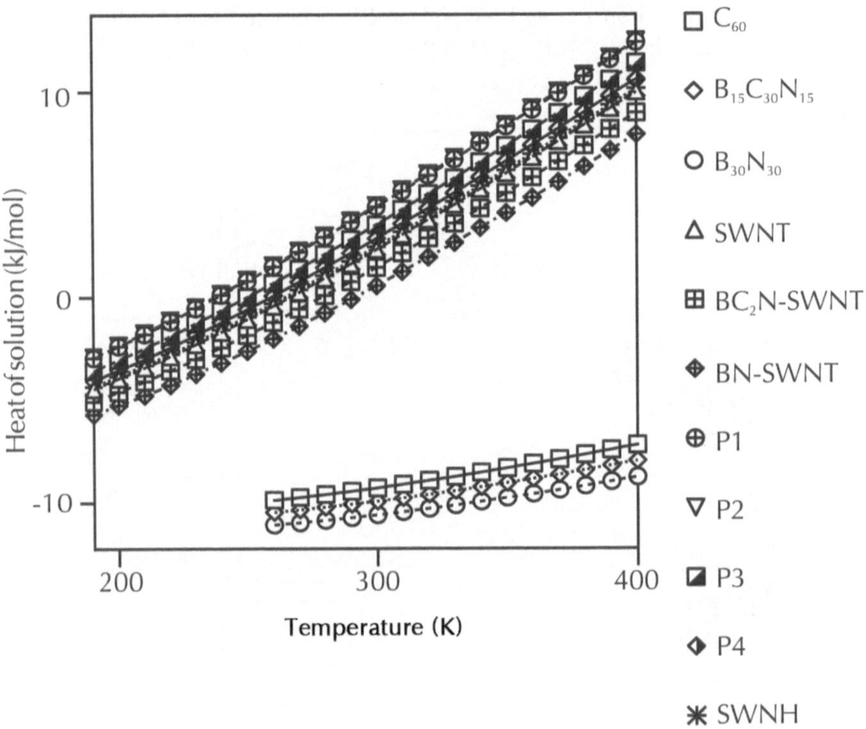

FIGURE 7 Heat of solution *vs.* temperature of $C_{60}/B_{15}C_{30}N_{15}/B_{30}N_{30}$–($BC_2N$/BN-)SWNT/SWNC in toluene/benzene/CS_2 for saturation.

Results of the dependence of the diffusion coefficient on the concentration in toluene, at $T = 298.15$K (*cf.* Fig. 8), show that data for SWNH are superimposed on SWNT, BC_2N-SWNH on BC_2N-SWNT, and BN-SWNH on BN-SWNT. The cluster formation in a solution close to saturation decreases the diffusion coefficients by 56%, 69% and 69–71% for C_{60}, SWNT and SWNCs, respectively, as compared with that for C_{60} molecule. For SWNT (bundlet) the diffusion coefficient drops by 29%, and for SWNCs (bundlet) the diffusion coefficients, by 29–33% with regard to C_{60} (droplet). The cluster formation close to saturation diminishes the diffusion coefficients by 56%, 68% and 68–70% for $B_{15}C_{30}N_{15}$, BC_2N-SWNT and BC_2N-SWNCs as compared with that for $B_{15}C_{30}N_{15}$ molecule. For BC_2N-SWNT (bundlet), the diffusion coefficient decays by 28%, and for BC_2N-SWNCs (bundlet), by 28–31% with regard to $B_{15}C_{30}N_{15}$ (droplet). The cluster formation close to saturation decreases the diffusion coefficients by 56%, 67% and 67–69% for $B_{30}N_{30}$, BN-SWNT and BN-SWNCs as compared with that for $B_{30}N_{30}$ molecule. For BN-SWNT (bundlet), the diffusion coefficient decays by 26%, and for BN-SWNCs (bundlet), by 26–29% with regard to $B_{30}N_{30}$ (droplet).

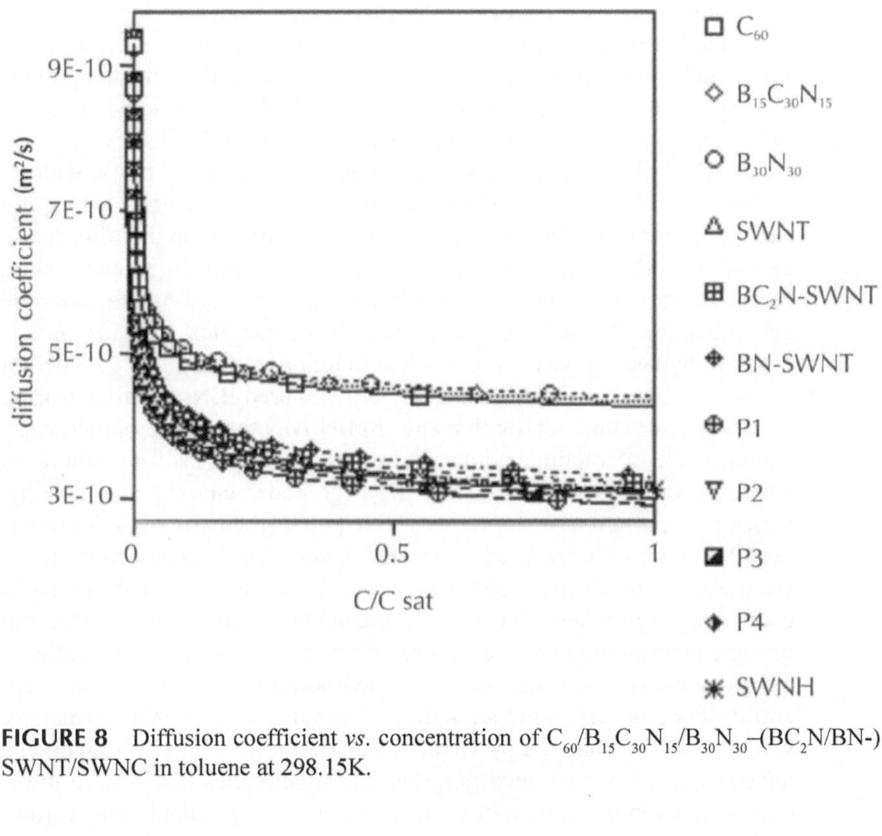

FIGURE 8 Diffusion coefficient *vs.* concentration of $C_{60}/B_{15}C_{30}N_{15}/B_{30}N_{30}$–($BC_2N$/BN-) SWNT/SWNC in toluene at 298.15K.

8.12 CONCLUSIONS

From the discussion of the present results the following conclusions can be drawn.

1. Atomic structures were deduced fitting voids of close-packed spheres. Several criteria reduced the analysis to manageable quantity of properties: closeness, dimension and efficiency. The model predicted the packing entity's property. A noncomputationally intensive approach, object clustering plus property prediction, assessed calculation reliability, solved problems and presented applications.
2. Packing efficiencies and interaction-energy parameters of nanocones are intermediate between C_{60} and tubes: in-between behavior was expected; however, they showed to be closer to tubes. Tube-like behavior is observed and cone properties are calculated closer to tubes. The packing efficiency and the interaction-energy parameters of horns are closest to tubes: most tube-like behavior is observed and properties are calculated closest to tubes. Large structural asymmetry in different cone types, characterized by number of pentagons (1–5), distinguished calculated properties especially for two pentagons P2. The heat of solution varies with the following behavior: P2 ≈ P1 > P3 > P4 > SWNH ≈ SWNT >> C_{60}.
3. BC_2N/BN will be especially stable species that are isoelectronic with C-analogs. Specific morphologies were observed for tube ends, which are suggested to result from B–N units. Chemical frustration that 60° disclinations would introduce in $B_{30}N_{30}$ governs structural difference with C tubes. Further work will explore similar nanostructure nature: possible generalization of conclusions to systems more complex; e.g., (a) there is a way of bypassing weak homonuclear bonding in closed B_xN_x, involving replacement of 5-membered rings by 4-membered B_2N_2 annuli, ensuring a perfect heteroatom alternation and (*b*) BN/AlN tubes/heterojunctions.
4. Nanoparticle interactions with vesicles as model systems for membranes covers a variable field, which is key for understanding nanotoxicity, which is important as nanoparticles were applied because of their properties. Nanoparticles are used as transport vehicles for selectively delivering active agents into cells, where membrane interaction is crucial in controlling their release. Nanoparticles can bind to membrane surface and become incorporated into/transported *via* bilayers; the latter is essential in nanotoxicology as the nanoparticles incorporation in cells is elementary and depends on p*H*, ionic strength and nanoparticles detailed structure, which could be changed by chem/physical modifications. Alternatively, nanoparticles selective incorporation into membranes is a way to introduce sensors into a system or control properties, e.g., membranes perme-

ability *via* light-induced heating. Nanoparticle/liposome interactions are key for understanding nanomedicine developments.

ACKNOWLEDGMENTS

The authors want to dedicate this manuscript to Dr. Luis Serrano-Andrés, who was greatly interested in this research and would have loved to see its conclusion. F. T. thanks support from the Spanish Ministerio de Ciencia e Innovación (Project No. BFU2010–19118).

KEYWORDS

- ***ab initio*** **calculations**
- ***helio-photocatalysis***
- ***isolated pentagon rule***
- **stoichiometric threshold**

REFERENCES

Anpo, M.; Kamat, P. V. Environmentally Benign Photocatalysts: Applications of Titanium Oxide-Based Materials; Springer: Berlin, 2010.

Asenjo, A.; Jaafar, M.; Navas, D.; Vázquez, M. J. Appl. Phys. **2006,** *100,* 023909–1–6.

Baker, J.; Kudrolli, A. Phys. Rev. E **2010,** 8*2,* 061304–1–5.

Balaban, A. T.; Klein, D. J. J. Phys. Chem. C **2009,** *113,* 19123–19133.

Balaban, A. T.; Klein, D. J.; Liu, X. Carbon **1994,** 357–359.

Barnard, A. S. Nat. Nanotechnol. **2009,** *4,* 332–335.

Berhanu, W. M.; Pillai, G. G.; Oliferenko, A. A.; Katritzky, A. R. ChemPlusChem **2012,** 7*7,* 507–517.

Betke, U.; Henk, M. Comput. Geom. **2000,** 1*6,* 157–186.

Betke, U.; Henk, M. Comput. Geom. **2000,** 1*6,* 157–186.

Bezmel'nitsyn, V. N.; Eletskii, A. V.; Okun,' M. V. Physics–Uspekhi **1998,** 4*1,* 1091–1114.

Bourgeois, L.; Bando, Y.; Han, W. Q.; Sato, T. Phys. Rev. B **2000,** 6*1,* 7686–7691.

Bourgeois, L.; Bando, Y.; Shinozaki, S.; Kurashima, K.; Sato, T. Acta Crystallogr., Sect. A **1999,** 5*5,* 168–177.

Carroll, D. L.; Redlich, P.; Ajayan, P. M.; Charlier, J.-C.; Blase, X.; de Vita, A.; Car, R. Phys. Rev. Lett. **1997,** 2811–2814.

Cervenka, J.; Katsnelson, M. I.; Flipse, C. F. J. Nat. Phys. **2009,** *5,* 840–844.

Chatterjee, R. Environ. Sci. Technol. **2008,** *42,* 339–343.

Chen, E. R.; Engel, M.; Glotzer, S. C. Discrete Comput. Geom. **2010,** *44,* 253–280.

Chopra, N. G.; Luyken, R. J.; Cherrey, K.; Crespi, V. H.; Cohen, M. L.; Louie, S. G.; Zettl, A. Science **1995,** *269,* 966–967.

Cioffi, C.; Campidelli, S.; Brunetti, F. G.; Meneghetti, M.; Prato, M. Chem. Commun. **2006,** 2129–2131.

Cioffi, C.; Campidelli, S.; Sooambar, C.; Marcaccio, M.; Marcolongo, G.; Meneghetti, M.; Paolucci, D.; Paolucci, F.; Ehli, G.; Rahman, G. M. A.; Sgobba, V.; Guldi, D. M.; Prato, M. J. Am. Chem. Soc. **2007,** *129,* 3938–3945.

Conway, J. H.; Torquato, S. Proc. Natl. Acad. Sci. U.S.A. **2006,** *103,* 10612–10617.

D'Oliveira, J.-C.; Minero, C.; Pelizzetti, E.; Pichat, P. J. Photochem. Photobiol. A **1993,** *72,* 261–267.

Dean, L. Chem. Int. **2012,** 34(4), 6–9.

Dîrtu, M. M.; Neuhausen, C.; Naik, A. D.; Rotaru, A.; Spinu, L.; Garcia, Y. Inorg. Chem. **2010,** *49,* 5723–5736.

Dreher, K. L. Toxicol. Sci. **2004,** *77,* 3–5.

Escrig, J.; Altbir, D.; Jaafar, M.; Navas, D.; Asenjo, A.; Vázquez, M. Phys. Rev. B **2007,** *75,* 184429–1–5.

Faraday, M. Philos. Trans. R. Soc. London **1857,** *147,* 145–181.

Fourches, D.; Pu, D. Q. Y.; Tassa, C.; Weissleder, R.; Shaw, S. Y.; Mumper, R. J. Tropsha, A. ACS Nano **2010,** *4,* 5703–5712.

Fuertes, G.; García-Sáez, A. J.; Esteban-Martín, S.; Giménez, D.; Sánchez-Muñoz, O. L.; Schwille, P.; Salgado, J. Biophys. J. **2010,** *99,* 2917–2925.

Fuertes, G.; Giménez, D.; Esteban-Martín, S.; Sánchez-Muñoz, O. L.; Salgado, J. Eur. Biophys. J. **2011,** *40,* 399–415.

Gasser, U.; Weeks, E. R.; Schofield, A.; Pusey, P. N.; Weitz, D. A. Science **2001,** *292,* 258–262.

Ghorbani, M. MATCH Commun. Math. Comput. Chem. **2012,** *68,* 381–414.

Haluska, C. K.; Riske, K. A.; Marchi-Artzner, V.; Lehn, J.-M.; Lipowsky, R.; Dimova, R. Proc. Natl. Acad. Sci. U. S. A. **2006,** *103,* 15841–15846.

Hamilton, E. J. M.; Dolan, S. E.; Mann, C. M.; Colijn, H. O.; McDonald, C. A.; Shore, S. G. Science **1993,** *260,* 659–661.

Hamilton, E. J. M.; Dolan, S. E.; Mann, C. M.; Colijn, H. O.; Shore, S. G. Chem. Mater. **1995,** *7,* 111–117.

Han, J.; Jaffe, R. J. Chem. Phys. **1998,** *108,* 2817–2823.

Heimburg, T. Biophys. Chem. **2010,** *150,* 2–22.

Herrmann, J.-M. Catal. Today **1999,** *53,* 115–129.

Herrmann, J.-M. J. Photochem. Photobiol. A **2010,** *216,* 85–93.

Herrmann, J.-M. Photocatalysis. In Kirk-Othmer encyclopedia of chemical technology; Wiley: New York, 2006a.Herrmann, J.-M. Top. Catal. **2006b**, *39,* 3–10.

Herrmann, J.-M. Top. Catal. **2005,** *34,* 49–65.

Herrmann, J.-M.; Guillard, C.; Pichat, P. Catal. Today **1993,** *17,* 7–20.

Hoffmann, M. R.; Martin, S. T.; Choi, W.; Bahnemann, D. W. Chem. Rev. **1995,** *95,* 69–96.

Horsell, D. W.; Hale, P. J.; Savchenko, A. K. Microsc. Anal. **2011,** *25(1),* 15–17.

Jaafar, M.; Gómez-Herrero, J.; Gil, A.; Ares, P.; Vázquez, M.; Asenjo, A. Ultramicroscopy **2009,** *109,* 693–699.

Jaafar, M.; Iglesias-Freire, O.; Serrano-Ramón, L.; Ibarra, M. R.; de Teresa, J. M.; Asenjo, A. Beilstein J. Nanotechnol. **2011,** *2,* 552–560.

Jaafar, M.; Martínez-Martín, D.; Cuenca, M.; Melcher, J.; Raman, A.; Gómez-Herrero, J. Beilstein J. Nanotechnol. **2012,** *3,* 336–344.

Jiao, Y.; Torquato, S. J. Chem. Phys. **2011,** *135,* 151101–1–4.

Kahn, O.; Martinez, C. J. Science **1998,** *279,* 44–48.

Kallus, Y.; Elser, V.; Gravel, S. Discrete Comput. Geom. **2010,** *44,* 245–252.

Kim, P.; Odom, T. W.; Huang, J.-L.; Lieber, C. M. Phys. Rev. Lett. **1999,** 1225–1228.

Klein, D. J. J. Chem. Educ. **1992,** 691–694.

Klein, D. J. Phys. Chem. Chem. Phys. **2002,** 4, 2099–2110.

Klein, D. J.; Balaban, A. T. J. Chem. Inf. Model. **2006,** 307–320.

Klein, D. J.; Balaban, A. T. Open Org. Chem. J. **2011,** 5(Suppl. 1-M3), 27–61.

Knorr, N.; Vinzelberg, S. Microsc. Anal. **2012,** *26(5),* 7–12.

Kohler-Redlich, P.; Terrones, M.; Manteca-Diego, C.; Hsu, W. K.; Terrones, H.; Rühle, M.; Kroto, H. W.; Walton, D. R. M. Chem. Phys. Lett. **1999,** *310,* 459–465.

Krishnan, A.; Dujardin, E.; Treacy, M. M. J.; Hugdahl, J.; Lynum, S.; Ebbesen, T. W. Nature (London) **1997,** *388,* 451–454.

Kroto, H. W. Nature (London) **1987,** *329,* 529–531.

Ksibi, M.; Zemzemi, A., Boukchina, R. J. Photochem. Photobiol. A **2003,** *159,* 61–70.

Kuznetsov, V. N.; and Serpone, N. J. Phys. Chem. B **2006,** *110,* 25203–25209.

Kuznetsov, V. N.; Serpone, N. J. Phys. Chem. C **2007,** *111,* 15277–15288.

Legrini, O.; Oliveros, E.; Braun, A. M. Chem. Rev. **1993,** *93,* 671–698.

Leite, F. L.; Bueno, C. C.; Da Róz, A. L.; Ziemath, E. C.; Oliveira, O. N., Jr. Int. J. Mol. Sci. **2012,** *13,* 12773–12856.

Loiseau, A.; Willaime, F.; Demoncy, N.; Hug, G.; Pascard, H. Phys. Rev. Lett. **1996,** *76,* 4737–4740.

Machado, M., Piquini, P. Mota, R. Chem. Phys. Lett. **2004,** *392,* 428–432.

Machado, M.; Mota, R.; Piquini, P. Microelectron. J. **2003c**, *34,* 545–547.

Machado, M.; Piquini, P.; Mota, R. Eur. Phys. J. D **2003a**, *23,* 91–93.

Machado, M.; Piquini, P.; Mota, R. Mater. Charact. **2003b**, *50,* 179–182.

Machado, M.; Piquini, P.; Mota, R. Nanotechnology **2005,** *16,* 302–306.

Madden, J. D. W. Science **2009,** *323,* 1571–1572.

Margulis, L.; Salitra, G.; Tenne, R.; Talianker, M. Nature (London) **1993,** *365,* 113–114.

Martínez-Martín, D.; Jaafar, M., Pérez, R., Gómez-Herrero, J., Asenjo, A. Phys. Rev. Lett. **2010,** *105,* 257203–1–4.

Misra, A.; Klein, D. J.; Morikawa, T. J. Phys. Chem. A 2009a, *113,* 1151–1158.

Misra, A.; Schmalz, T. G.; Klein, D. J. J. Chem. Inf. Model. 2009b, 2670–2676.

Miyamoto, Y.; Rubio, A.; Cohen, M. L.; Louie, S. G. Phys. Rev. B **1994,** *50,* 4976–4979.

Mota, R.; Machado, M.; Piquini, P. Phys. Status Solidi C **2003,** 799–802.

Murphy, C. J.; Thompson, L. B.; Alkilany, A. M.; Sisco, P. N.; Boulos, S. P.; Sivapalan, S. T.; Yang, J. A.; Chernak, D. J.; Huang, J. J. Phys. Chem. Lett. **2010,** 1, 2867–2875.

Neu, J. C.; Cañizo, J. A.; Bonilla, L. L. Phys. Rev. E **2002,** *66,* 61406–1–9.

Notman, R.; Noro, M.; O'Malley, B.; Anwar, J. J. Am. Chem. Soc. **2006,** *128,* 13982–13983.

Oppenländer, T. Photochemical Purification of Water and Air; Wiley-VCH: Weinheim (Ger.), 2003.

Pagona, G.; Fan, J.; Maigne, A.; Yudasaka, M.; Iijima, S.; Tagmatarchis, N. Diamond Relat. Mater. **2007b**, *16,* 1150–1153.

Pagona, G.; Fan, J.; Tagmatarchis, N.; Yudasaka, M.; Iijima, S. Chem. Mater. **2006a**, *18,* 3918–3920.

Pagona, G.; Sandanayaka, A. S. D.; Araki, Y.; Fan, J.; Tagmatarchis, N.; Charalambidis, G.; Coutsolelos, A. G.; Boitrel, B.; Yudasaka, M.; Iijima, S.; Ito, O. Adv. Funct. Mater. **2007a**, 1705–1711.

Pagona, G.; Sandanayaka, A. S. D.; Araki, Y.; Fan, J.; Tagmatarchis, N.; Yudasaka, M.; Iijima, S.; Ito, O. J. Phys. Chem. B **2006b**, *110,* 20729–20732.

Pino, E.; Encinas, M. V. J. Photochem. Photobiol. A **2012,** *242,* 20–27.

Puzyn, T.; Rasulev, B.; Gajewicz, A.; Hu, X. K.; Dasari, T. P.; Michalkova, A.; Hwang, H. M.; Toropov, A.; Leszczynska, D.; Leszczynski, J. Nat. Nanotechnol. **2011,** *6,* 175–178.

Rallo, R.; France, B.; Liu, R.; Nair, S.; George, S.; Damoiseaux, R.; Giralt, F.; Nel, A.; Bradley, K.; Cohen, Y. Environ. Sci. Technol. **2011,** *45,* 1695–1702.

Roubeau, O.; Alcazar Gomez, J. M.; Balskus, E.; Kolnaar, J. J. A.; Haasnoot, J. G.; Reedijk, J. New J. Chem. **2001,** *25,* 144–150.

Rubio, A.; Corkill, J. L.; Cohen, M. L. Phys. Rev. B **1994,** *49,* 5081–5084.

Serpone, N. J. Phys. Chem. B **2006,** *110,* 24287–24293.

Silaghi-Dumitescu, I.; Haiduc, I.; Sowerby, D. B. Inorg. Chem. **1993,** *32,* 3755–3758.

Som, C.; Berges, M.; Chaudhry, Q.; Dusinska, M.; Fernandes, T. F.; Olsen, S. I.; Nowack, B. Toxicology **2010,** *269,* 160–169.

Tagmatarchis, N.; Maigne, A.; Yudasaka, M.; Iijima, S. Small **2006,** 2, 490–494.

Tamura, R.; Tsukada, M. Phys. Rev. B **1995,** 6015–6026.

Tenne, R.; Margulis, L.; Genut, M.; Hodes, G. Nature (London) **1992,** *360,* 444–446.

Terauchi, M.; Tanaka, M.; Suzuki, K.; Ogino, A.; Kimura, K. Chem. Phys. Lett. **2000,** *324,* 359–364.

Terrones, M.; Benito, A. M.; Manteca-Diego, C.; Hsu, W. K.; Osman, O. I.; Hare, J. P.; Reid, D. G.; Terrones, H.; Cheetham, A. K.; Prassides, K.; Kroto, H. W.; Walton, D. R. M. Chem. Phys. Lett. **1996,** *257,* 576–582.

Thesing, L. A.; Piquini, P.; Kar, T. Nanotechnology **2006,** *17,* 1637–1641.

Tiberj, A.; Camara, N.; Godignon, P.; Camassel, J. Nanoscale Res. Lett. **2011,** *6,* 478–1–9.

Torrens, F.; Castellano, G. Comput. Lett. **2005,** *1,* 331–336.

Torrens, F.; Castellano, G. Curr. Res. Nanotechn. **2007a**, *1,* 1–29.

Torrens, F.; Castellano, G. Int. J. Chemoinf. Chem. Eng. **2012,** *2(1),* 48–98.

Torrens, F.; Castellano, G. Int. J. Quantum Chem. **2010,** *110,* 563–570.

Torrens, F.; Castellano, G. J. Comput. Theor. Nanosci. **2007c**, *4,* 588–603.

Torrens, F.; Castellano, G. Microelectron. Eng., **2013**, 108, 127–133.

Torrens, F.; Castellano, G. Microelectron. J. **2007b**, *38,* 1109–1122.

Torrens, F.; Castellano, G. Nanoscale **2011,** *3,* 2494–2510.

Torrens, F.; Castellano, G. Nanoscale Res. Lett. **2007d**, *2,* 337–349.

Weng-Sieh, Z.; Cherrey, K.; Chopra, N. G.; Blase, X.; Miyamoto, Y.; Rubio, A.; Cohen, M. L.; Louie, S. G.; Zettl, A.; Gronsky, R. Phys. Rev. B **1995,** *51,* 11229–11232.

Xia, X.; Jelski, D. A.; Bowser, J. R.; George, T. F. J. Am. Chem. Soc. **1992,** *114,* 6493–6496.

Zhu, J.; Kase, D.; Shiba, K.; Kasuya, D.; Yudasaka, M.; Iijima, S. Nanoletters **2003,** *3,* 1033–1036.

Zhu, Y.; Li, W.; Li, Q.; Li, Y.; Li, Y.; Zhang, X.; Huang, Q. Carbon **2009,** *47,* 1351–1358.

Zhu, Y.; Zhang, X.; Zhu, J.; Zhao, Q.; Li, Y.; Li, W.; Fan, C.; Huang, Q. Int. J. Mol. Sci. **2012,** *13,* 12336 12348.

CHAPTER 9

COMPUTATIONAL STRATEGIES FOR NONLINEAR OPTICAL PROPERTIES OF CARBON NANO-SYSTEMS

SHABBIR MUHAMMAD and MASAYOSHI NAKANO

CONTENTS

Abstract ... 310
9.1 Introduction ... 310
9.2 Computational Methodology ... 312
9.2.1 The Hamiltonian ... 313
9.2.2 The Bon-Oppenheimer Approximation ... 314
9.2.3 Methodology of Nonlinear Optical (NLO) Polarizabilities ... 315
9.3 Zero Dimensional Fullerenes ... 316
9.3.1 Electro-Optical Properties ... 319
9.4 One Dimensional Carbon Nanotubes ... 321
9.4.1 Electro-Optical Properties ... 322
9.5 Two Dimensional Graphenes ... 324
9.5.1 Electro-Optical Properties ... 324
9.6 Conclusions and Prospective ... 328
Keywords ... 329
References ... 329

ABSTRACT

Carbon nanomaterials consisting of graphenes, nanotubes and fullerenes are the hi-tech and the smartest materials of the present era. Elemental carbon in the sp2 hybridization can form a variety of amazing structures. Apart from the well-known graphite, carbon can build an infinite number zero dimensional closed cages consisting entirely of carbon, which are called fullerenes. Besides this, carbon can also form different one-dimensional (1D) crystalline geometries in the form of seamless cylinders known as nanotubes. The discovery of fullerenes and carbon nanotubes has aroused remarkable excitement not only in experimental chemistry but also in computational chemistry. A lack of experimental knowledge about the electronic processes in nano-systems has expedited the progress in computational chemistry. In recent years, many computational studies have been performed on nano-systems for their mechanical and electrical properties including their conductivity, superconductivity, ferromagnetism, charge transport, optical and nonlinear optical properties. Of particular interest to the present work are their strong electrical and nonlinear optical properties, as measured by several quantum chemical methods. Nano-systems have been potential candidates for nonlinear application because of their extensive π-conjugation. Fullerenes and carbon nanotubes have been challenging molecules for first-principles calculations because of their relatively larger sizes. Recent advances in parallel and supercomputing have brought a substantial improvement in capabilities of quantum chemical methods for predicting the optical and nonlinear optical properties of large molecules like fullerenes and carbon nanotubes. Quantum chemical methods are becoming more important and more practical for the interpretation of structure property relationships at molecular level. Computations have been performed on these nano-systems to functionalize them by either covalent attachment of chemical groups or the non-covalent adsorption or wrapping of various functional molecules. The role of several functionalized nano-systems to modify the electrical and optical properties in the field of carbon nanomaterial will be reviewed in this chapter, in which an overview of different structure property relationships applied to assess the electrical and nonlinear optical properties will be summarized.

9.1 INTRODUCTION

Over the last two decades, several thousand chapters have been written to explain many features of carbon nano-materials. During these two decades, the carbon nano-materials consisting of fullerenes, graphenes and nanotubes have emerged as the hi-tech and the smartest material of the era with their exceptional electro-optical properties. Elemental carbon in the sp^2 hybridization can form a variety of wonderful structures. Apart from the well-known graphite, carbon can form an infinite number zero dimensional closed cages consisting entirely of carbon, which are called fullerenes (Ijima, 1991). Besides this, carbon can also build different one-dimensional (1D) crystalline geometries in the form of seamless cylinders known as nanotubes (see Fig. 1) (Kroto, 1987). The discovery of fullerenes and carbon nanotubes has aroused remarkable excitement not only in experimental chemistry but also in computational chemistry. The lack of several experimental aspects about the electronic processes in nano-systems has expedited the progress in computational chemistry. In recent years, many computational studies have been performed on nano-systems for their mechanical and electrical properties including their conductivity, superconductivity, ferromagnetism, charge transport, optical and nonlinear optical properties. Of particular interest in this chapter are the optical and nonlinear optical properties, as measured by several quantum chemical methods.

Nano-systems can be considered potential candidates for nonlinear optical application because of their extensive π-conjugation. Fullerenes and carbon nanotubes have been challenging molecules for first-principles calculations because of their relatively larger sizes. Recent advances in parallel and supercomputing have brought a substantial improvement in capabilities of quantum chemical methods for predicting the optical and nonlinear optical properties of large molecules like fullerenes and carbon nanotubes. Quantum chemical methods are becoming more important and more practical for the interpretation of structure property relationships at molecular level. Computations have been performed on these nano-systems to functionalize them by either covalent attachment of chemical groups or the noncovalent adsorption or wrapping of various functional molecules. The role of several functionalized nano-systems to modify the electrical and optical properties in the field of carbon nanomaterial will be highlighted in this chapter, in which an overview of different structure property relationships applied to assess the electrical and nonlinear optical properties will be summarized.

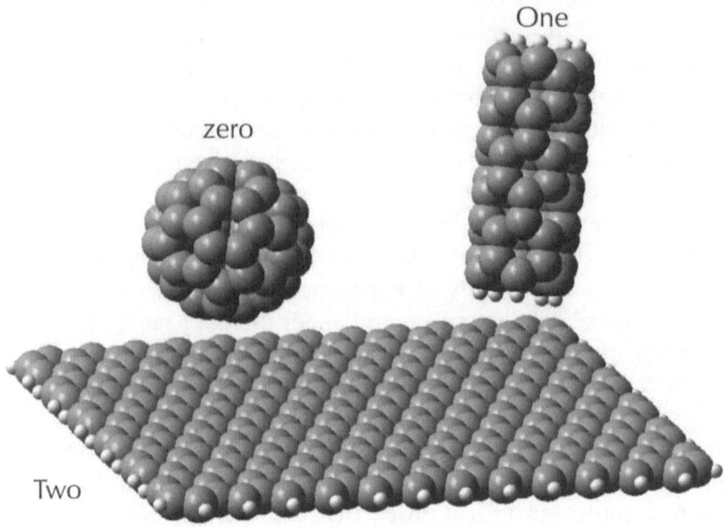

FIGURE 1 Different carbon nano-systems including zero-dimensional fullerenes, one-dimensional nano-tubes, and two-dimensional graphene sheets.

9.2 COMPUTATIONAL METHODOLOGY

Many aspects of molecular structure and dynamics can be modeled using classical methods in the form of molecular mechanics and dynamics. The classical force field is based on empirical results, averaged over a large number of molecules (Anisimov et al., 2004). Because of this extensive averaging, the results can be good for standard systems, but there are many important questions in chemistry that can not at all be addressed by means of this empirical approach. If one wants to know more than just the structure or other properties that are derived only from the potential energy surface, in particular properties that depend directly on the electron density distribution, one has to resort to a more fundamental and general approach which is quantum chemistry. The same holds for all nonstandard cases for which molecular mechanics is simply not applicable. Quantum chemistry is based on the postulates of quantum mechanics. In this chapter we shall recall some basic aspects of the theory of quantum chemistry with an emphasis on their practical implications for the molecular modeler. Quantum chemical methods, in which no empirical or semiempirical parameters are included in their equations except for the basic physical constants, are usually called *ab initio* (from the beginning) methods. *Ab initio* quantum chemical methods have nowadays gained much popularity for their efficiency and accuracy. Semi-empirical methods are based on the same theoretical formalism as *ab initio* methods, but with more approximations and parameters fitted from experimental data (Ahmen and Boudreaux, 1973). The present day quantum chemistry contains several methods like the Hartree-Fock (HF), postHartree-Fock with the MPn (Head Gordon and Pople, 1988) couple cluster (Cizek, 1966) and density functional theory (DFT) methods (Koch and Holthausen, 2000) The last group of DFT methods emerges as the most adaptable in quantum chemistry with apt accuracy and computational cost. The detail about these methods can be found elsewhere, in this chapter we will give a brief introduction about the Hamiltonian of a wave function.

9.2.1 THE HAMILTONIAN

In quantum chemistry, the system is described by a wavefunction, which can be found by solving the Schrödinger equation. This equation relates the stationary states of the system and their energies to the Hamiltonian operator, which can be viewed as the recipe for obtaining the energy associated with a wavefunction describing the positions of the nuclei and electrons in the system. In practice the Schrödinger equation cannot be solved exactly and approximations have to be made, as we shall see below. The approach is called "*ab initio*" when it makes no use of empirical information, except for the fundamental constants of nature such as the mass of the electron, Planck's constant etc., which are required to arrive at numerical predictions. The Schroëdinger's Equation:

$$H\psi = E\psi \tag{1}$$

where, ψ = psi, the wavefunction; E = the energy of the sytem; H = The "Hamiltonian", a nasty mathematical operator.

A complete Hamiltonian, even for an isolated molecular system, is much too complicated to use in the calculation. Neglecting the relativistic mass effects and all magnetic interactions due to coupling between spin and orbital motions of nuclei and electrons, a nonrelativistic total molecular Hamiltonian of an isolated molecule includes only operators for the nuclear kinetic energy ($\hat{V}_{NN}$), the nuclear-nuclear repulsion ($\hat{V}_{NN}$), the electronic kinetic energy ($\hat{T}e$), the nuclear-electron attraction ($\hat{V}_{Ne}$) and the electron-electron repulsion ($\hat{V}ee$). For an isolated molecule with M nuclei and N electrons, the total Hamiltonian can be written in atomic units as,

$$\hat{\mathrm{H}} = -\sum_{A=1}^{M} \frac{1}{2M_A}\nabla_A^2 + \sum_{A=1}^{M-1}\sum_{B>A}^{M} \frac{Z_A Z_B}{R_{AB}} - \sum_{i=1}^{N} \frac{1}{2}\nabla_i^2 - \sum_{i=1}^{N}\sum_{A=1}^{M} \frac{Z_A}{r_{iA}} + \sum_{i=1}^{N-1}\sum_{j>i}^{N} \frac{1}{r_{ij}} \tag{2}$$

In this expression, M_A and Z_A are, respectively, the mass and charge of nucleus A. R_{AB}, r_{iA} and r_{ij} are distances between charged particles (nuclei A and B, electron i and nucleus A, and electrons i and j, respectively). The Laplacian operator r_2 i involves differentiation with respect to coordinates of the ith electron and r_2 A involves differentiation with respect to those of the Ath nucleus.

9.2.2 THE BON-OPPENHEIMER APPROXIMATION

A system of interacting atoms is really made up of nuclei and electrons, which interact with each other. The true Hamiltonian for this system may be written as in Eq. (2). In 1923, Born and Oppenheimer noted that nuclei are much heavier than electrons and they made a famous approximation called Born-Oppenheimer (BO) approximation. The BO approximation assumes that at a given instant the electrons are moving around the frozen nuclei. Thus, the nuclear kinetic energy term $\hat{T}_{NN}$ in the total molecular Hamiltonian expression (1) can be neglected and the nuclear-nuclear repulsion term $\hat{V}_{NN}$ becomes a scalar operator. Consequently, the last four operators in expression (1) depend directly on the coordinates of the electrons and are usually called the electronic Hamiltonian $\hat{H}_e$,

$$\hat{H}_e = \sum_{A=1}^{M-1}\sum_{B>A}^{M}\frac{Z_A Z_B}{R_{AB}} - \sum_{i=1}^{N}\frac{1}{2}\nabla_i^2 - \sum_{i=1}^{N}\sum_{A=1}^{M}\frac{Z_A}{r_{iA}} + \sum_{i=1}^{N-1}\sum_{j>i}^{N}\frac{1}{r_{ij}} \tag{3}$$

Solving the electronic Schrödinger equation,

$$\hat{H}_e \left|\Psi_e(\vec{r};\vec{R})\right\rangle = E_e \left|\Psi_e(\vec{r};\vec{R})\right\rangle \tag{4}$$

at the fixed nuclear geometry $\vec{R}$, one can get the corresponding electronic energy eigenvalue E_e and eigenwavefunctions (or orbitals) $\left|\Psi_e(\vec{r};\vec{R})\right\rangle$. Here, the variable $\vec{r}$ stands for all electronic coordinates. Obviously, the electronic energy E_e and wavefunctions $\left|\Psi_e(\vec{r};\vec{R})\right\rangle$ depend on the chosen positions $\vec{R}$ of the nuclei parametrically. Changing the nuclear positions $\vec{R}$ in adiabatically small steps and repeatedly solving the electronic Schrödinger equations, one will obtain the potential energy surface (PES): E_e ($\vec{R}$), on which the nuclei move. Then, the BO approximation reintroduces the nuclear kinetic energy $\hat{T}_n$ into the nuclear Schrödinger equation,

$$\left(-\sum_{A=1}^{M}\frac{1}{2M_A}\nabla_A^2 + E_e(\vec{R})\right)\left|\Psi_{nuc}(\vec{R})\right\rangle = E\left|\Psi_{nuc}(\vec{R})\right\rangle \tag{5}$$

and rotations. This separation of the electronic and nuclear degrees of wavefunction is known as the Born-Oppenheimer approximation,

$$\left|\Psi_{tot}(\vec{r},\vec{R})\right\rangle = \left|\Psi_{e}(\vec{r};\vec{R})\right\rangle\left|\Psi_{nuc}(\vec{R})\right\rangle \tag{6}$$

9.2.3 METHODOLOGY OF NONLINEAR OPTICAL (NLO) POLARIZABILITIES

NLO active materials are capable of generating new electromagnetic fields with new frequencies, phase, or other physical properties by the interacting with intense laser light (see Fig. 2). These materials have several potential applications in modern optical data storage, holographic imaging, frequency mixing, telecommunications, and so forth (Champagne and Kirtman, 2001) Macroscopic polarization P and microscopic polarization p are given by:

$$P = \chi^{(1)}F + \chi^{(2)}F^2 + \chi^{(3)}F^3 \cdots \tag{7}$$

$$p = \alpha F + \beta F^2 + \gamma F^3 \cdots \tag{8}$$

where $\chi^{(1)}$, $\chi^{(2)}$, $\chi^{(3)}$ are linear susceptibility, second order nonlinear susceptibility, and third order nonlinear susceptibility and α, β, γ are polarizability, first hyperpolarizability, second hyperpolarizability. *Ab initio* techniques coupled with finite field (FF) approach are typically used to calculate first and second hyperpolarizabilities, β and γ, which are the origins of the macroscopic second- and third-order nonlinear optical (NLO) susceptibilities, $\chi^{(2)}$ and $\chi^{(3)}$, respectively.

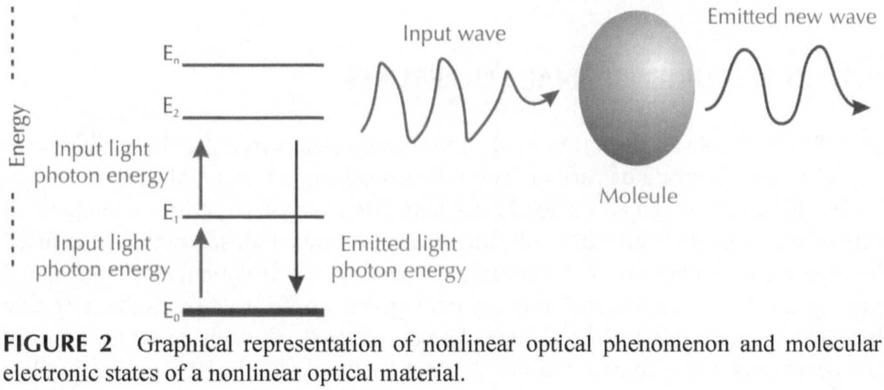

FIGURE 2 Graphical representation of nonlinear optical phenomenon and molecular electronic states of a nonlinear optical material.

In FF method when a molecule is subjected to the static electric field (F), the energy (E) of the molecule is expressed by the following equation

$$E = E^{(0)} - m_i F_i - \frac{1}{2} a_{ij} F_i F_j - \frac{1}{6} b_{ijk} F_i F_j F_k - \frac{1}{24} g_{ijkl} F_i F_j F_k F_l - \ldots \tag{9}$$

where E(0) is the energy of molecule in the absence of an electronic field, μ is the component of the dipole moment vector, *α* is the linear polarizability tensor, *β* and *γ* are the first and second hyperpolarizability tensors whereas *i*, *j* and *k* label the with *x*, *y* and *z* components, respectively. For a molecule the average dipole moment μ_0 and polarizability (α_0) are defined as follows:

$$\mu_0 = \left(\mu_x^2 + \mu_y^2 + \mu_z^2\right)^{\frac{1}{2}} \tag{10}$$

$$\alpha_0 = \frac{1}{3}\left(\alpha_{xx} + \alpha_{yy} + \alpha_{zz}\right) \tag{11}$$

The first hyperpolarizability (β_0) is defined as:

$$\beta_0 = \left(\beta_x^2 + \beta_y^2 + \beta_z^2\right)^{\frac{1}{2}} \tag{12}$$

where,

$$\beta_i = \frac{3}{5}\left(\beta_{iii} + \beta_{ijj} + \beta_{ikk}\right) \qquad \text{i, j, k=x, y, z} \tag{13}$$

and

$$\gamma_0 = \frac{1}{5}\left(\gamma_{xxxx} + \gamma_{yyyy} + \gamma_{zzzz}\right) + 2\left(\gamma_{xxyy} + \gamma_{xxzz} + \gamma_{yyzz}\right) \tag{14}$$

9.3 ZERO DIMENSIONAL FULLERENES

In 1985, W. Kroto et al. (Kroto et al., 1985) have discovered the first fullerene of C_{60} as a new allotrope of carbon. The fullerenes form a unique class of spherical molecules containing a conjugated *π*-system. Each carbon atom is bonded to three others and is sp^2 hybridized. Each fullerene represents a closed network of fused hexagons and pentagons. This building principle is a consequence of the isolated pentagon rule, which states that every pentagonal ring should be isolated by five hexagonal rings in a fullerene cage. In other words, in such structure none of the pentagons make contact with each other (Schmalz et al., 1988). The smallest fullerene that obey the isolated pentagon rule is C_{60} with icosahedral structures. The next stable homologue is C_{70}, followed by higher fullerenes (Thilgen and Diederich, 1999) like C_{76}, C_{78}, C_{80}, C_{82}, C_{84}, etc. The other smaller cage-like molecules containing less than 60 carbon atoms are classified as lower fullerene family. Some theoretically studied fullerenes including C_{20} (Sackers et al., 2006) C_{26} (Wang et al., 2001) C_{30} (Song et al., 2010) C_{36} (Piskoti et al., 1998) C_{40} (Xiao et al., 1998) C_{42} (Sun et al., 2005) C_{48} (Sun et al., 2005) C_{60} (Kroto et al., 1985) and C_{70} (Talyzin, 1997) have been shown in Fig. 3. It has been well known that usually the lower fullerenes are built up of only pentagons and hexagons. Unlike the conventional fullerenes, the lower fullerenes inevitably contain fused pentagon rings, they violate the "isolated pentagon rule" and, therefore, these are considered to be highly strained fullerenes. The relative stability of lower fullerenes family usually ascertained with the pentagon adjacency penalty rule (Diaz-tendero et al., 2005) i.e., more stable fullerenes with the minimum number of adjacent pentagons.

The lowest molecular weight structure that has some stability under ambient conditions is C_{36}, described by Zettl and co-workers (Piskoti et al., 1998). It has an icosahedral symmetry, similar to that of C_{60}. Due to the strongly curved C–C bonds, the electron–phonon interaction is expected to be stronger than in higher fullerenes, and superconductors with higher Tc were predicted from doped C_{36} based compounds (Cote et al., 1998). The first measurements, however, indicate that the C_{36} molecule is strongly reactive and in the solid state, it spontaneously forms covalently bonded structures. Furthermore, the calculated endohedral binding energies demonstrate that C_{36} is perhaps the smallest fullerene size, which can easily trap a range of atoms. The NMR chemical shifts, evaluated for the two lowest energy fullerene isomers, show that they are sufficiently different chemically to be distinguishable experimentally.

We will limit our discussion on the bonding nature of C_{60}, as it is the most stable and at the same time the most abundant fullerene. The icosahedral cage of C_{60} fullerene with I_h symmetry forms a spherical shape of the nonplanar conjugated system that is merely composed of sp^2 hybridized carbon atoms.

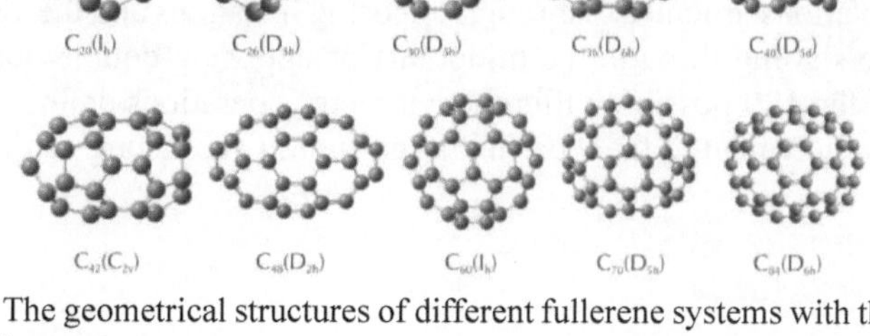

FIGURE 3 The geometrical structures of different fullerene systems with their respective symmetry point groups.

The molecule is comprised of 20 hexagon rings and 12 pentagon rings, such that all the atoms are identical. Each atom is at the vertex of one pentagon and two hexagons. There are two independent bond types: the shorter C-C bonds are between two pentagons and longer C-C bonds are between hexagon and pentagon. Some computed properties of C_{60} are compared with experimental results as given in Table 1. As seen from Table 1, these are technology- dependent results and semiquantitatively vary from each other not only in theoretical calculations but also in experimental values.

In C_{60} molecule the carbon nuclei exist on a sphere of about 7 Å diameter, with the electronic wavefunctions extending inside and outside by about 1.5 Å. The diameter of the molecule is approximately 10 Å, and there is a 4 Å diameter cavity inside. The atoms are actually positioned at the 60 vertices of a truncated icosahedron structure, with 90 edges, 12 pentagons and 20 hexagons (Forro and Mihaly, 2001). The two different C–C bond lengths in C60 (1.40 and 1.46 Å), indicate that the π electrons are not delocalized evenly over all bonds.

TABLE 1 Calculated and experimental [Pentagon-Hexagon (long bond)] and [Hexagon-Hexagon (Short bond)] bond distances [Å] of C60 fullerene.

Method	Long-bond	Short-bond	Ref.
HF	1.448	1.370	(Nakazawa et al., 2006)
MP2	1.446	1.406	(Novoselov et al., 2004)
B3LYP	1.453	1.396	(Piskoti et al., 1998)
Avg.	1.449	1.390	–
X-ray	1.467	1.355	(Prinzbach et al., 2000)
Electron diffraction	1.458	1.401	(Robertson et al., 2003)
NMR	1.450	1.40	(Sackers et al., 2006)
Avg.	1.458	1.378	

Symmetry group of the C_{60} include the inversion symmetry, the mirror planes, the rotation by 72◦ or 144◦ around axs piercing the centers of two opposing pentagons, 120◦ rotations around axs through opposing hexagons and the 180◦ rotations around the axs going through the midpoints of opposing double bonds. The relationship of the 120 possible different symmetry operations defines a symmetry group that is equivalent to the icosahedral group *Ih* (Yeo-heung et al., 2006).

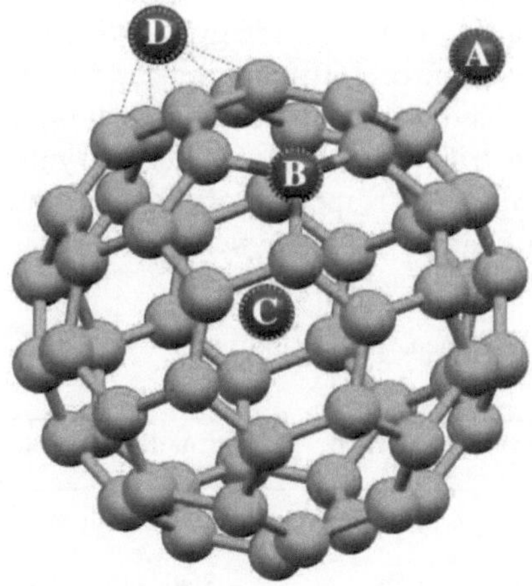

FIGURE 4 A schematic illustration of different possible doping options at C_{60} fullerene.

9.3.1 ELECTRO-OPTICAL PROPERTIES

The fullerenes have attracted much attention because of their unique structural and electronic properties. Hence, wide ranges of schemes for obtaining tunable optical properties from fullerenes are available with different doping approaches. For example, there are large empty spaces between the molecules, which are freely available for smaller atoms, ions or molecules. Due to the unsaturated character of the C-C bonds on the fullerene, there are plenty of electronic states to accept electrons from appropriate donors forming donor-π-acceptor combination suitable for NLO properties. Organic and inorganic molecules are also used to produce charge transfer salts with fullerenes (Dresselhaus et al., 1996). Doping shows the change in its the electronic structures, charge distribution patterns, and energy gaps ΔE between the highest occupied molecular orbital (HOMO) and the lowest unoccupied molecular orbital (LUMO) of the parent fullerenes. The investigations based on doping of fullerenes have indicated that dopant atoms can interact with the fullerene cages to form four kinds of heterofullerenes: (a) subsitutional fullerenes by the addition of a functional group, (b) substitutional fullerenes with atoms incorporated into the fullerene net-work, (c) endohedral fullerenes with atoms trapped inside the hollow fullerene cage, (d) exohedral fullerenes with atoms located outside the cage as shown in Fig. 4.

Due to the extensive π-conjugation, the fullerenes are the potential candidates for nonlinear optical properties. Different theoretical studies have shown that C_{60} have reasonably large amplitude of third-order nonlinear polarizability ranging from 4.94 to 7.00×10^{-36} esu. This amplitude has been further tuned by adopting different doping techniques. A tuning strategy for NLO properties of fullerenes is the formation of a series of novel [60]fullerene–ferrocene and [60]fullerene–porphyrin conjugates, in which a fullerene and an electron donating moiety are attached through a flexible triethylene glycol linker are synthesized. The third-order nonlinear optical (γ amplitude) responses are found to increase about double in magnitude than those of parent C_{60} fullerene.

TABLE 2 Third-order Nonlinear polarizabilities ($\times 10^{-36}$ esu) for C_{60} Fullerene and its derivatives.

Fullerenes	γ value	Method	Condition	Ref.
Fullerene C_{60}	5.50	LB94	Static	(Gisbergen et al., 1998)
Fullerene C_{60}	7.00	LDA	Static	(Gisbergen et al., 1998)
Fullerene C_{60}	4.94	INDO-TDHF	Static	(Talapatra et al., 1992)
[60]fullerene-ferrocene	8.01	PM3	Static	(Xenogiannopoulou et al., 2007)
[60]fullerene-porphyrin	8.68	PM3	Static	(Xenogiannopoulou et al., 2007)
$Fe(\eta^5\text{-}C_{60}H_5)_2$	-2820	BP86/TZP	Static	(Liu et al., 2008)
$Fe(\eta^5\text{-}C_{55}N_5)_2$	-7823	BP86/TZP	Static	(Liu et al., 2008)
$Fe(\eta^5\text{-}C_{55}B_5)_2$	-10410	BP86/TZP	Static	(Liu et al., 2008)
$C_{58}NN$	65.55	LDA TD-FF	Static	(Jensen et al., 2002)
$C_{58}BB$	78.29	LDATD-FF	Static	(Jensen et al., 2002)
$C_{58}BN$	105.01	LDATD-FF	Static	(Jensen et al., 2002)
$C_{60}F_{18}$	25.50	PBE/6–31G*	static	(Tang et al., 2010)
$C_{60}Cl_{30}$	136.10	PBE/6–31G*	static	(Tang et al., 2010)
$Sc_3N@C_{68}$	20	B3LYP/6–31G*	static	(Wang et al., 2012)
$Sc_3N@C_{70}$	25	B3LYP/6–31G*	static	(Wang et al., 2012)
$Sc_3N@C_{78}$	48	B3LYP/6–31G*	static	(Wang et al., 2012)
$Sc_3N@C_{80}$	23	B3LYP/6–31G*	static	(Wang et al., 2012)

In another computational study, the cyclopentadienyl rings of ferrocene have been replaced by $C_{55}x_5$ fullerenes where x = CH, B and N atoms (Liu et al., 2008) These ferrocene and fullerene hybrid molecules demonstrated a remarkably large third-order NLO response, especially for $Fe(\eta^{5}\text{-}C_{55}B_5)^2$ with the static third-order polarizability of -10410×10^{-36} esu as given in Table 2. The largest value of $Fe(\eta^{5}\text{-}C_{55}B_5)^2$ is found due to the charge transfer from ferrocene to C_{50} moiety along the z-axis (through Fe atom and the centers of two hybrid fullerenes) and play the key role in the NLO response. L. Jensen et.al. have investigated the effects of substituting carbon atoms with B and N on the second hyperpolarizability of C_{60} is different by using time-dependent density functional theory. They have calculated the second hyperpolarizability of the double substitute-doped fullerenes $C_{58}NN$, $C_{58}BB$ and $C_{58}BN$ (*see* Table 2). For C_{60} only small changes in the second hyperpolarizability were found when doping with either 2B or 2N. However, by doping C_{60} with both B and N, created a donor-acceptor system, where an increase in the second hyperpolarizability about 50% was found. The deformation of C_{60} fullerene by halogenation is also an important way to tune the γ amplitude. The effect of cage size is also important in several endohedral metallofullerenes where charge transfers from cage to encapsulated Sc_3N cluster.

9.4 ONE DIMENSIONAL CARBON NANOTUBES

Unlike the C_{60}, carbon nanotubes (CNTs) are one-dimensional carbon nanostructures with sp^2 hybridized carbon atoms and long enough tubic π-conjugations. The main two types of CNTs include multiwalled nanotubes (MWNTs) and single-walled nanotubes (SWNTs). Both experimental and theoretical studies suggested that the SWNTs are probably more important for future investigations on advance molecular materials and nanotechnology because their electronic properties can vary from semiconducting to metallic depending on their molecular structure. The molecular structure of single wall SWCNTs can be obtained by rolling up an infinite graphene sheet into a cylinder.

As illustrated in Fig. 5, SWCNTs are characterized by a chiral (or circumferential) vector XY, which is a linear combination of two unit lattice vectors a and b. In other words, XY = ma + nb where m and n are integers. The pair of indices (m, n) determines the diameter and chirality of the tube, as well as the basic electronic character. If $n = m$, the nanotube is classified as armchair and is metallic in nature (i.e., having a band gap of 0 eV). If $n = 0$, the nanotubes are called zigzag nanotubes. Fig. 5 also highlights the wrapping direction x and translation direction y for armchair SWCNTs. The carbon nanotubes can be considered as seamless graphene cylinders in which their sidewall curvatures have a strong influence on electronic structure. Importantly, this leads to a pyramidalization of the C atoms, hence weakening the π-conjugation of the SWCNT. Each individual C atom exhibits slightly bent sp^2 hybridization of carbon-carbon bonds.

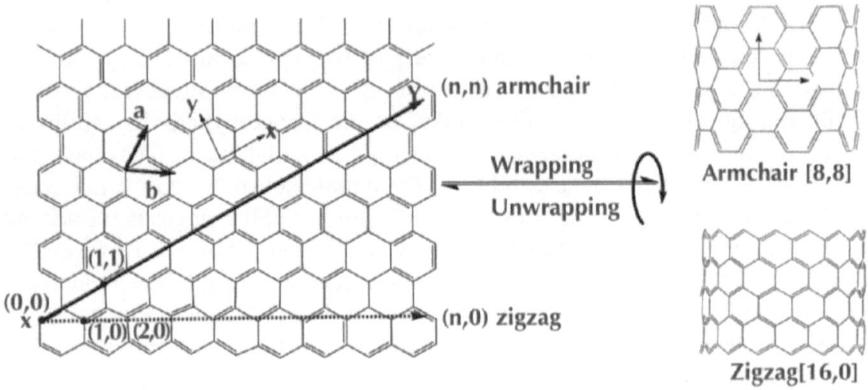

FIGURE 5 The representation of rolling principle to form SWCNTs from two-dimensional graphene sheet. The SWCNTs (8,8) and (16,0) are shown on right.

9.4.1 *ELECTRO-OPTICAL PROPERTIES*

Carbon nanotubes have always been promizing as novel nanomaterials in a wide variety of applications since Iijima first discovered them in 1991(Iijima, 1991). The small diameter (at scale of nanometers) and the long length (at the order of microns) lead to such large aspect ratios that the carbon nanotubes act as ideal one dimensional systems. Their rope like crystals offer host lattice for intercalation and storage. All these characteristics make carbon nanotubes the focus of extensive studies in nano-scale science and technology with potential applications in various materials and devices. Over the past 20 years, research efforts have directed to improve their efficiency and understanding of their chemical reactivity and extraordinary electronic and thermal properties (Karousis et al., 2010). Carbon nanotubes have significant potential for application in molecular electronics (Robertson and Mc Gowan, 2003), nanomechanics (Britz and Khlobystove, 2006) and optics (Nakazawa et al., 2006). In addition, the length and diameter are also important in determining the properties of CNT. For example, it has been theoretically predicted that chiral ($m = n$ or $m = 0$) SWNTs possess eight Raman-active and three infrared (IR) active phonon modes, whereas chiral ($0 < m \neq n$) SWNTs possess 14 Raman-active and six IR-active ones (Ikeda et al., 2007)

TABLE 3 First Hyperpolarizabilities (×10^{-30} esu) oscillator strength, f_0, difference of dipole moment, Δμ [D], transition energy, ΔE [eV] for perfect (6,0), defective (6,0) nanotubes, NO_{2}-NT(6,0)-NH_2 and Li5-CNT (5,0) nanotubes.

Nanotube	Method	β	ΔE	f_0	$\Delta\mu$	Ref.
Perfect (6,0)	LC-BLYP	0.00	1.9378	0.1081	0.000	(Liu et al., 2011)
DV585-z (6,0)	LC-BLYP	4.32	1.8471	0.070	0.23	(Liu et al., 2011)
SW5577-z (6,0)	LC-BLYP	4.58	1.7037	0.513	0.09	(Liu et al., 2011)
DV585-xz (6,0)	LC-BLYP	39.7	1.1316	0.074	0.88	(Liu et al., 2011)
DV555777 (6,0)	LC-BLYP	51.8	1.0092	0.031	0.81	(Liu et al., 2011)
SV-xz (6,0)	LC-BLYP	63.1	1.2576	0.035	3.61	(Liu et al., 2011)
SV-z (6,0)	LC-BLYP	95.00	1.2120	0.039	3.01	(Liu et al., 2011)
SW5577-xz (6,0)	LC-BLYP	285.0	0.9088	0.051	2.52	(Liu et al., 2011)
SW57–75 (6,0)	LC-BLYP	1900	0.5651	0.035	2.05	(Liu et al., 2011)
NO_{2}-NT(6,0)	HF-FF	299	–	–	–	(Xiao et al., 2008)
NO_{2}-NT(6,0)-NH_2	HF-FF	330	–	–	–	(Xiao et al., 2008)
NT(6,0)-NH_2	HF-FF	1140	–	–	–	(Xiao et al., 2008)
NT(6,0)—$(NH_2)_2$	HF-FF	485	–	–	–	(Xiao et al., 2008)
NT(6,0)$[Ru(tpy)_2]^{2+}$	HF-FF	1240	–	–	–	(Xiao et al., 2008)
Li5-CNT (5,0)	B3LYP-FF	1270	0.8416	0.0473	5.77	(Xu et al., 2011)

A perfect structure of a finite size (6,0) nanotube usually contains centro-symmetry element in its point groups, which in return nullifies the first hyperpolarizability (β) as given in Table 3. Several attempts have been made to find the finite size nanotubes without centro-symmetric point group involving topological defects (Liu et al., 2011) and one-end terminal substitutions (Xiao et al., 2008). The perfect structure of a zigzag (6,0) CNT pristine nanotube has a centro-symmetric point group and its β amplitude is zero. However a topological defect in pristine (6,0) CNT can interestingly modify its β amplitude. These defects are usually classified into the topological Stone-Wales (SW), single vacancy (SV) and double vacancy (DV) defects (for details see Wang, et al., 2001). The letters of SW, SV and DV in their names are the defect class. The number following the defect class means the defect type, such as 5577, 585 and 555777. The suffix after the defect type means the direction of the defect. For the SW and DV defects, the direction of the defect is an important factor to enhance the β value (Liu et al., 2011). For example, the β value of SW5577-xz structure is 285.00×10^{-30} esu (near xz direction), which is much large than 4.58×10^{-30} esu of SW5577-*z* (the *z* direction along the tube axis). It is also interesting that unlike the vacancy defects, the topological defect can bring the largest β values (1900×10^{-30} esu for the SW57–75) as shown in Table 3.

A typical NLO molecule usually has a donor-π-acceptor (D-π-A) configuration. A similar type of combination has been exploited by using (6,0) nanotube as π-conjugated bridge while NH_2 and NO_2 as donor, acceptor groups, respectively (Xiao et al., 2008). Interestingly, the nanotube (6,0) not only functions as π-conjugation bridge but also as an acceptor leading to a donor-π-conjugated paradigm (NT(6,0)-NH_2) with β value as large as 1140×10^{-30} esu. The effect of donor orientation relative to nanotube is also a crucial factor to further tune its β amplitude. Additionally, two more effects including the addition of lithium atom at one end of a (5,0) nanotube (Xu et al., 2011) and the substitution of heteroatom (N) in (5,0) nanotubes (Sun et al. 2012) have also proved the effective tuning of β values in the SWCNTs.

9.5 TWO DIMENSIONAL GRAPHENES

Graphene was discovered in 2004 (Novoselove et al., 2004) and has since sparked immense potential to be a key ingredient of new devices such as single molecule gas sensors, ballistic transistors, and spintronic devices. Graphene has been called the "mother of all graphitic forms (Geim and Novoselove, 2007) because it can be wrapped into fullerenes, rolled into carbon nanotubes and stacked into graphite. Graphene consists of hexagonal arrangement of carbon atoms in two-dimensional honeycomb crystals. Graphene differs from most conventional three-dimensional materials. Basic graphene is a semimetal or zero-gap semiconductor. The nature of the zigzag edges imposes localization of the electron density with the maximum at the border carbon atoms leading to the formation of flat conduction and valence bands near the Fermi level. The localized states are spin-polarized and in case of ordering of the electron spin along the zigzag edges, graphene can be established in ferromagnetic or antiferromagnetic phases. The antiferromagnetic spin ordering of the localized states at the opposite zigzag edges breaks the sublattice symmetry of graphene that changes its band structure and opens a gap.

9.5.1 ELECTRO-OPTICAL PROPERTIES

The electronic properties of graphene, which is two-dimensional crystal of carbon atoms, are intriguing especially in the field of electro-optics (Hod et al., 2008).

Different computing strategies have been applied to tune the energetic, electronic and magnetic properties of graphene by spin reordering of the edge states, by changing the size of graphene nanoribbons, edge geometry as well as orbital hybridization at the edges. The edges of graphene are chemically active and prone to structural modifications thereby influencing the properties of graphene as well. Therefore, we discuss the influence of the changes brought by the functionalization of different groups on graphene to tune its first hyperpolarizability and the prospects of these manipulations in a controllable way.

Similar to the nanotubes, the graphene fragment, which are called graphene nanoribbons (GNRs) have the same unique hexagonal carbon lattice and remarkable π-conjugated structures with abundant π-electrons. Hence these are natural candidates for excellent conjugated bridges. Despite the similar π-conjugation, there is a distinct difference between CNTs and GNRs. Unlike a tubular CNT, GNR is a planar geometrical structure that is a more effective conjugated bridge to design high-performance NLO materials. And the planar geometry of GNR allows for the application of standard lithographic techniques for the flexible design of a variety of experimental devices. Recently, some experimental studies reported the nonlinear optical behavior of large populations of graphenes in organic solvent dispersions, implying a potential broadband optical-limiting application (Wang et al., 2009). For frequency doubling applications, the abundance of polarizable π-electrons in graphenes could yield structures with large first hyperpolarizability if the centro-symmetry is broken (Marder, 2006). For this purpose a donor-graphene nanoribbon-acceptor (D-GNR-A) framework have proved a successful motif with large first hyperpolarizabilities (Zhou et al., 2011). The D-GNR-A model has been successfully used as Conjugation Bridge to tune the first hyperpolarizability by Z. J. Zhou et al.

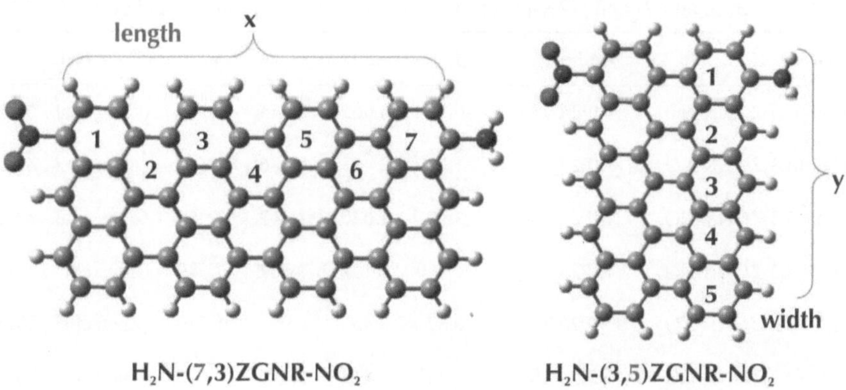

FIGURE 6 The representation of H_2N-(x, y)ZNGR-NO_2 systems with $x = 7$ and $y = 3$.

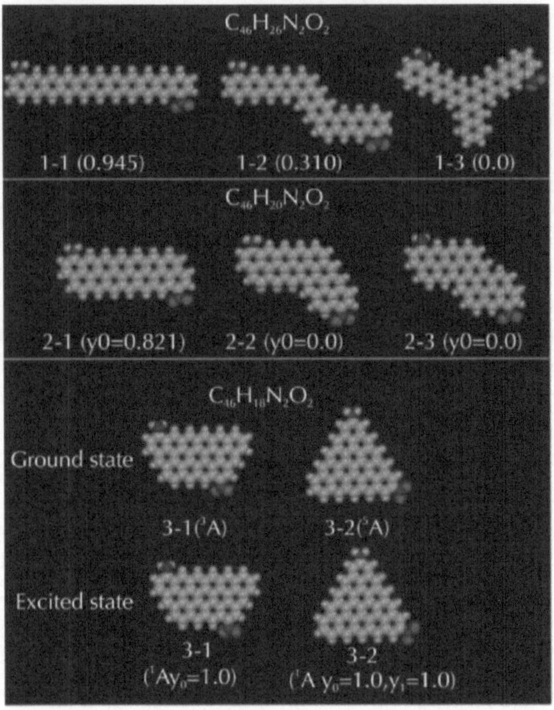

FIGURE 7 Geometrical structures of di-substituted GQDs, where light green, dark blue, red, and pink balls denote C, N, O, and H atoms, respectively. (Reprinted with permission from Zhou, 2011, Copyright American Chemical Society, 2011).

TABLE 4 First Hyperpolarizabilities ($\times 10^4$ a.u.)[a], oscillator strength, f_0, difference of dipole moment, $\Delta\mu$ [D], transition energy, ΔE [eV], at the CIS level for H2N–(x,3) ZGNR–NO2 (x=1, 3, 5, and 7) and H2N-(3, y) ZGNR-NO2 (y=2, 3, 4, and 5) and GQDs[b].

System	Method[c]	β	ΔE	f_0	$\Delta\mu$	Ref.
H_2N-(1,3)ZGNR-NO_2	MP2/CIS[d]	0.12	4.002	0.345	0.746	(Zhou et al., 2011)
H_2N-(3,3)ZGNR-NO_2	MP2/CIS	1.70	2.413	0.786	1.600	(Zhou et al., 2011)
H_2N-(5,3)ZGNR-NO_2	MP2/CIS	35.0	1.620	1.442	2.966	(Zhou et al., 2011)
H_2N-(7,3)ZGNR-NO_2	MP2/CIS	250.0	1.117	1.684	7.089	(Zhou et al., 2011)
H_2N-(3,2)ZGNR-NO_2	MP2/CIS	0.57	3.456	0.994	1.749	(Zhou et al., 2011)
H_2N-(3,3)ZGNR-NO_2	MP2/CIS	1.70	2.413	0.876	1.600	(Zhou et al., 2011)
H_2N-(3,4)ZGNR-NO_2	MP2/CIS	11.0	1.665	0.706	1.411	(Zhou et al., 2011)

TABLE 4 *(Continued)*

H_2N-(3,5)ZGNR-NO_2	MP2/CIS	85.0	1.161	0.553	1.487	(Zhou et al., 2011)
GQD **1–1** (^{1}A)	CAM-B3LYP	1.20	0.755	0.032	–	(Zhou et al., 2011)
GQD **1–2** (^{1}A)	CAM-B3LYP	0.69	2.857	0.193	–	(Zhou et al., 2011)
GQD **1–3** (^{1}A)	CAM-B3LYP	0.62	2.975	0.221	–	(Zhou et al., 2011)
GQD **2–1** (^{1}A)	CAM-B3LYP	5.0	1.029	0.310	–	(Zhou et al., 2011)
GQD **2–2** (^{1}A)	CAM-B3LYP	1.50	1.813	1.365	–	(Zhou et al., 2011)
GQD **2–3** (^{1}A)	CAM-B3LYP	2.0	1.714	1.832	–	(Zhou et al., 2011)
GQD **3–1** (^{3}A)	CAM-B3LYP	0.98	2.386	0.652	–	(Zhou et al., 2011)
GQD **3–1** (^{1}A)	CAM-B3LYP	5.60	1.708	0.313	–	(Zhou et al., 2011)
GQD **3–2** (^{5}A)	CAM-B3LYP	0.27	2.493	0.029	–	(Zhou et al., 2011)
GQD **3–2** (^{1}A)	CAM-B3LYP	17.0	0.556	0.106	–	(Zhou et al., 2011)

[a] (1 a.u. of $\beta = 8.6392 \times 10^{-33}$ esu), [b]System names are according to Fig. 7, [c] 6–31G* basis set is used for all methods, [d] f_0, $\Delta\mu$, and ΔE for H_2N-(x, y)ZGNR-NO_2 are calculated with CIS method

The planer conformation of GNR proved to be very effective π-conjugation bridge between donor (NH_2) and acceptor (NO2) groups substituted at its zigzag edges. The D-ZGNR-A with a planer structure showed β amplitude as large as 250×10^4 a.u. for H_2N-(7,3)ZNGR-NO_2 system as shown in Fig. 6. The size effect playes a crucial role to increase the β amplitude for H_2N-GNR-NO_2, in which effect of GNR perpendicular to D-A direction is superior to length effect along D-A direction. For example in H_2N-(x,3)ZGNR–NO_2 (x =1, 3, 5, and 7) with the constant bridge width (y = 3), the β amplitude greatly increases as the bridge length (x) increases (*see* Table 4). The bridging of ZGNR especially in H_2N-(7,3) ZGNR–NO_2 is also found to be superior with an order of magnitude larger β values than previously reported D-A-π conjugated systems like H_2N-$(CH{=}CH)_6$–NO_2 (Champagne et al., 2000) and H_2N-(6,0)ZCNT (Xiao et al., 2008), having similar lengths of their π-conjugations.

Apart from GNR, graphene quantum dots (GQDs) have been also implemented in donor-acceptor-π conjugation combination (Zhou et al., 2011). It has been found that with same number of carbon atoms, changing the shape of GQDs changes their first hyperpolarizability as well as multiplicity. As shown in Table 4 and Fig. 7, the GQDs with close-shell singlet conformations (y_i = 0.00) have lower value of first hyperpolarizability. While conformations with trapezoidal 3–1 and triangle 3–2 geometries have shown largest β values of 5.6 $\times 10^4$ and 1.7×10^4 a.u. in their lower singlet spin states, respectively.

It is very important to know the reason of larger first hyperpolarizabilities behind these all donor-π bridge-acceptor sytems. According to the perturbation theory, in the summation-overstates (SOS) expression, β is related to the excitation energies (E_{n0}), transition moments between states (μ_{mn}), and dipole moments differences between the ground and excitation states ($\mu_{nn}-\mu_{00}$) of a molecule as given by the following equation:

$$\beta_{ijk} = \frac{1}{2} P(i,j,k) \left\{ \sum_{n} \frac{\mu^{i}_{n0} \Delta\mu^{j}_{nn} \mu^{k}_{n0}}{E^{2}_{n0}} + \sum_{\substack{n,m \\ (m \neq n)}} \frac{\mu^{i}_{0n} \mu^{j}_{nm} \mu^{k}_{m0}}{E_{n0} E_{m0}} \right\} \tag{15}$$

where subscript "0" represents the ground state, μ_{n0} is transition moment between ground state (**0**) and ***n***th excited state (*n*), $\Delta\mu_{nn}$ is the difference between dipole moments of ground state (0) and *n*th excited state (*n*), E_{n0} is define as transition energy (E_{n0}=E_{n-} E_0) and $P(i, j, k)$ is a permutation operator. The excitation energies are found in the denominators of the SOS expression, while the transition moments and dipole moment differences appear in the numerators. The above expression indicates that (i) small excitation energies, (ii) large transition moments, and large dipole moment differences are required for large β amplitudes. As seen from Table 4, D-GNR-A and D-GQD-A systems with optimal combination of these three factors show larger β amplitudes than others with higher transition energies and lower transition moments and dipole moment differences.

9.6 CONCLUSIONS AND PROSPECTIVE

Carbon nano materials including zero-dimensional fullerenes, one-dimensional carbon nanotunes and two-dimensional graphene sheets are promizing to revolutionize several fields of material sciences and are major components of nanotecnology. Inspite of the huge amount of data available for carbon nano systems, we have spotlighted a relatively less explored feature of carbon nanosystems which is their possible use in nonlinear optical applications. In this chapter, we have briefly looked over different attempts to calculate the first and second hyperpolarizabilities in above-mentioned three types of carbon nanosystems. Nano-systems are found to be potential candidates for nonlinear optical application because of their extensive π-conjugation. Zero-dimensional fullerenes due to their higly symmetric structure (with inversion of symmetry) do not show first hyperpolarizability but have significant values of their sencond hyperpolarzability.

For C_{60} fullerene different doping stretegies have been adepted to increase the amplitude of second hyperpolarizability. Among these ferrocene and fullerene hybrid molecule especially $Fe(\eta_5\text{-}C_{55}B_5)^2$ compounds, which perhaps possess the largest π-conjugated systems with robustly large amplitude of second hyperpolarizability. The incorporation of a heteroatom inside the C_{60} fullerene also brings important changes in its electro-optical properties. Unlike the C_{60} fullerene, carbon nanotubes and graphenes have been attempted to integrate into D-π-A model for possible tuning of their first hyperpolarizabilties. The GNRs and GQDs having planer π-conjugations are found to be excellent conjugated bridges in D-conjugated bridge-A framework for designing high-performance NLO materials.

KEYWORDS

- ***ab initio***
- **carbon nanomaterials**
- **fullerenes**
- **quantum chemical methods**
- **wavefunction**
- **isolated pentagon rule**
- **carbon nanotubes**
- **sp^2 hybridization**
- **graphene**

REFERENCES

Ahmen, A. D.; Boudreaux, E. A. *Inorg. Chem.* **1973,** *12,* 1590–1597.

Ajayan, P. M. Chemical Reviews, **1999,** *99(7),* 1787–1799.

Anisimov V. M.; Lamoureux, G.; Vorobyov, I. V. *J. Chem. Theo. Comp.,* **2004,** *1,* 153–168.

Bakowies, D.; Thiel, W. *J. Am. Chem. Soc.*, **1991,** *113,* 3704.

Balch, A. L.; Olmstead, M. M. *Chem. Rev.* **1998,** *98,* 2123–2165.

Britz, D. A.; Khlobystov, A. N. *Chem. Soc. Rev.*, **2006,** *35(7),* 637–659.

Champagne, B.; Kirtman, B. In *Handbook of Advanced Electronic*, Photonic Materials, Devices, H. S. Nalwa Ed.; Academic Press: New York, USA, **2001,** Vol *9,* Chap.2, p. 63.

Champagne, B.; Perp_te, E. A.; Jacquemin, D.; van Gisbergen, S. J. A.; Baerends, E.-J.; Soubra-Ghaoui, C.; Robins, K. A.; Kirtman, B. *J. Phys. Chem. A* **2000,** *104,* 4755–4763.

Chiu, Y. N.; Ganelin, P.; Jiang, X.; Wang, B. C. J. Mol. Struct-THEOCHEM, **1994,** *312,* 215.

Cizek, J. *J. Chem. Phys.* **1966,** *45,* 4256.

Cote, M.; Grossman, J. C.; Cohen, M. L.; Louie, S. G. *Phys. Rev. Lett.* **1998,** *81,* 697–700.

David, W. I. F.; Ibberson, R. M.; Matthewman, J. C.; Prassides, K.; Dennis, T. J. S.; Hare, J. P.; Kroto, H. W.; Taylor, R.; Walton, D. R. M. *Nature* London, **1991,** *353,* 147.

Diaz-Tendero, S.; Martin, F.; Alcami, M. *Chem. Phys. Chem.,* **2005,** *6,* 92.

Dresselhaus, M.S.; Dresselhaus, G.; Eklund, P.C. Science of Fullerenes and Carbon Nanotubes, Academic Press, New York, 1996.

Forro L.; Mihaly, L. *Rep. Prog. Phys.* **2001,** *64,* 649–699.

Fowler, P.W.; Manolopoulos, D. E. *An Atlas of Fullerenes*, Oxford University Press, New York, 1995.

Geim, A. K.; Novoselov, K. S. *Nature Materials*, **2007,** *6,* 183.

Grossman, J. C.; Cote, M.; Louie, S. G.; Cohen, M. L. *Chem. Phys. Lett.,* **1998,** *284,* 344.

Haser, M.; Almlof, J.; Scuseria, G. E. *Chem. Phys. Lett.* **1991,** *181,* 497–500.

Head-Gordon, M.; Pople, J. A. M. J. *Chem. Phys. Lett.* **1988,** *153,* 503.

Hedberg, K.; Hedberg, L.; Bethune, D. S.; Brown, C. A.; Dorn, H. C.; Johnson, R. D.; de Vries, M. *Science* **1991,** *254,* 410–412.

Hod, O.; Barone, V.; Scuseria, G. E. *Phys. Rev. B.,* **2008,** *77,* 035–411.

Iijima, S. *Nature*, **1991,** *354,* **6348,** 56–58.

Iijima, S.; Ichihashi, T. *Nature*, **1993,** *363,* **6430,** 603–605.

Ikeda, K.; Saito, Y.; Hayazawa, N.; Kawata, S.; Uosaki, K. *Chem. Phys. Lett.,* **2007,** *438,* 109–112.

Ito, A.; Monobe, T.; Yoshii, T.; Tanaka. K. *Chem. Phys. Lett.,* **1999,** *315,* 348.

Jagadeesh, M. N.; Chandrasekhar, J. *Chem. Phys. Lett.,* **1999,** *305,* 298.

Jensen, L.; van Duijnen, P. T.; Snijders, J. G.; Chong, D. P. *Chem. Phys. Lett.,* **2002,** *359,* 524–529.

Karousis, N.; Tagmatarchis, N.; Tasis, D. *Chem. Rev.,* **2010,** *110,* 5366–5397.

Koch, W.; Holthausen, M. C. In *A Chemist's Guide to Density Functional Theory*, Wiley-VCH, Weinheim, Germany, 2000.

Kratschmer, W.; Lamb, L. D.; Fostiropoulos, K.; Huffman, D. R. *Nature*, London, **1990,** *347,* 354.

Kroto, H. W.; Heath, J. R.; Brien, S. C. O.; Curl, R. F.; Smalley, R. E. *Nature*, London, **1985,** *318,* 162.

Kroto, H.W. *Nature*, London. *329,* **1987,** 529.

Liu, S.; Lu, Y. J.; Kappes, M. M.; Ibers, J. A. *Science* **1991,** *254,* 408–410.

Liu, Y.-C.; Kan, Y.-H.; Wu, S.-X.; Yang, G.-C.; Zhao, L.; Zhang, M.; Guan, W.; Su, Z.-M. *J. Phys. Chem. A* **2008,** *112,* 8086–8092.

Liu, Z.-B.; Zhou, Z.-J.; Li, Z.-R.; Li, Q.-Z.; Jia, F.-Y.; Cheng, J.-B.; Sun, C.-C. *J. Mat. Chem.* **2011,** *21,* 8905–8910.

Marder, S. R. *Chem. Commun.* **2006,** 131–134.

Muhammad, S.; Fukuda, K.; Minami, T.; Kishi, R.; Shigeta, Y.; Nakano, M. *Chem. European J.* **2012,** DOI: 10.1002/chem.201203463.

Nakazawa, M.; Nakahara, S.; Hirooka, T.; Yoshida, M.; Kaino, T.; Komatsu, K. *Optics Letters*, vol. *31,* no. *7,* pp. 915–917, 2006.

Novoselov, K. S.; Geim, A. K.; Morozov, S. V.; Jiang, D.; Zhang, Y.; Dubonos, S. V.; Grigorieva, I. V.; Firsov, A. A. Science, **2004**, *306,* 666.

Piskoti, C.; Yager, J.; Zettl, A. *Nature*, **1998**, *393,* 771.

Prinzbach, H.; Weiler, A.; Landenberger, P.; Wahl, F.; Worth, J.; Scott, L. T.; Gelmont, M.; Olevano, D.; Issendorff, B. V. *Nature*, **2000,** *407,* 60.

Robertson, N.; McGowan, C. A. Chem. Soc. Rev., **2003,** *32(2),* 96–103.

Sackers, E.; Osswald, T.; Weber, K.; Keller, M.; Hunkler, D.; Woerth, J.; Knothe, L.; Prinzbach, H. *Chem. Eur. J.*, **2006,** *12,* 6242.

Schmalz, T. G.; Zeitz, W. A.; Klein, D. J.; Hite, G. E. *J. Am. Chem. Soc.* **1988,**110,111.

Schultz, D.; Droppa, R.; Alvarez, F. M. C. dos Santos, *Phys. Rev. Lett.*, **2003,** *90,* 15501.

Schultz, D.; Droppa, R.; Alvarez, F. M. C. dos Santos, *Phys. Rev. Lett.*, **2003,** *90,* 015501.

Schwedhelm, R.; Kipp, L.; Dallmeyer, A.; Skibowski, M. *Phys. Rev. B.*, **1998,** *58,* 13176.

Scuseria, G. E. *Chem. Phys. Lett.* **1991,** *176,* 423–427.

Song, J.; Parker, M. G.; Schoendorff, A.; Kus, M.; Vaziri, J. Mol. Struct- HEOCHEM, **2010,** *942,* 71.

Sun, G. Y.; Nicklaus, M. C.; Xie, R. H. *J. Phys. Chem. A,* **2005,** *109,* 4617.

Sun, S. L.; Y Hu, Y.; Xu, H. L.; Su, Z. M.; Hao, L. Z. *J. Mol. Model.* **2012,**18, 3219–3225.

Talapatra et al., G. B. *J. Phys. Chem.* **1992,** *96,* 5206.

Talyzin, A. V. *J. Phys. Chem. B.* **1997,** *101,* 9679.

Tang, S.-W.; Feng, J.-D.; Qiu, Y.-Q.; Sun, H.; Wang, F.-D.; Chang, Y.-F.; Wang, R.-S. *J. Comp. Chem.* **2010,** *31,* 2650–2657.

Thilgen, C.; Diederich, F. *Top. Curr. Chem.* **1999,** *199,* 135–171.

Valsakumar, M. C.; Subramanian, N.; Yousuf, M.; Ch Sahu, P.; Hariharan, Y.; Bharati, A.; Sastry, V. S.; Janaki, J.; Rao, G. V. N.; Radhakrishnan, T. S.; Sundar, C. S. Phys. Rev. B., **1993,** *48,* 9080.

Van Gisbergen, S. J. A.; Snijders, J. G.; Baerends, E. J. *Phys. Rev. Lett.* **1998,** *78,* 3097–3100.

Wang, J.-Y.; Lin, C.-S.; Zhang, M.-Y.; Chai, G.-L.; Cheng, W.-D. *Int. J. Quant. Chem* **2012,** *112,* 759–769.

Wang, J.; Hernandez, Y.; Lotya, M.; Coleman, J. N.; Blau, W. J. *Adv. Mater.* **2009,** *21,* 2430–2435.

Wang, Z. X.; Ke, X. Z.; Zhu, Z. Y.; Zhu, F. Y.; Ruan, M. L.; Chen, H.; Huang, R. B.; Zheng, L. D. *Phys. Lett. A*, **2001,** *280,* 351.

Xenogiannopoulou, E.; Medved, M.; Iliopoulos, K.; Couris, S.; Papadopoulos, M. G.; Bonifazi, D.; Sooambar, C.; Mateo-Alonso, A.; Prato, M. *Chem. Phys. Chem.* **2007,** *8,* 1056–1064.

Xiao, D.; Bulat, F. A.; Yang, W.; Beratan, D. N. *Nano Letters* **2008,** *8,* 2814–2818.

Xiao, J.; Lin, M.; Chiu, Y. N.; Fu, M.; Lai, S. T.; Li, N. N. J. Mol. Struct-THEOCHEM, **1998,** *428,* 149.

Xie, S. Y.; Gao, F.; Lu, X.; Huang, R. B.; Wang, C. R.; Zhang, X.; Liu, M. L.; Deng, S. L.;. Zheng, L. S. *Science*, **2004,** *304,* 699.

Xu, H.-L.; Zhong, R.-L.; Sun, S.-L.; Su, Z.-M. *J. Phys. Chem. C* **2011,** *115,* 16340–16346.

Yannoni, C. S.; Bernier, P. P.; Bethune, D. S.; Meijer, G.; Salem, J. *Am. Chem. Soc.* **1991,** *113,* 3190–3192.

Yeo-Heung, Y.; Miskin, A.; Kang, P. J. *Intel. Mater. Sys. Struc.*, **2006**, *17*, 107–116.

Zhou, X.; Liu, J.; Jin, Z.; Gu, Z.; Wu, Y.; Sun, Y. *Fullerene Sci. Technol.* **1997**, *5,* 285.

Zhou, Z. J.; Li, X.P.; Ma, F.; Liu, Z.B.; Li, Z. R.; Huang, X. R.; Sun, C. C. *Chem. Eur. J.* **2011,** *17,* 2414–2419.

Zhou, Z. J.; Liu, Z. B.; Li, Z. R.; Huang, X. R.; Sun, C. C. *J. Phys. Chem. C* **2011,** *115,* 16282–16286.

CHAPTER 10

MODERN DENSITY FUNCTIONAL THEORY: A USEFUL TOOL FOR THE COMPUTATIONAL STUDY OF NANOTECHNOLOGY

REINALDO PIS DIEZ

CONTENTS

Abstract ... 334
10.1 Introduction ... 334
10.2 The Early Days ... 335
10.3 The 1960s – The Birth of Modern Density Functional Theory ... 336
10.4 The 1980s – The Local Density Approximation, the First Rung in the Jacob's Ladder ... 339
10.5 The Generalized Gradient Approximation, the Second Rung in the Jacob's Ladder ... 342
10.6 The Meta Generalized Gradient Approximation, the Third Rung in the Jacob's Ladder ... 347
10.7 The Hybrid (Meta) Generalized Gradient Approximation, the Fourth Rung in the Jacob's Ladder ... 353
10.8 The Double-Hybrid (Meta) Generalized Gradient Approximation, the Fifth Rung in the Jacob's Ladder ... 359
10.9 The Performance of Modern Density Functionals to Predict Properties of Interest to Chemists ... 360
10.10 Conclusions ... 366
Acknowledgments ... 367
Keywords ... 368
References ... 368

ABSTRACT

Since the 1980s, the Density Functional Theory has evolved to become nowadays the most used tool for the calculation of electron structure of molecules, clusters and solids at a first-principles level of theory. In this chapter, the formal evolution of the theory from ancient local models to the modern hybrid meta generalized gradient approximations and double hybrid generalized gradient approximations is discussed. Moreover, the success and limitations of current Density Functional Theory in the prediction of a variety of properties, such as geometries, energies and thermodynamic functions, as well as in the interpretation of spectroscopic data are commented from a critical point of view.

10.1 INTRODUCTION

The first attempts to develop a theory of the electronic structure of atoms and molecules based exclusively on the electron density can be traced back to the 1920s and 1930s. The statistical nature of such attempts, however, led to very poor results for systems containing a few electrons, specially when they were compared with results obtained within the framework of Hartree-Fock and post Hartree-Fock methodologies.

Despite some efforts to improve the methods mentioned above by including the gradient of the electron density, it was in the mid 1960s that the formal framework for a theory of the electronic structure of atoms and molecules, in which the electron density is the basic variable, was developed. The pioneer work of Hohenberg, Kohn and Sham (Hohenberg and Kohn, 1964; Kohn and Sham, 1965) gave place to the birth of what it is known nowadays as Modern Density Functional Theory (DFT).

However, and due to the lack of efficient and reliable correlation functionals, the routine use of DFT by theoretical chemists was delayed up to early 1980s. From those days, DFT has exhibited an enormous progress both from methodological and practical point of views. It is completely safe to state that DFT is currently the most used computational tool both for theoretical chemists and for experimental researchers that look for an elegant explanation for their empirical results.

In 2000, Perdew (Perdew and Schmidt, 2001) used the concept of the Jacob's Ladder found in the Bible to make an analogy with the development of DFT from the earliest models, based only on the electron density, to the modern approaches making explicit use of orbitals. Thus, the most basic approach to the electronic structure problem within the DFT is equivalent to the first rung in the Ladder, and involves the electron density only. To ascend to the second rung, the gradient of the electron density must be added to the calculations. The inclusion of the kinetic energy density (instead of the Laplacian of the electron density) allows the Ladder's user to climb to the third rung. The approach that gives place to the fourth rung includes the occupied orbitals through a given amount of the exact exchange energy, that is, the one that is obtained from Hartree-Fock. Finally, the inclusion of unoccupied orbitals by means of perturbation theory allows the user to reach the Heaven of Chemical Accuracy after climbing the fifth rung.

In the present chapter, the Jacob's Ladder analogy is used as a guide to show the evolution of DFT in the last 50 years. Special emphasis is laid on the pros and cons of the theory in the prediction of those properties of interest for chemists. It is important to make clear at this point that the amount of existing density functionals is enormous. Thus, only a few functionals, representative of every rung in the Ladder, will be described in detail. The election of such functionals is rather arbitrary, though, but it is mainly based on the contribution of the most important researchers in the field according to the author's viewpoint.

10.2 THE EARLY DAYS

In 1927, Thomas (Thomas, 1927) and Fermi (Fermi, 1927), independently and simultaneously, developed the mathematical expression for the kinetic energy of a set of independent particles in terms of their density and in form of a *functional* of the density.

$$T_{TF}[\rho] = C_K \int \rho^{5/3}\, d\mathbf{r}, \tag{1}$$

where ρ is a shorthand notation for $\rho(\mathbf{r})$ and $C_K = 3(3\pi^2)^{2/3}/10$. The electrostatic terms adopt a classic form,

$$E_{Ne}[\rho] = \sum_{\alpha} \int Z_{\alpha} \frac{\rho}{|\mathbf{R}_{\alpha} - \mathbf{r}|} d\mathbf{r}, \tag{2}$$

and

$$J[\rho] = \frac{1}{2} \int \frac{\rho\rho'}{|\mathbf{r} - \mathbf{r}'|} d\mathbf{r}d\mathbf{r}', \tag{3}$$

where Z_{α} and R_{α} are the atomic number and coordinates of nuclei.

The three Eqs. given above form the first attempt to define a theory to calculate electronic structure: the Thomas-Fermi model.

$$E_{TF}[\rho] = T_{TF}[\rho] + E_{Ne}[\rho] + J[\rho]. \tag{4}$$

It is easily recognized that no exchange nor correlation terms are included in the Thomas-Fermi model. A few years later, Dirac (Dirac, 1930) proposed an exchange energy density functional.

$$K_D[\rho] = C_X \int \rho^{4/3}\, d\mathbf{r}, \tag{5}$$

where $C_X = -3(3/\pi)^{1/3}/4$. In this way, the electronic energy as a functional of the electron density could be written as

$$E_{TFD}[\rho] = T_{TF}[\rho] + E_{Ne}[\rho] + J[\rho] + K_D[\rho], \tag{6}$$

giving place to the Thomas-Fermi-Dirac (TFD) model. The TFD model performs very poor for atoms with a few electrons, with errors of about 40% with respect to Hartree-Fock results. Most important are the facts that no negative ions nor molecules exist within the TFD model due to the crude description of the kinetic energy.

The first correction to the TFD model was introduced by von Weizsäcker (von Weizsäcker, 1935), who improved slightly the kinetic energy functional with a term including the gradient of the electron density.

$$T_{TFvW}[\rho] = C_K \int \rho^{5/3}\, d\mathbf{r} + \frac{1}{8}\int \frac{|\nabla\rho|^2}{\rho}\, d\mathbf{r}. \tag{7}$$

Another important shortfall of the original TFD model is the absence of correlation effects. In spite of some efforts to include a correlation energy functional into the TFD or the TFvWD models, they have nowadays a historical interest only.

10.3 THE 1960S – THE BIRTH OF MODERN DENSITY FUNCTIONAL THEORY

In 1964, Hohenberg and Kohn (HK) (Hohenberg and Kohn, 1964) established by means of two theorems the basis of what it is known as Modern Density Functional Theory nowadays. In the first theorem, HK deal with a system formed by an arbitrary number of electrons moving under the influence of an external potential, $v(\mathbf{r})$, and the Coulomb repulsion. The external potential could be due to the presence of a nucleus or of a set of nuclei in the case of atoms or molecules, respectively, although it is not restricted to Coulombic potentials. The keystone in the

HK argument is that the electron density should be thought of as the independent variable in an electronic structure determination problem as it defines the external potential. Once the external potential is known, the Hamiltonian is fixed and all the properties of the system could be obtained after solving the Schrödinger equation. Thus, it is said that the external potential is a functional of the density.

In the second theorem, HK define an electronic energy density functional as:

$$E_{HK}[\rho] = F_{HK}[\rho] + \int v\rho\, d\mathbf{r}, \tag{8}$$

where F_{HK} includes the kinetic energy and the electron-electron interaction energy, and could be considered as a universal functional as it is independent of the external potential. HK state that the functional in Eq. (8) assumes its minimum value, say E_0, for the ground state density, ρ_0, which fixes in turn the external potential, v. Then, any trial density $\rho \neq \rho_0$, will deliver an electronic energy, E, that is an upper limit to E_0, $E \geq E_0$.

Unfortunately, Eq. (8) could not be used in practical calculations due to the uncertainty in the mathematical form of the universal functional, F_{HK}, beyond the obvious relationship:

$$F_{HK}[\rho] = T[\rho] + V_{ee}[\rho]. \tag{9}$$

One year later, Kohn and Sham (KS) (Kohn and Sham, 1965) proposed a solution to the problem on the uncertainty in F_{HK}. Indeed, KS proposed a smart solution to solve the problem of the uncertainty in the kinetic energy functional, $T[\rho]$. They invoked a reference system formed by N non-interacting electrons, described by an electron density ρ, under the influence of an external potential v. Thus, the HK electronic energy density functional could be written as:

$$E_{HK,s}[\rho] = T_s[\rho] + \int v\rho\, d\mathbf{r}. \tag{10}$$

The key ingredient in Eq. (10) is that due to the non-interacting nature of the system, the T_s functional could be written as:

$$T_s[\rho] = -\frac{1}{2}\sum_{i}^{N} \int \psi_i^* \nabla^2 \psi_i\, d\mathbf{r}, \tag{11}$$

where the set $\{\psi\}$ is a set of auxiliary functions that helps in the construction of the density ρ

$$\rho = \sum_{i}^{N} \psi_i \psi_i^* . \tag{12}$$

It should be stressed again that ψ is a shorthand notation for $\psi(\mathbf{r})$. The HK electronic energy density functional can now be written for the real system, formed by N interacting electrons, as:

$$E_{HK}[\rho] = T[\rho] + \int v\rho \, d\mathbf{r} + V_{ee}[\rho]. \tag{13}$$

If it is accepted that the density of the interacting system is the *same* as the density of the non-interacting system, then it is possible to rewrite Eq. (13) as:

$$E_{KS}[\rho] = T_s[\rho] + \int v\rho \, d\mathbf{r} + \frac{1}{2}\int \frac{\rho\rho'}{|\mathbf{r}-\mathbf{r}'|} d\mathbf{r}d\mathbf{r}' + E_{xc}[\rho], \tag{14}$$

where the subscript HK has been changed to KS to indicate that this new functional differs from the original one due to Hohenberg and Kohn. Furthermore, the electron–electron interaction term has been divided into the classical Coulomb term and the non classical term E_{xc} that describes exchange and correlation effects and also accounts for the error introduced when T_s replaces T,

$$E_{xc}[\rho] = V_{ee}[\rho] - \frac{1}{2}\int \frac{\rho\rho'}{|\mathbf{r}-\mathbf{r}'|} d\mathbf{r}d\mathbf{r}' + T[\rho] - T_s[\rho]. \tag{15}$$

The set of auxiliary functions that minimizes the KS functional, Eq. (14), can be found by writing the new functional,

$$\Omega[\{\psi\}] = E_{KS}[\rho] + \sum_{i}^{N}\sum_{j}^{N} \varepsilon_{ij} \left(\int \psi_i^* \psi_j \, d\mathbf{r} - \delta_{ij}\right), \tag{16}$$

where the set of Lagrange multipliers, $\{\varepsilon\}$, assures the orthonormality of the auxiliary functions. The condition:

$$\delta\Omega[\{\psi\}] = 0, \tag{17}$$

leads to a set of N one-particle Eqs. in canonical form:

$$\left[-\frac{1}{2}\nabla^2 + v_{ef}[\rho]\right]\psi_i = \varepsilon_i \psi_i, \; i = 1, \ldots, N \tag{18}$$

where the *effective* potential, v_{ef}, is:

$$v_{ef}[\rho] = v + \int \frac{\rho(\mathbf{r}')}{|\mathbf{r}-\mathbf{r}'|} d\mathbf{r}' + v_{xc}[\rho], \tag{19}$$

and $v_{xc}[\rho]$ is the *exchange-correlation* potential:

$$v_{xc}[\rho] = \frac{\delta E_{xc}[\rho]}{\delta \rho}. \tag{20}$$

It is evident from Eqs. (18) and (19) that the KS scheme must be solved self-consistently.

In summary, the merits of the proposal made by Kohn and Sham is that they solve the uncertainty in the kinetic energy density functional by invoking a reference, non interacting system of N electrons with density ρ, which in turn is the same as the density of the real, interacting system of N electrons. Furthermore, the KS Eqs. introduce both exchange and correlation effects at a computational cost similar to the Hartree or Hartree-Fock methods. However, KS did not propose any solution to the non-classical part of the electron-electron interaction. Moreover, they introduced a set of auxiliary functions that helps in modeling the electron density and must be obtained from a self-consistent procedure, loosing in this way the original appealing of a theory of the electronic structure of atoms and molecules based solely on the electron density.

The next step to put Eq. (14) to work is to give $E_{xc}[\rho]$ a practical mathematical form.

10.4 THE 1980S – THE LOCAL DENSITY APPROXIMATION, THE FIRST RUNG IN THE JACOB'S LADDER

It is common practice to separate the exchange contribution from the correlation contribution to $E_{xc}[\rho]$

$$E_{xc}[\rho] = E_x[\rho] + E_c[\rho], \tag{21}$$

to facilitate the modeling of the functionals needed to accomplish practical calculations. It is also accepted that those functional can be written in terms of the energy per particle o energy density

$$E_{xc}[\rho] = E_x[\rho] + E_c[\rho] = \int \varepsilon_x(\rho)\rho\, d\mathbf{r} + \int \varepsilon_c(\rho)\rho\, d\mathbf{r}. \tag{22}$$

The simplest form for the exchange energy density functional is the expression derived by Dirac, Eq. (5), that can be written in a more general form to account for spin polarized cases

$$E_x[\rho] = 2^{1/3} C_X \int \{\rho_\alpha^{4/3} + \rho_\beta^{4/3}\} d\mathbf{r}. \tag{23}$$

The first effort to model $E_c[\rho]$ can be traced back to 1972, when von Barth and Hedin (von Barth and Hedin, 1972) proposed a simple form for the correlation energy, in which the correlation energy per particle is:

$$\varepsilon_c^{vBH}(\rho,\xi) = \varepsilon^0(r_s) + [\varepsilon^1(r_s) - \varepsilon^0(r_s)] f(\xi), \tag{24}$$

where r_s is the Wigner-Seitz radius,

$$\frac{4}{3}\pi r_s^3 = \frac{1}{\rho}, \tag{25}$$

ξ is the spin polarization.

$$\xi = \frac{\rho^\alpha(\mathbf{r}) - \rho^\beta(\mathbf{r})}{\rho(\mathbf{r})}, \tag{26}$$

and the function $f(\xi)$ adopts the form:

$$f(\xi) = \frac{(1+\xi)^{4/3} + (1-\xi)^{4/3} - 2}{2^{4/3} - 2}. \tag{27}$$

In Eq. (24) the superscripts 0 and 1 indicates the spin compensated and ferromagnetic cases, respectively. von Barth and Hedin suggested the use of analytic expressions for $\varepsilon^0(r_s)$ and $\varepsilon^1(r_s)$ due to Hedin and Lundqvist (Hedin and Lundqvist, 1971).

In 1980, Ceperley and Alder (Ceperley and Alder, 1980) carried out a set of Quantum Monte Carlo simulations on the homogeneous electron gas to obtain the correlation energy for a wide range of densities. They presented their results in tabular form, however, and then, their use in real calculations was impractical due to the necessity of interpolation to obtain intermediate values of $E_c[\rho]$.

In the same year, Vosko, Wilk and Nusair (VWN) (Vosko et al., 1980) proposed an expression for the correlation energy density obtained from a random-phase approximation analysis:

$$\varepsilon_c^{VWN}(r_s,\xi) = \varepsilon_c^{VWN}(r_s,0) + \alpha_c(r_s)\frac{f(\xi)}{f''(0)}(1-\xi^4)$$

$$+[\varepsilon_c^{VWN}(r_s,1)-\varepsilon_c^{VWN}(r_s,0)]f(\xi)\xi^4, \tag{28}$$

where $\alpha_c(r_s)$ is the spin stiffness. VWN offer analytic expressions for $\varepsilon_c^{VWN}(r_s,0)$, $\varepsilon_c^{VWN}(r_s,1)$ and $\alpha_c(r_s)$, which were obtained from accurate fits to the numerical results of Ceperley and Alder.

$$G_c(r_s)=A\{\ln\frac{x^2}{X(x)}+\frac{2b}{Q}\tan^{-1}\frac{Q}{2x+b}-$$

$$\frac{bx_0}{X(x_0)}[\ln\frac{(x-x_0)^2}{X(x)}+\frac{2(b+2x_0)}{Q}\tan^{-1}\frac{Q}{2x+b}]\}, \tag{29}$$

where $G_c(r_s)$ is used to represent $\varepsilon_c^{VWN}(r_s,i)$ and $\alpha_c(r_s)$; $x=r_s^{1/2}$, $X(x)=x^2+bx+c$ and $Q=(4c-b^2)^{1/2}$ are functions, and A, x_0, b and c are parameters. The VWN correlation functional is accepted as one of the most accurate functional available for the homogeneous electron gas. The work of VWN presents various fits to the above equations. In particular, expressions III and V in reference (Vosko et al., 1980) are the ones more often found in different implementations of the functional, leading obviously to slightly different results. The authors recommend the use of formula V.

The combination of the Dirac exchange functional and the VWN correlation functional conforms what it is labeled nowadays as the Local Density Approximation (LDA) or Local Spin Density Approximation.

$$E_{xc}^{L(S)DA}=E_x^D+E_c^{VWN}. \tag{30}$$

The term *local* is used with a physical meaning in the sense that knowing the electron density at a given point then, it has the same value in any point around the reference one.

As expected, the L(S)DA fails when it is applied to inhomogeneous systems as atoms and molecules, providing energies larger than those obtained by post Hartree-Fock methods. Nevertheless, it is worth noting that equilibrium geometries are reasonably accurate, the bond lengths being slightly larger than experimental ones. Thus, L(S)DA geometries can be a good starting point for the study of large molecules.

As a rule of thumb, deviation of homogeneity can be considered non-negligible when $|\nabla\rho|/\rho$ is comparable to $(3\pi\rho)^{1/3}$. As this comparison is routinely found in atomic and molecular systems, it is mandatory to go beyond the L(S)DAto consider DFT on a competitive basis with respect to post Hartree-Fock methods.

Before analyzing possible improvements to the L(S)DA, it is very important to mention that the exact Hohenberg-Kohn functional remains unknown. Thus, all the functionals proposed to carry out practical calculations are mere approximations to it. This factindicates that the Variational Principle demonstrated by HK is no longer valid. In other words, the total electronic energy calculated within the KS scheme could approach the exact value well from above or from below. No systematic improvement can be achieved in modern DFT by the use of different functionals. The only possible enhancement concerns the improvement in the basis sets used in the calculations.

10.5 THE GENERALIZED GRADIENT APPROXIMATION, THE SECOND RUNG IN THE JACOB'S LADDER

It seems reasonable to account for the L(S)DA shortfall in describing inhomogeneous systems by including some function of the electron density gradient. The so-called *reduced density gradient* or *dimensionless density gradient*

$$s = \frac{|\nabla\rho|}{(24\pi^2)^{1/3}\rho^{4/3}}, \tag{31}$$

is used to this end.

It was shown at the end of Section 10.2 that the correction to the Thomas-Fermi kinetic energy density functional due to von Weiszäcker included the gradient of the electron density. It is then safe to say that von Wieszäcker was a pioneer in improving early local DFT by the inclusion of a density gradient.

The exchange energy density functional due to Dirac also possesses density gradient corrections.

$$E_x[\rho] = C_X \int \rho^{4/3} d\mathbf{r} + \beta \int \frac{|\nabla\rho|^2}{\rho^{4/3}} d\mathbf{r}, \tag{32}$$

where β is a constant that accepts both pure theoretical and empirical values (see reference (Parr and Yang, 1989) for a comparison of the different values). Unfortunately, the potential associated to the above correction diverges when $r \to \infty$.

In 1986, Perdew and Wang (PW) (Perdew and Wang, 1986, 1989) proposed for the first time the term Generalized Gradient Approximation (GGA) and presented a simple equation for the exchange energy density functional, in which the inhomogeneity is considered

$$E_x^{GGA}[\rho] = C_X \int \rho^{4/3} F^{GGA}(s) d\mathbf{r}, \tag{33}$$

where $F^{GGA}(s)$ is a function that accounts for some conditions that the exact exchange hole must satisfy and depends on the dimensionless density gradient. PW proposed a simple analytic form for $F^{GGA}(s)$

$$F^{GGA}(s) = (1 + 0.0864 s^2/m + bs^4 + cs^6)^m, \tag{34}$$

where *m*, and *b* and *c* are constants. This GGA exchange energy density functional is known as PW86.

In the same year, more precisely a few pages ahead in the same issue of the Physical Review B in which the PW86 exchange energy density functional was presented, Perdew (Perdew, 1986a, *b*) proposed some modifications to the correlation energy density functional due to Langreth and (Hu and Langreth, 1985, 1986; Langreth and Mehl, 1983; Langreth and Perdew, 1980). This correlation density functional is known as P86 and has the form:

$$E_c^{GGA}[\rho,\xi] = \int \varepsilon_c(\rho,\xi)\rho\, d\mathbf{r} + \int \frac{C(\rho)\exp(-\Phi)}{d}\frac{|\nabla\rho|^2}{\rho^{4/3}} d\mathbf{r}, \tag{35}$$

where $\varepsilon_c(\rho,\xi)$ is taken from another parameterization of Ceperley and Alder results for the correlation energy of the homogeneous electron gas (Perdew and Zunger, 1981). The other functions present in Eq. (35) are written as:

$$C(\rho) = 0.001667 + \frac{0.002568 + \alpha r_s + \beta r_s^2}{1 + \gamma r_s + \delta r_s^2 + 10^4 \beta r_s^3}, \tag{36}$$

$$\Phi = 0.19195 \frac{C(\infty)}{C(\rho)} \frac{|\nabla\rho|}{\rho^{7/6}}, \tag{37}$$

and

$$d = 2^{1/3}[(\frac{1+\xi}{2})^{5/3} + (\frac{1-\xi}{2})^{5/3}]^{1/2}, \tag{38}$$

where α, β, γ and δ are constants.

In 1988, Becke (Becke, 1988) concentrated his efforts in the development of an exchange energy density functional with the correct asymptotic behavior. His functional, known as B88, exhibits the following mathematical form:

$$E_x^{B88}[\rho] = C_X \int \rho^{4/3} F^{B88}(s)\, d\mathbf{r}, \tag{39}$$

where the inhomogeneity function is:

$$F^{B88}(s)=1-\beta\frac{s^2}{1+6\beta s\sinh^{-1}(s)}, \tag{40}$$

and β is a constant that best fits the Hartree-Fock exchange energy of the noble gases. The $\sinh^{-1}(x)$ function accepts a simple analytic expression, which facilitates its use in practical calculations.

$$\sinh^{-1}(x)=\ln[x+\sqrt{1+x^2}]. \tag{41}$$

The extension of Eqs. (39) and (40) to spin-polarized systems is straightforward.

In 1988, Lee, Yang and Parr (LYP) (Lee et al., 1988) used as starting point in their investigation an expression for the correlation energy due to Colle and Salvetti (CS) (Colle and Salvetti, 1975, 1983), which is based on the electron density and on the second-order reduced density matrix, both obtained from the Hartree-Fock wave function. LYP rewrote the original CS equations in terms of the local Hartree-Fock kinetic energy and a density-dependent term, which is similar to the von Weiszäcker correction, see Eq. (7). By expanding the Hartree-Fock kinetic energy about the Thomas-Fermi kinetic energy functional, LYP obtained three different formulas depending on the degree of the expansion. The most general expression is:

$$E_c^{LYP}[\rho]=-\int\frac{a}{1+d\rho^{-1/3}}(\rho+b\rho^{-2/3}[C_K\rho^{5/3}-2t_{vW}+$$

$$(\frac{1}{9}t_{vW}+\frac{1}{18}\nabla^2\rho)]\times\exp(-c\rho^{-1/3})d\mathbf{r}, \tag{42}$$

where a, b, c and d are constants and $t_{vW}=|\nabla\rho|^2/(8\rho)-\nabla^2\rho/8$. A similar equation was defined by LYP for open-shell systems.

Three possible formulas can be derived from Eq. (42). The first formula is called the second-order LYP correlation functional and agrees with Eq. (42). The zero-order formula can be easily obtained by discarding the two terms in parenthesis. Finally, a zero-order mean-path formula is obtained after replacing $(1/18)$ by $(-1/36)$ in Eq. (42).

In 1992, Perdew and Wang (Burke et al., 1997; Perdew et al., 1992) proposed a GGA version for both exchange and correlation energies, which is known as PW92 nowadays[1]. The exchange energy density functional is based on the B88 functional. Some modifications are introduced to fulfill bound and scaling conditions, giving place to the following inhomogeneity factor

$$F^{PW92}(s)=\frac{1+a\,s\sinh^{-1}(b\,s)+(c-d\exp(-100s^2))\,s^2}{1+a\,s\sinh^{-1}(b\,s)+e\,s^4}. \tag{43}$$

where *a, b, c, d* and *e* are constants. The correlation part of PW92 is far more involved as it defines a new inhomogeneity function, $H(t,r_s,\xi)$

$$E_c^{PW92}[\rho]=\int\rho(\varepsilon_c^{L(S)DA}(\rho,\xi)+H(t,r_s,\xi))\,d\mathbf{r}, \tag{44}$$

that depends on the Wigner-Seitz radius, the spin polarization and a new, scaled density gradient, *t*

$$t=\frac{|\nabla\rho|}{4g(\xi)(3/\pi)^{1/6}\rho^{7/6}}, \tag{45}$$

where *g* is a function of the spin polarization

$$g(\xi)=\frac{(1+\xi)^{2/3}+(1-\xi)^{2/3}}{2}. \tag{46}$$

The local part of the correlation energy per particle, $\varepsilon_c^{L(S)DA}(\rho,\xi)$, can be taken from any accurate fit to the Ceperley and Alder results for the homogeneous electron gas. Perdew and Wang proposed in fact an expression for $\varepsilon_c^{L(S)DA}(\rho,\xi)$ that becomes a simplification with respect to the VWN correlation functional. The *gradient-corrected* term is, in turn, divided into two contributions

$$H(t,r_s,\xi)=H_0(t,r_s,\xi)+H_1(t,r_s,\xi), \tag{47}$$

where $H_0(t,r_s,\xi)$ is developed to simulate the behavior in the limit of high densities, that is, when $r_s\to 0$

$$H_0(t,r_s,\xi)=g^3(\xi)\frac{\beta^2}{2\alpha}\ln\left[1+\frac{2\alpha}{\beta}t^2\frac{1+At^2}{1+At^2+A^2t^4}\right], \tag{48}$$

[1]Sometimes, it is also referred to as PW91.

where α is a constant, $\beta = 16(3/\pi)^{1/3}C(0)$, $C(\rho)$ has been defined in Eq. (36), and

$$A = \frac{2\alpha}{\beta}\frac{1}{\exp(-2\alpha\varepsilon_c(\rho,\xi)/g^3(\xi)\beta^2)-1}. \tag{49}$$

On the other hand, $H_1(t,r_s,\xi)$ is included to recover the correct result when $s \to 0$, and it is negligible unless $s \ll 1$,

$$H_1(t,r_s,\xi) = 16(\frac{3}{\pi})^{1/3}[C(\rho)-C(0)+0.0007144]g^3(\xi)t^2 \times$$

$$\exp(-400g^4(\xi)t^2/\left((3\pi^5)^{1/3}\rho^{1/3}\right)). \tag{50}$$

In 1996, Perdew, Burke and Ernzerhof (PBE) (Perdew et al., 1996a) recognize that the PW92 exchange-correlation functional has some problems, besides it incorporates inhomogeneity effects and obeys some conditions from the LDA. In particular, PW92 contains many parameters that are not joined smoothly, thus creating artifacts in the corresponding potentials in the regimes of small and large dimensionless density gradient. Then, instead of following the strategy of building up a density functional to fulfill asmany exact conditions as possible, PBE designed an exchange-correlation functional that satisfy only those conditions that are *energetically* significant.

The PBE exchange energy density functional is extremely simple in its mathematical form, the inhomogeneity factor given by

$$F(s) = 1+\kappa - \frac{\kappa}{1+\mu s^2/\kappa}, \tag{51}$$

where κ and μ are constants.

The PBE correlation functional shows the same form as the PW92 functional

$$E_c^{PBE}[\rho] = \int \rho(\varepsilon_c^{L(S)DA}(\rho,\xi) + H(t,r_s,\xi))\,d\mathbf{r}, \tag{52}$$

but the $H(t,r_s,\xi)$ function has only one term

$$H(t,r_s,\xi) = \gamma g^3(\xi)\ln[1+\frac{\beta}{\gamma}t^2\frac{1+At^2}{1+At^2+A^2t^4}], \tag{53}$$

where

$$A = \frac{\beta}{\gamma}\left[\frac{1}{\exp(-\varepsilon_c(\rho,\xi)/[\gamma g^3(\xi)]) - 1}\right], \tag{54}$$

where β and γ are constants.

The PW92 and PBE schemes are special cases within the GGA in the sense that can be invoked to model both the exchange energy and the correlation energy. The other functionals must be combined to get the desired DFT model, as they are not complete schemes. Thus, GGA DFT models such as B88P86 (or simply BP86) or B88LYP (or simply BLYP) were usually used some years ago and still are in use when cheap models are needed.

10.6 THE META GENERALIZED GRADIENT APPROXIMATION, THE THIRD RUNG IN THE JACOB'S LADDER

It has been mentioned before in this chapter that the consequence of the uncertainty in the form of the Hohenberg-Kohn functional is that no systematic improvement can be achieved as it is the case of post Hartree-Fock methodologies. Modern DFT is then based on the Kohn-Sham version of the DFT as it solves the problem of the uncertainty in the kinetic energy density functional. All the current efforts in methodological DFT are devoted to the modeling of accurate exchange-correlation energy density functionals.

The first two rungs in the Jacob's Ladder model the exchange-correlation energy density functional in terms of the electron density and its gradient

$$E_{xc}^{L(S)DA} = \int \varepsilon_{xc}^{L(S)DA}(\rho)\rho\, d\mathbf{r} \tag{55}$$

and

$$E_{xc}^{GGA} = \int \varepsilon_{xc}^{GGA}(\rho,\nabla\rho)\rho\, d\mathbf{r}, \tag{56}$$

where the explicit dependence of ε^{GGA} on both the electron density and its gradient in indicated.

It seems then reasonable to include higher-order derivatives of the electron density into the exchange-correlation energy per particle to go beyond the GGA, that is, to achieve a *meta* GGA (mGGA)

$$E_{xc}^{mGGA} = \int \varepsilon_{xc}^{mGGA}(\rho,\nabla\rho,\nabla^2\rho)\rho\, d\mathbf{r}. \tag{57}$$

However, Neumann and Handy (Neumann and Handy, 1997) have demonstrated that the inclusion of derivatives of the electron density beyond the density gradient to a general form like Eq. (57), leads to no significant improvements in the accuracy of the results.

In 1996, Becke (B95)[2] (Becke, 1996) proposed a correlation energy functional that included the term

$$D_\sigma = \sum_i^{occ} |\nabla \psi_{i,\sigma}|^2 - \frac{1}{4}\frac{|\nabla \rho_\sigma|^2}{\rho_\sigma}, \tag{58}$$

in the parallel-spin part, σ being both α and β. D_σ derives from the second-order expansion of the exact non interacting $\sigma\sigma$ pair density within the Kohn-Sham version of the DFT. Moreover, by recalling that Eq. (15), the definition of E_{xc} within the KS context, also contains information about the kinetic energy, it seems natural to include the kinetic energy density of the occupied KS auxiliary functions in ε_{xc}

$$E_{xc}^{mGGA} = \int \varepsilon_{xc}^{mGGA}(\rho, \nabla\rho, \tau)\rho \, d\mathbf{r}, \tag{59}$$

where $\tau = \sum_i^{occ} |\nabla \psi_i|^2$.

In 1998, Van Voorhis and Scuseria (VS98)[3] (Voorhis and Scuseria, 1998) developed an exchange energy density functional after expanding the reduced second-order density matrix, that enters the exact expression for the exchange energy, in terms of Bessel and Legendre polynomials. The VS98 exchange functional can be expressed as:

$$E_x^{VS98} = C_X \int \rho^{4/3} F^{VS98}(s,z)\, d\mathbf{r}, \tag{60}$$

where the inhomogeneity factor is:

$$F^{VS98}(s,z) = \frac{a}{\gamma(s,z)} + \frac{bs^2 + cz}{\gamma^2(s,z)} + \frac{ds^4 + es^2z + fz^2}{\gamma^3(s,z)}, \tag{61}$$

where $z = \tau/\rho^{5/3} - C_K$ and $\gamma(s,z) = 1 + \alpha(s^2 + z)$. The constants a to f and α were adjusted to fit an extensive set of molecular properties. Besides the development of the VS98 functional was originally devoted to the exchange energy, Van Voo-

[2] Although the paper was published in 1996, the functional is referred to as B95.
[3] This functional is also known as VSXC.

rhis and Scuseria proposed that the correlation energy functional could have an expression similar to that of Eq. (61), but accepting different terms for same-spin and opposite-spin electrons

$$E_c^{VS98} = E_c^{VS98,\alpha\beta} + \sum_\sigma E_c^{VS98,\sigma\sigma}, \tag{62}$$

and

$$E_c^{VS98,\sigma\sigma} = \int \rho \varepsilon_c^{L(S)DA} D_\sigma F^{VS98}(s,z)\, d\mathbf{r}$$

$$E_c^{VS98,\alpha\beta} = \int \rho \varepsilon_c^{L(S)DA} F^{VS98}(s,z)\, d\mathbf{r}, \tag{63}$$

where the function D_σ is similar to the one that appears in B95, see Eq. (58). In this way, the overall implementation of VS98 implies the evaluation of 21 parameters.

In 1999, Perdew, Kurth, Zupan and Blaha (PKZB) (Perdew et al., 1999) felt unhappy with *semiempirical* density functionals containing about 20 parameters. To remedy this, PKZB proposed to construct a mGGA exchange-correlation functional based on the philosophy of the PBE functional (Perdew et al., 1996a), that is, retaining all its good properties and adding some others to enhance its predictive power. The exchange part of the PKZB density functional has the form,

$$E_x^{PKZB} = C_X \int \rho^{4/3} F^{PKZB}(\rho,s,\tau)\, d\mathbf{r}, \tag{64}$$

where the inhomogeneity factor is:

$$F^{PKZB}(\rho,s,\tau) = 1 + \kappa - \frac{\kappa}{1+\frac{x}{\kappa}}, \tag{65}$$

and the function x is:

$$x = \frac{10}{81}s^2 + \frac{146}{2025}q^2 - \frac{73}{405}qs^2 + \left[D + \frac{1}{\kappa}\left(\frac{10}{81}\right)^2 \right] s^4, \tag{66}$$

where D is a constant set to 0.113 and the function q adopts the following form,

$$q = \frac{3\tau}{2(3\pi^2)^{2/3}\rho^{5/3}} - \frac{9}{20} - \frac{s^2}{12}. \tag{67}$$

For the correlation part of the PKZB becomes,

$$E_c^{PKZB} = \int \left\{ \rho \varepsilon_c^{PBE} \left[1 + C \left(\frac{\sum_\sigma \tau_\sigma^{vW}}{\sum_\sigma \tau_\sigma} \right)^2 \right] - (1+C) \sum_\sigma \rho_\sigma \varepsilon_{c,\sigma}^{PBE} \left(\frac{\tau_\sigma^{vW}}{\tau_\sigma} \right)^2 \right\} d\mathbf{r}, \tag{68}$$

where $\tau_\sigma^{vW} = \frac{1}{8} \frac{|\nabla \rho_\sigma|^2}{\rho_\sigma}$ and C is a constant set to 0.53. It can be seen that, despite the complexity of some formulas, the PKZB exchange-correlation density functional was formulated with only two adjustable parameters.

Interestingly, almost at the end of their work, PKZB state that although it is true that "there is no systematic way to construct density functional approximations… there are more or less systematic ways" to do so. The PBE and PKZB schemes demonstrate that.

In 2003, Tao, Perdew, Staroverov and Scuseria (TPSS) (Tao et al., 2003) proposed a revised version of the PKZB exchange-correlation density functional to overcome its failures in the prediction of some properties, notably bond lengths and hydrogen-bondedcomplexes. The philosophy to design the TPSS exchange-correlation energy density functional was to replace the empirical parameters D and C present in PKZB by other parameters chosen to enforce some exact conditions. Interestingly, the tests carried out by the authors on different sets of molecules indicated that the TPSS is indeed an improvement to the PKZB functional, with results similar to more sophisticated functionals containing a fraction of the exact exchange (see next section on hybrid GGA functionals). Also in their chapter, the authors coined the term *nested functionals* to mean that the L(S)DA is inside PBE and PBE is nested into TPSS. Consistent with the Jacob's Ladder concept, the authors, too, anticipate the evolution of TPSS to include a certain percentage of the exact exchange, thus climbing naturally to the fourth rung of the Ladder.

In 2005, Truhlar and co-workers presented the first member of a prolific family of exchange-correlation energy density functionals. They called this functional M05 (Zhao et al., 2005), but as it was designed to occupy the fourth rung in the Jacob's Ladder, the discussion is postponed to the next section. Now, the first mGGA member of that family is revised: the M06-L functional.

The exchange part of the M06-L functional (Zhao and Truhlar, 2006a) is based mainly on the VS98 functional

$$E_x^{M06-L} = \sum_\sigma \int \left\{ F_x^{PBE}(\rho_\sigma, \nabla \rho_\sigma) f(w_\sigma) + \varepsilon_x^{LSDA}(\rho_\sigma) h(s_\sigma, z_\sigma) \right\} d\mathbf{r}, \tag{69}$$

where $F_x^{PBE}(\rho_\sigma, \nabla \rho_\sigma)$ is given by Eq. (51) and $\varepsilon_x^{LSDA}(\rho_\sigma)$ is the Dirac exchange functional. The function $h(s_\sigma, z_\sigma)$ is identical to $F^{VS98}(s, z)$ given in Eq. (61) and

$$f(w_\sigma) = \sum_{i=0}^{m} a_i w_\sigma^i \tag{70}$$

$$w_\sigma = \frac{t_\sigma - 1}{t_\sigma + 1} \tag{71}$$

$$t_\sigma = \frac{2^{2/3} C_K \rho^{5/3}}{\tau}. \tag{72}$$

The correlation part of M06-L is divided in same-spin and opposite-spin contributions as in the VS98 case

$$E_c^{M06-L} = E_c^{M06-L,\alpha\beta} + \sum_\sigma E_c^{M06-L,\sigma\sigma}, \tag{73}$$

and

$$E_c^{M06-L,\alpha\beta} = \int e_{\alpha\beta}\left[g_{\alpha\beta}(s_\alpha, s_\beta) + h_{\alpha\beta}(s_{\alpha\beta}, z_{\alpha\beta})\right]d\mathbf{r} \tag{74}$$

$$E_c^{M06-L,\sigma\sigma} = \int e_{\sigma\sigma} D_\sigma \left[g_{\sigma\sigma}(s_\sigma) + h_{\sigma\sigma}(s_\sigma, z_\sigma)\right]d\mathbf{r}, \tag{75}$$

where

$$g_{\alpha\beta}(s_\alpha, s_\beta) = \sum_{i=0}^{n} c_i \left(\frac{\gamma_{\alpha\beta}(s_\alpha^2 + s_\beta^2)}{1 + \gamma_{\alpha\beta}(s_\alpha^2 + s_\beta^2)}\right)^i \tag{76}$$

$$g_{\sigma\sigma}(s_\sigma) = \sum_{i=0}^{n} c_i' \left(\frac{\gamma_{\sigma\sigma} s_\sigma^2}{1 + \gamma_{\sigma\sigma} s_\sigma^2}\right)^i, \tag{77}$$

the $h_{\alpha\beta}$ and $h_{\sigma\sigma}$ functions are the same as in the exchange functional, and $s_{\alpha\beta}^2 = s_\alpha^2 + s_\beta^2$ and $z_{\alpha\beta} = z_\alpha + z_\beta$. The D_σ function is a self-correlation factor

$$D_\sigma = 1 - \frac{s_\sigma^2}{4(z_\sigma + C_K)}. \tag{78}$$

Finally, the functions $e_{\alpha\beta}$ and $e_{\sigma\sigma}$ are taken from reference (Becke, 1996) and follow the analysis made by Stoll, Pavlidou and Press (Stoll et al., 1978).

$$e_{\alpha\beta} = \rho\varepsilon_c^{LSDA}(\rho_\alpha, \rho_\beta) - \rho_\alpha\varepsilon_c^{LSDA}(\rho_\alpha, 0) - \rho_\beta\varepsilon_c^{LSDA}(0, \rho_\beta) \quad (79)$$

$$e_{\sigma\sigma} = \rho_\sigma\varepsilon_c^{LSDA}(\rho_\sigma, 0). \quad (80)$$

The upper limit of the sum in Eq. (70) is set to 11, whereas the upper limit in the sums in Eqs. (72) and (75) is set to 4. Thus, the M06-L exchange-correlation energy density functional needs 42 parameters to beused in practical calculations. Besides the broad applicability of the functional to the chemistry of main-group elements, it is worth noting its good performance to describe the energetic of systems containing transition metals.

Very recently, Peverati and Truhlar (Peverati and Truhlar, 2012) proposed a new mGGA density functional, in which the exchange part is represented by a dual-range local functional. They called the functional M11-L.

The functional form of M11-L, in the spin-unpolarized version, can be written as:

$$E_{xc}^{M11-L} = E_x^{SR-M11-L} + E_x^{LR-M11-L} + E_c^{M11-L} \quad (81)$$

where the *short-range* part of the exchange energy density functional is:

$$E_x^{SR-M11-L} = \int \rho\varepsilon_x^{LDA}(\rho)G(\alpha)\left(f_1^{SR}(w)F_x^{PBE}(s) + f_2^{SR}(w)F_x^{RPBE}(s)\right)d\mathbf{r}, \quad (82)$$

and the inhomogeneity factors are the usual PBE exchange functional (Perdew et al., 1996a), and the revised version of the PBE functional, RPBE, proposed by Hammer, Hansen and Norskov (Hammer et al., 1999)to improve the description of adsorption processes. The range separation or attenuation function is expressed as:

$$G(\alpha) = 1 - \frac{8}{3}\alpha\left[\sqrt{\pi}\,\mathrm{erf}(\frac{1}{2\alpha}) - 3\alpha + 4\alpha^3 + (2\alpha - 4\alpha^3)\exp(-\frac{1}{4\alpha^2})\right], \quad (83)$$

where $\alpha = \omega/\left(2(6\pi^2\rho)^{1/3}\right)$ contains the *range-separation* parameter, ω (Chai and Head-Gordon, 2008). The *long-range* part of the exchange functional is similar in form to the short-range one.

$$E_x^{LR-M11-L} = \int \rho\varepsilon_x^{LDA}(\rho)(1 - G(\alpha))\left(f_1^{LR}(w)F_x^{PBE}(s) + f_2^{LR}(w)F_x^{RPBE}(s)\right)d\mathbf{r}. \quad (84)$$

The form of the correlation functional is quite similar to the PBE correlation functional, Eq. (52),

$$E_c^{M11-L} = \int \rho\left(f_3(w)\varepsilon_c^{L(S)DA}(\rho) + f_4(w)H(s,\rho)\right)d\mathbf{r}. \tag{85}$$

The six enhancement factors that depend on the kinetic energy density have the same functional form as the enhancement factors that appears in the development of M06-L, Eq. (70), and their argument, w, is given in Eq. (71). The upper limit in the sums that define the enhancement factors is set to 8. Considering the range-separation parameter as another adjustable parameter, the formulation of the M11-L contains 55 parameters.

10.7 THE HYBRID (META) GENERALIZED GRADIENT APPROXIMATION, THE FOURTH RUNG IN THE JACOB'S LADDER

The fourth rung in the Jacob's Ladder accounts for the occupied Kohn-Sham auxiliary functions by means of the inclusion of the exact exchange energy, the one that it is defined in the Hartree-Fock theory. Interestingly, the density functionals that constitute the fourth rung of the Jacob's Ladder were proposed a few years before the mGGA functionals appeared in the DFT realm.

The first formal discussion on the possibility to include an explicit orbital dependence on the exchange-correlation functional was raised in 1993 and is due to Becke (Becke, 1993a). Considering the KS version of the DFT as starting point, the HK universal functional $F[\rho]$ can be written in terms of the constrained search formulation due to Levy (Levy, 1979)

$$F^{\lambda}[\rho] = \int \left(\Psi_{\rho}^{min,\lambda}\right)^{*}\left(\hat{T}_s + \lambda\hat{V}_{ee}\right)\Psi_{\rho}^{min,\lambda}\,d\mathbf{r} \tag{86}$$

where $\Psi_{\rho}^{min,\lambda}$ is the N-electron wave function that minimizes $F^{\lambda}[\rho]$ and leads to the density ρ, which, in turns, remains constant when λ goes from 0 to 1. $\hat{T}_s$ and $\hat{V}_{ee}$ are the operators that define the corresponding density functionals $T_s[\rho]$ and $V_{ee}[\rho]$, respectively, where $V_{ee}[\rho] = J[\rho] + E_{xc}[\rho]$ as can be deduced from Eq. (15). Moreover,

$$E_{xc}[\rho] = E_{xc}^{'}[\rho] + T[\rho] - T_s[\rho], \tag{87}$$

where $E_{xc}^{'}[\rho]$ represents *pure* exchange-correlation effects. From the above equations, it is clear that both the non-interacting reference system and the interacting system of the KS scheme are closely related by the value of theparameter λ

$$F^{0}[\rho] = T_s[\rho] \tag{88}$$

$$F^1[\rho] = T_s[\rho] + J[\rho] + E_{xc}[\rho], \tag{89}$$

where the density, ρ, is constructed from the N-electron wave functions $\Psi_\rho^{min,0}$ and $\Psi_\rho^{min,1}$, respectively.

From the Hellmann-Feynman theorem it is easily seen that

$$\frac{\partial F^\lambda[\rho]}{\partial \lambda} = \int \left(\Psi_\rho^{min,\lambda}\right)^* \hat{V}_{ee} \Psi_\rho^{min,\lambda}\, d\mathbf{r}, \tag{90}$$

that can be written after integration as:

$$F^1[\rho] - F^0[\rho] = \int_0^1 \int \left(\Psi_\rho^{min,\lambda}\right)^* \hat{V}_{ee} \Psi_\rho^{min,\lambda}\, d\mathbf{r}\, d\lambda. \tag{91}$$

Inserting Eqs. (88) and (89) in Eq. (91) it is readily obtained.

$$E_{xc}[\rho] = \int_0^1 \int \left(\Psi_\rho^{min,\lambda}\right)^* \hat{V}_{ee} \Psi_\rho^{min,\lambda}\, d\mathbf{r}\, d\lambda - J[\rho], \tag{92}$$

that can be written more concisely as:

$$E_{xc}[\rho] = \int_0^1 E_{xc}^\lambda[\rho]\, d\lambda. \tag{93}$$

In his work, Becke proposed to use a sort of linear interpolation to solve the integral

$$E_{xc}[\rho] \approx \frac{1}{2}\left(E_{xc}^0[\rho] + E_{xc}^1[\rho]\right), \tag{94}$$

and, most important, related $E_{xc}^0[\rho]$ to the pure exchange energy that can be obtained from a single Slater determinant. Thus, Becke suggested to use the KS auxiliary functions for the calculation of the *exact* exchange, that is, using the Hartree-Fock expression. On the other hand, the term $E_{xc}^1[\rho]$ contains both exchange and correlation effects and can be estimated from the L(S)DA. Thus, the first *hybrid* density functional, known as *half-and-half*, has an exchange-correlation density functional of the form,

$$E_{xc}^{H\&H}[\rho] = \frac{1}{2}\left(E_x^{HF} + E_{xc}^{L(S)DA}[\rho]\right). \tag{95}$$

In the same year, the same author, Becke (B3) (Becke, 1993b), suggested to extend the half-and-half formula to include the electron density gradient. The new expression for the exchange-correlation energy functional contained three semiempirical coefficients, a_0, a_x and a_c

$$E_{xc}^{B3PW92} = a_0 E_x^{HF} + (1 - a_0) E_x^{L(S)DA} + a_x \Delta E_x^{B88} + E_c^{L(S)DA} + a_c \Delta E_c^{PW92}, \tag{96}$$

where ΔE_x^{B88} is the GGA correction to the exchange energy proposed by Becke, see Eqs. (39) and (40), and ΔE_x^{PW92} is the GGA correction to the correlation energy due to Perdew and Wang, see the discussion following Eq. (44).

In 1994, Stephens and co-workers (Stephens et al., 1994) suggested the use of the LYP correlation energy density functional in the hybrid GGA of Becke instead of the PW92 functional. Moreover, due to the impossibility to get a local component from LYP, Eq. (96) is slightly modified to give:

$$E_{xc}^{B3LYP} = a_0 E_x^{HF} + (1 - a_0) E_x^{L(S)DA} + a_x \Delta E_x^{B88} + (1 - a_c) E_c^{VWN} + a_c E_c^{LYP}, \tag{97}$$

where the values for the semiempirical coefficients are the same as in the B3PW92 case. B3LYP became one of the most popular density functionals adopted by the community of computational chemists. In the first years of the new century, B3LYP represented about 80% of the overall amount of citations involving any density functional (Souza et al., 2007). In the last two years, the popularity poll of density functionals indicated that B3LYP occupies the second place in a list of 40 functionals further separated in two groups (Swart et al., 2011).

In 1997, Becke (B97) (Becke, 1997) proposed a new hybrid GGA with only one parameter affecting the exact exchange.

$$E_{xc}^{B97} = c_x E_x^{HF} + E_x^{B97} + E_c^{B97}. \tag{98}$$

Moreover, he also suggested a simplification in the form of the exchange and correlation functionals, inaugurating the philosophy of functionals design based on a *massive* parameterization.

$$E_x^{B97} = \sum_\sigma \int \rho_\sigma \varepsilon_{x,\sigma}^{LSDA}(\rho_\sigma) g_{x,\sigma}(s_\sigma^2) d\mathbf{r}, \tag{99}$$

$$E_c^{B97,\alpha\beta} = \int e_{\alpha\beta}^{LSDA}(\rho_\alpha, \rho_\beta) g_{c,\alpha\beta}(s_\alpha^2, s_\beta^2) d\mathbf{r}, \tag{100}$$

$$E_c^{B97,\sigma\sigma} = \int e_{\sigma\sigma}^{LSDA}(\rho_\sigma) g_{c,\sigma\sigma}(s_\sigma^2) d\mathbf{r}, \tag{101}$$

and

$$E_c^{B97} = E_c^{B97,\alpha\beta} + \sum_\sigma E_c^{B97,\sigma\sigma}. \tag{102}$$

The functions $e_{\alpha\beta}^{LSDA}$ and $e_{\sigma\sigma}^{LSDA}$ adopt the form shown in Eqs. (79) and (80), and the functions $g_{x,\sigma}$, $g_{c,\alpha\beta}$ and $g_{c,\sigma\sigma}$ share the same functional form

$$g(s^2)=\sum_{i=0}^{m} c_i u^i(s^2), \tag{103}$$

where $u(s^2)=\gamma s^2/(1+\gamma s^2)$. When g depends on both s_α^2 and s_β^2, a single average is formed, $s_{avg}^2=(s_\alpha^2+s_\alpha^2)/2$, and the γ values are fixed from atomic calculations. Becke found that a value of $m=2$ provides reasonable results. Thus, the new hybrid GGA functional contains *only* 10 parameters.

In 1999, Adamo and Barone (Adamo and Barone, 1999) proposed a hybrid GGA density functional based on the PBE GGA density functional. As the new functional does not contain any new additional parameter, except for the parameters already present in the PBE functional, it was known as PBE0[4]

$$E_{xc}^{PBE0}=\frac{1}{4}E_x^{HF}+\frac{3}{4}E_x^{PBE}+E_c^{PBE}. \tag{104}$$

The coefficients for the exact exchange energy and the PBE exchange energy were obtained by Perdew, Ernzerhof and Burke (Perdew et al., 1996b) from the lowest order in perturbation theory that can yield reliable atomization energies for typical molecules.

Meta GGA density functionals also have their hybrid versions. As mentioned in Section 10.6, Truhlar and co-workers presented in 2005 the first member of a family of exchange-correlation energy density functionals. They called this functional M05 (Zhao et al., 2005), and it was classified as hybrid meta GGA. As usual, only a fraction of the exact exchange energy is mixed with the exchange energy density functional

$$E_x^{M05}=\left(1-\frac{X}{100}\right)E_x^{mGGA}+\frac{X}{100}E_x^{HF}, \tag{105}$$

where E_x^{mGGA} is a simplified version of E_x^{M06-L}, Eq. (69)[5],

$$E_x^{mGGA}=\sum_{\sigma}\int F_x^{PBE}(\rho_\sigma,\nabla\rho_\sigma)f(w_\sigma)d\mathbf{r}, \tag{106}$$

[4]This hybrid GGA functional is also known as PBE1PBE.

[5]Although the M05 family of functionals was prior to the M06-L functional, they belong to different rungs in the Jacob's Ladder. This is why the M06-L functional, a meta GGA one, was presented in the previous section.

and $f(w_\sigma)$ is given by Eqs. (70) to (72). The correlation energy density functional also takes a form very similar to that of the M06-L method, that is, there are different functionals for the opposite-spin and same-spin contributions. The self-correlation factor, D_α (see Eq. (78)), takes a slightly different form.

Only a few months later, Truhlar et al. proposed another hybrid meta GGA functional, very similar to M05, but containing a larger amount of the exact exchange energy. In particular, the X variable in Eq. (106) increases from 0.28 to 0.56. The authors called the functional M05–2X (Zhao et al., 2006). The justification for the increase in the amount of exact exchange in the exchange functional was to provide a better functional with broad applications in systems containing main-group elements only. The original M05 functional, on the other hand, could still be used to calculate several properties of systems containing both main-group elements and transition metals with reasonable accuracy.

Also in 2006, the same group at the University of Minessota presented the M06-HF functional (Zhao and Truhlar, 2006b)

$$E_{xc}^{M06-HF} = E_{xc}^{M06-L} + E_x^{HF}, \tag{107}$$

where the original M06-L functional, but with a different set of parameters, was used. The objective of including the exact exchange at a full extent was not only to eliminate long-range self-interaction errors, but to achieve an overall performance as good as or even better that the hybrid GGA B3LYP functional.

In 2008, Zhao and Truhlar completed the M06 family of functionals with two new members that belong to the fourth rung of the Jacob's Ladder: M06 and M06–2X (Zhao and Truhlar, 2008a). They return to the philosophy of the M05 and M05–2X functionals and give the new functionals the form

$$E_x^{M06,M06-2X} = \left(1 - \frac{X}{100}\right) E_x^{M06-L} + \frac{X}{100} E_x^{HF}, \tag{108}$$

where X is 27 and 54 for M06 and M06–2X, respectively. As in the case of the M05 functionals, M06 was parametrized to be used with systems containing both non-metals and transition metals, whereas M06–2X was parametrized only for non metals.

Also in 2008, Zhao and Truhlar presented two new hybrid mGGA functionals, which were developed with the aim of improving self-consistent convergence and to fulfill some exact conditions up to second order in the reduced density matrix (Zhao and Truhlar, 2008b). The functionals were called M08-HX and M08-SO. The exchange part of the functionals takes the form

$$E_x^{M08-HX,M08-SO} = \int \rho \varepsilon_x^{LDA}(\rho) \left(f_1(w) F_x^{PBE}(s) + f_2(w) F_x^{RPBE}(s) \right) d\mathbf{r}, \tag{109}$$

which is very similar to the exchange part of the mGGA M11-L functional, see Eq. (81). The correlation part of the M08 functionals is also similar to that of the M11-L one

$$E_c^{M08-HX,M08-SO} = \int \rho\left(f_3(w)\varepsilon_c^{L(S)DA}(\rho,\xi) + f_4(w)H(s,\rho,\xi)\right)d\mathbf{r}, \tag{110}$$

which, in turn, is taken from the correlation part of the PBE functional. The four enhancement factors, $f_{1-4}(w)$, are defined in such a way that the sums, see Eq. (70), goes from 0 to 11. Moreover, the hybrid scheme is defined for both functionals as

$$E_{xc}^{M08-HX,M08-SO} = \left(1-\frac{X}{100}\right)E_x^{M08-HX,M08-SO} + \frac{X}{100}E_x^{HF} + E_c^{M08-HX,M08-SO}. \tag{111}$$

Every functional has 49 parameters to optimize, some of them are fixed due to the constraints imposed to the functionals, though. The exchange-mixing parameter, X, is 52.23 and 56.79 for M08-HX and M08-SO, respectively.

In 2011, Peverati and Truhlar presented a dual-range hybrid meta GGA functional called M11 (Peverati and Truhlar, 2011a). The form of the functional is

$$E_{xc}^{M11} = \frac{X}{100}E_x^{HF} + \left(1-\frac{X}{100}\right)\left(E_x^{SR-M11-L} + E_x^{LR-HF}\right) + E_c^{M11-L}, \tag{112}$$

where the short-range part of the exchange energy density functional is that of M11-L, Eq. (82) and the long-range functional is the exact Hartree-Fock exchange using the long-range of a range-separated Coulomb operator

$$E_x^{LR-HF} = \frac{1}{2}\sum_{\sigma}\sum_{i,j}^{occ}\iint \psi_{i\sigma}^*(\mathbf{r}_1)\psi_{j\sigma}^*(\mathbf{r}_1)\frac{\mathrm{erf}(\omega r_{12})}{r_{12}}\psi_{i\sigma}(\mathbf{r}_2)\psi_{j\sigma}(\mathbf{r}_2)\,d\mathbf{r}_1 d\mathbf{r}_2, \tag{113}$$

where ω is the usual range-separation parameter. The correlation part of the M11 functional is the same as that of the M11-L functional, see Eq. (85).

The dual-range hybrid meta GGA M11 functional is defined by 45 parameters as the four enhancement factors in $E_x^{SR-M11-L}$ and E_c^{M11-L} are polynomials of degree 10 and X is considered a parameter to optimize, too. The optimum value of X is 42.8

10.8 THE DOUBLE-HYBRID (META) GENERALIZED GRADIENT APPROXIMATION, THE FIFTH RUNG IN THE JACOB'S LADDER

The fifth rung in the Jacob's Ladder accounts for the inclusion of the unoccupied Kohn-Sham auxiliary functions by means of low-order Perturbation Theory.

In 2004, Zhao, Lynch and Truhlar presented two-hybrid meta GGA density functionals containing a certain amount of correlation energy estimated through MP2 (Zhao et al., 2004). They called those functionals MC3 for Multi-Coefficient Three-Parameter. The form of the exchange-correlation energy proposed for the MC3 functionals is:

$$E_{xc}^{MC3DFT} = c_1 E_x^{HF} + (1-c_1) E_{xc}^{DFT} + c_2 E_c^{MP2}, \tag{114}$$

where E_{xc}^{DFT} is calculated, in turn, from a hybrid (meta) GGA scheme.

$$E_{xc}^{DFT} = \frac{X}{100} E_x^{HF} + \left(1 - \frac{X}{100}\right) E_x^{DFT} + E_c^{DFT}. \tag{115}$$

The three parameters c_1, c_2 and X give the functionals their name. The first functional, MC3BB, uses the B88 and B95 functionals for exchange and correlation, respectively. The second functional, MC3 mPW uses the PW92 functional for exchange and correlation, with the modification introduced by Adamo and Barone for the exchange (Adamo and Barone, 1999).

In 2006, Grimme and co-workers (Grimme, 2006; Schwabe and Grimme, 2006) suggested an expression for the exchange-correlation energy very similar to that in Eq. (114), but retaining the original philosophy of hybrid functionals, in which exchange and correlation bear different coefficients.

$$E_{xc}^{B2-PLYP} = a_x E_x^{HF} + (1-a_x) E_x^{DFT} + (1-c) E_c^{LYP} + c E_c^{MP2}, \tag{116}$$

where pure GGA calculations could be performed with B88 or mPW92 for exchange, and the LYP functional is used for correlation. MP2 correlation energy is calculated according to:

$$E_c^{MP2} = \frac{1}{4} \sum_{ia} \sum_{jb} \frac{\left[(ia \mid jb) - (ib \mid ja)\right]^2}{\varepsilon_i + \varepsilon_j - \varepsilon_a - \varepsilon_b}, \tag{117}$$

where i, j represent occupied KS functions and a, b are unoccupied KS functions, and the ε 's are the eigenvalues of the corresponding functions. The functionals proposed by Grimme et al. were called B2-PLYP and mPW92-PLYP.

In 2008, Martin and co-workers (Karton et al., 2008; Tarnopolsky et al., 2008) carried out a series of benchmark studies to improve the original B2- and mPW92-PLYP double-hybrid functionals. The authors found that modifying both a_x and c much better results could be obtained for a series of properties. They called their modified functionals B2K-PLYP, mPW2K-PLYP, and B2GP-PLYP, respectively. The authors concluded that the B2GP-PLYP functional is a reliable general-purpose method, although B2K-PLYP should be used for nucleophilic substitution and hydrogen transfer reactions, in which its performance is much better.

In 2011, Brémond and Adamo proposed a double-hybrid functional free from adjustable parameters called PBE0-DH (Brémond and Adamo, 2011). The authors used an adiabatic connection formula, which contains the Hartree exchange-correlation operator, to obtain the following form for the exchange-correlation energy.

$$E_{xc}^{PBE0-DH} = \frac{1}{2}\left(E_x^{HF} + E_x^{PBE}\right) + \frac{1}{2}\left(\frac{7}{4}E_c^{PBE} + \frac{1}{4}E_c^{MP2}\right). \qquad (118)$$

10.9 THE PERFORMANCE OF MODERN DENSITY FUNCTIONALS TO PREDICT PROPERTIES OF INTEREST TO CHEMISTS

In the last sections, the evolution of the Density Functional Theory was revised from the point of view of the increasing complexity in the exchange-correlation functionals needed in the Kohn-Sham version of the theory.

Now, it is time to see how well the functionals of different rungs perform in describing existing properties. As it was mentioned in section 1, the amount of existing functionals is extremely large and a thorough review of all of them is almost impossible. Thus, the present section will be devoted to those functionals shown and discussed in some extent in previous sections. Only in those cases in which a given functional, not mentioned in the present chapter, exhibits an outstanding performance for some specific property, it will be added to the discussion.

Moreover, benchmark studies are usually performed when a new functional is designed. Thus, comparative performance studies are biased to the performance of the new functional or to the performance of the family to which the new functional belongs. In other words, it is not easy to find benchmark studies including a large variety of functionals from every rung of the Jacob's Ladder.

The most relevant comparative studies are shown in reverse chronological order in the present chapter.

Peverati and Truhlar (Peverati and Truhlar, 2012) performed an extensive study on the MXX family of functionals, where XX goes from 05 to 11, on the occasion of the development of the dual-range hybrid meta GGA M11 functional. Their authors used their own database, called BC338, where BC means *Broad Chemistry*, formed by 338 data taken from subdatabases containing atomization energies (SB1AE97, LB1AE12), single-reference metal bond energies (SRMBE13) and multireference bond energies (MRBE10), isomerization energies (IsoL6), ionization potentials, electron affinities and proton affinities (IP13/03, EA13/03, PA8), bond dissociation reaction energies (ABDE4/05, ABDEL8, HC7/11), hydrogen-transfer and nonhydrogen-transfer barrier heights (HTBH38/08, NHTBH38/08), thermochemistry of systems containing π electrons (πTC13), noncovalent complexation energies (NCCE31/05), atomic energies (AE17) and a dozen difficult cases for DFT (DC10). The performance of a series of functionals for every subdatabase is shown in Table 1 and the overall performance for the BC338 database is depicted in Figure 1. It can be seen from the Table that the hybrid meta GGA M06–2X functional performs very well in almost all cases, except for bond energies in systems with appreciable multireference character (MRBE10) and for those cases described as difficult for DFT (DC10). Interestingly, the only functional with a mean unsigned error (MUE) well below 10 kcal/mol for the MRBE10 subdatabase is the dual-range meta GGA M11-L. This fact is the main responsible for the lowest MUE exhibited by M11-L as shown in Fig. 1. Only the M06, M11 and M11-L functionals present MUE's below 3 kcal/mol for the overall BC338 set, whereas B97, M06-L, M06–2X, M08-HX andM08-SO have MUE's between 3 kcal/mol and 4 kcal/mol. Surprizingly, the B3LYP functional is the second worse functional behind PBE.

TABLE 1 Performance of different exchange-correlation energy density functionals for various subdatabases, see text and reference (Peverati and Truhlar, 2012). The mean unsigned error, in kcal/mol is shown. The top performer for every subdatabase is indicated in boldface.

Database	PBE	M06-L	M11-L	B3LYP	B97	M06	M06–2X	M06 HF	M08-HX	M08-SO	M11
SB1AE97	2.77	0.83	0.68	0.96	0.66	0.59	**0.42**	0.58	0.66	0.60	0.45
LB1AE12	12.96	2.25	2.94	**2.14**	2.22	2.28	2.18	5.67	2.43	2.36	2.58
SRMBE13	3.87	3.41	3.21	2.88	**2.64**	2.77	3.90	6.98	3.36	2.87	4.55
MRBE10	19.73	10.64	**6.14**	22.47	21.99	18.60	44.93	75.50	50.87	47.70	43.54
IsoL6	1.98	2.76	1.57	2.61	1.93	1.27	1.53	2.46	**0.59**	1.19	1.10
IP13/03	3.62	3.08	3.11	4.76	3.23	3.28	**2.56**	3.80	3.42	3.58	3.64
EA13/03	2.27	3.83	5.54	2.33	1.90	1.85	2.14	3.81	1.32	2.72	**0.89**

TABLE 1 *(Continued)*

Database	PBE	M06-L	M11-L	B3LYP	B97	M06	M06–2X	M06-HF	M08-HX	M08-SO	M11
PA8/06	1.34	1.88	2.17	**1.02**	1.53	1.84	1.65	2.28	1.08	1.64	1.03
ABDE4/05	4.09	5.54	5.14	8.73	4.99	2.84	**2.12**	4.43	2.67	2.51	2.45
ABDEL8	7.16	8.85	6.98	10.40	7.71	4.72	**2.69**	4.56	2.87	3.88	3.48
HC7/11	3.97	3.35	2.42	16.80	8.46	2.78	**2.15**	2.29	4.89	4.60	3.74
πTC13	6.01	6.52	5.47	6.06	7.01	4.08	**1.51**	1.92	1.98	1.97	2.12
HTBH38/08	9.31	4.15	1.44	4.23	4.16	1.98	1.14	2.07	**0.72**	1.07	1.30
NHTBH38/08	8.42	3.81	2.86	4.55	3.31	2.33	**1.22**	2.53	**1.22**	1.23	1.28
NCCE31/08	1.24	0.58	0.56	0.96	0.70	0.41	0.29	0.41	0.35	0.37	**0.26**
DC10	39.22	19.76	10.38	20.66	15.34	9.45	10.46	15.63	9.60	10.87	**8.03**
AE17	54.44	3.86	6.42	13.6	9.25	6.85	6.18	**4.28**	6.54	10.54	5.15

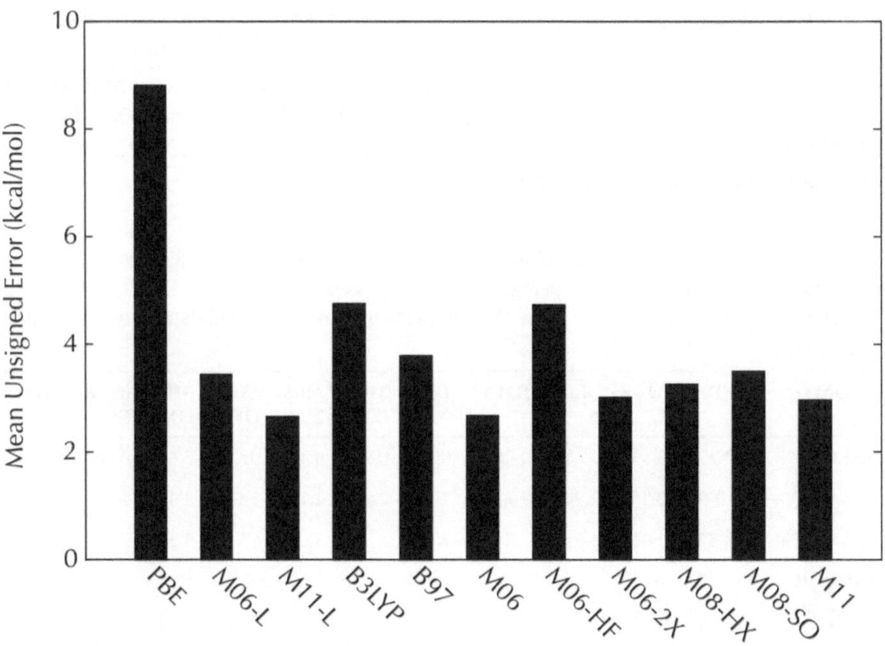

FIGURE 1 Overall performance of exchange-correlation energy density functionals for the BC338 database (Peverati and Truhlar, 2012).

Two other subdatabases were used to test the performance of the same functionals, except for M11-L (Peverati and Truhlar, 2011a). The subdatabases contain charge-transfer electronic transitions (CTS8) and non-covalent binding energies (S22A). Results are shown in Fig. 2. It can be seen in the Figure that M06-HF and M06–2X are the top performers for CTS8 and S22A, respectively. Besides the good performance of some others functionals, it is interesting to see that B3LYP is the worst functional for S22A and the second worse behind M06-L.

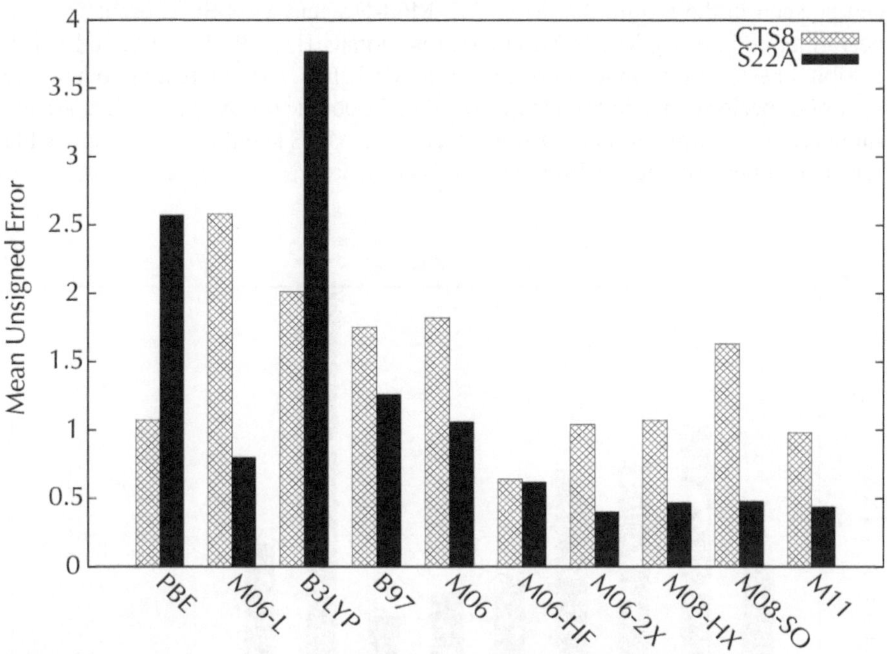

FIGURE 2 Performance of exchange-correlation energy density functionals for the CTS8 and S22A subdatabases, see text and reference (Peverati and Truhlar, 2011a). MUE's are given in eV and kcal/mol for CTS8 and S22A, respectively.

Peverati, Zhao and Truhlar published another extensive study on the performance of a set of functionals on a smaller database, BC322 (Peverati and Truhlar, 2011b). Besides the absence of hybrid GGA and hybrid meta GGA functionals, the interest of the benchmark study lies in the inclusion of various GGA and meta GGA functionals. Thus, within the GGA group formed by B88PW92, B88LYP, PW92, B88P86 and PBE, the best functional is B88PW92 with a MUE of 4.32 kcal/mol, whereas PBE becomes the worst with a MUE of 7.27 kcal/mol. The meta GGA TPSS functional has a MUE of 4.71 kcal/mol for the BC322 database.

In 2009, Zheng, Zhao and Truhlar performed a benchmark study on barrier heights of various reactions comprised in four subdatabases that, in turn, form the DBH24/08 database. The salient features of that work is that the computational cost, relative to the MP2/6–31+G(d, p) level of theory, is reported together with the partial and overall MUE's. The product of the overall MUE and the relative computational cost, both obtained with the 6–31+G(d, p) basis set, could be taken as a measure of the overall performance of a given method. Thus, the smaller the product, the better the performance index. Figure 3 summarizes the results. It can be seen in the Figure that M06–2X, M08-HX and M08-SO present the best performance indexes. The hybrid GGA functionals B3LYP, B3PW92 and PBE0 exhibit a performance index very similar to MP2. It is very interesting to note the very poor performance index shown by M06. As both M06 and M06–2X share the same relative computational cost, a rather large MUE should be the responsible for the bad performance of M06 when it is compared with M06–2X.

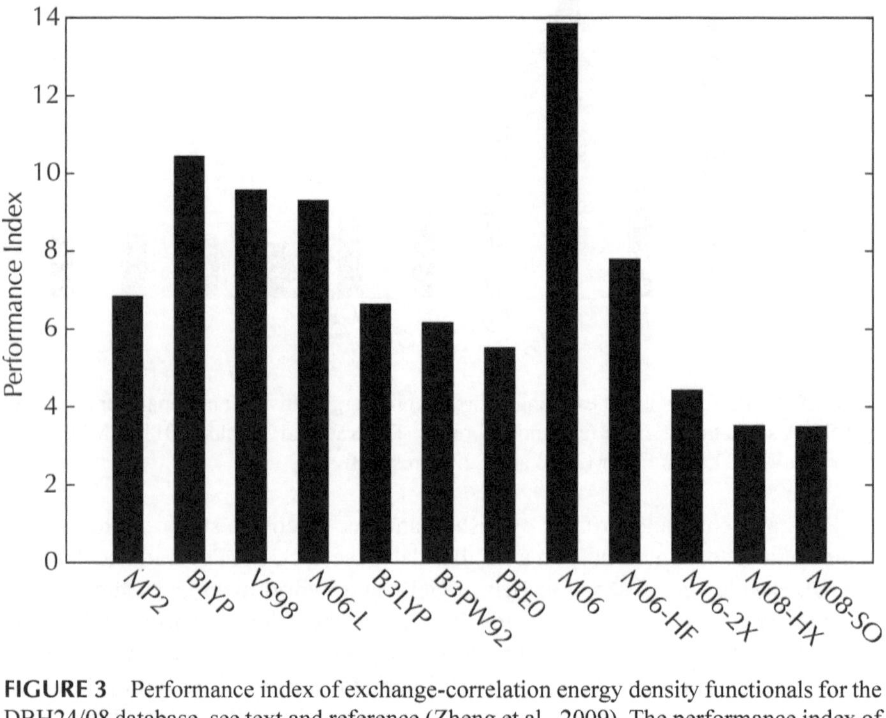

FIGURE 3 Performance index of exchange-correlation energy density functionals for the DBH24/08 database, see text and reference (Zheng et al., 2009). The performance index of the MP2 method is also shown for reference.

In 2008, Zhao and Truhlar carried out an extensive benchmark study of various functionals and several databases (Zhao and Truhlar, 2008a). One of such databases, TMRE48, contains information on the thermochemistry of systems containing transition metals. In particular, atomization energies (TMAE9/05) and reaction energies ($3d$ TMRE18/06) of systems containing transition metals, and metal-ligand bond energies (MLBE21/05) were calculated with a series of functionals. Results are summarized in Fig. 4. It can be seen in the Figure that despite the good performance of BLYP for the TMAE9/05 database, and B3LYP and B97 for MLBE21/05, both M06 and M06-L show very good performances with overall MUE's of 5.6 and 5.7 kcal/mol, respectively. It is also very interesting to note the very bad performance of M06–2X for systems containing transition metals. This behavior should not be surprizing, as it is a well-known fact that the Hartree-Fock theory fails to describe open-shell systems, as those containing transition metals. The increasing contribution of the exact Hartree-Fock exchange energy should be the main responsible for that failure.

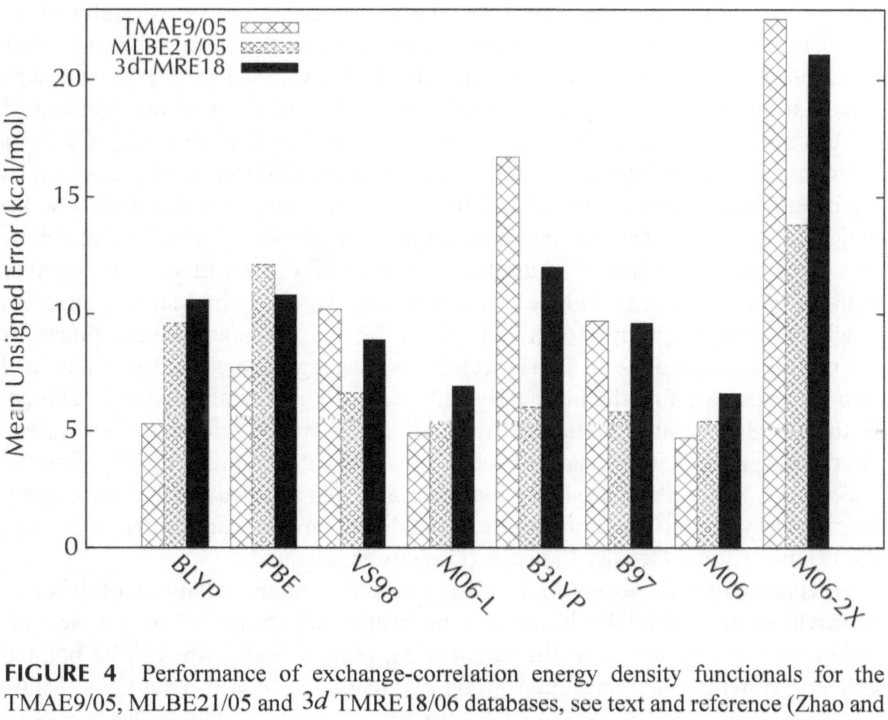

FIGURE 4 Performance of exchange-correlation energy density functionals for the TMAE9/05, MLBE21/05 and $3d$ TMRE18/06 databases, see text and reference (Zhao and Truhlar, 2008a).

In the same work, Zhao and Truhlar show that bond lengths of systems containing main group elements are very well described by DFT, with an average MUE of 0.008 Å. When transition-metal systems are included in the benchmark, the average MUE notoriously increases up to about 0.055 Å, indicating the well-known difficulty of dealing with transition metals. Surprizingly, the GGA PBE exchange-correlation functional is the top performer for bond lengths.

10.10 CONCLUSIONS

The Density Functional Theory has greatly evolved from the 1920s, when the first attempts to devise a theory, in which the electron density is the basic variable were made.

Thanks to the work of Hohenberg, Kohn and Sham in mid 1960s, the Density Functional Theory was given a formal background, from which further evolution was possible.

The evolution of modern Density Functional Theory can be understood in terms of some basic ingredients. The first ingredient is the set of Kohn-Sham auxiliary functions, $\{\psi(\mathbf{r})\}$, which are needed to define the electron density, $\rho(\mathbf{r})$. The electron density is the basic ingredient of the Local (Spin) Density Approximation. The next step in the evolution is the inclusion of the gradient of the electron density through the dimensionless gradient $s = |\nabla\rho(\mathbf{r})| / \left((24i^2)^{1/3}\rho^{4/3}(\mathbf{r})\right)$. Functionals with both the electron density and the electron density gradient as arguments belong to the so-called Generalized Gradient Approximation. The kinetic energy density can be calculated from the Kohn-Sham auxiliary functions as $\tau = \sum_i^{occ} |\nabla\psi_i(\mathbf{r})|^2$. When τ is used as argument of a given functional, together with the electron density and the electron density gradient, the Meta Generalized Gradient Approximation is obtained. When the Kohn-Sham auxiliary functions are used to calculate the exact Hartree-Fock exchange energy and it is included into the exchange functional, the Hybrid Density Functional Theory is defined. Both Hybrid Generalized Gradient Approximations and Hybrid Meta Generalized Gradient Approximations can be used in practical calculations. Finally, when the *unoccupied* Kohn-Sham auxiliary functions are used to calculate the exact correlation energy up to second order and it is included into the correlation functional, the Double-Hybrid Density Functional Theory is obtained.

It is difficult to achieve a conclusive evidence about the reliability of different approximations to the DFT based only on benchmark studies. However, according to what it was shown in the last section, several useful tips can be pointed out. For systemscontaining main group elements, the hybrid meta GGA M06–2X functional performs very well for a variety of areas, such as thermochemistry, energy barriers and non covalent interactions. When transition metals are involved, the meta GGA M06-L functional isan excellent alternative as M06–2X is a poor choice due to the large amount of Hartree-Fock exchange. For more difficult cases, in which an appreciable multireference character is presented, such as bond dissociation or reactions involving different isomers, meta GGA, hybrid GGA and hybrid meta GGA are useless due to their single-range character. Thus, dual-range functionals, such as the meta GGA M11-L or the hybrid meta GGA M11 should be used instead. The more recent double-hybrid functionals seem to bean appealing alternative for practical calculations within the DFT. The relative computational cost of those functionals should be carefully considered, though, as the MP2 part of the calculation, if carried out self consistently, is rather demanding. Insummary, a small subset of functionals formed by M06-L, M06–2X, M11 and M11-L could be well considered for routine calculations in various fields of Chemistry.

Finally, it is interesting to lay some emphasis on the mathematical form of modern functionals. Some researchers are devoted to design functionals with as few parameters as possible, whose values, in turn, are obtained after applying some well known boundary conditions that exchange and correlation functionals must fulfill. Other researchers prefer to develop functionals with the aim of chemical accuracy in mind. This philosophy involves mathematical forms with up to 50 adjustable parameters. Functionals designed following the former philosophy are unable to provide accurate results in certain fields of Chemistry. Functionals designed following the second philosophy, on the other hand, provide results accurate enough in several fields to consider modern DFT a reliable methodology for electronic structure calculations. This, however, seems to move modern DFT away from *ab-initio* or first principle methods and tend to *push* it to the side of semiempirical methods. Unfortunately, with this sort of *semiempirical* DFT the universal functional of Hohenberg and Kohn will surely remain unknown.

ACKNOWLEDGMENTS

The author acknowledges CONICET, Argentina, for financial support under the PIP program. He is also member of the Scientific Researcher Career of CONICET.

KEYWORDS

- **attenuation function**
- **Generalized Gradient Approximation**
- **Hartree-Fock theory**
- **Hellmann-Feynman theorem**
- **Kohn-Sham auxiliary functions**
- **massive parameterization**
- **nested functionals**
- **opposite-spin electrons**
- **reduced density gradient**
- **same-spin electrons**
- **spin-polarized systems**
- **Wigner-Seitz radius**

REFERENCES

Adamo, C.; Barone, V. *J. Chem. Phys.*, **1999,** *110,* 6158.

Becke, A. D. **1993a,** *98,* 1372.

Becke, A. D. *J. Chem. Phys.*, **1993b,** *98,* 5648.

Becke, A. D. *J. Chem. Phys.*, **1996,** *104,* 1040.

Becke, A. D. *J. Chem. Phys.*, **1997,** *107,* 8554.

Becke, A. D. *Phys. Rev. A*, **1988,** *38,* 3098.

Brémond, E. A. G.; Adamo, C. *J. Chem. Phys.*, **2011,** *135,* 024106.

Burke, K.; Perdew, J. P.; Wang, Y.; Dobson, J. F.; Vignale, G.; Das, M. P. (eds), *Electron Density Functional Theory: Recent Progress and New Directions*. New York: Plenum. **1997**, 18.

Ceperley, D. M.; Alder, B. J. *Phys. Rev. Lett.*, **1980,** *45,* 566.

Chai, J.-D.; Head-Gordon, M. *J. Chem. Phys.*, **2008,** *128,* 084106.

Colle, R.; Salvetti, D. *J. Chem. Phys.*, **1983,** *79,* 1404.

Colle, R.; Salvetti, D. *Theor. Chim. Acta*, **1975,** *37,* 329.

Dirac, P. A. M. *Proc. Camb. Phil. Soc.*, **1930,** *26,* 376.

Fermi, E. *Rend. Accad. Lincei*, **1927,** *6,* 602.

Grimme, S. *J. Chem. Phys.*, **2006,** *124,* 034108.

Hammer, B.; Hansen, L.; Norskov, J. *Phys. Rev. B*, **1999,** *59,* 7413.

Hedin, L.; Lundqvist, B. I. *J. Phys. C: Solid State Phys.*, **1971,** *4,* 2064.

Hohenberg, P.; Kohn, W. *Phys. Rev.*, **1964,** *136,* B864.

Hu, C. D.; Langreth, D. C. *Phys. Rev. B*, **1986,** *33,* 943.

Hu, C. D.; Langreth, D. C. *Phys. Scr.*, **1985,** *32,* 391.

Karton, A.; Tarnopolsky, A.; Lamère, J. F.; Schatz, G. C.; Martin, J. M. L. *J. Phys. Chem. A*, **2008,***112,* 12868.

Kohn, W.; Sham, L.J. *Phys. Rev.*, **1965,** *140,* A1133.

Langreth, D. C.; Mehl, M. J. *Phys. Rev. B*, **1983,** *28,* 1809.

Langreth, D. C.; Perdew, J. P. *Phys. Rev. B*, **1980,** *21,* 5469.

Lee, C.; Yang, W.; Parr, R. G. *Phys. Rev. B*, **1988,** *37,* 785.

Levy, M. *Proc. Natl. Acad. Sci. USA*, **1979,** *76,* 6062.

Neumann, R.; Handy, N. C. *Chem. Phys. Lett.*, **1997,** *266,* 16.

Parr, R. G.; Yang, W. **1989,** *Density Functional Theory of Atoms and Molecules*. New York: Oxford University Press.

Perdew, J. P. *Phys. Rev. B*, **1986a**. *33,* 8822.

Perdew, J. P. *Phys. Rev. B*, **1986b**. *34,* 7406.

Perdew, J. P.; Burke, K.; Ernzerhof, M. *Phys. Rev. Lett.*, **1996a**. *77,* 3865.

Perdew, J. P.; Chevary, J. A.; Vosko, S. H.; Jackson, K. A.; Pederson, M. R., Singh, D. J.; Fiolhais, C. *Phys. Rev. B*, **1992,** *46,* 6671.

Perdew, J. P.; Ernzerhof, M.; Burke, K. *J. Chem. Phys.*, **1996b**. *105,* 9982.

Perdew, J. P.; Kurth, S.; Zupan, A.; Blaha, P. *Phys. Rev. Lett.*, **1999,***82,* 2544.

Perdew, J. P.; Schmidt, K. In: Doren, V. Van, Alsenoy, C. Van, and Geerlings, P. (eds), *Density Functional Theory and Its Application to Materials*. Melville, NY: AIP, **2001.**

Perdew, J. P.; Wang, Y. *Phys. Rev. B*, **1986,** *33,* 8800.

Perdew, J. P.; Wang, Y. *Phys. Rev. B*, **1989,** *40,* 3399.

Perdew, J. P.; Zunger, A. *Phys. Rev. B*, **1981,** *23,* 5048.

Peverati, R.; Truhlar, D. G. *J. Phys. Chem. Lett.*, **2011a,** *2,* 2810.

Peverati, R.; Truhlar, D. G. *J. Phys. Chem. Lett.*, **2011b,** *2,* 2810.

Peverati, R.; Truhlar, D. G. *J. Phys. Chem. Lett.*, **2012,** *3,* 117.

Schwabe, T.; Grimme, S. *Phys. Chem. Chem. Phys.*, **2006,** *8,* 4398.

Souza, S. F.; Fernandes, P. A.; Ramos, M. J. *J. Phys. Chem. A*, **2007,** *111,* 10439.

Stephens, P. J.; Devlin, F. J.; Chabalowski, C. F.; Frisch, M. J. *J. Phys. Chem.*, **1994,** *98,* 11623.

Stoll, H.; Pavlidou, C. M. E.; Preuss, H. *Theoret. Chim. Acta*, **1978,** *49,* 143.

Swart, M.; Bickelhaupt, F. M.; Duran, M. **2011,** *Density Functionals Poll.* http://www.marcelswart.eu/dft-poll.

Tao, J.; Perdew, J. P.; Staroverov, V. N.; Scuseria, G. E. *Phys. Rev. Lett.*, **2003,** *91,* 146401.

Tarnopolsky, A.; Karton, A.; Sertchook, R.; Vuzman, D.; Martin, J. M. L. *J. Phys. Chem. A*, **2008,** *112,* 3.

Thomas, L. H. *Proc. Camb. Phil. Soc.*, **1927,** *23,* 542.

von Barth, U.; Hedin, L. *J. Phys. C: Solid State Phys.*, **1972,** *5,* 1629.

von Weizsäcker, E. *Z. Physik*, **1935,** *96,* 431.

Voorhis, T. Van, and Scuseria, G. E. *J. Chem. Phys.*, **1998,** *109,* 400.

Vosko, S. H.; Wilk, L.; Nusair, M. *Can. J. Phys.*, **1980,** *58,* 1200.

Zhao, Y.; Lynch, B. J.; Truhlar, D. G. *J. Phys. Chem. A*, **2004,** *108,* 4786.

Zhao, Y.; Schultz, N. E.; Truhlar, D. G. *J. Chem. Phys.*, **2005,** *123,* 161103.

Zhao, Y.; Schultz, N. E.; Truhlar, D. G. *J. Chem. Theory Comput.*, **2006,** *2,* 364.

Zhao, Y.; Truhlar, D. G. *J. Chem. Phys.*, **2006a,** *125,* 194101.

Zhao, Y.; Truhlar, D. G. *J. Chem. Theory Comput.*, **2008b,** *4,* 1849.

Zhao, Y.; Truhlar, D. G. *J. Phys. Chem. A*, **2006b,** *110,* 13126.

Zhao, Y.; Truhlar, D. G. *Theor. Chem. Account*, **2008a,** *120,* 215.

Zheng, J.; Zhao, Y.; Truhlar, D. G. *J. Chem. Theory Comput.*, **2009,** *5,* 808.

CHAPTER 11

MOLECULAR DYNAMICS SIMULATIONS: APPLICABILITY AND SCOPES IN COMPUTATIONAL BIOCHEMISTRY

KSHATRESH DUTTA DUBEY and RAJENDRA PRASAD OJHA

CONTENTS

Abstract 372
11.1 Introduction 372
11.2 Basic Scheme of Molecular-Dynamics 373
11.3 Simulation Protocols 375
11.3.1 System Preparation and Parametrization 376
11.3.2 Annealing 377
11.3.3 Stability of Integration 378
11.3.4 Production Dynamics 379
11.4 Applications of Molecular Dynamics Simulations 382
11.4.1 Study of Biological Properties at Atomic Scales 382
11.4.2 Prediction of Conformational Changes Using MD Simulations 384
11.4.3 Unrevealing the Folding and Unfolding Process Using MD Simulations 384
11.4.4 Calculation of Free Energy Changes by MD Simulation 386
11.4.5 Simulation Approaches for Protein-Ligand Binding 387
11.5 Conclusion and Future Aspects 394
Acknowledgments 395
Keywords 395
References 395

ABSTRACT

Macroscopic properties of biomolecular systems are a straightforward consequence of time dependent microscopic interactions. Molecular Dynamics (MD) Simulations provide a time dependent microscopic properties of biomolecules, which could not be explained by experimental methods like X-ray crystallography. These specifications enable MD simulations as most widely used computational techniques for the study of dynamical properties of proteins, DNAs and other bio-macromolecules. In the present chapter, we focus the scopes, applicability and major case studies of MD simulations. The chapter also provides some lucid discussions about the pros and cons of free energy pathways like MM-PB (GB)/SA, LIE and other alchemical methods which are frequently used to study the binding mode of several biomolecular complexes.

11.1 INTRODUCTION

Computational chemistry is a branch of chemistry that links the experimental data with theoretical models by resolving chemical problems with computational methods. Here, a model of the real world is constructed, the measurable and unmeasurable properties are then computed and finally they are compared to experimental determined properties. This comparison with experimental data validates the computational model used to study the chemical problems.

The availability of the first protein structures determined by X-ray crystallography led to the initial view that these molecules were very rigid, an idea consistent with the lock-and-key model of enzyme catalysis. Detailed analysis of protein structures, however, indicated that proteins had to be flexible in order to perform their biological functions. The first detailed microscopic view of atomic motions in a protein was provided in 1977 via a molecular dynamics (MD) simulation of bovine pancreatic trypsin inhibitor by McCammon et al. (1977). This study was performed on bovine pancreatic trypsin inhibitors for a very small simulation time of 9.2 ps. However, the experimental facts, say, hydrogen bond exchange were already explained before computational molecular dynamics (Berger et al., 1957). Two years after the first MD simulation, the role of thermal (B) factor in internal motion of protein, calculated during X-ray crystallographic refinement was also studied (Brunger et al., 1985; Smith et al., 1984). As a consequence of these studies, the estimation of mean square fluctuations versus residue number became a usual part of almost all analysis during MD simulations. In subsequent ten years, a wide range of motional phenomena was investigated by molecular dynamic simulations of protein and nucleic acids. These studies were focused on conformational flexibility and interpretation of experimental results. Such studies were mostly about analysis of fluorescence depolarization (Frauenfelder et al., 1987), dynamics of NMR parameters (Brunger et al., 1987), effect of solvent and temperature on protein's stability (Colonna-Ceseri et al., 1986; Nilson and Clore, 1987). During these time periods, simulated annealing was widely used for *x*-ray structure refinement (Harvey et al., 1984) and NMR structure determinations (Case and Karplus, 1979). Meanwhile, several applications were also performed for demonstration of internal motion in biological functions such that hinge bending mode for opening and closing of binding site (Brooks and Karplus, 1983). In this way, molecular dynamics simulation, which provides the methodology for detailed microscopical modeling on the atomic scale, became a powerful and widely used tool in chemistry, physics, and materials science. This technique is a scheme for the study of the natural time evolution of the system that allows prediction of the static and dynamic properties of substances directly from the underlying interactions between the molecules.

11.2 BASIC SCHEME OF MOLECULAR-DYNAMICS

The time dependent behavior of molecules can be best described by time dependent quantum mechanical equations (Schrodinger wave equation), but it is extremely hard to solve these equations for a system containing large number of atoms (advances in supercomputing facilitates us to solve Schrodinger equations for small system but for larger system it is almost impossible to solve). Therefore, a simpler classical mechanical description is often used to approximate the motion executed by the molecule's heavy atoms. In a molecular dynamic simulation, the classical equations of motion for the system of interest, say, biomolecules in solution are integrated numerically by solving Newton's equation of motion (Harvey and McCammon, 1987):

$$m_i \frac{d^2 r}{dt^2} = -\nabla_i \left[U\left(r_{1,} r_{2,} .. r_N\right)\right] \quad \mathrm{l} = 1,2,3......\mathrm{N} \tag{1}$$

From the solution of these equations, the atomic positions and velocities as a function of time are obtained (here m_i and r_i represents the mass and position of particle *i* and *U* is the potential-energy surface, which depends on the positions of the N particles in the system). Knowledge of the time history or trajectory of the atoms permits the computation of properties such as structure, folding pathways, diffusion, and thermodynamics to be studied. The key steps in the numerical solution of the classical equation of motion may be divided into two parts: the evaluation of energies and forces and the propagation of atomic positions and velocities (Harvey and McCammon, 1987).

The forcefield equation for biological molecules includes energy terms to represent chemical bonds, angles, and improper torsions as well as rotations about bonds (dihedrals) and pair wise additive nonbonded interactions (van der Waals and Coulombic). One form for the overall potential, which is often used, is indicated below: (Brooks et al., 1985, 1988).

$$U(r_{1,}r_{2,}...r_N)=\sum\frac{1}{2}k_b(b-b_0)^2+\sum\frac{1}{2}k_\theta(\theta-\theta_0)^2+\sum k_\omega(\omega-\omega_0)^2+\sum k\left[1+\cos(n\chi-\delta)\right]$$

$$+\sum\left\{4_{ij}\left[\left(\frac{\sigma_{ij}}{r}\right)^{12}-\left(\frac{\sigma_{ij}}{r}\right)^{6}\right]+\frac{q_iq_j}{r}\right\} \tag{2}$$

In this equation, the constants k_b, k_θ, k_ω and k_φ are the force constants for deformation of bonds, angles, impropers, and dihedrals, respectively. The equilibrium values of bond distance, valence angle, and improper torsion correspond to b_0, θ_0, and ω_0. In the dihedral terms, n is the periodicity of the underlying torsional potential, e.g., $n = 3$ representing the 3-fold rotational minima around a C-C single bond, and 6 is a phase factor. The constants ε_{ij}, σ_{ij}, and q_i represent the atomic (i, j) pair Lennard-Jones well depth and diameter and the partial electrostatic charges.

It is clear from above equation that we must have to differentiate it with respect to the cartesian positions of each atom in order to obtain a numerical solution. A standard method for solving ordinary differential equations, such as Newton's equation of motion, is the finite-difference approach. In this approach, the molecular coordinates and velocities at a time $t+\Delta t$ are obtained (to a sufficient degree of accuracy) from the molecular coordinates and velocities at an earlier time t. The equations are solved on a step-by-step basis. The Taylor expansion about time t of the position at time $t + \Delta t$ can be treated as starting point of finite distance approach, which is fundamental of integration algorithm like Verlet (Verlet, 1967). The Taylor's expansion of position at $t + \Delta t$ time can be expressed as:

$$r(t+\Delta t) = \mathrm{r}(t) + \dot{r}(t)\Delta t+\frac{1}{2}\ddot{r}(t)\Delta t^2+.....$$

Alternatively, it can be written as:

$$r(t+\Delta t)=r(t)+v(t)\Delta t+\frac{1}{2}\left(\frac{F}{m}\right)\Delta t^2+O\left(\Delta t^3\right)+...$$

where v (t) is velocity vector and (F/m) represents accelerations. Similarly for backward time, (t – Δt):

$$r(t-\Delta t)=r(t)-v(t)\Delta t+\frac{1}{2}\left(\frac{F}{m}\right)\Delta t^2-O\left(\Delta t^3\right)+...$$

The sum of these two expressions yields:

$$r(t+\Delta t) = 2r(t) - r(t-\Delta t) + \left(\frac{F}{m}\right)\Delta t^2 + ...$$

or

$$r(t+\Delta t) = 2r(t) - r(t-\Delta t) - \left(\frac{\Delta U}{m}\right)\Delta t^2 + ... \quad (3)$$

where, U is function of the coordinates and can be obtained by force fields or potential-energy function. The above algorithm is the basic for all integral and is commonly known as Verlet algorithm (Verlet, 1967), which has excellent stability properties appropriate for long integrations. In addition, there are other integration algorithms such as Leap-Frog (Hockney. 1970; Potter, 1972) and Velocity-Verlet (Swope et al., 1992), where small modifications in Verlet algorithm were done to achieve velocity scaling and small numerical errors. Step sizes Δt are generally chosen to be around 10^{-15} s. The resulting trajectories describe the time course of the system, and it is currently feasible to simulate proteins in solution for times on the order of nanoseconds, i.e., for about 10^6–10^7 integration steps.

The simulations with water increase the bulk size of the system and integration becomes very computational expensive. Therefore, an additional set of coupled constraint equations may be solved via an iterative self-consistent method known as SHAKE to keep all water molecules at their equilibrium geometry. The most common form of the applied constraints used in simulations of pure solvents and biopolymer solutions represents the distance constraints between a pair of atoms as:

$$r_{ij}^2 - d_{ij}^2 = 0 \quad (4)$$

where r_{ij} is the instantaneous separation between atoms i and j and d_{ij} is the reference constraint value. These constraints are implemented in molecular dynamics by first taking an unconstrained step using Eq. (3), then adding a displacement vector, which represents the displacement due to the forces of constraint, to satisfy the constraints specified by Eq. (4).

11.3 SIMULATION PROTOCOLS

Here we will briefly describe the step-wise protocols to be followed for conventional MD simulations.

11.3.1 SYSTEM PREPARATION AND PARAMETRIZATION

The initial coordinates are usually obtained from experimentally determined molecular structures, mainly from X-ray crystallography and NMR experiments. If initial structures are not known by any experimental techniques, comparative modeling can be performed (a flowchart of comparative protein modeling is shown in Fig. 1).

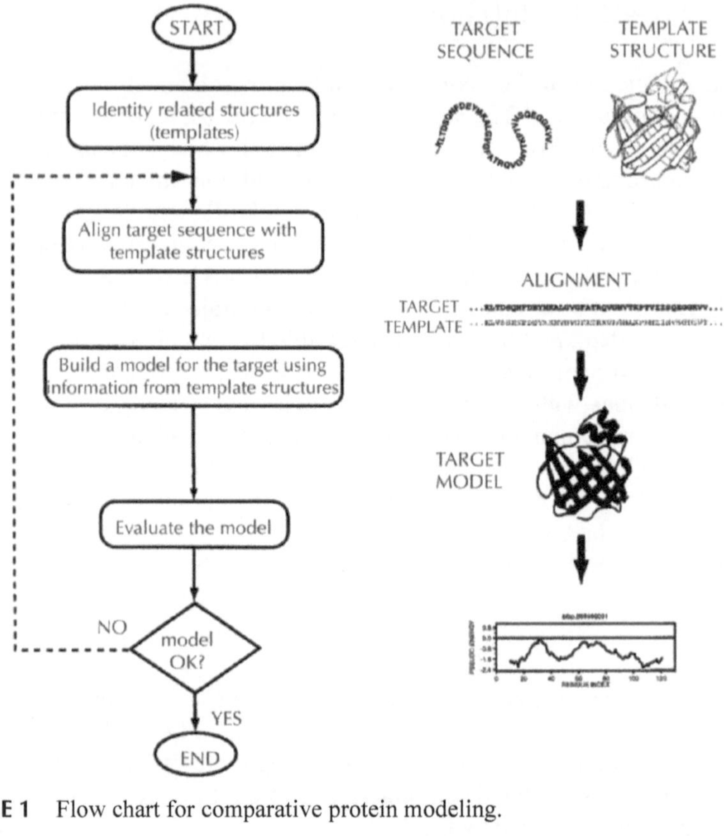

FIGURE 1 Flow chart for comparative protein modeling.

The experimental also contains several issues, which should be rectified before proceeding to computations. First, it is not possible to determine hydrogen atom positions by X-ray crystallography. Thus the coordinates for the many hydrogen atoms in the molecule are missing from X-ray coordinate files. These coordinates must be added to the initial structure before the simulation is started. In some cases, whole parts of the protein are missing from the experimentally determined structure. At times, these omissions reflect flexible parts of the molecule that do not have a well-defined structure (such as loops). At other times, they reflect parts of the molecule (e.g., terminal sequences) that were intentionally removed to facilitate the crystallization process. In both cases, structural models may be used to complete the structure, before starting the computations. In addition, experimental structures also possess several steric clashes due to nonbonded overlaps. Therefore, it is good practice to refine the initial structure by submitting it to energy minimization. The role of this minimization is to relieve local stresses due to nonbonded overlaps, as well as to relax bond length and bond angle distortions in the experimental structure.

After the minimization and other structural modifications, the system to be simulated must be proceeded for parameterizations according to molecular dynamics softwares. Often, the library for standard amino acids and nucleic acids is expected to be included in the library of the used molecular dynamics softwares (AMBER, CHARMM, GROMACS, NAMD etc.). However, sometimes one has to manually prepare library for modeled drugs. It is a good practice to run *ab-initio* calculations for the correct estimation of bonded and nonbonded parameters of force field equations.

11.3.2 ANNEALING

Since molecular dynamics simulations deal with the dynamic nature of the system, the assignment of velocities plays a crucial role during simulations. Unlike initial coordinates, which can be obtained experimentally, velocity can be assign only using temperature. Moreover, initial assignments of velocities are at random due to Maxwellian velocity distribution at a temperature T, hence initial velocities are not at equilibrium. Instead, the random assignment process may accidentally assign high velocities to a localized cluster of atoms that makes the simulation unstable. To overcome this problem, it is common practice to start a simulation with a "heat-up" phase. Velocities are initially assigned at a low temperature, which is then increased gradually allowing for dynamic relaxation. This slow heating continues until the simulation reaches the desired temperature.

As velocity is a direct function of temperature, annealing process is generally performed by increasing the atomic velocities. The increment in velocity may be performed – (i) By reassigning new velocities from a Maxwellian distribution

$$P(v)dv = \left(\frac{m}{2\pi k_B T}\right)^{\frac{1}{2}} \exp\left[\frac{-mv^2}{2K_B T}\right] dv$$

However, the large total linear momenta and angular momenta cause a drift of the molecule's center of mass relative to the reference frame, which is a major drawback of this method of velocity assigning. (ii) By scaling the velocities by a uniform factor, which is a readily used technique. An obvious way to alter the temperature of the system is velocity scaling. If the temperature at time t is T(t) and the velocities are multiplied by a factor, then the associated temperature change can be calculated as:

$$\Delta T = \frac{1}{2}\sum_{i=1} 2\frac{m_i(\lambda v_i)^2}{N_{df}k_B} - \frac{1}{2}\sum_{i=1} 2\frac{m_i v_i^2}{N_{df}k_B}$$

$$\Delta T = (\lambda^2 - 1)T(t)$$

$$\lambda = \sqrt{\frac{T_0}{T(t)}}$$

The simplest way to control the temperature is thus to multiply the velocities at each time step by the factor $\lambda = \sqrt{\frac{T_0}{T(t)}}$ where *T (t)* is the current temperature as calculated from the kinetic energy and T_0 is the desired temperature. Berendsen temperature coupling (Brendsen et al., 1984) and Nose-Hoover temperature coupling (Nose, 1984) are other algorithms used for annealing process.

11.3.3 STABILITY OF INTEGRATION

The accuracy and stability of the simulation is an important issue in any numerical study. Due to large fluctuations and errors, the simulation becomes unstable and does not follow standard classical trajectories. Thus, the stability of the simulation must be checked at all times. This process is usually done by running a long simulation ranging from 10 picoseconds to several hundreds of picoseconds. This simulation is maintained until the energy, momenta, density, pressure and other parameters become constant. It is also referred as 'Equilibration' dynamics. A typical example of variations in several parameters like energy, pressure and density is shown in Fig. 2.

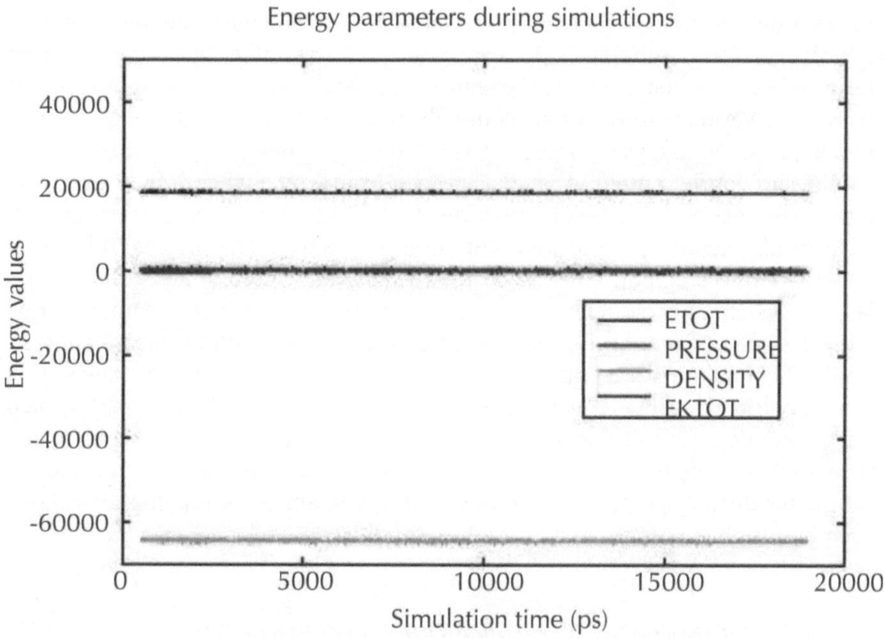

FIGURE 2 Graphs representing the results of equilibration dynamics. We see that all energy parameters are almost constant. The energy, pressure and density fluctuations are almost constant that is a necessary condition for a reliable MD simulation.

The heating process may also be treated as initial stage of equilibration. The stability of integration is an important process during simulations. If simulated system is not properly equilibrated, it may result in large numerical error in MD simulation that may cause inaccurate prediction of dynamical properties. It is a good practice to use constrained dynamics during equilibration to maintain the initial shape of system.

11.3.4 PRODUCTION DYNAMICS

When the simulation is equilibrated, the dynamic simulation is considered reliable. From this point on, the trajectory generated is stored for further analysis. Typical "production runs" take from several hundred picoseconds up to tens of nanoseconds (depending on the size of the system and the available computer power). It is the most fertile part of the molecular dynamic simulations, where system is relieved for its natural motion. Therefore, constraints should be removed from the system such that it can execute its natural motion. At the start of the production phase all counters are set to be zero and the system is permitted to evolve. If the simulation is performed at NVT ensemble, there should be no scaling of velocity and therefore the temperature becomes a calculated property of the system. Various properties are routinely calculated and stored during the production phase for analysis purpose. It is usual to store the positions, energies and velocities of configurations at regular interval from which other properties can be determined once the simulations have finished. Moreover, the parameters used in production dynamics depend upon the motive of study. The original molecular dynamics (MD) technique was used only to study the natural time evolution of a classical system of N particles in a volume V. In such simulations, the total energy, E, is a constant of motion, so that the time averages obtained by such a conventional MD simulation are equivalent to microcanonical ensemble averages. It is often scientifically more relevant to perform simulations in other ensembles, such as the canonical (NVT) that is associated with a Helmholtz free energy or the isothermal-isobaric (NPT) ensemble that is associated with a Gibbs free energy. Two rather different types of solutions to the problem of simulating alternative ensembles with the MD method have been proposed.

11.3.4.1 CONSTANT TEMPERATURE DYNAMICS

A constant temperature may be required if someone is interested to determine how the behavior of the system changes with temperature, such as the unfolding of a protein or glass formation. When the temperature of the system is gradually decreased to study folding and unfolding process, it is commonly known as simulated annealing. The relation between the temperature of the system and kinetic energy can be given as:

$$\langle K \rangle_{NVT} = \frac{3}{2} N k_B T$$

A famous approach to change the temperature of the system is to scale velocities, which have already been discussed in the Section 11.3.3. An alternative way to maintain the temperature is to couple the system to an external heat bath fixed at the desired temperature (Berendsen et al., 1984).

11.3.4.2 CONSTANT PREASSURE DYNAMICS

If someone is interested to specify the temperature in molecular dynamics simulations, so it may be desired to maintain the system at constant pressure. This enables the analysis of the behavior of the system as a function of the pressure enabling one to study phenomena such that onset of the pressure induced transitions. Several experiments are made under the condition of constant temperature and pressure, and so simulations at the isothermal-isobaric ensemble are most directly relevant to experimental data. Some structural rearrangement may be achieved more easily in an isobaric simulation than in a simulation at constant volume. Constant pressure condition may also be important when the number of particle in the system changes. Various schemes for prescribing the pressure of a molecular dynamics simulation have also been proposed and applied (Beglov and Roux, 1994; Berkowitz and McCammon, 1982; Evans and Morriss, 1984; Feller et al., 1995). In all of these approaches it is inevitable that the system box must change its volume. Among these schemes Andersen (Berkowitz and McCammon, 1982) originally proposed a method for constant-pressure MD that involves coupling the system to an external variable, *V*, the volume of the simulation box. This coupling mimics the action of a piston on a real system. Using the Lagrangian of extended system we may have:

$$\ddot{V} = \frac{1}{M_v}\left[P(t) - P_0\right]$$

where *V* is the volume, P(*t*) is the instantaneous pressure, P_0 is the desired pressure. Andersen proved that the solution to these equations produces trajectories in the isobaric-isoenthalpic (NPH) ensemble where the particle number, pressure, and enthalpy of the system are constant. Here the choice of the piston mass determines the decay time of the volume fluctuations. It has been proven that equilibrium quantities are independent of M_v, but in practice M_v influences the dynamic properties in the simulations. Even though there is no precise formulation for choosing M_v, Andersen suggests that the piston mass may be chosen by trial and error to satisfy the length of time required for a sound wave to travel through the simulation cell. These constant-pressure MD methods can be combined with a suitable temperature control mechanism, as discussed in the previous section, to produce a more useful method to control both pressure and temperature simultaneously during the simulation. There are several approaches. The simplest approach is to use the scaling factors. The details of the algorithm are given by Berendsen et al. (1987). Another approach (Hoover, 1985; Nose, 1984) is to define the appropriate extended Lagrangian for simultaneously coupling pressure and temperature to produce the isothermal-isobaric (NPT) ensemble.

In the above sections we have elaborated only fundamentals of molecular dynamics simulations. However, there are several other tricks and techniques which are frequently used to get an accurate prediction of properties using molecular dynamic simulations. There are several excellent reviews on these techniques which provide very useful information about the fundamentals and formulations of several schemes (Allen and Tildsley, 1987; Brenedsen and Gunsteren, 1986; Borhani and Shaw, 2012; Brooks et al., 1988; Brooks and Case, 1993; Haile, 1992; Helfand, 1984; Karplus and McCammon, 2002; MacCammon and Harvey, 1987).

11.4 APPLICATIONS OF MOLECULAR DYNAMICS SIMULATIONS

11.4.1 STUDY OF BIOLOGICAL PROPERTIES AT ATOMIC SCALES

The macroscopic properties of biological molecules are direct consequence of time dependent interactions at atomic level. Molecular dynamic simulation explains the time dependent properties of molecules; therefore it has utmost importance in study of biological system at atomic scale. Here, we are discussing some of the applications of MD simulations on biological systems.

11.4.1.1 STUDY OF ATOMIC INTERACTIONS IN DIMERIC AND TRIMERIC PROTEIN USING MD SIMULATIONS

A recent study explains the use of molecular dynamic simulations in explanation of dimeric interaction between two units of dimeric protein (Dubey et al., 2011). In this study the authors have performed molecular dynamics simulations at nanosecond scale. They prepared two separate complexes of dengue envelope protein (PDB ID: 1OAN) at neutral and low pH. The complexes were prepared using standard MD protocol and interactions between both units of envelope proteins were calculated. The experimental data suggest that dengue envelope protein exhibit dimeric form at neutral pH and this dimeric form dissociated into intermediate monomers after change in endosomal pH (Modis et al., 2003). In this study the authors have successfully used MD simulations to support above assumptions of mechanism of dengue envelope protein. They found that there is a large number of hydrogen bonds between the dimeric partner at low pH. They also calculated salt-bridges between dimeric partners for the simulated complexes. They found that at normal pH the dimer has significant salt-bridges. The molecular dynamics simulation at low pH suggests that the number of hydrogen bonds between the dimeric partners decrease remarkably and also there were no salt-bridges at low pH. This result shows that due to decrease in hydrogen bonds and other interactions, dimeric complex is less stable at low pH, which induces the dissociation of dimeric partners into monomeric intermediates.

Furthermore, the experimental structure of dengue envelope protein (DENV) after fusion tells that DENV protein exhibits a trimeric form at low pH (Modis et al., 2004). The same group has also performed several other molecular dynamic simulations of trimeric DENV protein in several pH conditions and salt concentrations to check the stability of trimeric complexes at different pH conditions at several nanosecond scale (Dubey et al., 2012). They found that the trimeric form is the most stable form at pH ~6.5. They found that at neutral pH the trimeric complex is at a less stable state. Since the protonation of histidines yields pH ~6.5, hence they proposed that histidine protonation plays an important role in the stability of the trimeric complex. These results were in good agreement with previous studies of histidine protonation (Kampmann et al., 2006; Mueller et al., 2008).

11.4.1.2 STUDY OF NONCOVALENT INTERACTION IN MEMBRANE PROTEINS USING MD SIMULATIONS

Non-covalent interactions like hydrogen, plays important role in structural and functional mechanism of proteins and other bio-macromolecules. These interactions can be easily recognized using several experimental techniques, but the dynamic behavior of such interactions through experimental studies have not been investigated due to difficulties in carrying out such experiments.

A recent study has been carried out to explore the dynamic character of these interactions using molecular dynamic simulations for membrane protein (Alok and Sankararamkrishnan, 2010). In this study, the authors have carried out simulations on four globular and two membrane proteins with different secondary structural contents. The dynamic nature of six different noncovalent interactions was analyzed to identify their behavior over time within and across the different classes (all-α versus all-β) and different types (globular versus membrane) of proteins. Some of the properties analyzed were a fraction of each type of interaction that was maintained throughout the simulation, maximum residence time (MRT) and the lifetime of the interactions.

This study reveals that conventional H-bonds are dominant (>60%) interactions and is mostly due to main-chain functional groups. They are predominantly stable with a MRT of at least 10 ns, owing to their role in maintaining the secondary structure of proteins. Their analysis reveals that C-H···O interactions involving the main-chain Cα and main-chain carbonyl oxygen atoms are the second most dominant interactions in the all-β-proteins. A large proportion of them are relatively more stable. Cation···π interactions are less frequently observed (<2%), but surprizingly it is one of the most stable interactions. The longest MRT of many such interactions exceeds 20 ns. Such strong and stable interactions have implications in the biological activity of proteins, protein-ligand interactions and protein folding studies.

11.4.2 PREDICTION OF CONFORMATIONAL CHANGES USING MD SIMULATIONS

A large part of the current knowledge of conformational flexibility in biomolecules is derived from experimental data, in particular X-ray crystallography and NMR. There are several examples of proteins structurally characterized when trapped in different functional states, and the time resolution of structural studies improves steadily. Nevertheless, details on the pathways between different known conformations often remain obscure.

In this scenario, computer simulation techniques became very important tools for understanding the physical basis of the structure and function of biological macromolecules. In particular, molecular dynamics (MD) simulations can provide the ultimate detail concerning individual particle motions as a function of time. Thus, they can be used to address specific questions about the properties of a modeled system, often more easily than experiments on actual systems. Of course, experiments play an essential role in validating the simulation methodology: comparisons of simulation and experimental data serve to test the accuracy of the calculated results and to provide criteria for improving the technique. Recent experimental advances allowed an increasingly accurate description of short time protein dynamics, that is more accessible to atomically detailed computer simulations and therefore open the way for more meaningful comparison to experiments.

The conformational study of biological molecules is one of the fundamental applications of MD simulations, therefore hundreds of articles can be found on conformational study using MD simulations. Consequently, we are not citing works related to conformational study using simulations and one may easily find several articles by single click on the web.

11.4.3 UNREVEALING THE FOLDING AND UNFOLDING PROCESS USING MD SIMULATIONS

The simulation of protein folding is the most computational expensive process that's why it was almost impossible to simulate during early age of MD simulations. However, due to rapidly enhancement in computational efficiency, folding simulation is being performed, nowadays. The main problem facing the attempt to study room temperature folding by direct molecular dynamics simulations of an all-atom model is that of time scales. Whereas protein folding takes place on the millisecond time scale and up, the time scale accessible to molecular dynamics is on the order of nanoseconds. Initially, using a massively parallel computer, Duan and Kollman (1998) performed a 1 μs simulation of the villin headgroup subdomain protein, a 36-residue peptide, in water. Starting from a fully unfolded extended state, including approximately 3000 water molecules, the simulation was able to follow the dynamics of this protein as it adopted a partially folded conformation. Such long-timescale molecular dynamics (MD) simulations require exceptionally large computational resources. Furthermore, the usefulness of these simulations is limited by the fact that they cannot provide the level of statistics required for studying folding kinetics and thermodynamics. Another problem associated with a direct MD approach to the folding process is that it is unclear how well the MD potential energy functions used fare in the unfolded regime. It is supposed that analysis of the unfolding process will contribute to the understanding of the folding process. Thus, instead of using molecular dynamics to simulate the folding process, many researchers turned their attention to using MD simulations as a tool for studying the inverse process of protein unfolding from the native state. It is in good practice to use high temperature molecular dynamics to speed up the unfolding process. Some reviews on unfolding process are stated elsewhere (Daggett and Levitt, 1994; Karplus and Sali, 1995; Karplus and Carflish, 1994).

Recently, Shaw group has engaged into the simulations of protein folding using high speed supercomputer where they have performed simulation on WW domain protein FiP35, which comprise a three-stranded *b*-sheet arranged as two *b*-hairpins, bind proline-rich sequences (Lindorff-Larsen et al., 2011). In simulations of Fip35—initiated from the extended state—the protein achieved the folded state, with a backbone root-meansquared deviation (RMSD) of ~1 A° from the crystal structure. The simulations were carried out under conditions where the folded and unfolded states exist in reversible equilibrium; repeated folding/unfolding barrier crossings followed a single well-defined pathway, with kinetics that closely matches experiment. Elucidation of the folding transition state allowed an even faster-folding Fip35 variant to be designed, which was subsequently confirmed experimentally. These folding results have been extended to encompass 12 small proteins of diverse structure—*a*-helical, *b*-sheet, and mixed *a*/*b* —with 8 of the 12 proteins reaching RMSD values less than 2 A° from the respective crystal structure. It is noteworthy that all 12 of these folding simulations used single, physics-based molecular mechanics force fields—a modified version of the CHARMM force field—indicating an increased level of accuracy that enables simulation of large conformational changes.

In addition to the simulation, several other protein folding simulations can be found on the web page of Theoretical and Computational Biophysics Group i.e. http://www.ks.uiuc.edu/Research/folding/.

11.4.4 CALCULATION OF FREE ENERGY CHANGES BY MD SIMULATION

From an MD trajectory the statistical equilibrium averages can be obtained for any desired property of the molecular system for which a value can be computed at each point of the trajectory. A number of thermodynamic properties can be derived from such averages. In principle, free energy perturbation and molecular dynamics (FEP/MD) simulations based on atomic models are the most powerful and promizing approaches to estimate binding free energies of ligands to macromolecules (Beveridge and Dicapua, 1989; Brandsak and Smalas, 2000; Florian et al., 2000; Lu and Kofke, 2001 A and B; Straatsma and McCammon, 1992). It is supposed that calculations based on FEP/MD simulations for protein-ligand interactions could become a useful tool in drug discovery and optimization (Ajay and Murcko, 1995; Boresch et al., 2003; Gilson et al., 1997; McCammon, 1998; Kollman, 1993; Lizaridis et al., 2002; Simonson et al., 2002; Woo and Roux, 2005).

To simulate accurately the behavior of molecules, one must be able to account for the thermal fluctuations and the environment-mediated interactions arizing in diverse and complex systems (e.g., a protein binding site or bulk solution). In FEP/MD simulations, the computational cost is generally dominated by the treatment of solvent molecules. Computational approaches at different level of complexity and sophistication have been used to describe the influence of solvent on biomolecular systems (Roux and Simonson, 1999). Those range from MD simulations based on all-atom models in which the solvent is treated explicitly (Brooks et al., 1988; Honig and Nicholls, 1995), to Poisson-Boltzmann (PB) continuum electrostatic models in which the influence of the solvent is incorporated implicitly (Edinger et al., 1997; Roux and Simonson, 1999). There are also semianalytical approximations to continuum electrostatics, such as generalized Born (Dominy et al., 1999; Ghosh et al., 1998; Jayram et al., 1998 A and B; Straatsma and McCammon, 1992), as well as empirical treatments based on solvent-exposed surface area (Colonna-Cesari and Sander, 1990; Cummings et al., 1995; Eisenberg and McClachlan, 1986; Lazaridis and Karplus, 1997; Scheraga, 1979; Stouten et al., 1993; Wesson and Eisenberg, 1992). However, even though such approximations are computationally convenient, they are often of unknown validity when they are applied to a new situation.

Some excellent reviews on statistical formulations of free energy from simulations can be found elsewhere (Gilson et al., 1997; Gilson and Zhou, 2007; Shirts, 2012; Woo and Roux, 2005; Zhou and Gilson, 2009).

11.4.5 SIMULATION APPROACHES FOR PROTEIN-LIGAND BINDING

11.4.5.1 MM-PBSA METHODS

The MM-PBSA method along with its GB variant (MM-GBSA), uses MD simulations of the free ligand, free protein, and their complex as a basis for calculating their average potential and solvation free energy. In this approach ΔG is written in terms of gas phase contribution, energy difference due to translational and rotational motion, desolvation energy and entropic contributions. Therefore:

$$G = G^{gas} + G^{tran/rot} + G^{sol} - TS^{ideal}$$

Here, the first term of right hand side of equation are the gas phase contributions which is the sum of the electrostatic energy, van der Waals energy and internal energy i.e. $G^{gas} = E^{vdw} + E^{ele} + E^{int}$ The second term of the RHS is the energy due to translational/rotational motion. In classical mechanics this term is equal to 3RT, which is generally omitted in MM-PB/SA calculations. The third term is the solvation energy term, which is further, composed of nonpolar solvation and polar solvation terms. The last term is the entropic contribution, which depends upon the degree of freedom of the translational, rotational and vibrational motion of the molecular system. Theoretically, the above formula is applicable for the ligand, protein and complex. The resultant binding free energy is expressed as:

$$\Delta G = G^{comp-} (G^{lig} + G^{protein})$$

which can be explained in details as (Predih et al., 2009).

Unbound ligand free energy: The free energy for the ligand is given by:

$$G(L) = E^{gas}(L^u) + G^{sol}(L^u) - TS^{ideal}(L^u)$$

where the superscript u indicates the unbound conformation of the ligand.

Unbound protein free energy: The protein is treated by classical mechanics during the simulation for unbound protein. Its free energy is composed of the total intramolecular energy, E^{gas} [van der Waal interaction (E^{vdw}) + Coulombic interaction (E^{coul}) + bonding interaction (E^{bond})], solvation free energy (E^{solv}) and entropic contributions ($-TS^{ideal}$).

$$G(P) = E^{gas}(P^u) + G^{solv}(P^u) - TS^{ideal}(P^u)$$

Complex free energy: It contains the sum of energy terms corresponding to the protein and ligand in bound state. The free energy of complex is composed of following terms:

$$G(C) = E^{gas}(P^b) + E^{gas}(L^b) + G^{solv}(C^b) + -TS^{ideal}$$

Here, E^{gas} is decomposed as the sum of E^{bond}, E^{coul} and E^{vdw} and superscript 'b' denotes the bound form.

The binding free energy for the noncovalent association of two molecules, may be written as:

$$\Delta G_{bind} = \Delta G\,(L+P \rightarrow C) = G(C) - G\,(L) - G\,(P) \text{ or}$$

$$\Delta G_{bind} = \Delta E^{int} + \Delta G^{solv} - T\Delta S^{ideal}$$

Here,

$$\Delta E^{int} = E^{int}(P^b) - E^{int}(P^u);$$

$$\Delta G^{solv} = G^{solv}(C^b) - G^{sol}(L^u) - G^{solv}(P^u), \text{ and}$$

$$-T\Delta S^{ideal} = -TS^{ideal} + TS^{ideal}(L^u) + TS^{ideal}(P^u)$$

Here, ΔE^{int} is the change in protein intramolecular energy, which is calculated by MM-PB/SA method. ΔG^{solv} (Fogolari et al., 2003; Gohlke and Case, 2004) is composed of a nonpolar contribution and a polar contribution. Nonploar solvation energy accounts for the unfavorable cavity formation and favorable van der Waals interaction between solute atoms and the solvent (Fogolari et al., 2003; Gohlke and Case, 2004)

$$E^{solv,\,np} = \gamma A + b$$

where A stands for solvent accessible surface area (SASA). γ and b are empirical constant which may have different values. The polar solvation energy is calculated using the linear Poisson- Boltzmann (PB) equation (Grochowaski and Trylska, 2008) that relates to the charge density $\rho(r)$ and to the electrostatic potential $\varphi(r)$, in a medium with non uniform dielectric permittivity $\varepsilon(r)$, which was used as 1 and 80 for the solute and the solvent, respectively.

$$\Delta\,\varepsilon(r)\,\Delta\,\varphi(r) = -4\pi\,\rho(r) + \kappa^2\,\varepsilon(r)\,\Delta\,\varphi(r)$$

where κ is the Debye-Huckel screening parameter to take into account the electrostatic screening effect.

$T\Delta S$ arises from the change in translational, rotational and vibrational degrees of freedom in the system upon the protein ligand binding. A protocol for free energy calculation is shown in Fig. 3.

The entropy term, due to the loss of degrees of freedom upon association, is decomposed into translational, S_{trans}, rotational, S_{rot}, and vibrational, S_{vib}, contributions. These terms are calculated using standard equations of statistical mechanics (Schwarzl et al., 2002), S_{rot} is a function of the moments of inertia of the molecule, whereas S_{trans} is a function of the mass and the solute concentration. S_{trans} is the only term in the free energy of an ideal solution that depends on solute concentration, leading to the concentration-dependence of the binding reactions. The vibrational entropy term is calculated with the quantum formula from a normal mode analysis (NMA) (Tidor and Karplus, 1994). A quasi-harmonic analysis of the MD simulations is also possible. However, it has been found that it does not always yield convergent values, even using very long MD simulation trajectories, and also led to large deviations from the results obtained with NMA, giving an overall unreasonable entropic contribution (Pearlman, 2005).

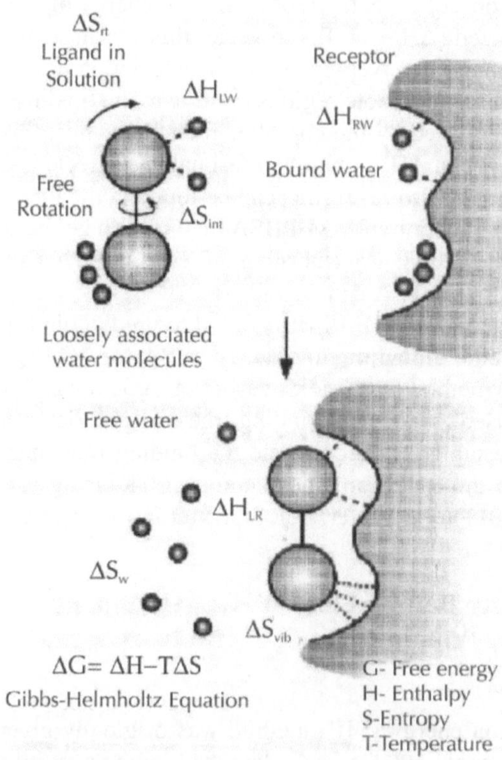

FIGURE 3 A simple protocol for calculations of free energy.

Generally in MM-PBSA methods the trajectories obtained from long MD simulations are averaged and single trajectories are analyzed for free energy calculations. However, its GB variant, MMGBSA, is often performed on single snapshots especially in industrial uses. The ions and the explicit solvent were striped before MM-PBSA analysis and periodic boundary conditions are also applied. A study in implicit solvent has been also performed by Pearlman and Rizzo (Pearlman, 2005; Rizzo et al., 2004). In the study by Pearlman, it has been found that MM-PBSA yields better results with MD simulations restrained around the X-ray structure, compared to unrestrained simulations. There are two possibilities regarding the MD simulation for MM-PBSA. In the first approach one should make three trajectories, one for the complex and each of the isolated partners, and calculate the energy terms using the adequate simulation. However, a popular alternative consists in performing only one MD simulation for the complex. In this variant, the terms relative to one isolated partner are calculated after removing the atoms of the other partner in the frames extracted from the MD simulation of the complex. As a consequence, the reorganization energy of the molecules upon association is neglected ($\Delta H_{intra} = 0$). However, this variant is less CPU demanding and leads to increased convergence due to cancelation of errors, reduction of noise arizing from flexible remote regions relative to the binding site, and conformational restraints imposed by the complex geometry. Thus, this one-simulation variant is attractive when ΔH_{intra} may be reasonably neglected. Comparisons between one- and three trajectories results can be found in the literature (Shao et al., 2007). The recent application of MM-PBSA method can be found in some recent study (Agrawal et al., 2008; Balasubrimanian et al., 2007; Dubey et al., 2011, 2012; Dubey and Ojha, 2011 A and B, and 2012).

For some systems, where MM/PBSA failed to predict binding free energy by an order of magnitude, a new method MM-QMSA (or MM/QM-COSMO as it was initially called) method from the Cavasotto group, in which QM calculations on full protein are used to calculate binding free energy (Victor et al., 2011). This method is already established to calculate the binding free energy of a series of phosphopeptides to the SH2 domain of human LCK using same methodology (Anisimov et al., 2011). Parameters for COSMO intrinsic solvation model have been also derived.

11.4.5.2 LIE METHODS

The linear interaction energy (LIE) method was originally given by Aqvist and co-workers (Aqvist et al., 1994) to calculate the binding affinity of protein ligand complexes. It is also known as a linear response method and it is a semi empirical method for estimating absolute free energies and requires just two simulations, one for the ligand in solution and other for the ligand in the protein binding site. The snapshots saved from the simulations represent Boltzmann ensembles of conformations and are used to compute the Boltzmann-averaged electrostatic and van der Waals interaction energies of the ligand with its environments in the bound and Free states. The binding free energy is estimated as:

$$\Delta G = \beta\,(<U^{elec}_{ligand\text{-}protein}> - <U^{elec}_{ligand\text{-}solvent}>) + \alpha\,(<U^{vdw}_{ligand\text{-}protein}> - <U^{vdw}_{ligand\text{-}solvent}>) + \gamma$$

As usual, the angle bracket indicates Boltzmann average, α and β are two parameters. To determine ΔG one thus needs to perform just two simulations, one of the ligand in the solvent and other of the ligand in the bounded form with the protein. The first term describes the electrostatic contributions in the ΔG according to the linear response approximation (LRA) theory (Aqvist et al., 1994; Lee et al., 1992). The second term in the above equation holds for the nonpolar contributions to ΔG. Its linear relationship with the surrounding van der Waals energy is based on the observation that solvation energies of nonpolar compounds are linearly correlated with the surrounding van der Waals energies (Aqvist and Hansson, 1996; Hansson et al., 1998). γ is a constant that can be added to get a correct energy. In the initial implementation, β was fixed to ½ following the LRA approximation, while α was fitted empirically to a value of 0.16 to reproduce the experimental activity of four structurally related endothiapepsin inhibitors (**Ljungberg et al., 2003**). γ was kept to 0 to limit the overparameterization. Although these parameters gave satisfying results for protein-ligand systems, it was found using FEP calculations that β could be considered as a function of the ligand nature. Values of 0.5, 0.43, 0.37, and 0.33 were suggested for ionic molecules and neutral compounds with one, two, or more hydroxyl groups, respectively (Carlson and Jorgenson, 1995; Zhou et al., 2001). A value of 0.18 was found to be optimal for α. Non-zero values of γ can be necessary to reproduce ΔG for some systems (Aqvist et al., 2002). More recently, it was suggested that γ could be expressed as a function of the buried solvent accessible surface area (SASA) of the ligand that is buried upon complexation, leading to the modified equation:

$$\Delta G = \beta\,(<U^{elec}_{ligand\text{-}protein}> - <U^{elec}_{ligand\text{-}solvent}>) + \alpha\,(<U^{vdw}_{ligand\text{-}protein}> - <U^{vdw}_{ligand\text{-}solvent}>) + \gamma\,(<SASA_{ligand\text{-}protein}> - <SASA_{ligand\text{-}solvent}>)$$

However, in the study of Aqvist et al. (Aqvist and Marelius, 2001; Carlsson et al., 2006) the addition of this term is questioned since buried SASA is correlated with van der Waals term. Some efforts have been done to reduce the computational costs where the electrostatic terms are replaced by the function of Coulombic interaction between the ligand and the protein and the solvation reaction field energy. This variant is attractive due to reduced computational costs.

A study of Alam et al. (Alam and Naik, 2009) shows the reliability of LIE method. In this study, a training set of 76 podophyllotoxin analogs was used to build a binding affinity model for estimating the free energy of binding for 36 inhibitors (test set) with diverse structural modifications. They found that the average root mean square error (RMSE) between the experimental and predicted binding free energy values was 0.56 kcal/mol, which is comparable to the level of accuracy achieved by the most accurate methods, such as free energy perturbation (FEP) or thermodynamic integration (TI). The squared correlation coefficient between experimental and SGB-LIE estimates for the free energy for the test set compounds were also significant. Similar applications were found by many studies (Kayani and Goliaei, 2009; Singh et al., 2005).

Several important contributions to molecular recognition are neglected in LIE, such as the conformational rearrangement upon complexation of the ligand and the receptor, the receptor desolvation energy, and the entropies. However, it has been argued that these terms are implicitly taken into account by the LRA approximation and the adjustable parameters of the model. Nonetheless the quality of the results obtained by LIE methods is somewhat surprizing, because the LIE method does not account explicitly for standard configurational entropy or the internal energy of the ligand. This method may be successful in part, because it is generally used to compare ligands within a single chemical series.

11.4.5.3 FREE ENERGY PERTURBATION METHODS

Among all the known methods of binding free energy calculations, this method gives the most promizing results. However, due to large computational time, these methods are less applicable for huge system. A brief principle of this method is discussed below.

The change in free energy between two states of a system, such as before and after binding, can be written as $-RT\text{In}<e^{-\Delta U/RT}>$, where ΔU is the change in the energy function between the initial and final states, and the angle bracket indicates a Boltzmann average taken in the initial state. The change in the energy can be found by the interaction energy term and the Boltzmann average taken by a classical MD run. In reality, such a simulation is extraordinarily difficult to converge unless the initial and final states are simple. The FEP methods solves this problem by breaking the change into small steps δU (perturbations) and running a separate simulations for each resulting energy function U_i to obtain the stepwise free energy changes. Pathways (perturbation) methods can also be used to compute the standard binding free energy of a protein and ligand.

The free energy difference between two states A and B can formally be obtained from Zwanzig's formula (Zwanzing, 1954)

$$\Delta G = G_B - G_A = \beta^{-1}\text{In}<\exp(-\beta\Delta V)>_A$$

where $\beta = 1/KT$ and angular bracket denotes a MD generated ensemble average.

The main criterion for the above equation to be practically useful is that the configurations sampled on the potential V_A should have a reasonable (at least nonvanishing) probability of occurring also for V_B. This essentially means that thermally accessible regions of the two potentials should have a significant degree of overlap. If not, the result will be a very slow convergence of the average. That convergence can be assessed by interchanging the labels A and B and changing the sign of ΔG in above equation, thus applying the formula "backwards."

In order to solve the above convergence problem associated with the implementation of the above equation, a multistage approach is normally adopted. A path between the states A and B is defined by introducing a set of intermediate potential energy functions that are usually constructed as linear combinations of the initial (A) and final (B) state potentials

$$V_m = (1 - \lambda_m) V_A + \lambda_m V_B$$

where λ_m varies from 0 to 1. In practice this path is thus discretized into a number of points (m=1… n), each represented by a separate potential energy function that corresponds to a given value of λ. This coupling parameter approach rests upon the fact that the free energy difference is uniquely defined by the initial and final states (i.e., a state function) and can be computed along any reversible path connecting those states. Now the total free energy change can be obtained by summing over the intermediate states along the λ variable

$$\Delta G = G_B - G_A = \beta^{-1} \text{In} < \exp(-\beta (V_{m+1} - V_m)] >_m$$

This approach is generally referred to as the free energy perturbation (FEP) method.

FEP methods can also be used to compute the standard binding free energy of a protein and ligand. Many successful applications of these methods can be found elsewhere (Chandani et al., 2005). Although the equation used in FEP method is exact, many studies have demonstrated that except in the case of rather small changes, it converges as a function of the amount of data collected is far from ideal (Shirts and Pande, 2005).

11.4.5.4 THERMODYNAMIC INTEGRATION (TI)

The energy expression of thermodynamical integration can be obtain by taking derivative of the free energy with respect to some continuous parameter λ describing a series of intermediate alchemical states-

$$dG/d\lambda = d/d\lambda \int e^{-\beta H(\lambda)}\, dx = <dH(\lambda)/d\lambda>_\lambda$$

$$\Delta G = \int \langle dH(\lambda)/d\lambda \rangle_{\lambda}$$

where the pathways of intermediates between the states of interest is parameterized between λ=0 and λ=1. When the end states have different masses, the momenta will have λ dependence as well. The thermodynamic integration trades variance for bias. Averaging over $\langle dH/d\lambda \rangle$ will require fewer uncorrected samples to reach a given level of relative error than averaging $e^{-\beta\Delta H(x)}$, as as long as $\langle dH/d\lambda \rangle$ is well behaved. However, to compute the total free energy from a series of individual simulations, we must use some type of numerical integration of the integral, which, by definition, introduces bias for which a number of different numerical techniques have been applied (Lu et al., 2003; Reset and Mezei, 1993).

For alchemical changes that result in smooth, monotonic curves for $\langle dH/d\lambda \rangle$, TI can be quite accurate using a relatively small number of points. However, if the curvature becomes large, as can frequently be the case in alchemical simulations where Lennard–Jones potentials are turned on or off, then the bias introduced by discretization of the integral can become large (Lopex and Kollman, 1993; Piterra and Gunsteren, 2002). Even in the case of small curvature (i.e., charging of SPC water in water) reasonably large errors can be introduced (i.e., 5–10% of the total free energy with 5 values) (Lopex and Kollman, 1993).

For relatively long simulations, TI becomes bad approximation and it should be avoided.

11.5 CONCLUSION AND FUTURE ASPECTS

With the increase of computational efficiency and robotic techniques, MD simulations have become a frequently used method that provide a tangible link between theories and experiment, by quantifying at the thermodynamic level the physical phenomena modeled by statistical simulations. With 20 years of hindsight gained from methodological development and characterization, a variety of problems of both chemical and biological relevance can now be tackled with confidence. Hopefully, we will be able to perform simulations for folding of proteins of large size at milli- to micro second levels. Using MD simulations, the structure of the target (or target complex), if not available, might be obtained using folding simulations. Alternatively, simulations may be used to prepare homology models as accurate as an experimental structure. Similarly, the free energy methods, in near future, will become an avoidable element of screening pipelines, discriminating between candidates, selected from cruder approaches to retain only the best leads towards a given target. The implementation of quantum mechanical methods in free energy calculations will drastically enhance the accuracy of binding free energy and we can easily elaborate the polarization and electronic effect in protein-ligand binding as well as protein-protein interactions in upcoming years.

ACKNOWLEDGMENTS

KDD acknowledge Council for Scientific and Industrial Research, New Delhi for SRF.

KEYWORDS

- ***ab initio* calculations**
- **force fields**
- **linear interaction energy**
- **Maxwellian velocity distribution**
- **Schrodinger wave equation**
- **SHAKE**
- **Verlet algorithm**

REFERENCES

Agrawal, S.; Ojha. R. P.; Maiti, S. J. Phys. Chem. B., **2008,** *112,* 6828–6836.

Ajay, A.; Murcko, M. A. J. Med. Chem. **1995,** *38,* 4953–4967.

Alam, M. A.; Naik, P. K. J. Mol. Graph. Mod., **2009,** *27,* 930–947.

Allen, M.; Tildsley, P. Computer simulation of liquids. **1987,** Oxford, Oxford University Press.

Anisimov, V. M.; Cavasotto, C. M. J. Phys. Chem. B. **2011,** *115,* 7896–7905.

Anisimov, V. M.; Cavasotto, C. N. J. Comput. Chem., **2011,** *32,* 2254–2263.

Åqvist, J.; Hansson, T. J. Phys. Chem., **1996,** *100,* 9512.

Åqvist, J.; Luzhkov, V. B.; Brandsdal, B. O. Acc. Chem. Res. **2002,** *35,* 358

Åqvist, J.; Marelius, J. J. Combin. Chem. High. Through. Scr., **2001,** *4,* 613.

Åqvist, J.; Medina, C.; Samuelsson, J.-E. Protein Eng., **1994,** *7,* 385.

Balasubrimaniam, C.; Ojha, R. P.; Maiti, S. Biochem. Biophys. Res. Com., **2007,** *355,* 1081–1086.

Beglov, D.; Roux, B. J. Chem. Phys. **1994,** *100,* 9050.

Berger, A; Linderstrom-Lang, K. Arch. Biochem. Biophys. **1957,** *69,* 106–118.

Berkowitz, M. L.; McCammon, J. A. Chem. Phys. Lett. **1982,** *90,* 215.

Beveridge, D. L.; Dicapua, F. M. Annu. Rev. Biophys. Biol. **1989,** *18,*431–492.

Boresch, S.; Tettinger, F.; Leitgeb, M.; Karplus, M. J. Phys. Chem. B. **2003,** *107,* 9535–9551.

Borhani, D. W.; Shaw, D. E.; J. Comp. Aid. Drug. Design, **2012,** *26,* 15–26.

Bortolato, A.; Moro, S. J. Chem. Inf. Model., **2007,** *47,* 572–82.

Brandsdal, B. O.; Smalas, A. O. Protein Eng. **2000**, *13,* 239–245.

Brenedsen, H. C.; van Gunsteren, W. F. Molecular dynamics simulations: Techniques and approaches. In Barnes AJ, Orville- Thomas WJ and Yarwood J (Editors). Molecular liquids, dynamics and interactions. NATO ASI Series C*135,* New York, Reidel, 475–600.

Brenedsen, H. C.; van Gunsteren, W. F. Practical algorithms for dynamic simulations. Moleculardynamics simulations of statistical mechanical systems. Proceedings of the Enrico Fermi Summer School Varenna Soc. Italian di Fiscia. Bolgana, **1986,** 43–65.

Brenedsen, H. J. C..; Postma J. P. M..; van Gunsteren, W. F.; Di Nola, A.; Haak, J. R., J. Chem. Phys. **1984,** *81,* 3684–3690.

Brooks III, C. L.; Karplus, M.; Pettitt, B. M. Proteins. A theoretical perspective of dynamics, structure and thermodynamics. In Advances in Chemical Physics, Vol. LXXI. I. Prigogine and S. A. Rice, editors. John Wiley & Sons, New York, 1988.

Brooks, B.R; Karplus, M. Proc. Nat. Acad. Sci. **1983,** *80,* 6571–6575.

Brooks, C. L. III.; Case, D. A. Chem. Rev. **1993,** *93,* 2487–2502.

Brooks, C. L., III; Karplus, M.; Pettitt, B. M. Adv. Chem. Phys. **1988,** *71,* 1–249.

Brooks, C.; Brthger, A.; Karplus, M. Biopolymers **1985,***24,* 843–865.

Brunger, A.T; Brooks, C.L III; Karplus, M. Proc. Nat. Acad. Sci. **1985,** *82,* 8458–8462.

Brunger, A.T; Kuriyan, J; Karplus, M. Science. **1987,** *235,* 458–460.

Caflisch, A.; Karplus, M. Molecular dynamics studies of protein and peptide folding and unfolding. In: Merz, K. Jr, Grand, S. Le. eds. The Protein Folding Problem and Tertiary Structure Prediction. Boston: Birkhauser, **1994,** pp 193–230.

Carlson, H. A.; Jorgensen, W. L. J. Phys. Chem., **1995,** *99,* 10667–10673.

Carlsson, J.; Ander, M.; Nervall, M.; Åqvist, J. J Phys Chem B., **2006,** *110,* 12034–12041.

Case, D.A; Karplus M. J. Mol. Biol. **1979,** *132,* 343–368.

Chandani, S.; Lee, C. H.; Loecher, E. L. Chem. Res. Tox., **2005,** *18,* 1108–1123.

Chaubey, A. K.; Dubey, K. D.; Ojha, R. P. J. Comp. Aided. Drug. Design., **2012.**

Colonna-Cesari, F.; Sander. Biophys. J. **1990,** *57,* 1103–1107.

Colonna-Ceseri, F. et al. J. Biol. Chem., **1986,** *261,* 15273–15280.

Cummings, M. D.; Hart, T. N.; Read, R. J. Protein Sci. **1995,** *4,* 2087–2089.

Daggett, V.; Levitt, M. Curr. Opin. Struct. Biol., **1994,** *4,* 291–295.

Dominy, B. N.; Brooks, C. L. III. J. Phys. Chem. B. **1999,** *103,* 3765–3773.

Duan, Y.; Kollman, P. A. Science, **1998,** *282,*740–744.

Dubey, K. D.; Chaubey, A. K.; Ojha, R. P. Biochim. Biophys. Acta, **2012,** DOI: 10.1016/j.bbapap.2012.08.014.

Dubey, K. D.; Chaubey, A. K.; Ojha, R. P. Biochim. Biophys. Acta., **2011,1814,** 1796–1801.

Dubey, K. D.; Ojha, R. P. J. Biol. Phys.,**2012,** *37,* 69–78.

Dubey, K. D.; Ojha, R. P. J. Mol. Model., **2012,** 1679–1689.

Dubey, K. D.; Ojha, R. P. Mol. Simult., **2012,** *37,* 1151–1163.

Edinger, S. R.; Cortis, C.; Shenkin, P. S.; Friesner, R. A. J. Phys. Chem. B., **1997,** *101,* 1190–1197.

Eisenberg, D.; McClachlan, A. Nature. **1986,** *319,* 199–203.

Evans, D. J.; Morriss, G. P. Comp. Phys. Repts. **1984,** *1,* 297.

Feller, S. E.; Zhang, Y.; Pastor, R. W.; Brooks, B. R. J. Chem. Phys. **1995,** *103,* 4613.

Florian, J.; Goodman, M. F.; Warshel. A. J. Phys. Chem. B. **2000**, *104,* 10092–10099.

Fogolari, F.; Brigo, A.; Molinari, H. Biophys. J. **2003,** *85,*159–166.

Fraternali, F.; Gunsteren, W. F. V. J. Mol. Biol. **1996**, *256,* 939–948.

Frauenfelder, H. et al. Biochemistry. **1987,** *26,* 254–261.

Ghosh, A.; Rapp, C. S.; Friesner. R. A.; J. Phys. Chem. B. **1998**, *102,* 10983–10990.

Gilson, M. K.; Given, J. A.; Bush, B. L.; McCammon, J. A. Biophys. J, **1997,** *72,* 1047–1069.

Gilson, M. K.; Zhou, H. X. Ann. Rev. Biophys. **2007,** *36,* 21–42.

Gohlke, H.; Case, D. A. J. Comput. Chem. **2004,** *25,* 238–250.

Grochowski, P.; Trylska. J. Biopolymers., **2008,** *89,* 93–113.

Haile, J. M. Molecular dynamics simulation. Elementary methods. New York, John Wiley & Sons. **1992**.

Hansson, T.; Marelius, J.; Åqvist, J. J Comput Aided Mol Des, **1998,** *12,* 27–35.

Harvey, S.; McCammon, J. A. Dynamics of Proteins and Nucleic Acids; Cambridge University Press: Cambridge, **1987**.

Harvey, SC; Prabhakaran, M; Mao, B; MacCammon, J. A. Science, **1984,** *223,* 1189–1191.

Helfand E. Science, **1984,** *226,* 647–650.

Hockney, R. W. Methods. Comput. Phys., **1970,** *9,* 136.

Honig, B.; Nicholls. A. Science. **1995,** *268,* 1144–1149.

Hoover, W. G. Phys. Rev. A., **1985,** *31,* 1695.

Jain, A.; Sankararamkrishnan, R. Biophys. J, Suppl. **2010,** *98,* 635a-636a.

Jayaram, B.; Liu, Y.; Beveridge, D. L. J. Chem. Phys. **1998**, *109,* 1465–1471.

Jayaram, B.; Sprous, D.; Beveridge, D. L. J. Phys. Chem. B. **1998**, *102,* 9571–9576.

Kampmann, T.; Mueller, D. S.; Mark, A. E.; Young, P. R.; Kobe, B. Structure., **2006,** *14,* 1481–1487.

Kang, Y. K.; Gibson, K. D.; Nemethy, G.; Scheraga, H. J. Phys. Chem. **1988,** *92,* 4739–4742.

Karplus, M.; McCammon, J. A. Nat. Struct. Biol., **2002,** *9,* 646–652.

Kollman, P. Chem. Rev. **1993,** *93,* 2395–2417.

Kyani A.; Goliaei. B. J. Mol. Struct., **2009,** *913,* 63–69.

Lazaridis, T.; Karplus, M. Science, **1997,** *278,* 1928–1931.

Lazaridis, T.; Masunov, A.; Gandolfo. F. Proteins, **2002,** *47,* 194–208.

Leach, A.R; Molecular modeling: Principle and application, Pearsons Prentice Hall.

Lee, F. S.; Chu, Z. T.; Bolger, M. B.; Warshel. A. Protein. Eng., **1992,** *5,* 215–228.

Lindorff-Larsen, K.; Piana, S.; Dror, R. O.; Shaw, D. E. Science., **2011,** *334,* 517–520.

Ljungberg, K. B.; Marelius, J.; Musil, D.; Svensson, P.; Norden, B.; Åqvist, J. Eur. J. Pharm. Sci. *12*, 441–446.

Lopex, M. A.; Kollman, P. A. Prot Sci, **1993,** *2,* 1975–1986.

Lu, N. D.; Kofke, D. A. J. Chem. Phys., **2001,** *114,* 7303–7311.

Lu, N. D.; Kofke, D. A. J. Chem. Phys., **2001**, *115,* 6866–6875.

Lu, N. D.; Singh, J. K.; Kofke, D. A. J. Chem. Phys. **2003,** *118,* 2977–2984.

M Karplus, A Sali. Curr Opin Struct Biol **1995**, *5*, 58–73.

McCammon, J. A. Curr. Opin. Struct. Biol., **1998**, *8,* 245–249.

McCammon, J. A.; Gelin, B. R.; Karplus, M. Nature, **1977**, *267, 585.*

McCammon, J. A.; Harvey, S. C. Dynamics of proteins and nucleic acids. Cambridge, CambridgeUniversity Press, 1987.

Modis, Y.; Ogata, S.; Clements, D.; Harrison, S. C. Nature., **2004,** *427,* 313–319.

Modis, Y.; Ogata, S.; Clements, D.; Harrison, S. C. Proc. Natl. Acad. Sci., **2003,** *100,* 6986–6991.

Mueller, D. S.; Kampamann, T.; Yennamalli, R.; Young, P. R.; Kobe, B.; Mark, A. E. Biochem. Soc. Trans. **2008,** *36,* 43–45.

Nilson, L; Clore, G.M; Gronenborn, A.M; Brunger, A.T; Karplus M. J Mol Biol., **1986,** *188,* 455–475.

Nose, S. J. Chem. Phys., **1984,** *81,* 511–519.

Pearlman, D. A. J. Med. Chem, **2005,** *48,* 7796–7807.

Perdih, A.; Bren, U.; Solemajer, T. J Mol. Mod., **2009,** *15,* 983–996.

Pitera, J. W.; vanGunsteren, W. F. Mol. Simulat. **2002,** *28,* 45–65.

Potter, D. Computational Physics. New York: Wiley, **1972,** Chap 2.

Resat, H.; Mezei, M. J. Chem. Phys. **1993,** *99,* 6052–6061.

Rizzo, R. C.; Toba, S.; Kuntz, I. D. J Med Chem., **2004,** *47,* 3065–3074.

Roux, B.; Simonson, T. Biophys. Chem., **1999**, *78,* 1–20.

Scheraga, H. A. Acc. Chem. Res., **1979,** *12,* 7–14.

Schwarzl, S. M.; Tschopp, T. B.; Smith, J. C.; Fischer, S. J. Comp. Chem., **2002,** *23,* 1143–1149.

Shao, J.; Tanner, S. W.; Thompson, N.; Cheatham, T. E. I.II. J Chem. Theory. Comput., **2007,** *3,* 2312–2334.

Shirts, M. R. Meth. Mol. Biol. **2012,** *819,* 425–467.

Shirts, M. R.; Pande, V. S. J. Chem. Phys. **2005,** *122,* 144107.

Simonson, T.; Archontis, G.; Karplus, M. Acc. Chem. Res., **2002,** *35,* 430–437.

Singh, P.; Mhaka, A. M.; Christensen, S. B.; Gray, J. J.; Denmeade, S. R.; Isaacs, J. T. J. Med.Chem., **2005,** *48,* 3005–3014.

Smith, J.; Causack, S.; Pezzeca, U.; Brooks, B. R.; Karplus, M. J. Chem. Phys., **1984,** *85,* 3636–3654.

Stouten, P.; Frommel, C.; Nakamura, H.; Sander, C. **1993**. Mol. Sim. *10,* 97–120.

Straatsma, T. P.; McCammon, J. A. Annu. Rev. Phys. Chem., **1992,** *43,* 407–435.

Swope, W.; Andersen, H. C.; Berens, H.; Wilson, KR. J. Chem. Phys. **1992**, *76,* 637.

Tidor, B.; Karplus, M. J. Mol. Biol., **1994,** *238,* 405–414.

van Gunsteren, W. F. Molecular dynamics and stochastic dynamic simulations: A Primer. In van Gunsteren W. F.; Weiner P. K.; Wilkinson A. J. (Editors). Computer simulations of biomolecular systems, Vol 2, Leiden, ESCOM.

van Gunsteren, W. F.; Brenedsen, H. C. Angew. Chemie. Intern. Ed. Eng, *29,* 992–1023.

Verlet, L. Phys. Rev. **1967,** *159,* 98–103.

Victor, A.; Ziemys, A.; Kizhake, S.; Yuan, Z.; Natarajan, A.; Cavasotto, C J. Comp. Aided. Drug. Desg. **2011,** *25,* 1071–1084.

Wesson, L.; Eisenberg, D. Protein. Sci., **1992,** *1,* 227–235.

Woo, H. J.; Roux, B. Proc. Nat. Acad. Sci., **2005,** *102,* 6825–6830.

Zhou, H. X.; Gilson, M. K. Theory.Chem. Rev. **2009,** *109,* 4092–4107.

Zhou, R. H.; Friesner, R. A.; Ghosh, A.; Rizzo, R. C.; Jorgensen, W. L.; Levy, R. M. J Phys Chem B., **2001,** *105,* 10388–10397.

Zwanzig, R. W. J. Chem. Phys. **1954,** *22,* 1420–1426.

CHAPTER 12

STRUCTURAL INFORMATION FROM MOLECULAR DESCRIPTORS BASED ON THE MONTE CARLO METHOD: APPLICATIONS TO NANOSCALE STRUCTURES

DAN CIUBOTARIU, VALENTIN GOGONEA and CIPRIAN CIUBOTARIU

CONTENTS

Abstract 402
12.1 Introduction..402
12.2 The Essence of the Monte Carlo method...403
12.3 Monte Carlo Method as a General Integration Technique.............406
12.4 Molecular Descriptors Calculated by Means of the Monte Carlo method 408
12.4.1 Molecular Volume...408
12.4.2 Molecular Surface...413
12.4.3 Synthetic Descriptors of Molecular Size and Shape.........417
12.4.4 Steric Static Integral Parameter 421
12.5 Descriptors for the Accessibility of Molecular Surface and the Dynamics of the Molecular Structure...429
12.6 Simulations of the Copolymerization Reactions based on the Monte Carlo Method...433
12.6.1 Introduction..433
12.6.2 Implementing Monte Carlo Models on Computers..........436
12.6.4 Simulation of Copolymerization Reactions......................441
12.7 Description of the IRS Computer Program 442
12.8 Conclusions 444
Keywords 446
References..446

ABSTRACT

In this chapter, we intend to review our contributions to the field of molecular descriptors calculated by Monte Carlo method. We begin with a short description of the fundamentals of Monte Carlo method and its general application as a stochastic integration technique for objects in an N-dimensional space. We describe the structural descriptors of molecular shape and size, which have been developed by us in the approximation of hard-spheres (i.e., the molecule is considered as a collection of relatively impenetrable atomic spheres, positioned at their Cartesian coordinates and having radii equal to van der Waals (vdW) radii). The simulation of the copolymerization reactions is also included, as another example of usefulness of the Monte Carlo method in other areas of theoretical research. The applications of these molecular descriptors are discussed in each chapter.

We emphasize that the Monte Carlo algorithms described here are general and applicable to any shape or/and size of a molecular structure or arbitrary object.

12.1 INTRODUCTION

In this chapter, we present a series of molecular descriptors (MDs) developed at molecular nanoscale dimensions. MDs were developed in the past years based on geometrical three-dimensional (3D) molecular models. The standard (Ciubotariu, 1987) or optimized geometry (Ciubotariu et al., 1991; Gogonea et al., 2010) of molecules and their van der Waals (vdW) spaces were used to calculate these MDs. MDs have been introduced as measures of molecular size and shape, and were applied for modeling various chemical and biological properties of organic molecules by means of quantitative structure activity relationship (QSAR) studies.

We also present here some characteristics of the Monte Carlo (MC) method, which has been used by us to develop the algorithms for calculating these MDs by stochastic integration and to simulate the polymerization reactions.

The *Monte Carlo* method (or methods) emerged during the Second World War, and they were used for simulating the disintegration of uranium atoms under bombardment with neutrons, an experiment initially difficult to perform under real conditions. The name "*Monte Carlo*" appears explicitly in 1949 in the work of Metropolis and Ulam (Metropolis and Ulam, 1949), owing its origin to the use of random number tables displayed on the Monte Carlo casino roulette.

By applying this method in chemistry it was possible to estimate the free energy, to simulate kinetic processes and those in the collision theory, (Binder, 1955; Laidler, 1969; Szabo and Ostlund, 1989) as well as to quantitatively assess the steric effects by using van der Waals (vdW) indicators (Ciubotariu, 1987; Voiculescu-Duvaz et al., 1991). Beginning with the second half of the 60s,

Monte Carlo techniques started to be applied in macromolecular chemistry as well (Baeurle, S. A., 2009).

12.2 THE ESSENCE OF THE MONTE CARLO METHOD

The *Monte Carlo* method is a set of techniques simulating a system with multiple degrees of freedom. It has a wide applicability, conditioned, however, by the use of electronic computers. The method is based on using the intimate relationship between solutions to various problems and some statistical characteristics of random experiments. Consequently, the use of these techniques is conditioned by the possibility to move from a given problem to a *random model*, which can be easily implemented, on a computer, which allows solving the respective problem.

The usual way to solve a problem consists of developing an algorithm in order to obtain an unknown quantity y either exactly or with a given accuracy. In this case, if $y_1, y_2, ..., y_n, ...$ are the results corresponding to an iterative process of successive accumulated approximation, we are able to write:

$$y = \lim_{n \to \infty} y_n \tag{1}$$

In this case, the process expressed by relationship (1) is deterministic, and it may stop after a finite number of steps.

But, there are some specific problems for which the use of such algorithms is practically impossible or, in the case that they might be applicable, the algorithms are prohibitively complicated. For such problem one can model the mathematical or physical essence of the problem using the law of large numbers of the probability theory by means of a random model (Demidovich and Maron, 1981). In this random model the average value, y^*, of a certain random variable is usually equal to the unknown quantity y. The practical application is based on the fact that the mean value of the given random variable (y^*) can be estimated with the aid of the mean selection value (y_s^*), which depends, obviously, on the selection volume. The corresponding estimations $y_1^*, y_2^*, ..., y_n^*, ...$ of the quantity y will be obtained after the statistical processing of the data provided by the random experiment under study. It results that the random variable y^*, with its values y_1^*, $y_2^*, ..., y_n^*, ...$, must be convergent in probability to the value y when $n \to \infty$, namely for any $\varepsilon > 0$ the following relationship applies:

$$\lim P\left(\left|y - y_n^*\right| < \varepsilon\right) = 1 \tag{2}$$

where P stands for probability. The choice of the quantity y_n* is determined by the specific particularities of the modeled phenomenon. In this case, the process is not deterministic, since it is defined by a random model (experiment).

Problem-solving procedures using random models have been called *Monte Carlo* methods (or models). A particularity of this method is that the analysis of the developed model (experiment) and the obtained results are replaced by the results of calculations made with random numbers. The estimation of the desired quantity, y^*, is obtained statistically and has a probabilistic nature (Knuth, 1983).

The exact definition of random numbers (or of a set of random numbers) is rather complicated (Knuth, 1983), the main difficulties arising from the meaning of the word *random*. To a certain extent, random numbers do not exist; to be more precise, references must be made to a set of random, independent numbers, with a specified distribution. Here, we adopt the definition of von Mises (von Mises, 1919). Thus, the numbers x_1, x_2, ..., x_n within an interval I represent a set of random numbers provided that the following two conditions are fulfilled:

(i) $(x_i)_{i=1,\,n}$ satisfy a certain distribution law, and
(ii) the distribution law is nonvariant, in the conditions of a common selection rule, for any subset of the $(x_i)_{i=1,\,n}$ set.

The fundamental set of random numbers is the set of random numbers uniformly distributed on the interval (0,1), namely $x_i \in (0,1)$, $i=1, 2, \ldots, n$ and the probability density function is $F(x)=1$ if $x \in (0,1)$ and $F(x)=0$ if $x \notin (0,1)$. Once this set is obtained, other sets of random numbers may be derived from it, with any distribution on any interval I.

In order to obtain random numbers uniformly distributed on a common interval, the following types of methods can be used: manual method (roulette, dice, etc.), random number tables, analogical methods (representing random physical phenomena), and arithmetic methods (Feuvrier, 1972).

For practical purposes, the most beneficial method consists in obtaining random numbers on computers, with the help of arithmetic methods called *random number generators*; the numbers obtained are called *pseudorandom*, because the algorithms used are deterministic.

The most used algorithms are congruential, such as:

$$x_{n+1} = f(n);\, x_0 \text{ being given} \tag{3}$$

Before using them, it is advisable to test the quality of the random number generators implemented on computers. This is usually done by checking if the following two criteria are fulfilled: the uniformity of the generated set $(x_i)_{i=1,\,n}$ on the interval (0,1) and the independence of its elements. These conditions are checked by calculating the moments $M_k, k=1,4$ and applying the test χ^2 (Knuth, 1983; Feuvrier,

1972). Nowadays all computers have implementations of excellent quality of random number generators, uniformly distributed on the interval (0,1).

As previously mentioned, the unknown quantity evaluated by the random model attached to a certain problem usually appears as mean value of a certain random variable. In this case, the estimation given by the mean value of selection for the mean theoretical value is consistent, unbiased and efficient (Demidovici and Maron, 1981; Stoka, M. and Teodorescu, R., 1966).

Suppose the observable y is calculated with the help of a well-defined *Monte Carlo* model. The expected value of y is $E(y)$. Based on the model, it is possible to obtain, one by one, the different values y_i, $i=1, N$, namely $y_i \neq y_j$ if $i \neq j$ by using various sets of random numbers. After N simulations, the average value of y will be $\bar{y}$:

$$\bar{y} = \frac{1}{N}\sum_{i=1}^{N} y_i \tag{4}$$

According to the law of large numbers, the following equation results:

$$\sigma^2(\bar{y}) = \frac{\sigma^2(y)}{N} = 0 = E\{[\bar{y} - E(y)]^2\} \tag{5}$$

where σ^2 stands for variation. By reaching the limit in Eq. (5), we obtain:

$$\sigma^2(\bar{y}) = \frac{\sigma^2(y)}{N} = 0 = E\{[\bar{y} - E(y)]^2\} \tag{6}$$

Relationship (6) illustrates that $\bar{y}$ approaches the true value $E(y)$ as N grows. Hence, it can be *concluded* that the accuracy of *Monte Carlo* calculations depends on the value of N.

In order to estimate the value of N for a given accuracy of y, one may use Bernoulli's theorem and apply Chebyshev's inequality (Demidovici and Maron, 1981):

$$P\left[\left|\bar{y} - E(y)\right| \le \varepsilon\right] \ge 1 - \frac{\sigma^2(y)}{\varepsilon^2 N} \tag{7}$$

where $\sigma^2(y)$ is estimated by the selection variation, s^2:

$$s^2 = \frac{1}{N-1}\sum_{i=1}^{N}(y_i - \bar{y})^2 \tag{8}$$

An equivalent form of relationship (7), where $\sigma^2(y)$ is estimated by s^2 in Eq. (8), is the following:

$$P\left[\bar{y}-\varepsilon<E(y)<\bar{y}+\varepsilon\right]>1-\frac{\sigma^2(y)}{\varepsilon^2 N} \tag{9}$$

Relationship (9) indicates that the true value $E(y)$ belongs to the interval $(y-\varepsilon, y+\varepsilon)$ with a (guaranteed) probability higher than $1-\delta$, where $\delta=\sigma^2(y)/(\varepsilon^2 N)$. Hence, it can be inferred that the estimation accuracy, ε, for a maximum given probability is inversely proportional to the square root of the number of experiments, N (Moskowitz and Caflisch, 1996). This fact shows (relatively, in the context of present day computer speed) the poor convergence of the method: for instance, in order to reduce the error of the result 10-fold, the number of experiments has to increase 100-fold.

By using the method described previously, with the aid of the relationships presented, it is possible both to estimate the accuracy of *Monte Carlo* results and to establish the value of random experiment numbers N which should ensure the expected accuracy with a given guarantee probability.

12.3 MONTE CARLO METHOD AS A GENERAL INTEGRATION TECHNIQUE

Consider the problem of numerical estimation (Barker, 1987; Sobol and Tutunnikov, 1996; Ökten, 1999) of a convergent multidimensional integral.

$$V=\iint\underset{M}{\cdots}\int f\left(x_1,x_2,\ \dots\ x_k\right)dx_1dx_2\ \dots\ dx_k \tag{10}$$

where $f(x_1x_2 \dots x_k)$ is a function continuous in a closed bounded domain. The number V in Eq. (10) represents the volume of a right cylindroid developed on the base M in an (k+1)-dimensional space $Ox_1, x_2, \dots, x_k f(x_1, x_2, \dots, x_k)$.

The integral (10) should be transformed in such a manner that the new integration domain will be contained in a k-dimensional hypercube (Barker, 1987). This is realized by a change of variables, supposing that the domain M can be placed in the following k-dimensional parallelepiped

$$p_i \le x_i \le P_i\ ;\quad i=1,2,\dots,k \tag{11}$$

The new k-dimensional hypercube is described as follows,

$$0 \leq \chi_i \leq 1 \; ; \; i = 1,2,\ldots,k \tag{12}$$

The new variables χ_i are obtained by the relations:

$$\chi_i = \frac{x_i - p_i}{P_i - p_i} \tag{13}$$

By calculating the Jacobian of this transformation, the integral (10) becomes:

$$V = \iint \cdot \underset{M}{\cdot \cdot} \int f'(\chi_1, \chi_2, \ldots \chi_k) d\chi_1 d\chi_2 \ldots d\chi_k \tag{14}$$

where

$f'(\chi_1, \chi_2, \ldots, \chi_k) = (P_1 - p_1)(P_2 - p_2)\ldots(P_k - p_k) \times f(p_1 + (P_1 - p_1)\chi_1, p_2 + (P_2 - p_2)\chi_2, \ldots, p_k + (P_k - p_k)\chi_k)$

If we note $\chi = (\chi_1, \chi_2, \ldots, \chi_k)$ and $d\lambda = d\chi_1 d\chi_2 \ldots d\chi_k$, the integral (14) can be written as follows,

$$V = \int \cdot \underset{\lambda}{\cdot \cdot} \int f'(\chi) d\lambda \tag{15}$$

Since the function f' is nonnegative, the integral (15) may be viewed as a volume of a solid in a k-dimensional space (hypercube). The value of this integral, V, can be evaluated by the Monte Carlo method using random trials. Thus, generating k independent sequences of (pseudo)random numbers uniformly distributed on the interval (0, 1), the points $X_i(\chi_i^{(1)}, \chi_i^{(2)}, \ldots, \chi_i^{(k)}), i = 1, 2, \ldots, N$ may be considered random. If from the total number of points X_i (N being a sufficiently large number), n points belong to the domain λ, then the integral (10) is given by the relation

$$V = \frac{\lambda}{N} \sum_{i=1}^{n} f(X_i) \tag{16}$$

where λ is a k-dimensional volume in the domain of integration λ. Below n is also a sufficiently large number, so that

$$X_i \in \lambda \quad \text{for} \quad i = 1,2,\ldots,n \tag{17}$$

$$X_i \notin \lambda \quad \text{for} \quad i = n+1, n+2, \ldots, N \tag{18}$$

The analytic representation of the boundary Γ of the domain λ is taken into consideration for verifying the conditions in Eqs. (17) and (18).

The accuracy of the estimation, ε, i.e., the estimation of the integral Eq. (10) with Eq. (16) for a given maximum probability (*1*–δ) is inversely proportional to the square root of the total number of random trials (points):

$$\varepsilon = \frac{1}{2\sqrt{\delta N}} = O\left(\frac{1}{\sqrt{N}}\right) \tag{19}$$

The integration by the Monte Carlo method shows a relatively slow convergence. According to the relation (19), for ε=0.05 and δ=0.01, it results that N=20000. For details about the algorithm of calculation see (Ciubotariu, 1990).

12.4 MOLECULAR DESCRIPTORS CALCULATED BY MEANS OF THE MONTE CARLO METHOD

The requirement to define vdW numerical descriptors in order to properly quantify both the shape and size of molecules results from theoretical considerations regarding the 3D-nature of steric effects and from the fact that molecules fill the tri-dimensional (3D) Cartesian space. Among calculated vdW MDs we selected and detailed below only the vdW volume (V^W) and surface (S^W), the ratio V^W/S^W (r^{WV}), the semiaxs of the ellipsoid (E_X, E_Y, E_Z), which embeds a given molecule, and two globularity measures – G^{LEL} and G^{LOB} (Ciubotariu, 1987, 1990, 2001, 2010, 2011; Niculescu-Duvăz, 1991; Todeschini and Consonni, 2009). In each case the molecule is viewed as a collection of atomic spheres distributed in 3D-space, each atomic sphere having a radius equal with its vdW radius. These descriptors have been calculated with the Monte Carlo algorithms implemented on the computer in IRS software (http://irs.cheepee.homedns.org).

12.4.1 MOLECULAR VOLUME

Let M be a molecule (or a substituent) composed from a collection of atomic spheres $i=1, m$ centered at the equilibrium positions of the corresponding atomic nuclei i, and having a radius equal to the vdW radius r_i^W. A molecular vdW envelope Γ can be defined (in the "hard sphere" approximation) as the outer surface resulted from the intersection of the all vdW atomic spheres. The points (x, y, z) inside this envelope Γ satisfy at least one of the following inequations:

$$(X_i - x)^2 + (Y_i - y)^2 + (Z_i - z)^2 \le \left(r_i^W\right)^2 \quad i=1, m \tag{20}$$

where (X_i, Y_i, Z_i) is the center of atomic sphere i (i=1, m), and m represents the atomic number.

Consequently, the molecular vdW volume of the substituent M (V_M^W) is the total volume embedded by Γ. The following integral

$$V_M^W = \iiint_M dV \quad dV = dxdydz \tag{21}$$

can be intuitively justified as the volume of M. To estimate the integral (21), the substituent M is inserted into a bounding parallelepiped and the random points are generated into this parallelepiped.

$$V_M^W = \frac{n_S}{n_T} \cdot V_P \ (\text{in Å}^3) \tag{22}$$

n_{S} – the random points satisfying at least one of the conditions in Eq. (20)
n_T – the random points generated in the parallelepiped
V_P – the volume of bounding parallelepiped

The optimized geometry of alkane and cycloalkanes molecules, obtained by MM+ molecular mechanics and AM1 semiempirical quantum-mechanical parameterizations in the HyperChem Package (www.hyper.com), were loaded into in home developed IRS software (http://irs.cheepee.homedns.org) and the corresponding vdW volumes were calculated. The values of vdW molecular volume (V^W) of a series of 84 alkanes and cycloalkanes (63 alkanes and 21 cycloalkanes) calculated as described above, together with their vapor pressure (log VP) selected of literature (Katritzky, 2007) are presented in Table 1.

TABLE 1 Vapor pressure for a series of (a) 63 alkanes and (*b*) 21 cycloalkanes and their vdW volumes. (The values of vapor pressure, log VP (pa, 25°C), are taken from Katritzky, 2007)*

(a) Cn, where n represents the number of carbon atoms in an alkane molecule.

No	Alkane – Cn	log VP (Pa, 25°C)	V^W (Å^3)
1	C2	6.620	45.613
2	C3	5.970	62.421
3	C4	5.380	79.254

TABLE 1 *(Continued)*

4	2 M-C3	5.540	79.193
5	C5	4.830	96.249
6	2 M-C4	4.960	96.065
7	22 MM-C3	5.230	96.013
8	C6	4.300	113.053
9	3 M-C5	4.400	112.415
10	2 M-C5	4.440	113.071
11	23 MM-C4	4.490	112.443
12	22 MM-C4	4.820	112.655
13	C7	3.780	130.109
14	3 M-C6	3.910	129.551
15	2 M-C6	3.940	129.860
16	23 MM-C5	3.960	128.878
17	33 MM-C5	4.040	128.850
18	223 MMM-C4	4.135	128.996
19	24 MM-C5	4.020	129.463
20	22 MM-C5	4.140	129.137
21	C8	3.270	146.738
22	3E-C6	3.428	146.313
23	3 M-C7	3.417	146.831
24	34 MM-C6	3.481	145.441
25	3E3 M-C5	3.486	145.318
26	4 M-C7	3.436	146.446
27	2 M-C7	3.439	146.663

TABLE 1 *(Continued)*

28	3E2 M-C5	3.503	145.603
29	23 MM-C6	3.495	146.292
30	233 MMM-C5	3.556	145.504
31	234 MMM-C5	3.550	145.228
32	33 MM-C6	3.581	145.715
33	223 MMM-C5	3.631	145.538
34	24 MM-C6	3.607	146.298
35	25 MM-C6	3.606	146.951
36	22 MM-C6	3.657	146.144
37	224 MMM-C5	3.818	145.831
38	C9	2.770	163.667
39	3 M-C8	2.921	163.651
40	4 M-C8	2.959	163.250
41	2 M-C8	2.927	163.525
42	2233 MMMM-C5	3.103	161.341
43	3E24 MM-C5	3.126	162.438
44	26 MM-C7	3.094	163.053
45	3E22 MM-C5	3.177	161.778
46	225 MMM-C6	3.347	163.093
47	2244 MMMM-C5	3.427	162.285
48	2233 MMMM-C6	2.730	177.529
49	22 MM-C8	2.686	179.886
50	2 M-C9	2.400	181.218
51	335-MMMC7	2.746	178.774
52	33EE-C5	2.988	161.747

TABLE 1 *(Continued)*

53	3 M-C9	2.421	179.98
54	4 M-C9	2.490	180.106
55	5 M-C9	2.468	180.559
56	C1	7.790	28.6748
57	C10	2.270	180.398
58	C11	1.745	197.54
59	C12	1.252	214.688
60	C13	0.755	231.487
61	C14	0.270	247.959
62	C15	-0.183	264.924
63	C16	-0.700	281.996

(b) CyCn, where *n* represents the number of carbon atoms in the cycle (Cy) of a cycloalkanes molecule.

No	Cycloalkane	log VP (Pa, 25°C)	V^W (Å^3)
1	CyC3	5.850	54.7898
2	CyC4	5.195	70.2555
3	CyC5	4.620	85.6123
4	CyC6	4.110	101.69
5	CyC7	3.450	117.729
6	CyC8	2.876	134.666
7	E-CyC5	3.727	119.003
8	E-CyC6	3.230	135.202
9	IPR-CyC5	3.332	136.005
10	NPR-CycloC6	2.747	152.132
11	1,1 MM-CyC5	4.006	119.152
12	1,1 MM-CyC6	3.480	135.061
13	BU-CyloC6	2.243	169.077
14	*cis*-1,2 MM-CyC5	3.799	118.916
15	*trans*1,2 MM-CyC5	3.931	118.958
16	*cis*-1,3 MM-CyC6	3.457	135.308
17	*trans*1,3 MM-CyC6	3.371	134.928
18	*cis*-1,3 MM-Cy5	3.945	119.315
19	*trans*1,3 MM-CyC5	3.934	119.042
20	*cis*-1,4 MM-CyC6	3.379	135.638
21	*trans*1,4 MM-CyC6	3.481	135.081

[*] The substituents in Table 1 are denoted as following: M-methyl, E-ethyl, IPR-isopropyl, PR-propyl, BU-butyl.

The linear quantitative structure–property relation (QSPR) models Eqs. (23) and (24) were obtained using MobyDigs computer program (http://www.talete.mi.it; see references for details).

$$\log VP = 7.761\ (\pm 0.229) - 0.0298\ (\pm 0.0015)\cdot V^{W} \quad (23)$$

(n=84; r=0.976; r^2_{adj}=0.953; s=0.293; F=1515.0; q^2_{LOO}=0.930)

$$\log VP = 8.089\ (\pm 0.073) - 0.0312\ (\pm 0.0005)\cdot V^{W} \quad (24)$$

(n=63; r=0.993; r^2_{adj}=0.986; s=0.166; F=4409.0; q^2_{LOO}=0.930)

In the Eqs. (23) and (24) r is the correlation coefficient, r^2_{adj} represents the adjusted correlation coefficient (the explained variance, EV), s and F stand for standard error of points (from the regression line) and statistical Fisher test, respectively. The predictive capacities of the above QSPRs were measured by the cross validation coefficient (q^2) obtained by "leave one-out" method, q^2_{LOO}.

The statistical quality of the model (24), developed only for the alkanes, is better than those of (23), where the entire series of 84 alkanes and cycloalkanes was used. This fact may be explained by the difference of 3D geometry of these molecules. Actually, this work is in progress.

12.4.2 MOLECULAR SURFACE

The vdW envelope, Γ, defined in the above section, is, obviously, a surface. Several methods were developed to compute the area of this surface (Ciubotariu, 1987, 1991; Gogonea, 1991; Pearlmann, 1983). Some of them are based on a Monte Carlo method (Ciubotariu, 1987, 1991), others on an analytical algorithm (Gogonea, 1996). The computed surfaces were especially used to characterize the shape and the similarity of the molecules, their graphical representation, and so on (Gogonea, 1996).

The Monte Carlo algorithm (Ciubotariu, 1987, 1990, 1991; Gogonea, 1991, 1996) in IRS computer program implies the generation of an uniform distribution of random points on each sphere of the molecule, followed by the detection of the number of points generated on the entire surface (n_t) and those (n_e) that do not satisfy the inequalities in (20). For every "*hard sphere*" i, one computes the outer part of each sphere's surface area, S_i^W:

$$S_i^W = \frac{(n_e)_i}{n_t} \times 4 \times \pi \times (r_i^W)^2 \tag{25}$$

The final surface area is computed as a sum of the exterior surface area of each sphere, S_i^W:

$$S^W = \sum_{i=1}^{m} S_i^W \tag{26}$$

See refs. (Ciubotariu, 1987; Gogonea, 1991) for details about how to generate a uniform grid by means of Monte Carlo method and, also, for the developed algorithms using this stochastic method of integration.

The S^W descriptor (Todeschini and Consonni, 2009b) was used in a QSPR study of boiling points (Bp) of alkanes (Ciubotariu, 2004). The values of Bp and S^W are given in Table 2.

TABLE 2 The values of boiling points (Bp) and surface area (S^W) of a series of 72 alkanes.

No	Alkanes*	Bp [°C]	S^W
1	C2	-88.5	70.90
2	C3	-44.5	92.85
3	C4	-0.5	114.83
4	2 M-C3	-10.5	114.11
5	C5	36.5	136.62
6	2 M-C4	27.9	134.18
7	22 MM-C3	9.5	134.48
8	C6	68.7	158.59
9	3 M-C5	63.2	153.57
10	2 M-C5	60.2	155.89
11	23 MM-C4	58.1	151.86
12	22 MM-C4	49.7	152.20
13	C7	98.4	180.46
14	3E-C5	93.5	173.32
15	3 M-C6	91.8	175.30
16	2 M-C6	90.1	177.74
17	23 MM-C5	89.8	170.73
18	33 MM-C5	86.0	168.74
19	223 MMM-C4	80.9	167.74
20	24 MM-C5	80.5	174.47

TABLE 2 *(Continued)*

No	Alkanes*	Bp [°C]	S^W
21	22 MM-C5	79.2	173.79
22	C8	125.8	202.23
23	3E-C6	118.9	195.35
24	3 M-C7	118.8	197.04
25	34 MM-C6	118.7	189.82
26	3E3 M-C5	118.2	185.25
27	4 M-C7	117.7	196.97
28	2 M-C7	117.6	199.68
29	3E2 M-C5	115.6	192.53
30	23 MM-C6	115.3	193.49
31	233 MMM-C5	114.6	184.91
32	234 MMM-C5	113.4	185.75
33	33 MM-C6	112.0	190.40
34	223 MMM-C5	110.5	186.76
35	24 MM-C6	109.4	194.69
36	25 MM-C6	108.4	197.02
37	22 MM-C6	107.0	195.69
38	2233 MMMM-C4	106.0	181.19
39	224 MMM-C5	99.3	190.77
40	C9	150.6	224.10
41	33EE-C5	146.2	200.62
42	3E-C7	143.0	216.39
43	3 M-C8	143.0	219.05
44	4 M-C8	142.5	218.84
45	2 M-C8	142.5	221.45
46	3E23 MM-C5	141.6	200.33
47	2334 MMMM-C5	141.5	201.55
48	4E-C7	141.2	216.73
49	3E3 M-C6	140.6	207.01
50	23 MM-C7	140.5	214.33
51	334 MMM-C6	140.5	203.35
52	2233 MMMM-C5	140.3	196.87
53	34 MM-C7	140.1	211.40
54	234 MMM-C6	139.0	207.74

TABLE 2 *(Continued)*

No	Alkanes*	Bp [°C]	S^W
55	233 MMM-C6	137.7	205.60
56	33 MM-C7	137.3	212.37
57	3E24 MM-C5	136.7	207.95
58	35 MM-C7	136.0	213.36
59	25 MM-C7	136.0	216.37
60	26 MM-C7	135.2	218.95
61	44 MM-C7	135.2	211.95
62	4E2 M-C6	133.8	212.85
63	3E22 MM-C5	133.8	204.41
64	24 MM-C7	133.5	217.09
65	2234 MMMM-C5	133.0	201.53
66	22 MM-C7	132.7	217.66
67	223 MMM-C6	131.7	209.36
68	235 MMM-C6	131.3	210.55
69	244 MMM-C6	126.5	207.63
70	224 MMM-C6	126.5	210.06
71	225 MMM-C6	124.0	215.04
72	2244 MMMM-C5	122.7	207.63

* M = Methyl, E = Ethyl, C_n represents the alkane C_nH_{2n+2}, n=1,2, …, 8.

QSPR linear model Bp *vs* S^W is as follows,

$$Bp = -168.93\ (\pm 7.228) + 1.46\ (\pm 0.038)\ S^w \quad (27)$$

$$(n=72;\ r=0.977;\ r^2_{adj}=0.954;\ s=9.966;\ F=1482.0;\ q^2_{LOO}=0.951)$$

The resulting QSPR model of Eq. (27) as well as the models of Eqs. (23) and (24) were measured for goodness of fit by the correlation coefficient (r) and the coefficient of determination adjusted for the degree of freedom (r^2_{adj}). The uncertainty in the model was quantified by standard error (s), and the reliability by the F (Fisher) and t (Student) statistics. The t-test was used to determine the 95% confidence limits of the QSAR models. The predictive ability of QSARs was noted as the cross-validation coefficient q^2 determined by the leave-n-out method (LOO if n=1). The quantity q^2 is also known as *coefficient of prediction*.

The model (27) indicates that the vdW surface area may be a successful structural descriptor in QSPR studies. The statistical calculations were made with MobyDigs computer program (http://www.talete.mi.it; see references for details).

12.4.3 SYNTHETIC DESCRIPTORS OF MOLECULAR SIZE AND SHAPE

The shape of molecules is doubtlessly the main element of most chemical interactions. Quantitative treatment of molecular shape, that is the development of appropriate molecular descriptors able to synthesize the characteristics of 3D extension of molecules, is a very difficult problem. Most procedures are based either on comparing molecules with a reference structure, or on dividing them and defining the sectors by means of Euclidean distance between certain atoms or with the aid of Cartesian coordinates of those sectors.

Using a *hard-spheres* model, we developed a series of van der Waals indicators of the molecular shape (Ciubotariu, 1987, 1991, 2001, 2010, 2011). This model allowed the introduction of several synthetic descriptors of molecular shape, which are presented below.

A first set of indicators was developed starting from the fact that a molecule can be characterized by the surface of molecular envelope described by Eq. (20).

The Eq. (20) represents a 2nd -degree equation describing a general surface (Safarevich, 1976):

$$a_{11} \cdot x^2 + a_{22} \cdot y^2 + a_{33} \cdot z^2 + 2a_{14} \cdot x + 2a_{24} \cdot y + 2a_{34} \cdot z + a_{44} = 0 \quad (28)$$

By transformations of coordinates (translation), the equation (28) is simplified and reduced to one of 15 equations composed of four terms (Ciubotariu, 1987).

For obvious physical reasons related to spatial extension of substituents, we neglected both singular quadrics and the equations that do not have real solutions – and, therefore, do not represent geometrical figures. From the five nonsingular surfaces of 2nd degree, which remain and represent geometrical figures (ellipsoid, ellipsoidal and hyperbolic paraboloid, and one-sheet and two-sheet hyperboloid), only the ellipsoid fulfills the physical conditions so that by assimilating the molecule with this geometrical figure the physical meaning of the calculated parameters is maintained (Ciubotariu, 1987, 1991).

It is known that the relationship:

$$\frac{x^2}{E_X^2} + \frac{y^2}{E_Y^2} + \frac{z^2}{E_Z^2} = 1 \quad (29)$$

represents an ellipsoid, namely a spheroid (or conoid). If $E_X < E_Y = E_Z$ equation (29) represent a prolate ellipsoid. If $E_X = E_Y > E_Z$ the relations (29) represent an oblate ellipsoid of revolution, and if $E_X = E_Y = E_Z$ we have a sphere.

The molecules are oriented along the Ox axis of the Cartesian coordinate system and the volume of the ellipsoid (29) and its vdW center are estimated by a

Monte Carlo algorithm implemented in the IRS computer program (Ciubotariu, 2006; http://irs.cheepee.homedns.org). Then, the semiaxs of the ellipsoid are calculated.

Starting from the concept of *packing density* and from the fact that the experimental determination of the cross-section area of a molecule is performed by assimilating it to a sphere, and, supplementary, assuming a maximal packing of molecule spheres, one can consider the descriptor R^{WV} as a quantitative measure of the steric characteristics of molecules, defined as follows:

$$R^{WV} = \frac{V^W}{S^W} \tag{30}$$

In Eq. (30), V^W and S^W are the vdW volume and surface, respectively. R^{WV} was used earlier for characterization of globularity of various substituents (Ciubotariu, 1987, 1991; Todeschini and Consonni, 2009c).

On the basis of the molecular vdW descriptors described in this chapter, two other parameters were be defined. The first one (G^{LOB}) was introduced as a measure of globularity for acyclic molecules. It is given by the relation (Ciubotariu, 1987; Todeschini and Consonni, 2009c):

$$G^{LOB} = \frac{R^{WV}}{R_s} \tag{31}$$

where R^{WV} is defined by relation (30) and R_s represents the ratio between the volume and the surface of an equivalent sphere, which surrounds the molecule, with the radius equal to the half of the longest dimension of the parallelepiped that embeds the molecule. The above relation cannot be used for cyclic molecules, because the volume of the equivalent sphere includes the internal empty space, which is not included in the van der Waals volume.

The second one is defined by the following equation (Ciubotariu, 1987; Todeschini and Consonni, 2009c):

$$G^{LEL} = \frac{V^E}{V^S} \tag{32}$$

where V^E is the volume of the ellipsoid surrounding the whole molecule, and V^S is the volume of a sphere with a radius equal to half of the longest ellipsoid ax. This parameter should be more useful for characterizing globularity because it includes the volume of all holes, which may appear.

These two parameters can be used to describe the shape of acyclic molecules. The globularity measure decreases with the growth of the linear chains and increases toward unity when the molecule is highly branched or compacted.

These synthetic descriptors of molecular size (E_X, E_Y, E_Z) and shape (R^{WV}, G^{LOB}, and G^{LEL}) were used in QSPR analysis of boiling points of alkanes in Table 2. The true nature of the intermolecular forces involved in process of boiling and the entropy change in the transition from liquid to gas phase are not considered in detail. We are interested here to test only the correlation ability of the considered MDs in Table 3.

TABLE 3 The values of synthetic MDs of size (E_X, E_Y, E_Z) and shape (R^{WV}, G^{LOB}, and G^{LEL}) of the alkanes in Table 2.

No	E_X	E_Y	E_Z	R^{WV}	G^{LOB}	G^{LEL}
1	3.021	3.145	2.881	0.643	0.613	0.880
2	3.516	3.780	2.881	0.672	0.533	0.709
3	3.763	4.200	2.881	0.692	0.494	0.615
4	3.889	3.781	3.536	0.695	0.536	0.884
5	4.252	4.820	2.881	0.704	0.438	0.527
6	4.182	4.248	3.561	0.712	0.503	0.825
7	3.875	3.766	4.168	0.712	0.512	0.840
8	4.503	5.241	2.881	0.713	0.408	0.472
9	4.307	4.899	3.599	0.728	0.446	0.646
10	4.687	4.811	3.594	0.720	0.449	0.728
11	4.503	4.220	4.175	0.732	0.488	0.869
12	4.215	4.235	4.168	0.729	0.516	0.979
13	4.988	5.861	2.881	0.721	0.369	0.418
14	4.661	4.931	4.055	0.738	0.449	0.777
15	4.503	5.352	3.664	0.733	0.411	0.576
16	4.940	5.265	3.572	0.726	0.413	0.637
17	4.681	4.859	4.168	0.746	0.460	0.826
18	4.344	4.921	4.168	0.750	0.457	0.748
19	4.587	4.231	4.169	0.752	0.492	0.842
20	4.746	5.027	3.635	0.734	0.438	0.683
21	4.663	4.799	4.168	0.736	0.460	0.844
22	5.240	6.281	2.881	0.724	0.346	0.383
23	4.662	5.378	4.033	0.741	0.414	0.650
24	5.026	5.961	3.660	0.737	0.371	0.518
25	4.536	5.351	4.156	0.756	0.424	0.658
26	4.792	4.950	4.314	0.760	0.461	0.844
27	4.972	5.854	3.696	0.737	0.378	0.536
28	5.432	5.861	3.591	0.732	0.375	0.568

TABLE 3 *(Continued)*

No	E_X	E_Y	E_Z	R^{WV}	G^{LOB}	G^{LEL}
29	4.946	4.910	4.201	0.760	0.461	0.843
30	4.852	5.351	4.165	0.749	0.420	0.706
31	4.769	4.944	4.164	0.767	0.466	0.813
32	4.780	4.962	3.998	0.758	0.458	0.776
33	4.523	5.353	4.168	0.754	0.422	0.658
34	4.740	4.854	4.169	0.761	0.470	0.839
35	5.286	5.260	4.232	0.734	0.417	0.797
36	4.961	5.270	4.168	0.742	0.422	0.744
37	4.563	4.239	4.175	0.780	0.513	0.850
38	4.714	5.036	4.130	0.748	0.446	0.768
39	5.723	6.901	2.881	0.728	0.316	0.346
40	4.769	5.035	4.566	0.776	0.463	0.859
41	5.028	6.010	4.206	0.745	0.372	0.585
42	5.259	6.396	3.680	0.741	0.348	0.473
43	5.214	6.354	3.739	0.741	0.350	0.483
44	5.679	6.300	3.572	0.735	0.350	0.511
45	5.057	5.038	4.269	0.777	0.461	0.841
46	4.812	5.003	4.155	0.779	0.467	0.799
47	5.054	5.713	4.029	0.746	0.392	0.624
48	4.804	5.410	4.338	0.761	0.422	0.712
49	5.408	5.934	4.158	0.751	0.380	0.639
50	4.587	5.505	4.160	0.771	0.420	0.629
51	4.676	4.902	4.176	0.795	0.486	0.813
52	5.081	5.921	4.150	0.759	0.385	0.601
53	4.964	5.428	4.078	0.765	0.423	0.687
54	4.831	5.398	4.167	0.768	0.427	0.691
55	5.085	5.962	4.168	0.752	0.379	0.596
56	5.057	5.064	3.982	0.761	0.451	0.785
57	5.433	5.307	4.420	0.752	0.415	0.795
58	5.393	5.886	4.272	0.746	0.380	0.665
59	5.443	5.995	3.596	0.741	0.371	0.545
60	4.898	5.869	4.168	0.756	0.386	0.593
61	5.125	5.414	3.757	0.749	0.415	0.657
62	5.003	5.113	4.168	0.765	0.449	0.798
63	5.480	5.752	3.844	0.749	0.391	0.636

TABLE 3 *(Continued)*

No	E_X	E_Y	E_Z	R^{WV}	G^{LOB}	G^{LEL}
64	4.731	5.047	4.106	0.779	0.463	0.763
65	5.411	5.865	4.168	0.741	0.379	0.656
66	4.911	5.286	4.197	0.760	0.431	0.738
67	5.275	5.384	4.093	0.757	0.422	0.745
68	5.128	5.402	4.127	0.764	0.424	0.725
69	5.085	5.392	4.104	0.757	0.421	0.718
70	5.365	5.269	4.184	0.749	0.419	0.766
71	4.639	5.042	4.168	0.765	0.455	0.761
72	5.132	5.353	4.294	0.760	0.421	0.743

Linear correlations with boiling points (at normal pressure) for all 72 alkanes with $N = 2–9$ carbon atoms were tested for these six MDs (E_x, E_y, E_z, R^{WV}, G^{LOB}, G^{LEL}) with a linear equation of the following type:

$$BP = \alpha \cdot (\pm\Delta\alpha) + \beta \cdot (\pm\Delta\beta) \cdot MD \tag{33}$$

The statistical characteristics of these QSPR models (33) are systematized in Table 4. In this table, *r* is the correlation coefficient, *s* is the standard deviation, *EV* is the explained variance, *t* is the Student test for *r* and *F* is the Fisher test.

TABLE 4 The vales of statistics calculated for QSPR (33).

MD	R	α	Δα	β	Δβ	s	F	EV
E_X	0.849	−291.59	29.28	82.60	6.05	24.600	186.4	0.718
E_Y	0.790	−178.72	26.28	54.54	4.99	28.575	119.5	0.619
E_Z	0.523	−112.71	42.28	55.91	10.73	39.718	27.1	0.264
G^{LOB}	0.710	383.22	32.62	612.13	75.10	32.829	73.1	0.497
G^{LEL}	0.285	176.94	28.50	-101.46	40.22	44.673	6.4	0.069
R^{WV}	0.833	−1060.80	91.34	1568.38	122.68	25.773	163.4	0.690

12.4.4 STERIC STATIC INTEGRAL PARAMETER

The first generally successful quantitative characterization of steric effects in organic reactions was that of Taft (1952). Following a suggestion of Ingold (Ingold, 1930), Taft defined the steric constant E_S as (Taft, 1952, 1953):

$$E_S = \log (k_X / k_H)_A \tag{34}$$

where k refers to the rate constant for the acid hydrolysis (denoted by A) of esters of type A and/or the esterification reaction of carboxylic acids B.

$$X - CH_2 - COOR \ (A) \quad X - CH_2 - COOH \ (B)$$

The acid-catalyzed esterification and/or hydrolysis of esters are reversible. The first step in the mechanism for acid-catalyzed ester hydrolysis (and/or esterification reaction) is protonating of the carbonyl oxygen by acid (see Fig. 1). This is rapidly realized, and thus increases the susceptibility of the carbonyl carbon to nucleophilic attack of water (and/or alcohol).

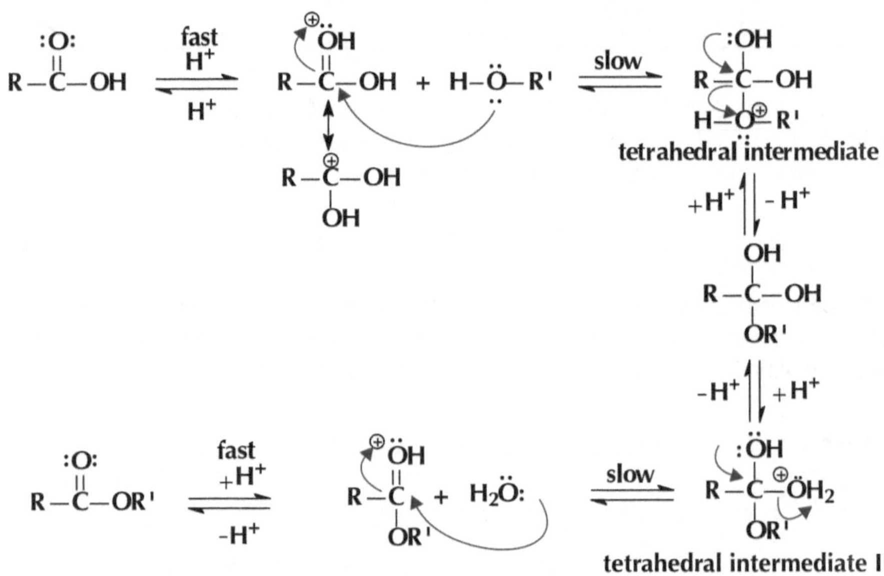

FIGURE 1 Mechanism of acid-catalyzed esterification and hydrolysis reactions.

Hydrolysis of an ester (and/or esterification reaction of carboxylic acids with alcohols) involves two relatively slow steps: formation of tetrahedral intermediates I (and/or II) and collapse of a tetrahedral intermediate II (and/or I). The acid (H^+) increases the rates of both slow steps of these reactions by lowering the activation energy of the corresponding transition states – see Figure 2a for acid-catalyzed esterification reaction and Figure 2b for the acid-catalyzed hydrolysis of esters. The size of X will affect attainment of the transition state by alcohol (Fig. 2a) and/or by water (Fig. 2b).

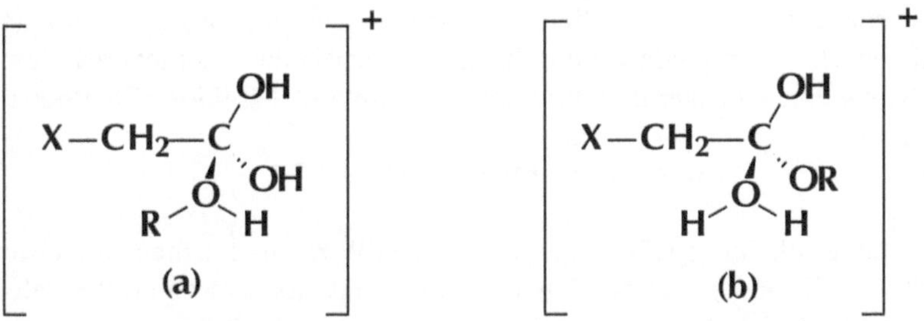

FIGURE 2 The transition states for the esterification (a) and for hydrolysis of esters (b).

The size of X will affect the attainment of the transition states by alcohol in the case of esterification (Fig. 2a), by water (Fig. 2b) or by some other ligand in the processes that are to be modeled by E_S constant defined in relation (34). This definition assures that the electronic effects of X can be neglected on attainment of the transition state.

Unfortunately, variations of the structures (A) and (B) cannot be used to obtain E_S values for many common substituents (e.g., when $X - CH_2 = NO_2$, CN, halide, OR, etc.) because they are unstable under conditions of acid hydrolysis.

Charton opened up a route to determine E_S values for such groups when showed that E_S is related to the van der Waals (vdW) radii of substituents (Charton, 1975).

Hansch et al. (Hansch and Leo, 1995) used this finding to extend E_S values to a variety of new substituents via the following correlation equation:

$$E_S = -1.839 \cdot r_\upsilon(\text{av}) + 3.484 \quad (n{=}6;\ r{=}0.996;\ s{=}0.132) \tag{35}$$

In this equation, E_S holds only for symmetrical top functions such as H, Br, CF_3, $C(CH_3)_3$ and $r_\upsilon(\text{av})$ is the average of the minimum and maximum vdW radii of the substituent estimated according to Charton (Charton, 1975). The linear equation (35) is based only on six substituents; r is the correlation coefficient and s represents the standard deviation (error).

This equation was used to extend the scale of steric constants E_S for various substituents. But, it must be noted that in Eq. (35) only the radius of the first atom has been used to calculate E_S for substituents such as OH, OCH_3, SH, SCH_3 and NH_2. Two values of E_S have been calculated for groups such as NO_2 and C_6H_5: in one, the width of the substituent was employed and in the other, the thickness was used. To explain this approach, Hansch and Leo argue that "since one has little or no idea of what expect in steric effects in biologic correlation analysis, one must generally try both parameters to discover which yields the best fit" (Hansch, 1980).

Charton has made a more direct approach to defining the steric hindrance of a substituent. He attempted to avoid the use of a particular chemical reaction to establish new extended steric scales with the following definition (Charton, 1975):

$$\upsilon_X = r_\upsilon(X) - r^W(H) = r_\upsilon(X) - 1.2 \quad (36)$$

In the above relation, $r_\upsilon(X)$ is the minimum vdW radius for the symmetrical top substituents X, and $r^W(H) = 1.2$ is the vdW radius for hydrogen; the values are measured in Å (1Å = 10^{-8} cm = 10^{-10} m = 10^{-1} nm). Charton (Charton, 1976) has used the primary E_S values obtained through Eq. (37) to estimate many new values for substituents with various configurations by means of the following linear correlation equation:

$$\log k_X = a \cdot \upsilon_X + b \quad (37)$$

The principal disadvantage of the Taft-type constants, which are determined using a certain reaction as experimental model and a reference substituent X_0, is that they are not steric "pure" parameters; they do not measure a single type of substituent effect. This explains the existence of many constants of this type: E_S, E_S', E_S^C, E_S^e, E_S^0, etc. Even if the physical meaning of those constants is not clear (as in the case of topological indices), many examples in E_S have been used to rationalize steric effects in organic reactions and biochemical systems where the interaction with macromolecules (biological enzymes or receptors) is of primary importance (Katritzky, 2010).

Starting from the difficulties appearing in a coherent quantification of steric effects and taking into account the requirement of a physical meaning of the parameters in a range of a certain scale (as far as this possible), we have introduced the steric static integral parameter I_S as a measure of steric effects. (Ciubotariu, 1987) This parameter was developed taking into consideration the fact that the proximal steric effects are usually attributed to repulsive forces, which appear when two atoms (or groups) are closer than the sum of their vdW radius. The developed model allows the quantitative evaluation of the steric effects in every molecular context and for any type of substituents interacting in chemical and biological processes, by means of Monte Carlo that has been used as a mathematical integration technique. I_S has a clear physical meaning and, also, it can give a physical significance to E_{S-}Taft steric scale.

The steric static integral parameter I_S has been constructed on the basis of the following assumptions:

1. The substituent is treated as a collection of atomic hard spheres whose centers are in the equilibrium positions of atomic nuclei. An atom has a radius equal to its vdW radius, r^W. A vdW envelope, Γ, may be uniquely defined as the outer surface resulted through the intersection of all vdW

spheres of a substituent (or molecule). The points of Cartesian coordinates (x, y, z) located within the envelope Γ have to satisfy at least one of the Eq. (20) inequalities.

2. The steric behavior of the substituent or molecule is due to the homogeneous and isotropic molecular vdW space delimited by the envelope Γ.
3. The steric effect is of vectorial nature. In the adopted hard sphere approximation, every substituent (molecule) occupies a certain volume in the 3D space, relatively impenetrable to external influences.
4. Quantitatively, each volume element of the substituent, dV, exerts on the reaction center a steric effect which is proportional to its extent and to $1/R^n$, where R is the distance between each element dV and the reaction center, and n represents a power, which should be determined; generally, $1 \leq n \leq 12$.
5. The steric effect of the hydrogen atoms can be ignored.

On the basis of the above hypotheses, we admitted that the steric effect of each volume element dV depends on the size of dV, and this effect will increase as the volume element dV is nearer to the reaction center. Clearly, the steric effect tends to zero if the separation distance between the position of dV and reaction center tends to infinite. Neglecting the steric influence of hydrogen atoms, i.e. the volume of the substituent disposed at a distance less than 1.2Å, we defined the steric static integral parameter as follows (Ciubotariu, 1987,):

$$I_S = \iiint_{V^W} \frac{dV}{R^n} \tag{38}$$

Let V^W be the vdW space of the substituent or molecule (M) corresponding to the vdW volume of M located at a distance greater than 1,2Å to the reaction center (neglecting the influence of hydrogen atoms). An element of this volume, α_k, sufficiently small, can be obtained by the following relation:

$$\alpha_k = dV = \frac{V^W}{n_i}; \; k=1,2, \ldots, n_i \tag{39}$$

Based on the previous assumptions, it can be considered that the variation in the influence of each α_k on the reaction center depends on its magnitude, which is constant for a given situation, and on the distance to that center raised at a given constant power λ,

$$dI_k = -(\lambda - 1)\alpha_k \left(\frac{1}{R^\lambda}\right) dr_k \quad \lambda > 1 \tag{40}$$

By integration of rearranged relation (40)

$$\int dI_k = -\alpha_k(\lambda - 1)\int \frac{dR_k}{R_k^{\lambda}} \tag{41}$$

we can obtain the influence of each α_k:

$$I_k = \frac{\alpha_k}{R_k^{\lambda-1}} \tag{42}$$

Taking $n = \lambda - 1$, the relation (42) can be rewritten as follows

$$I_k = \frac{\alpha_k}{R_k^{n}} \tag{43}$$

For a volume element α_k sufficiently small, and by extension, over the entire considered molecular space V^W, it is possible to define a steric static integral parameter I_S as:

$$I_S(n) = \sum_{k=1}^{n_i} I_k = \int_{V^W} \frac{dV}{R^n} \tag{44}$$

The optimal value of n may be obtained by trial and error, using experimental data of chemical reactivity for reaction modeling the steric effects.

In this static model described here we used the atomic coordinates based on the standard bond lengths, valence and dihedral angles, and the Bondi vdW radii. The reaction center was placed in the origin of the Cartesian coordinate system with the first atom of the M located on the Ox axis and the second in the xOy plane.

The I_S parameter was applied to the acid-catalyzed esterification and hydrolysis reactions of esters, the same reaction used by Taft for modeling the steric effects. The exact values of n were obtained by trial and error method, using the coefficient of correlation as selection criteria. Table 5 systematizes the experimental data – the values of rate constants for a series of 17 carboxylic acids determined in their acid-catalyzed esterification reaction with methanol, the solvent being also the methanol. The global reaction is the following,

$$\text{M–COOH} + \text{MeOH} \xrightarrow[\text{HCl}]{\text{MeOH; 50°C}} \text{M–COOMe} + H_2O$$

TABLE 5 Rate constants, k, and values of $I_S(n)$ corresponding to substituents M; n=4 and n=6.

No	M*	$k \cdot 10^{2\#}$	$I_s(6)$	$I_s(4)$
1	Me	21.9	0.698	1.740
2	Et	19.3	0.734	1.950
3	*n*-Pr	10.3	0.710	1.914
4	*n*-Bu	10.1	0.724	2.023
5	Am	10.2	0.691	1.898
6	*i*-Pr	7.27	0.719	2.078
7	*i*-Bu	2.48	0.748	2.218
8	*s-Bu*	2.19	0.824	2.433
9	*i*-Am	10.4	0.725	2.050
10	$PhCH_2$	9.44	0.666	1.965
11	$PhCH_2CH_2$	9.20	0.677	1.921
12	*s*-$BuCH_2$	2.40	0.817	2.429
13	*t*-Bu	0.858	0.865	2.543
14	Et_2CH	0.253	0.911	2.751
15	cy-$HxCH_2$	2.79	0.739	2.236
16	cy-Hx	4.38	0.770	2.299
17	PhEtCH	1.05	0.884	2.636

* Me=Methyl; Et=Ethyl; *n*-Pr=Propyl normal; *i*-Pr=isopropyl; Bu=butyl; Am=amyl (pentyl); *t*-butyl=*tert*-butyl; s-Bu=*sec*-butyl; cy-Hex=cyclohexyl.

the values of rate constants, k, were taken from the chapter of Charton (1975).

The optimal I_S molecular descriptors were determined by trial and error method, making correlations between the rate constants of the reaction and these parameters. For each substituent in two conformations, staggered and eclipsed, 120 values of $I_S(n)$ were computed, $\Delta n = 0.1$, $0 \leq n \leq 12$. In every case we made a correlation $\log(k)$ *vs.* $I_S(n)$, the statistical quality of the model being estimated by means of the correlation coefficient, r. For the eclipsed conformations, the best value was obtained for n=2.2 (r=0.652), but this has a poor statistical significance. For staggered conformations the best value was obtained for n=4 (r=0.958). The corresponding linear model is the following:

$$\log(k) = 10.094(\pm 0.675) - 3.943(\pm 0.307) \times I_S \quad (45)$$

$$(r=0.958;\ s=0.359;\ F(1,15)=165.3;\ EV=0.905)$$

In the Eq. (45) the others statistical indicators are: *s* – standard deviation (error), F – Fisher test, and EV – explained variance. This linear model explains about 90% of the variance of experimental data – log(k) values in Table 5.

Another set of data used to determine the value of *n* in $I_S(n)$ was obtained from the acid-catalyzed hydrolysis of ethyl esters X-COOEt in a mixture of 70% MeAc-H_2O at different temperatures. The experimental data (Charton, 1975) and the values of $I_S(4)$ are presented in Table 6.

TABLE 6 Rate constants and values of $I_S(n)$ corresponding to acid-catalyzed hydrolysis.

No.	X*	10^5k at 24.8°C	10^5k at 35.0°C	10^5k at 44.7°C	$I_S(4)$
1.	Me	4.47	10.9	24.7	1.740
2.	Et	3.70	9.24	20.7	1.950
3.	*n*-Pr	1.96	4.83	10.8	1.914
4.	*n*-Bu	1.79	4.45	10.2	2.023
5.	*n*-Am	1.77	4.30	9.76	1.898
6.	*i*-Pr	1.46	3.43	7.46	2.078
7.	*i*-Bu	0.572	1.46	3.30	2.218
8.	*t*-Bu	0.128	0.363	1.10	2.543
9.	$PhCH_2$	1.58	3.84	8.84	1.965

* Acronyms for the substituents X are the same as in Table 5.

The correlation equations obtained using the data in Table 6 are the following:

$$\log(10^5\times k) = -0.945-1.924\cdot I_S \tag{46}$$

$$(r=0.960;\ s=0.138;\ EV=0.896)$$

$$\log(10^5\times k) = -0.717-1.840\cdot I_S \tag{47}$$

$$(r=0.958;\ s=0.136;\ EV=0.890)$$

$$\log(10^5\times k) = -0.652-1.692\cdot I_S \tag{48}$$

$$(r=0.954;\ s=0.131;\ EV=0.881)$$

The results obtained with the steric static integral molecular descriptor I_S are encouraging. The fact that $I_S(4)$ computed for standard geometry is capable to reflect the steric effects in a chemical reaction represents an advantage because it can be calculated for any kind of substituents, independent of their complexity. The physical meaning of this parameter is clear: the steric effects can be seen as a steric gradient depending on the size of the volume element and on the distance from the reaction center raised to a certain power, *n*. In the case of one of the reactions used by Taft to define the steric parameter E_S (log *k* values in Tables 5 and 6), the steric effect is inversely proportional with the distance of each volume element to the reaction center, raised at the power 4. The extension on the entire space that is significant in the steric interaction is realized by integration. Using the regression analysis models, it seems that the molecular descriptor I_S can estimate the conformation preferred by the substituent, as a general behavior of the whole molecule. This should be very useful in QSAR (quantitative structureactivity relationships) studies to explain receptor-ligand interactions.

The method described here also gives the possibility of a dynamically approach, using different geometry for the transition state of reactions or biological interactions.

12.5 DESCRIPTORS FOR THE ACCESSIBILITY OF MOLECULAR SURFACE AND THE DYNAMICS OF THE MOLECULAR STRUCTURE

Traditionally, molecular descriptors have been used to quantify the size, shape and structure of molecules (Ciubotariu, et al., 2001). While, some experimental techniques like electron, X-ray, NMR and neutron diffraction, can provide structural information about molecules in various aggregation states, other techniques (e.g. various spectroscopic techniques, like incoherent neutron scattering (Gabel, et al., 2002)), are capable of probing the dynamics of atoms in molecules and of the molecules as a whole. Thus, sometimes it is useful to use traditional molecular descriptors to capture dynamic properties of molecules. The development of the descriptors for molecular surface accessibility and dynamics (MSAD) of the molecular structure (Gogonea, et al., 2010) was made possible by the advent of a combination of two experimental techniques used in tandem, namely, hydrogen-deuterium exchange and mass spectrometry (H/D-MS/MS) (Englander 2006).

In this kind of experiment, performed on proteins, the amide hydrogen of the peptide bond is exchanged with deuterium. The rate of the exchange reaction is controlled by a combination of steric and electronic factors (Bai, et al., 1993). Thus descriptors that quantify both electronic (inductive, mesomeric, charge transfer, H-bonding) and steric effects (size and shape) can be used in combination to construct MSAD descriptors. Here we present one MSAD descriptor constructed from a molecular descriptor that quantifies an electronic effect (H-bonding), and two descriptors that describe a steric effect (the accessibility of the amide hydrogen and the accessibility of the amino acid residue to which the amide hydrogen belongs). The outcome of the experiment is a series of hydrogen-deuterium exchange (HDX) incorporation factors (D_0) determined by mass spectrometry for individual peptic peptides, obtained by digesting the protein with various proteases (e.g., pepsin, XIII, etc.) (Englander, 2006). The experiment does not provide an amino acid resolution, that is, it only gives a D_0 for an entire peptide. The purpose of constructing the MSAD descriptor is to use structural information obtained from a crystal structure or a molecular model of the protein to determine the deuterium incorporation (D^i_0) for individual amino acid residues. Because the rate of the HDX reaction relates with the rate of the unfolding of amino acid resides in the protein, the MSAD descriptor quantifies both the accessibility of deuterium oxide to the amide hydrogen of the protein backbone (i.e., surface accessibility) and the local dynamics of the residues captured through the unfolding and refolding of the protein backbone (Gogonea, et al., 2010).

In the following we present the theoretical formalism used to derive the MSAD descriptor and exemplify its use in calculations performed on apoA1 protein, the major protein component of high density lipoproteins (HDL) (Curtiss, et al., 2006).

The hydrogen-deuterium exchange incorporation factors (D^i_0) were calculated for all residues of apoA1 using experimental data from overlapping apoA1 peptic peptides of nascent HDL analyzed by H/D-MS/MS (Wu, et al., 2007, 2009). The MSAD descriptor was validated by previously applying it to a set of protein crystal structures with published NMR derived HDX rate constants (k_{xc}^i), and/or H/D-MS/MS derived D_0 values for peptides (Wu, et al., 2009). Experimentally measured D_0 values for overlapping peptides determined by H/D-MS/MS (in the case of apoA1 of reconstituted nascent HDL there was >95% coverage), and a molecular model is used to produce a set of per amino acid residue MSAD descriptors (XP_i) that are used to predict per residue deuterium incorporation factors (D_0^i), residue unfolding equilibrium constants (K_u^i), and HDX rate constants (k_{xc}^i). Gauging the difference between experimental and calculated D_0 values assesses the crystal structure or the initial model, which can be subsequently refined to increase the match between experimental and calculated D_0.

To start with, we note that the HDX rate constant, k_{xc}^i, for a residue i relates to the experimentally derived D_0^i and the HDX time, t, as follows:

$$-k_{xc}^i t = \ln\left(1 - D_0^i\right) \qquad (49)$$

The unfolding equilibrium constant, K_u^i, of an amino acid residue in the protein, is the ratio of k_{xc}^i and the intrinsic HDX rate constant (k_{ch}^i) of individual residues in random coil conformation (Bai, et al., 1993; Chetty, Mayne, et al., 2009):

$$K_u^i = \frac{k_{xc}^i}{k_{ch}^i} \qquad (50)$$

For an ensemble of protein conformations, the MSAD descriptor, XP_i, defines the ratio of the number of protein molecules (N_D^i) with the amide H of residue *i* exchanged to D, to the total number of molecules:

$$XP_i = \frac{N_D^i}{N_D^i + N_H^i} \tag{51}$$

For a single conformation (and not an ensemble), XP_i for each residue *i* can be calculated as the product of two molecular descriptors (RAI_i, BAI_i) encoding structural information about residue *i*'s size, solvent accessibility and its electronic effects (chemical composition and interactions with the rest of the system):

$$XP_i = RAI_i \cdot BAI_i \tag{52}$$

Equation 52 incorporates information about the solvent accessibility of both backbone amide H (*BAI* descriptor) and the whole residue *i* (*RAI* descriptor), and the unfolding/refolding dynamics of this residue. BAI_i and RAI_i are the backbone and residue accessibility/dynamics molecular descriptors, respectively, with values between 0 and 1.

The correction factor (PCF_i), is the ratio of the experimentally determined per residue deuterium incorporation factor D_0^i (obtained by partitioning the experimentally measured D_0 for peptic peptides) and the calculated MSAD descriptor, XP_i:

$$PCF_i = \frac{D_0^i}{XP_i} \tag{53}$$

PCF gages the difference in protection to HDX of a backbone amide H atom in the crystal structure or molecular model with respect to the same amide H in the solution structure. *PCF* is used to refine the conformation of individual residues in the crystal structure/model by adjusting the protection of amide H's through altering their 3D position. The molecular descriptors *BAI* and *RAI* are calculated using an approach similar to COREX (Hilser and Freire 1996).

The molecular descriptor *BAI* is the ratio of the van der Waals surface area of the backbone amide H atom in the crystal structure/model $(A_H^i)_{\text{model}}$ and the random coil conformation $(A_H^i)_{\text{rc}}$, respectively.

$$BAI_i = \frac{(A_H^i)_{\text{model}}}{(A_H^i)_{\text{rc}}} \exp\left(-c\sum_j \left(d_{HB} - d_{HB}^o\right)\right) \tag{54}$$

If the amide H is involved in H-bonding, the descriptor *BAI* is modified by an electronic effect contribution, which is expressed as an exponential factor that includes the length of the H-bond (d_{HB}).

The molecular descriptor RAI_i of residue i is the ratio of the solvent accessible surface area of the residue in the crystal structure/model $(A^i)_{model}$ and in the random coil $(A^i)_{rc}$ conformation, respectively:

$$RAI_i = \frac{(A^i)_{model}}{(A^i)_{rc}} f_i^{PD} \tag{55}$$

The descriptor RAI is modulated by a factor f_i^{PD} that accounts for the dynamics of the protein (Wu, et al., 2007; Wu, et al., 2009), present in experimental data but not accounted for in a single frozen conformation provided by a crystal structure or a model. f_i^{PD} can be calculated from a molecular dynamics or Monte Carlo ensemble or can be extracted from the experimental deuterium incorporations by an optimization procedure.

Table 7 lists calculated values for the MSAD descriptor XP for a few selected amino acid residues of apoA1 protein, the main protein component of HDL. To obtain the solution structure of the protein that produces the same deuterium incorporations as those obtained experimentally, the crystal structure/initial model is altered repetitively until PCF for each residue becomes unity. During the process of altering the 3D structure of the protein, the BAI and RAI descriptors change accordingly until PCF reaches one for each amino acid residue. The values of the MSAD descriptor (XP_{opt}) for the residues corresponding to the protein structure in solution are given in the last column of Table 7. The difference between the initial and optimized values of the MSAD descriptor provides and estimates the f_i^{PD} factor used in Eq. (55).

TABLE 7 Molecular descriptors (MSAD, surface area) for selected amino acid residues in apoA1 protein of high-density lipoprotein.

Residue	*SASA*[a]	*BAI*[b]	*RAI*[c]	*PCF*[d]	*XP*[e]	XP_{opt}[f]
Leu_{14}	1.0	0.61	1.0	0.4	0.61	0.24
Asp_{20}	0.31	1.0	0.92	0.8	0.92	0.74
Phe_{71}	0.31	0.79	0.56	0.87	0.44	0.39
Leu_{75}	0.32	0.71	0.68	0.51	0.37	0.25
His_{161}	0.82	0.52	1.0	1.07	0.52	0.56
Leu_{218}	0.47	0.84	0.85	0.68	0.72	0.48
Asn_{241}	0.89	0.54	1.0	0.98	0.54	0.53

[a] Solvent accessible surface area for the amino acid residue (nm^2).

[b] Backbone amide hydrogen accessibility descriptor.

[c] Residue accessibility descriptor.

[d] Initial correction factor.

[e] MSAD descriptor calculated from crystal structure or molecular model of the protein.

[f] Optimized MSAD descriptor calculated from the molecular model of the protein obtained by altering the 3-D position of the amino acid residues in the crystal structure or the initial molecular model of the protein. The position of the residues is stepwise modified until the theoretical and experimentally derived hydrogen deuterium incorporations agree.

12.6 SIMULATIONS OF THE COPOLYMERIZATION REACTIONS BASED ON THE MONTE CARLO METHOD

12.6.1 INTRODUCTION

Any theoretical model correlating copolymer structure with monomer reactivity must consider two fundamental aspects, namely:

(i) Determining the structure of copolymers resulted under given reaction conditions and

(ii) Quantitatively evaluating the relative monomer reactivity during the binary copolymerization process.

These two issues will be discussed presently in this chapter.

(i) The knowledge of copolymer structure is essential both for understanding their chemical, physical and mechanical behavior and for examining the mechanisms of polymer formation or modification reactions. Extensive theoretical studies have focused on the relationship between the structure of the polymer macromolecule and its properties (Harwood, 1981). Computers are essential in such studies for estimating the various structural characteristics of polymers, for at least the following three reasons:

(1) In most cases it is necessary to evaluate a very large number of characteristics. Thus, if we consider a copolymer derived of the symmetrical monomers M_1 and M_2, which can be incorporated into the macromolecular chain in only one way, we shall be able to identify, from an experimental point of view, three dyads M_iM_j, six triads $M_iM_jM_k$, 10 tetrads $M_iM_jM_kM_n$, etc. (see Fig. 3).

Monomers:	M_1			M_2
Dyads	M_1M_1	$(M_1M_2$	$M_2M_1)$	M_2M_2
Triads:	$M_1M_1M_1$	$(M_2M_1M_1$	$M_1M_1M_2)$	$M_2M_2M_2$
	$M_2M_1M_2$	$(M_1M_2M_2$	$M_2M_2M_1)$	$M_1M_2M_1$
Tetrads:	$M_1M_1M_1M_1$	$(M_2M_1M_1M_1$	$M_1M_1M_1M_2)$	$M_2M_2M_2M_2$
	$M_2M_1M_1M_2$	$(M_2M_2M_2M_1$	$M_1M_2M_2M_2)$	$M_1M_2M_2M_1$
	$(M_1M_1M_2M_1$	$M_1M_2M_1M_1)$	$(M_1M_1M_2M_2$	$M_2M_2M_1M_1)$
	$(M_1M_2M_1M_2$	$M_2M_1M_2M_1)$	$(M_2M_1M_2M_2$	$M_1M_2M_1M_1)$

FIGURE 3 Structural characteristics present in binary copolymers (symmetrical monomers M_1 and M_2).

If monomers are not symmetrical, complications may arise because of the "head – tail" or "tail – head" additional pattern, and if the macromolecular chain contains chiral carbon atoms, things get even more complicated through the apparition of stereo-sequences.

(2) Complex mathematical relationships have been developed for evaluating the various structural characteristics. For example, the general calculation procedure when using stochastic methods (Harwood and Kodaira, 1977) implies calculating probabilities conditioned by the apparition, within the chain, of monomers and/or (stereo) sequences based on monomer concentrations, on reactivity ratios (copolymerization constants) and corresponding equilibrium constants (if is the case). Using the compositions thus evaluated, with the aid of conditioned probabilities, it is possible to calculate other structural characteristics. The equations necessary for studying a given problem are relatively easy to build, but solving them is usually difficult, and solutions have a complicated form, error-prone and extremely difficult to interpret in practice.

In many situations, when the polymer characteristics cannot be calculated by stochastic modeling, it is possible to use numerical methods or, even more simply and efficiently, the Monte Carlo method which also allows obtaining a representation of the polymer macromolecular chain structure. This immediate representation can be extremely useful for understanding polymer structure.

(3) The computer allows simplifying the calculation of structural characteristics of polymers obtained at high conversions. Many theoretical approaches used for modeling binary or ternary copolymerization are restricted to the instantaneous behavior of polymerization systems. They are applied to polymers synthesized at low conversions, when the composition of the reaction mass does not vary with conversion. The integration of equations representing the basis for these models permits to obtain solutions for cases of high conversions as well, but their analytical resolution is possible only in a few situations. Fortunately, their numerical integration can be performed easily and simply with the help of computers.

(ii) Estimation of monomer relative reactivity to various radicals is characteristic to any copolymerization study (Mayo and Lewis, 1944). Experimental determination of the values of binary copolymerization reactivity ratios (r_1, r_2), significant from a quantitative point of view, has represented one of the most difficult issues of macromolecular chemistry (Young, 1961; Kelen and Tüdös, 1975).

If the kinetic model that describes binary copolymerization is expressed by the following chemical equations (Ciubotariu, Holban et al., 1975; Motoc and Ciubotariu, 1975):

$$\succ M_1^* + M_1 \xrightarrow{k_{11}} \succ M_1^* \text{, homopolymerization'} \tag{56a}$$

$$\succ M_1^* + M_2 \xrightarrow{k_{12}} \succ M_2^* \text{, copolymerization'} \tag{56b}$$

$$\succ M_2^* + M_1 \xrightarrow{k_{21}} \succ M_1^* \text{ , copolymerization'} \quad (56c)$$

$$\succ M_2^* + M_2 \xrightarrow{k_{22}} \succ M_2^* \text{ , homopolymerization'} \quad (56d)$$

then the basis of experimental data processing can be a differential or integral equation of copolymer composition, the choice between these two forms depending on the conversions at which copolymerization reaction have been stopped.

Taking into account the consumption rates of the two monomers:

$$\frac{-d[M_1]}{dt} = k_{11}[\succ M_1^*]\cdot[M_1] + k_{21}[\succ M_2^*]\cdot[M_1] \quad (57a)$$

$$\frac{-d[M_2]}{dt} = k_{22}[\succ M_2^*]\cdot[M_2] + k_{12}[\succ M_1^*]\cdot[M_2] \quad (57b)$$

and the fact that the ratio of the two rates (Eqs. (57a) and (57b)) expresses the copolymer composition (the mole ratio of monomers contained in the copolymer), and introducing the simplifying condition related to the stationary concentration of each type of radical undergoing an irreversible propagation, namely,

$$k_{21}[\succ M_2^*]\cdot[M_1] = k_{12}[\succ M_1^*]\cdot[M_2]$$

one obtains instantaneous binary copolymer composition equation, also known as the Mayo – Lewis equation (Mayo and Lewis, 1944).

$$\frac{d[M_1]}{d[M_2]} = \frac{[M_1]}{M_2}\cdot\frac{r_1[M_1]+[M_2]}{[M_1]+r_2[M_2]} \quad (58)$$

In Eq. (58) $r_1=k_{11}/k_{12}$ and $r_2=k_{22}/k_{21}$ represent the reactivity ratios and "[]" represents the concentrations. One may remark that r_1 and r_2 express the ratio between the homopolymerization and copolymerization rate constants for each monomer species, M_1 and M_2. Being constants for a given monomer system and a specified temperature, they are also called *copolymerization constants*.

The exact evaluation of these reactivity ratios, for a given copolymer system, is exquisitely important because they include significant information for the macromolecular system being obtained. Table 8 presents briefly, but, at the same time, synthetically, the information which can be revealed by the quantitative value, measured experimentally, of quantities r_1 and r_2.

TABLE 8 The synthetic presentation of the information included by the reactivity ratios (r_1, r_2). (Harwood et al., 1977).

$r \in (0, 1)$	Monomers copolymerize
$r_1 > 1$	Inclination to homopropagation
$r_1 \sim 0$	Inclination to alternation
$r_1 \sim r_2 \sim 1$	Perfectly random copolymer
$r_1 \cdot r_2 = 1$	"Ideal" system, by analogy with liquid-vapors equilibrium

By knowing these values, it is possible to qualitatively predict, which is not at all insignificant, the development of the binary copolymerization process in connection with both the composition of the resulting copolymer and its compositional poly dispersion. Moreover, it is also possible to choose compositions of the monomer system which can lead to copolymers with the expected composition and, by extension, with known physical, chemical and mechanical properties.

Generally, reactivity ratios are based on experimental data, by analyzing the composition of the resulting copolymers or by determining the residual monomer content in the under-layer, by applying a certain form of the Mayo – Lewis composition equation (Mayo and Lewis, 1944). Most methods developed initially, until 1965, were based on the linearization of equation (58) and the application of the least square method, nonlinear methods being used after 1975 (Kelen and Tüdös, 1977). Reactivity ratios (r_1, r_2) for the binary copolymerization systems were systematized and registered by L. J. Young (1961).

In order to study these aspects of copolymerization reactions, theoretical, analytical and stochastic models have been developed; they permit to determine copolymer structure, more specifically their composition. The use of the *Monte Carlo* method, allows to simulate the copolymerization process and to evaluate both the composition and the compositional poly dispersion of the synthesized copolymer. Besides, the use of adequate *Monte Carlo* models allows the determination of reactivity ratios as well. In what follows we shall present *Monte Carlo* simulation models developed for binary copolymerization reactions (Ciubotariu, Holban et al., 1975; Ciubotariu, 1987; Motoc, Ciubotariu et al., 1975).

12.6.2 *IMPLEMENTING MONTE CARLO MODELS ON COMPUTERS*

As a matter of fact, *Monte Carlo* techniques represent numerical methods of solving stochastic models without resorting to the analytical representation of the modeled system. Therefore, it is necessary to build an algorithmic model based on the Monte Carlo method, suitable for computers.

A possible formulation of a certain Monte Carlo model involves the following stages (Ciubotariu, 1987; Shreider, 1965):

(1) Identifying the states $E_1, E_2, \ldots, E_n$ of the real system taken into consideration.
(2) Evaluating transition probabilities $p_1, p_2, \ldots, p_n$. The system exists in state E_j with probability p_j.
(3) Generating random number $x \in (0,1)$. If inequality (59) is met,

$$\sum_{j=1}^{k} p_j < x \leq \sum_{j=1}^{k+1} p_j \tag{59}$$

the system is in state k.

This stage is reiterated as long as necessary. In connection with the relationship (59) other restrictions may be introduced, which can easily find their place in the context of *Monte Carlo* models.

Obviously, a *Monte Carlo* model does not directly reflect the behavior of the real system. The relationship between the *Monte Carlo* model and the real system is the following:

The *Monte Carlo* model helps to build a possible chronological *history* of the real system states. The chronological succession of the simulated model states is considered to be the *history* of the real system. The simulation results are validated by comparing a partial *history* of the states generated by the model with the same partial *history* of the real system states using experimental techniques. The model can be further extended for simulating the real system.

12.6.3 THE MONTE CARLO MODEL OF THE BINARY COPOLYMERIZATION

It is important to determine the distribution of sequences and copolymer structure in order to understand polymer properties and to estimate monomer reactivity, by offering, at the same time, extremely valuable information for discriminating between the possible reaction mechanisms.

Synthesis and formation processes of macromolecular complexes have a statistical character, which is reflected both by the way in which monomer units are distributed along the macromolecular chain and by the dependence between copolymer composition and conversion. This explains the use of the Monte Carlo method because it seems natural for simulating polymerization reactions.

A binary copolymerization reaction taking place according to the terminal kinetic model (with ultimate effect, meaning that only the last monomer unit of the growing macroradical affects the rate constants) implies four propagation reactions:

$$\succ M_i^* + M_j \xrightarrow{k_{ij}} \succ M_j^* \quad ; i, j = 1, 2 \tag{60}$$

Quantities $r_i = k_{ii}/k_{ij}$, $i, j = 1,2,\ i \neq j$, expressing the relative addition rates of the two monomers to the growing macroradical, represent *the reactivity ratios (copolymerization constants).*

Furthermore, we suppose that (Ciubotariu, 1987; Motoc, Ciubotariu et al., 1975):

(i) the length of the growing macroradical does not affect the rate constants,
(ii) there are no depropagation reactions, and
(iii) a negligible quantity of monomers is consumed during the initiation reactions, in the chain transfer reactions or in the case of interruption of the propagation chain reaction.

Consequently, the transition probabilities P_{IJ} from state $\succ M_I^*$ to $\succ M_J^*$ can be calculated (Motoc and Ciubotariu, 1975; Ciubotariu, 1987) with the relation:

$$P_{II} = \frac{k_{II}\left[\succ M_I^*\right]\cdot\left[M_I\right]}{k_{II}\left[\succ M_I^*\right]\cdot\left[M_I\right] + k_{IJ}\left[\succ M_I^*\right]\cdot\left[M_J\right]} \tag{61}$$

The concentration of macroradicals appears both in the numerator and denominator of relation (61). It can be equally simplified by taking into account the definition of reactivity ratios relation. Consequently, the relation (61) becomes as follows:

$$P_{II} = \frac{k_{II}\left[M_I\right]}{k_{II}\left[M_I\right] + k_I\left[M\ \right]} = \frac{r_I\cdot\left[M_I\right]}{r_I\cdot\left[M_I\right] + \left[M_J\right]} = \frac{r_I}{r_I + C}; \text{ I = 1,2; I = J} \tag{62}$$

$$P_{IJ} = 1 - P_{II} = \frac{C}{r_I + C} \tag{63}$$

where $C = [M_J]/[M_I]$stands for monomer concentrations in the reaction mass.

Taking into account that the states of the given system have been identified, A_1, A_2, B_1, B_2 (see Eq. 60 and Fig. 4), and the transition probabilities between them are known (see the Eqs. (62) and (63), one can build a computer Monte Carlo model. To do this, we introduced a function F(x), defined as follows:

$$F(x): (0, 1) \rightarrow (A_1, A_2, B_1, B_2) \tag{64}$$

where x has the property that $\forall\ x_1, x_2, \ldots \in (0,1) \Rightarrow P(x_1) = P(x_2) = \ldots$, and P() represents the probability, that is the numbers x_i, i=1, 2, … are independently, uniformly distributed on the interval (0,1), and the following notations:

$$\alpha = P_{11} = r_1 / (r_1 + C) \text{ and } \beta = P_{22} = r_2 / (r_2 + C^{-1}) \tag{65}$$

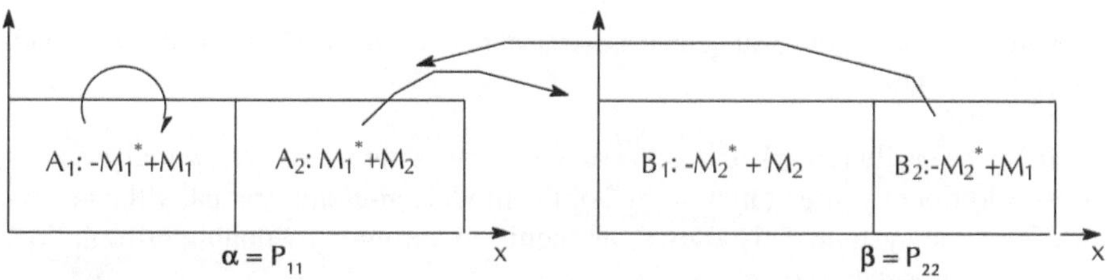

FIGURE 4 Representation of the function F(x) in Eq. (64); the transition probabilities are α and β.

The function F(x) defined in Eq. (64) presents the following properties:

$$\begin{gathered}\text{(i1) } \forall x : x \in (0,\alpha) \Rightarrow A_1 : x \Leftrightarrow A_1 \\ \ldots \\ \text{(i4) } \forall x : x \in (\beta,1) \Rightarrow B_2 : x \Leftrightarrow B_2\end{gathered} \tag{66}$$

Given the definition of function F and taking into account the properties (66), it obviously results that:

$$x_1 \Leftrightarrow A_1 \cup x_2 \Leftrightarrow A_1 \Rightarrow x_1 \approx x_2$$

$$\ldots\ldots.$$

$$x_7 \Leftrightarrow B_2 \cup x_8 \Leftrightarrow B_2 \Rightarrow x_7 \approx x_8$$

where x_1, …, x_8 Є (0,1) and "≈" is the notation for the equivalence relation.

The stochastic model described above allows a topological representation (Ciubotariu, 1987; Ciubotariu et al., 1976) which may be used as a block diagram for implementing on computers the *Monte Carlo* model proposed previously for modeling and simulating binary copolymerization reactions. This topological graph is illustrated in Fig. 5.

It should be pointed out that the lengths of segments ab, cd, ef and gh are, respectively, proportional to α, 1-α, 1-β and β.

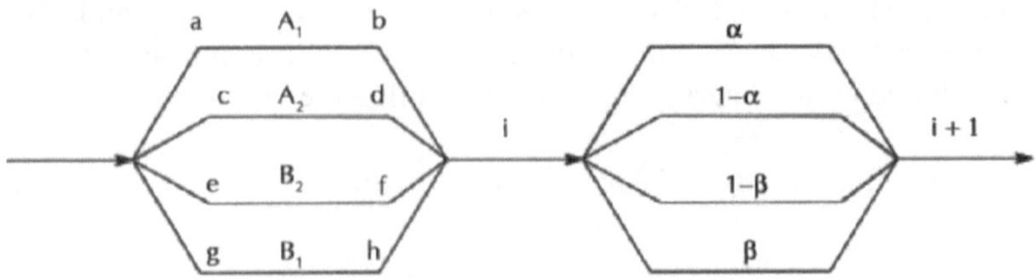

FIGURE 5 The topological graph associated to the *Monte Carlo* model of binary copolymerization.

It can be easily remarked that the model evolves naturally from initial assumptions. Additionally, the "functioning" of the model *simulates* gradually the behavior of the real system. This allows for acquiring extremely valuable information, otherwise impossible to get.

Moreover, an oriented graph can be attached to the copolymerization reaction to describe more clearly the (inter)relations between elementary reaction steps (Ciubotariu, 1987). Such graphs are extremely useful for building *Monte Carlo* algorithms, as they can be considered a sort of block diagrams of the model. Thus, if we take into consideration the Eq. (60) and the definition of the function F by relations 64–66, the propagation processes of the macroradical chain for various kinetic situations can be described by the graphs in Fig. 6.

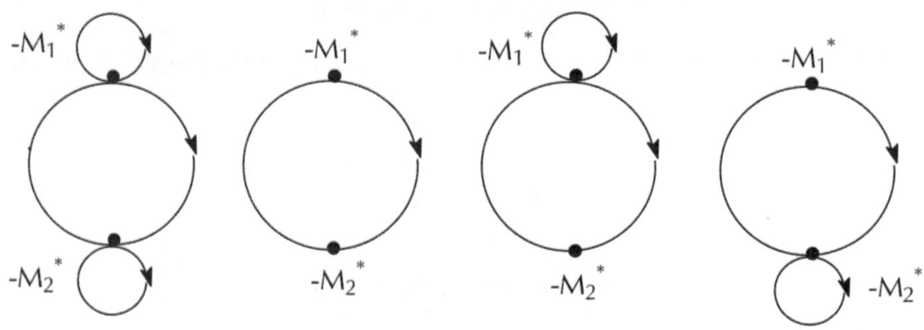

FIGURE 6 Oriented graphs associated to the process of binary copolymerization with ultimate effect.

Considering that the copolymerization reaction is stationary, it is necessary to introduce two indicator variables (*I, J*), which represent, respectively, the number of mer units of M_I and M_J type consumed during the copolymerization (simulation) process. This way, if $C_0 = [M_J]_0/[M_I]_0$ represents the initial concentration, then the concentration at a certain moment, *n*, during Monte Carlo simulation experiment will be done by the relationship:

$$C_n = \frac{\{[M_J]_{n-k} - J\}}{\{[M_I]_{n-k} - I\}} \tag{67}$$

By introducing a function of δ type:

$$\delta_I = \left\{ \frac{0, \text{ if the mer unit additioned at the previous stage is } M_J}{1, \text{ if the mer unit additioned at the previous stage is } M_I} \right\}$$

it is possible to determine the value of the ratio $\boldsymbol{C}$ at addition step $\boldsymbol{n}$, as follows,

$$C_n = \frac{\{[M_J]_{n-2} - \delta_J\}}{\{[M_I]_{n-2} - \delta_I\}} = \frac{[M_J]_{n-1}}{[M_I]_{n-1}} \tag{68}$$

By introducing relationships (67) and (68) in the relations (62) and (63), transition probabilities between the four system states become a conversion function.

12.6.4 SIMULATION OF COPOLYMERIZATION REACTIONS

The *Monte Carlo* model for simulating copolymerization reactions was implemented on computers as a software package called MEMORY. Binary copolymerization "experiments" performed with MEMORY provided a set of information about the dependence of copolymer composition on copolymerization degree. The obtained results are systematized in Table 9.

TABLE 9 Dependence of copolymer composition on polymerization degree, N (Ciubotariu and Holban et al., 1975), expressed as % molar of mer unit, $\boldsymbol{M_I}$.

N	% M_I	N	% M_I	N	% M_I
10	90,00	300	56,67	700	56,17
50	60,00	400	56,00	800	56,25
100	57,00	500	56,00	900	56,33
200	27,00	600	56,17	1000	56,31

The analysis of this data indicates that, for a polymerization degree $\boldsymbol{N} > \mathbf{500}$, the precision of the *Monte Carlo* method is enough to guarantee the stability of the model. Thus, it is reasonable to perform simulations for polymerization degrees $\boldsymbol{N} = \mathbf{1000}$.

Because the *Monte Carlo* method is sensitive to the statistical quality of the random number sequence, different random number sequences were used in order to simulate 100 macromolecules with $N = 1000$. Given that the model thus created proved stable in terms of random "noise", the choice of a random number sequence with a satisfactory statistical quality made it possible to acquire reliable results after only one simulation.

The simulation model previously proposed was tested for a wide variety of copolymerization reactions (Motoc and Ciubotariu, 1976; Motoc and Ciubotariu, et al., 1975, 1976, 1977). As an illustration, Table 10 presents the results obtained by simulating the copolymerization reaction of the ethyl methacrylate with the *para*-ethoxystirene, reaction performed at 50°C by using the azoisobutyronitrile as initiator. In Table 10, x_1 represents the molar fraction of monomer $\boldsymbol{M_1}$ (ethylic ester of methacrylic acid) in the reaction feed. Y_2 is the molar fraction of monomer $\boldsymbol{M_2}$ (*para*-ethoxystirene) in the resulting copolymer, which was determined by classical methods (Margerison et al., 1975), and Y_{MC} is the molar fraction resulting from the *Monte Carlo* model described above.

TABLE 10 The simulation results (Y_{MC}) of the copolymerization reaction of the methyl methacrylate with the *para*-ethoxystirene, reaction performed at 50°C by using the azoisobutyronitrile as initiator.

x_1	$Y_2^{(1)}$	$Y_2^{(2)}$	$Y_2^{(3)}$	Y_{MC}
0,873	0,223	0,235	0,224	0,240
0,876	0,264	0,255	0,222	0,276
0,230	0,605	0,597	0,600	0,610
0,256	0,641	0,633	0,593	0,642

(1) Analysis C; (2) Analysis O; (3) Analysis UV.

(2) Experimental data (Margerison et al., 1975): r_1=0,37; r_2=0,24; x_1 = molar fraction of the ethyl methacrylate (M_1) in the feed; Y_2 = molar fraction of the *para*-ethoxystirene (M_2) in the copolymer.

The *Monte Carlo* model of irreversible binary copolymerization with ultimate effect was formalized within an axiomatic model (Motoc and Ciubotariu, 1974) and as a structural model of pattern recognition (Ciubotariu, 1987). This stochastic linguistic model allowed for the development of an original method for determining reactivity ratios ($\boldsymbol{r_1}$, $\boldsymbol{r_2}$).

12.7 DESCRIPTION OF THE IRS COMPUTER PROGRAM

IRS (Investigated Receptor Space) represents an effort to consolidate our work in the field of computational chemistry. The project is intended to assist researchers in developing, testing and applying various QSAR and, possibly, QSPR methods, but it can be expanded further as a general computational chemistry workbench.

IRS relies on some basic data types for three-dimensional computations, as well as more specific abstractions for atoms, molecules and radicals (or substituents). IRS supports most of the usual chemistry formats by using the OpenBabel library (http://openbabel.org/). Tabular data can be imported using CSV and be used as input data (such as experimental values for performing linear correlation analysis).

At its core, IRS is a framework for developing and integrating computational chemistry algorithms, such as Monte Carlo integrators or statistical correlation. Each algorithm is abstracted as a black box with a set of *input data* and *output data*, as well as a set of *parameters* that define the preconditions of the user's particular calculus. The input and output data are always associated with *semantic concepts*, allowing a new approach in machine-processing of data. A work session is based on the *blackboard* architecture, where an experimenter places necessary data and calculi on the provided "blackboard" (see Fig. 7)

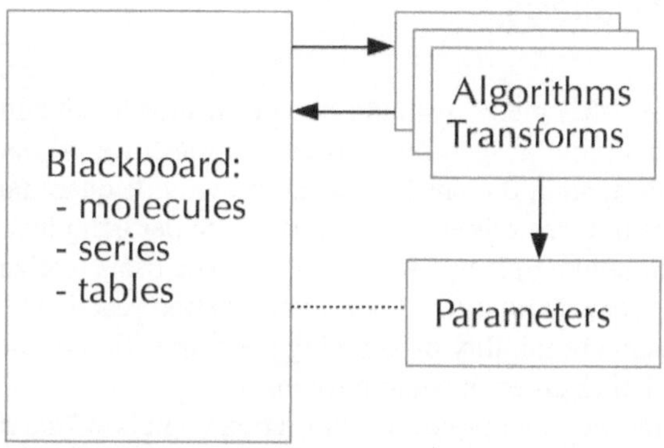

FIGURE 7 Blackboard architecture of IRS.

The architecture described above enhances the possibilities for semantically processing the available data. For instance, it makes it possible for the framework to automatically determine and execute intermediary steps in obtaining derived data. Depending on available data and algorithms, the system determines a computation path in order to provide the user with the requested data.

From a computer science perspective, the blackboard constructs a directed acyclic graph (DAG) where the nodes represent the data the user manipulates, and the arcs are the transformations that produced it (apart from data loaded from an external sources, which are nodes with no incident arcs). The user can grow this graph by selecting which type of node he wants to create, and the system finds an arrangement of algorithms (links) that can produce the desired node type.

IRS also allows users to compose existing data into series of data of the same type (as depicted in the tables in the current chapter) for the purpose of statistical analysis of the results. All tabular data can be exported in "comma separated values" format (CSV), allowing more complex investigation by specific programs.

Researchers use multiple sets of parameters when performing calculations. Some of these affect the quality of the calculus (i.e. the number of trials in a stochastic algorithm, or the number of points used in a Monte Carlo integration), while others change the results quantitatively. In the case of computing a molecule's volume and surface one can choose from a set of atom radii, each selection resulting in a different numerical value of the computed molecular descriptor. In order to accommodate this situation each data node computed by IRS is associated with the set of external parameters used by the user.

The final purpose of IRS is to provide researchers with the ability to create a database consisting of input data, results of the numerical analysis performed on them, together with the methods applied in obtaining their results.

12.8 CONCLUSIONS

The Monte Carlo (MC) method should be very useful for obtaining some information about the structural features of organic molecules or macromolecules of polymer materials at nano-dimensions. It should provide quantitative information about the features of three-dimensional (3D) molecular complexes, using its ability to solve multiple integrals by means of stochastic trials realized with (pseudo) random numbers. The simulation of copolymerization reactions by means of MC algorithms offers the possibility of knowing the disposition of monomers along a macromolecule, being given the reactivity ratios.

In this chapter we have presented the fundamentals of the MC method and the way by which it has been used to develop some shape and size descriptors at molecular nanoscale dimensions. We described here a MC simulation model of copolymerization reactions that was based on using the relationships between the stochastic character of these chemical reactions and some statistical characteristics of random experiments.

One of the fundamental concepts of chemistry is the molecular structure, since the physical, chemical and biological properties of molecules are determined by it. The quantitative treatment of various characteristics of molecular structure by means of so-called molecular descriptors is a great task of the actual research and explains the efforts in this field. The structural descriptors of molecular shape and size have been developed in the approximation of hard spheres, taking into consideration the molecular van der Waals (vdW) space. Thus, a molecule has been considered as a collection of relatively impenetrable atomic spheres, situated at their equilibrium positions defined by their Cartesian coordinates and having radii equal to the corresponding vdW radii. We selected and detailed here the vdW volume (V^W) and surface (S^W), the ratio V^W/S^W (r^{WV}), the semiaxes of the ellipsoid (E_X, E_Y, E_Z), which embeds a given molecule, and two globularity measures – G^{LEL} and G^{LOB}. The use of these vdW molecular descriptors in QSPR studies in modeling the vapor pressure of a series of alkanes and cycloalkanes (V^W) and boiling points of a series of 72 alkanes (S^W, R^{WV}, E_X, E_Y, E_Z, G^{LEL} and G^{LOB}) proves the utility of quantifying the various structural characteristics of the vdW molecular space in order to identify the structural requirement that conditions a given property.

Taking into consideration the fact that the proximal steric effects were usually attributed to repulsive forces, which appear when two atoms (or atomic groups) are closer than the sum of their vdW radius, the steric static integral parameter I_S was introduced before, and it was also presented in this chapter. I_S has a clear physical meaning. It represents a steric gradient directly proportional with the size of each volume element (dV) and inversely proportional with the distance of that volume element to the reaction center raised to a certain power, n (n>0). The extension on the entire space was realized by Monte Carlo integration. The optimal values of I_S (specified by calculated n value) can be determined by regression analysis, using the trial and error method. It can be mentioned that I_S can be calculated in every molecular context and for any type of substituents interacting in chemical and biological processes. Applying this steric indicator to the acid-catalyzed esterification and hydrolysis of esters as was described in this chapter, we have showed that the steric effect as measured by E_S-Taft steric parameter is due to the molecular vdW space, being directly proportional to the size of each dV and inversely proportional to the distance of that dV to the reaction center raised to the fourth power. I_S should be very useful in QSAR studies to measure and explain the steric receptor-ligand interactions.

The development of the descriptors for molecular surface accessibility and dynamics (MSAD) of the molecular structure was made possible by the introduction of a combination of two experimental techniques used in tandem, namely, hydrogen-deuterium exchange and mass spectrometry (H/D-MS/MS). We presented here one MSAD descriptor constructed from a molecular descriptor that quantifies an electronic effect (H-bonding), and two descriptors that describe a steric effect (the accessibility of the amide hydrogen and the accessibility of the amino acid residue to which the amide hydrogen belongs).

The MC method was used for simulating the copolymerization reaction that follows the kinetic model with ultimate effect because the synthesis of macromolecular assembly have a statistical character, commonly reflected by the way in which the monomer units are distributed along the macromolecular chain. The simulation MC model has been tested for a variety of copolymerization reactions and compared with experimental results. The model performed properly.

A short description of our in-house computer package named IRS (Investigated Receptor Space) has been presented in the finale section.

The development of nanoscience requires a sustained research effort focused on various structural characteristics of molecular and macromolecular systems. The progress in this field should be achieved, obviously, not by opposing the methods in competition, but by the union of all techniques in a combined assault on the problem in hand.

KEYWORDS

- **coefficient of prediction**
- **copolymerization constant**
- **hard sphere**
- **head–tail or tail–head**
- **leave one-out method**
- **Monte Carlo method**
- **packing density**
- **pseudorandom**
- **random number generators**

REFERENCES

Baeurle, S. A. "Multiscale modeling of polymer materials using field-theoretic methodologies: A survey about recent developments." J. Math. Chem. **2009,** *46,* 363–426.

Bai, Y.; Milne, J. S. et al. "Primary structure effects on peptide group hydrogen exchange." Proteins: Struc. Funct. Genet. **1993,** *17,* 75–86.

Barker, J. R. "Sums of Quantum States for Nonseparable Degrees of Freedom: Multidimensional Monte Carlo Integration" J. Phys. Chem. **1987,** *91,* 3849–3854.

Binder, K. "The Monte Carlo Method in Condensed Matter Physics." Springer: New York, **1995**.

Charton, M. "Steric Effects: I. Esterification and Acid-Catalyzed Hydrolysis of Esters." J. Am. Chem. Soc. **1975,** *97,* 1552–1556.

Charton, M. Steric Effects. 7. "Additional υ Constants." J. Org. Chem. **1976,** *41,* 2217–20.

Chetty, P. S.; Mayne, L. et al. "Helical structure and stability in human apolipoprotein A-I by hydrogen exchange and mass spectrometry." Proc. Natl. Acad. Sci. USA **2009,** *106(45),* 19005–10.

Ciubotariu, C.; Medeleanu, M.; Ciubotariu, D. "IRS – a Computer Program Package for QSAR and QSPR Studies." Chem. Bull. "Politehnica" Univ. (Timisoara). **2006,** *51(65),* 13–16.

Ciubotariu, D. "StructureReactivity Relationships in the Class of Carbon Oxide Derivatives", PhD Thesis: Polytechnic Institute of Bucharest, 1987.

Ciubotariu, D.; Deretey, E.; Medeleanu, M.; Gogonea, V.; Iorga, I. "New Shape Descriptors for Quantitative Treatment of Steric Effects: 1. The Molecular van der Waals Volume." Chem. Bull. Pitt.; Polytechnic Institute Timisoara **1990,** *35,* 83–92.

Ciubotariu, D.; Gogonea, V.; Medeleanu, M. "Van der Waals Molecular Descriptors. Minimal Steric Difference." In: QSPR/QSAR Studies by Molecular Descriptors; Diudea, M. V. Ed.; NOVA Science: Huntington, 2001; 281–361.

Ciubotariu, D.; Holban, S.; Motoc, I. A Computer Study of Copolymers. I. Ethylene / Propylene Copolymers, Preprint: Univ. Timişoara, Fac. St. Nat.; Ser. Chimie, **1975,** No. 10.

Ciubotariu, D.; Medeleanu, M.; Vlaia, V; Olariu, T.; Ciubotariu, C; Dragoş, D.; Seiman, C. "Molecular van der Waals Space and Topological Indices from the Distance Matrix." Molecules **2004,** *9,* 1053–1078.

Ciubotariu, D.; Motoc, I.; Holban, S.; Zur Kopolymerization, V. Die Simulierung der bildung eines makromoleküls durch binare kopolymerization. Das program MEMORY3, Rev. Roum. Chim. **1976,** *21,* 1253–1261.

Ciubotariu, D.; Vlaia, V.; Ciubotariu, C.; Olariu, T.; Medeleanu, M.; "Modeling the Toxicity of Alcohols. Topological Indices versus van der Waals Molecular Descriptors." In Quantum Frontiers of Atoms and Molecules, Putz, M.V. Ed.; Nova Science Publishers, Inc.; Huntington, New York, **2011,** 629–668.

Ciubotariu, D.; Vlaia, V.; Ciubotariu, C.; Olariu, T.; Medeleanu, M.; "Molecular Shape Descriptors: Applications to StructureActivity Studies." In Carbon Bonding and Structures Advances in Physics and Chemistry, Putz, M. V. Ed.; Springer Verlag: Dordrecht Heidelberg London New York, **2011,** 337–377.

Curtiss, L. K.; D. T. Valenta, et al. "What is so special about apolipoprotein AI in reverse cholesterol transport?" Arterioscler Thromb Vasc Biol. **2006,** *26,* 12–19.

Demidovich, B. P.; Maron, I. A. "Computational Mathematics" Mir: Moscow, **1981,** 649–674.

Englander, S. W. "Hydrogen exchange and mass spectrometry: A historical perspective" J. Am. Mass. Spectrom. **2006,** *17,* 1481–9.

Feuvrier, C. "La simulation des systemes" Dunod: Paris, 1972; 70–88.

Fishman, G. S. "Monte Carlo: Concepts, Algorithms, and Applications" Springer: New York; 1995.

Gabel, F.; Bicout, D.; et al. "Protein dynamics studied by neutron scattering." Quarterly Reviews of Biophysics **2002,** *35,* 327–67.

Gogonea, V. "An Approach to Solvent Effect Modelling by the Combined Scaled-Particle Theory and Dielectric Continuum-Medium Method", PhD. Thesis, Toyohashi Univ. of Technology, Japan, 1996.

Gogonea, V.; Ciubotariu, D.; Deretey, E.; Popescu, M.; Iorga, I.; Medeleanu, M. "Surface Area of Organic Molecules: a New Method of Computation" Rev. Roum. Chim. **1991,** *36,* 465–471.

Gogonea, V.; Wu, Z. et al. "Congruency between biophysical data from multiple platforms and molecular dynamics simulation of the double-super helix model of nascent high-density lipoprotein." Biochemistry **2010,** *49,* 7323–43.

Hansch, C.; Leo, A; Hoekman, D. "Exploring QSAR – Hydrophobic, Electronic and Steric Constants" ACS Professional Reference Book, Am. Chem. Soc.: Washington, D.C. 1995.

Hansch, C.; Leo, A. "Substituent Constants for Correlation Analysis in Chemistry and Biology" Wiley-Interscience: New York, **1980**.

Harwood, H. J. In Data processing in Chemistry, Hippe, Z. Ed.; Elsevier: Amsterdam; **1981,** 133–150.

Harwood, H. J.; Kodaira, Y.; Newman, D. L. In Computers in Polymer Sciences, Mattson, J. S.; Mark; H. B., Jr.; MacDonald, H. C., Eds.; M. Dekker: New York, **1977,** Chapter 2.

Hilser, V. J. and Freire E. "Structurebased Calculation of the Equilibrium Folding Pathway of Proteins. Correlation with Hydrogen Exchange Protection Factors." J. Mol. Biol. **1996,** *262,* 756–772.

I. Motoc; D. Ciubotariu, Zur Kopolymerization. I. Matematisches modell der bildung eines binaren kopolymerizationsproduktes, Rev. Roum. Chim. **1976,** *21,* 949–954.

I. Motoc; D. Ciubotariu; Holban, S. "Zur Kopolymerization. II. Die Siemulierung der bildung eines makromoleküls." Das program MEMORY*1,* Rev. Roum. Chim. **1976,** *21,* 769–773.

I. Motoc; D. Ciubotariu; Holban, S. "Zur Kopolymerization. III. Die Simulierung der bildung eines makromoleküls." Das program MEMORY*2,* Rev. Roum. Chim. **1976,** *21,* 775–780.

Ingold, C. K. "CXXIX-The mechanism of, and constitutional factors controlling, the hydrolysis of carboxylic esters. Part I. The constitutional significance of hydrolytic stability maxima" J. Chem. Soc. **1930,** 1032–1039.

Kalos, M. H.; Whitlock, P. A. "Monte Carlo Methods" Wiley-VCH: New York, **2008**.

Katritzky, A. R.; Kuanar, M.; Slavov, S.; Hall, C. D. "Quantitative Correlation of Physical and Chemical Properties with Chemical." Chem. Rev. **2010,** *110,* 5714–5789.

Katritzky, A. R.; Slavov, S. H.; Dobchev, D. A.; Karelson, M. "Rapid QSPR model development technique for prediction of vapor pressure of organic compounds." Comp. Chem. Eng. **2007,** *31,* 1123–1130.

Kelen, T.; Tüdös, F.; "Analysis of the Linear Methods for Determining Copolymerization Reactivity Ratios. I. A New Improved Linear Graphic Method." J. Macromol. Sci. **1977,** *9,* 1–27.

Knuth, D. E.; "The Art of the Computer Programming, Seminumerical Algorithms" (translated in Romanian language); Edit. Tehn.: Bucureşti, **1983**, 166–192.

Kroese, D. P.; Taimre, T.; Botev, Z.I. "Handbook of Monte Carlo Methods" John Wiley & Sons: New York; **2011**.

Laidler, K. J. "Theories of Chemical Reaction Rates" McGraw-Hill: New York, **1969**.

Margerison, D.; Bain, D. R.; Nogan, N. R.; Taylor, L.; "Reactivity ratios for the copolymerization systems stirenep-ethoxystirene and methyl methacrylatep-ethoxystirene" Polymer, **1975,** *16,* 278–280.

Mayo, F. R.; Lewis, F. M.; "Copolymerization. I. A Basis for Comparing the Behavior of Monomers in Copolymerization; The Copolymerization of Stirene and Methyl Methacrylate." J. Am. Chem. Soc. **1944,** *66,* 1594–1598.

Metropolis, N.; Ulam, S. "The Monte Carlo Method." J. Am. Statist. Assoc. **1949,** *44,* 335–341.

MobyDigs v.1.1 was obtained from Talete srl.; via V. Pisani, 13-20124, Milano, Italy, **2009,** Internet page http://www.talete.mi.it; at the time of this writing it seems discontinued on the web site.

Moskowitz, B.; Caflisch, R. E. "Smoothness and Dimension Reduction in Quasi-Monte Carlo Methods" Mathl. Comput. Modelling. **1996,** *23,* 37–54.

Motoc, I.; Ciubotariu, D.; Holban S. "The axiomatic model of the copolymerization reaction. The FORTRAN Program MEMORY-3" in Romanian: Modelul axiomatic al copolimerizării binare. Programul (FORTRAN) MEMORY_*3,* The fourth National Conference of theoretical and applied physical chemistry, Bucharest, **1974,** September, 2–*4,* Abstr. Vol.; p. 12.

Motoc, I.; Ciubotariu, D.; Holban, S. "Zur Kopolymerization: Axiomatische Formulierung eines Matematischen Modells. Das MEMORY3 Programm" J. Polym. Sci. **1977,** *15,* 1465–1470.

Motoc, I.; Ciubotariu, D.; Holban, S. "Zur Kopolymerization. IV. Das mathematische modell der bildung eines binaren kopolymers: die verbesserte version." Rev. Roum. Chim. **1976,** *21,* 1247–1251.

Motoc, I.; Ciubotariu, D.; Holban, S. A Computer Study of Copolymers. II. The Influence of Conversion on the Ethylene/Propylene Copolymer Composition, Preprint: Univ. Timişoara, Fac. St. Nat.; Ser. Chimie, **1975,** 11.

Niculescu-Duvăz, I.; Ciubotariu, D.; Simon, Z.; Şi Voiculetz, N. "QSAR (SAR) Models and Their Use for Carcinogenic Potency Prediction." In Modeling of Cancer Genesis and Prevention; Voiculetz, N.; Balaban, A. T.; Şi Simon, Z.; Eds.; CRC: Boca Raton, FL, **1991**, 157–214.

Ö kten, G. "Error Reduction Techniques in Quasi-Monte Carlo Integration" Mathl. Comput. Modelling. **1999,** *30,* 61–69.

Pearlmann, R. S. SAREA Program. QCPE No. *413,* **1983**.

Safarevich, R. I.; "Fundamentals of algebraic geometry." (translation form Russian in Romanian language). Edit. St. Enciclop.: Bucuresti, **1976**.

Shreider, Y. A. Ed. "The Monte Carlo Method." Pergamon: New York, **1965**.

Sobol, I. M.; Tutunnikov, A. A. "A Variance Reducing Multiplier for Monte Carlo Integrations" Mathl. Comput. Modelling. **1996,** *23,* 87–96.

Stoka, M.; Teodorescu, R.; "Probability and Geometry" (in Romanian language), Edit. Ştiinţifică: Bucureşti, **1966,** 200–203.

Szabo, A.; Ostlund, N.S. "Modern Quantum Chemistry. Introduction to Advanced Electronic Theory" McGraw-Hill: New York, **1989**.

Taft, R. W. "Linear Free Energy Relationships from Rates of Esterification and Hydrolysis of Aliphatic and Ortho-substituted Benzoate Esters." J. Am. Chem. Soc. **1952,** *74,* 2729–2732.

Taft, R. W. "Linear Steric Energy Relationships." J. Am. Chem. Soc. **1953,** *75*, 4538–4539.

Taft, R. W."Polar and Steric Substituent Constants for Aliphatic and o-Benzoate Groups from Rates of Esterification and Hydrolysis of Esters." J. Am. Chem. Soc. **1952,** *74,* 3120–3128.

Todeschini, R.; Consonni, V. "Molecular Descriptors for Chemoinformatics" Wiley-VCH, Weinheim, 2009; (a) p.141; (b) p. 545; (c) p. 688.

Von Mises, R. (1919) Math. Z. 5: 52 – ref. cited in Knuth, 1983.

Wu, Z.; Wagner, M. A. et al. "The refined structure of nascent HDL reveals a key functional domain for particle maturation and dysfunction." Nat. Struct. Mol. Biol. **2007,** *14,* 861–868.

Wu, Z.; Gogonea, V. et al. "The double super helix model of high density lipoprotein." J. Biol. Chem. **2009,** *284,* 36605–19.

Young, L. J.; "Copolymerization parameters." J. Polym. Sci. **1961,** *54,* 411–455.

INDEX

A

ABC transporters, circumvention
  impairment, 17
  LDH enzyme, leakage, 17
  P-gp efflux transporter, 17
  transepithelial electrical resistance (TEER), 17
  transporters modulation, 16
Active targeting, 17–18
Aerosols, 44
Allotropes, 43–44
Alternant conjugated organic oligomers
  carbon-based conducing polymers
    bond alternation, 167
    cis/transcontent, 167
    electrical conductivity, 168
    encapsulation, 168
    excitation energy, 168
    platinium wires and electroday, 168
    polymer infiltration, 168
    transisomers, 167
  CNTs, encapsulation
    aggregates, 169
    branched and cross-linked, 171–172
    carbon-chain encapsulation approach, 169
    carbon-molecular nanowires, 169
    charge-transfer, 171
    double-anode arc-discharge process, 169
    as encapsulators and protectors, 172
    field-effect nanotransistors, 171
    metal wires, application, 172
    phononscattering effects, 171–172
    small molecular magnets, 170
    spontaneous-insertion/synthesis, 169
    stability, 172
    supercritical CO2, 171
    SWCNT doped, 171
    synthesis, 171
    zero de-doping effects, 171
  five 4N/4N+2 oligomers, computational quantum chemistry (see Five oligomer-molecules)
  nanowires
    benzenoid compounds, 173
    biodegradable, 173
    Cadmium sulfide usage, 173
    cyclobutadiene/benzene oligomer, 174
    electronic analysis, 174–175
    energy harvesting, 173
    linear benzenoid molecules, 173
    optoelectronics, 173
    versatility, 172–173
  result
    A–E oligomers, electronic properties, 179
    anion and cation states and ground state, 179–180
    Coulomb analysis, 181
    E oligomer, 177–178
    HOMO/LUMO gaps, 177–180
    LUMO symmetry rule, 184–186
    regression analysis, 177, 180
    wavefunction analysis, 181–184
Amino acid zwitterion, 251–252

Ammonia and amonia-borane complex, 203
Attenuation function, 352

B

B88, 343
Biopharmaceutical considerations
ABC transporters, circumvention, 16–17
active targeting, 17–18
administration, routes for, 16–17
EPR effect, 14–15
MPS-pegylation, 13–14
$B_{12}N_{12}$ cage
calculated negative NICS values, 220
and hydrogen-loaded analogs, 219
optimized geometries, 219
reaction electrophilicity values, 220
stability and reactivity, 220
BN nanocones, 274
BN-nanotubes (BNNTs), 210, 274
BNNTs. *See* BN-nanotubes (BNNTs)
Born-Oppenheimer (BO) approximation, 314–315
Boron-lithium clusters, 231–232

C

Cage-like cluster, 224
Carbohydrates, hydrogen storage materials, 201–202
hydrogen from starch, 201
Carbon-based polymers, 168, 173
Carbon nanohorns (CNHs), 49–50
Carbon nanomaterials, 46, 209
Carbon nanoparticles
aerosols, 44
allotropes, 43–44
CNHs, 49–50
MAA, 44
metallofullerenes, 48–49
NDs, 46–47
NGO, 48
SWCNTs, 44–46
Carbon nano-systems. *See* Nonlinear optical properties, computational strategies
Carbon nanotubes
encapsulation of molecules, 169–172
Mn12Ac encapsulation, 170
multiwalled, 91
one dimensional, 321–324
oxidized multiwalled, 47
Si-doped carbon nanotubes (SiCNTs), 209
SWCNTs (see Single-wall C-nanotube (SWNT))
Cascade-like behavior, 159
CAS-SCF. *See* Complete-active-space selfconsistent field (CAS-SCF)
CBGs. *See* Covalently bonded graphenes (CBGs)
CDFT. *See* Conceptual density functional theory (CDFT)
Ceria nanoparticles, 54
Chemotopology, 63
Chitosan, nanoparticles coated, 40
Clathrate hydrates, 209
Clathratus, 211
CLIO. *See* Cross- Linked Iron Oxide (CLIO)
$C_{12}N_{12}$ cage
binding energy per hydrogen, 220
dispersion, 220
electric field, 222
gas phase heat, 221
hardness calculation, 222
HEDMs, 222
NICS(0) values, 222
optimizations and frequency calculations, 220
optimized geometries, 221
stability orders, 221
T-P phase diagram, 222–223
unit cell, optimized geometries, 222–223
Coefficient of prediction, 416
COFs. *See* Covalent organic frameworks (COFs)
Complete-active-space selfconsistent field (CAS-SCF), 128
Conceptual density functional theory (CDFT), 212

Conducting nanowires, 166
Constant pressure dynamics, 380–382
Constant temperature dynamics, 380
Copolymerization constant, 434, 438
Copolymerization reactions
analytical and stochastic models, 436
binary copolymers, structural characteristics, 433
complex mathematical relationships, 434
copolymerization constants, 435
copolymer structure, 433
Mayo-Lewis equation, 435
monomer relative reactivity, estimation, 434
reactivity ratios, 436
stochastic modeling, 434
Cornell dots (C-dots), 56
Coulomb Integral pr oligomer, 177
Covalently bonded graphenes (CBGs), 208
Covalent organic frameworks (COFs), 206
Cross-linked iron oxide (CLIO), 58

D

DAM. *See* Drive-amplitude modulation (DAM)
Data decay, 31–32
Delocalization rule, 173–174
Dendrimers
asymmetric, 8
biodegradable, 7
branching cycles, 5
Click chemistry, 7
convergent synthesis, 6–7
design, 5
divergent and convergent, schemes, 7
functional groups, 6
3-generation, schematic representation, 5
generation zero (G0), core, 5
hydrophobic interactions, 8
layers, 5
limitations, 8
parts, 4–5
polyamidoamine (PAMAM), 5
propylene imine (PPI), 5
structural properties, variation, 6
synthesis, 6
and traditional polymers, 4
Dendrogram, 63
Dendron. *See* Dendrimers
Density functional theory (DFT), 128, 217, 313, 334, 335
Dextran nanoparticles
composition, 35–36
microspheres, 36
size, 36
DFT. *See* Density functional theory (DFT)
Dimeric and trimeric protein
dengue envelope protein, 382–383
histidines protonation, 383
intermediate monomers, 382
Dispersion
arbitrary filler and polymer, interactions, 90
fillers inability, 90
graphene melt compounding, 92
MWCNTs, SEM pictures, 91
polymer infiltration, 92
DOTA, 36
Drive-amplitude modulation (DAM), 280
Drug delivery, nanotechnology applications
dendrimers (see Dendrimers)
features, 4
inorganic nanosystems (see Inorganic nanosystems)
lipid-based nanoparticles (see Lipid-based nanoparticles)
nanotoxicology, 18–19
nonspecific distribution, 2
specialized barrier tissues, 2
2-D stars, 228–230
3-D stars, 230–231

E

EFM. *See* Electrostatic force microscopy (EFM)
Electronic Hamiltonian, 314
Electrostatic force microscopy (EFM), 281

Electrostatic potential (ESP), 177
Elliot-Yafet mechanism, 134
Endohedral metallofullerene, 48
Enhanced permeation and retention (EPR) effect, 14–15
EOM-CC. *See* Equation-of-motion-coupled-cluster (EOM-CC)
EPR effect. *See* Enhanced permeation and retention (EPR) effect
Equation-of-motion-coupled-cluster (EOM-CC), 128
Euler theorem, 273

F

Faraday effect, 134
Fischer Tropsch (F-T) process, 202–203
Five oligomer-molecules
  methods, 175–176
  preliminary optimization, 176
  statistical analysis, 176
  structure optimization and preparation, 176
  wavefunction analysis, 177
Force fields, 375, 385
Formic acid and imidazolium ionic liquid, 203–204
Free energy perturbation methods, 392–393

G

Gamma rays, 32
Gauge-independent atomic orbital (GIAO) method, 214–215
Generalized gradient approximation (GGA)
  B88, 343
  dimensionless density gradient, 343
  double-hybrid (meta)
    B2- and mPW92-PLYP double-hybrid functionals, 360
    B2-PLYP and mPW92-PLYP, 359
    multi-coefficient three-parameter (MC3), 359
    PBE0-DH, 360
  exchange energy density, 342
  gradient-corrected term, 345
  Hartree-Fock exchange energy, 344
  hybrid (meta)
    B3LYP, 355
    functional M05, 356
    GGA M11 functional, 358–359
    half-and-half, 354
    Hartree-Fock theory, 353
    HK universal functional, 353
    massive parameterization, 355
    meta GGA density functionals, 356
    M08-HX and M08-SO, 357
    pure exchange-correlation effects, 353
    Slater determinant, 354
  meta
    energy density functional, 352
    enhancement factors, 353
    functional M05, 350
    functional M11-L, 352
    Hartree-Fock methodologies, 347
    inhomogeneity factor, 348
    kinetic energy density, 348
    M06-L exchange-correlation energy density functional, 352
    M06-L functional, 350–351
    PBE exchange functional, 352
    PKZB density functional, 349
    range-separation parameter, 352
    self-correlation factor, 351
    semiempirical density functionals, 349
    TPSS exchange correlation energy density, 350
    VS98 exchange functional, 348
  PW92 and PBE schemes, 347
  reduced density gradient, 342
  second-order LYP correlation functional, 344
  zero-order mean-path formula, 344
Generation zero (G0), 5
Gerade, 148, 154

GIAO method. *See* Gauge-independent atomic orbital (GIAO) method
GNRs. *See* Graphene nanoribbons (GNRs)
GNS. *See* Graphene nanosheets (GNS)
GODC. *See* Graphene oxide derived carbons (GODC)
GOFs. *See* Graphene-oxide frameworks (GOFs)
Gold nanoparticles
  absorption bands, 11
  in diagnostics and therapeutics, 10
  fluorescent materials, 11
  interacting electromagnetic field, 11
  plasmons, 11
Graphene
  Avrami plots, polymer, 112
  GNS, 108
  isotactic PP (iPP)
    CNT-induced crystallized systems, 110
    crystallization rate, 109–110
    synergistic effects, 109
  matrix, thermal stability, 106
  thermal expansion coefficient, 106
Graphene nanoribbons (GNRs), 325
Graphene nanosheets (GNS), 108
Graphene oxide derived carbons (GODC), 208
Graphene-oxide frameworks (GOFs), 207–208

H

Half life, 32
Hamiltonian, 313–314
Hard sphere, 408, 413, 417, 424, 425, 444
Hartree-Fock molecular orbitals, 128
Hartree-Fock theory, 353, 365
HCA. *See* Hierarchical cluster analysis (HCA)
Head–tail or tail–head, 434
HEDMs. *See* High-energy density materials (HEDMs)
Helio-photocatalysis, 284
Hellmann-Feynman force components, 215
Hellmann-Feynman theorem, 354
Hierarchical cluster analysis (HCA), 63–64
High-energy density materials (HEDMs), 222
Highly oriented pyrolytic graphite (HOPG), 280
HOMO/LUMO gaps
  aromatic content, 177
  band gaps, 182
  electronic excitability, 177
  modulations, 186
  reduced aromatic content, 177
HOPG. *See* Highly oriented pyrolytic graphite (HOPG)
Hydrocarbons and liquid organic hydrogen carriers
  Fischer Tropsch (F-T) process, 202–203
  liquid organic hydrogen carriers (LOHC), 203
  propane and methane, 202–203
  "synfuel," 202
Hydrogen hydrates
  binding energies per hydrogen sphere, 218
  DFT based reactivity descriptors, 217
  encapsulated, 217
  hydrogen trapping agents, 217
  Lennard-Jones interaction parameters, 218
  MEP, 217
  MHP, 217
  molecular hydrogen-trapping agent, 216–217
Hydrogen storage materials
  chemical
    ammonia and amonia-borane complex, 203
    carbohydrates, 201–202
    formic acid and imidazolium ionic liquid, 203–204
    metal hydrides, 199–201
    synthesized hydrocarbons and liquid organic hydrogen carriers, 202–203

computational details
Brillouin zone, 216
Coulombic interactions, 216
electric field, 215
GGA, 215
Hellmann-Feynman force components, 215
Lorentz-Berthelot rule, 216
PAW, 215
PBE, 215
PES, 215
SPC water model, 216
VMD program, 216
exohedral and endohedral hydrogen trapping
B12N12 cage, 219–220
C12N12 cage, 220–223
hydrogen hydrates, 216–219
Mgn and Can cage-like clusters, 224
lithium doped molecular clusters
associated 3-D functional material, 236–237
boron-lithium clusters, 231–232
2-D stars, 228–230
3-D stars, 230–231
Li3Al4− cluster, 228
Li-doped anular systems, 233–234
Li-doped borazine derivatives, 234–235
Li-doped boron hydride, 232–233
N4Li2 and N6Ca2 clusters, 235
metal clusters (see Metal clusters, hydrogen storage)
physical
COFs, 206
MOFs, 204–205
nanotubes and cage-like molecules, 208–212
sheet-like networks, 206–208
super-alkali clusters
hydrogen-trapped analogs and, 238
variation of interaction, 239
theoretical background
CDFT, 212
density functional theory, 212
electronegativity and hardness, 213
GIAO method, 214–215
global reactivity descriptors, 212
local philicity, calculation, 214
local reactivity descriptor, 213
MEP, 214
MHP, 214
MPP, 214
NICS, 214–215
Hypercore, 7

I

Inorganic nanoparticles
gold, 60–61
hydroxide, 55
iron oxide, 57–60
silica, 55–57
ZnNWs, 62
Inorganic nanosystems
biopharmaceutical considerations (see Biopharmaceutical considerations)
gold nanoparticles, 10–11
porous inorganic nanoparticles, 13
superparamagnetic nanoparticles, 12
Investigated Receptor Space (IRS) computer program
blackboard architecture, 443
comma separated values format (CSV), 444
directed acyclic graph (DAG), 443
OpenBabel library, 443
purpose, 444
Tabular data, 443
three-dimensional computations, 443
IPR. *See* Isolated pentagon rule (IPR)
Isolated pentagon rule (IPR), 273, 316–317

K

Kapitza resistance, 101
Kelvin probe force microscopy (KPFM), 279–280
Kohn-Sham auxiliary functions, 353, 359, 366
Koopmans' theorem, 213
KPFM. *See* Kelvin probe force microscopy (KPFM)

L

Lactobacillaceae family, 35
Laser-induced demagnetization experiments, 134–135
Latex nanoparticles, 36–37
Leave one-out method, 413
Li_3Al_4– cluster, 228
Li-doped anular systems, 233–234
Li-doped borazine derivatives, 234–235
Li-doped boron hydride, 232–233
LIE method. *See* Linear interaction energy (LIE) method
Linear benzenoid molecules, 173
Linear interaction energy (LIE) method
  Boltzmann-averaged electrostatic, 390–391
  linear response approximation (LRA) theory, 391
  podophyllotoxin analogs, 391–392
  solvent accessible surface area (SASA), 391
Linear response method, 390
Lipid-based nanoparticles
  clinical applications, 8
  drawbacks, 10
  hydrophobic constituents, 8
  liposomes, 8–9
  SLN manufacturing, 9–10
  solid lipids, emulsions preparation, 9
  spherical vesicles, 8
Lipid nanoparticles
  liposomes, 41–43
  SLNs, 43
Liposomes, 34
  ATP, 41
  in cancer therapeutic applications, 41
  classification, 41
  haemosomes, 41
  internal aqueous core, 41
  Kaposi's sarcoma, 42
  magnetic, 41
  PEG-lipid conjugates, 42
  in pharmacokinetic studies, 42–43
  photon radionuclides, 43
  radionuclides uasge, 41
  surface modifications, 42
  virosomes, 41
Liquid organic hydrogen carriers (LOHC), 203
Lithium doped molecular clusters. *see* Hydrogen storage materials
LOHC. *See* Liquid organic hydrogen carriers (LOHC)
Lorentz-Berthelot rule, 216
LUMO symmetry rule
  band-gap calculations, 185
  conductive oligomers, symmetry rule, 185
  five oligomers, HOMO and LUMO orbitals, 184
  improper arrangement, analysis, 185
  profiling, 185
  spin-orbits and spin-polarization, 186
  visual analysis, 185

M

MAA. *See* Macroaggregated albumin (MAA)
Macroaggregated albumin (MAA), 44
Magnetic force microscopy (MFM)
  DAM, 280
  EFM, 281
  graphene mechanical measurements, 279
  HOPG, 280
  KPFM, 279–280
  magnetic impurities, 280
  phase-locked loop, 280
  scanning force microscopy (SFM), 279
  SQUID magnetometer, 281

topography/magnetic signals, 279
Magnetic phases, 150–151, 160
Massive parameterization, 355
Maximum hardness principle (MHP), 214
Maximum residence time (MRT), 383
Maxwellian velocity distribution, 377
Membrane proteins, noncovalent interaction, 383
MEMORY, 441
MEP. *See* Minimum electrophilicity principle (MEP)
Mesoporous silica nanoparticles, 13
Metal clusters, hydrogen storage
aromatic clusters, 226–227
hydrogen trapping agents, 227–228
transition metal and doped ethylene complex
Kubas interaction, 225–226
M-ethylene complex and H2-trapped analogs, 225
reaction electrophilicity calculation, 225
Metal-diborides, 209–210
Metal hydrides, hydrogen storage materials
advantages, 199
destabilization, 200
hydrogenated magnesium nickel boride, 201
hydrogen sorption, phenomenon, 200
intermetallic compounds and properties, 200
NaAlH4 system, hydrogenation and dehydrogenation, 199
per mass and per volume, pictorial presentation, 200–201
United States DOE, 199
Metallofullerenes, 48–49
Metal organic frameworks (MOFs)
1, 4 benzene dicarboxylate (BDC), 204
effective hydrogen storing candidates, 204
for hydrogen adsorption, 205
LDI, 204
metal doping, 205
Mg_n and Ca_n cage-like clusters, 224
MHP. *See* Maximum hardness principle (MHP)
Minimum electrophilicity principle (MEP), 214
Minimum polarizability principle (MPP), 214
MM-PBSA methods, 387–390
binding free energy, 387
binding free energy prediction, 390
complex free energy, 388
Debye-Huckel screening parameter, 388
free energy, calculations protocol, 389
gas phase contributions, 387
GB variant, 387
nonpolar solvation, 387
normal mode analysis (NMA), 389
protein intramolecular energy, 388
solvent accessible surface area (SASA), 388
unbound ligand free energy, 387
Modern density functional theory
double-hybrid (meta) generalized gradient approximation
B2- and mPW92-PLYP double-hybrid functionals, 360
B2-PLYP and mPW92-PLYP, 359
multi-coefficient three-parameter (MC3), 359
PBE0-DH, 360
electron-electron interaction, 337
electronic energy density functional, 337
exchange-correlation potential, 339
external potential, 337
generalized gradient approximation (GGA)
B88, 343
dimensionless density gradient, 343
exchange energy density, 342
gradient-corrected term, 345
Hartree-Fock exchange energy, 344

PW92 and PBE schemes, 347
reduced density gradient, 342
second-order LYP correlation functional, 344
zero-order mean-path formula, 344
Hartree-Fock methods, 339
Hohenberg and Kohn (HK) theorems, 336–337
hybrid (meta) generalized gradient approximation
B3LYP, 355
functional M05, 356
GGA M11 functional, 358–359
half-and-half, 354
Hartree-Fock theory, 353
HK universal functional, 353
massive parameterization, 355
meta GGA density functionals, 356
M08-HX and M08-SO, 357
pure exchange-correlation effects, 353
Slater determinant, 354
Ladder analogy, 335
local density approximation
correlation energy per particle, 340
deviation of homogeneity, 341
Dirac exchange functional, 341
exchange energy density functional, 340
Quantum Monte Carlo simulations, 340
spin stiffness, 341
Variational Principle, 342
Vosko, Wilk and Nusair (VWN), 340
Wigner-Seitz radius, 340
meta generalized gradient approximation
energy density functional, 352
enhancement factors, 353
functional M05, 350
functional M11-L, 352
Hartree-Fock methodologies, 347
inhomogeneity factor, 348
kinetic energy density, 348
M06-L exchange-correlation energy density functional, 352
M06-L functional, 350–351
PBE exchange functional, 352
PKZB density functional, 349
range-separation parameter, 352
self-correlation factor, 351
semiempirical density functionals, 349
TPSS exchange correlation energy density, 350
VS98 exchange functional, 348
properties of interest, performance
BC322, 363
BC338, 361
charge-transfer electronic transitions (CTS8), 363
database, 361
exchange-correlation energy density functionals, 361–365
IsoL6, 361
mean unsigned error (MUE), 361, 364
meta GGA TPSS functional, 363
MRBE10, 361
MXX family of functionals, 361
non-covalent binding energies (S22A), 363
SB1AE97, LB1AE12, 361
SRMBE13, 361
subdatabases, 363
TMRE48, 365
self-consistent procedure, 339
Thomas-Fermi-Dirac (TFD) model, 336
MOFs. *See* Metal organic frameworks (MOFs)

Molecular descriptors (MD), structural information
copolymerization reactions, simulations (see Copolymerization reactions)
IRS computer program, 442–444
molecular surface and dynamics, 429–432
Monte Carlo method (see Monte Carlo method)
Molecular dynamics (MD) simulations
annealing, 377–378
biological properties, atomic scales
dimeric and trimeric protein, 382–383
membrane proteins, noncovalent interaction, 383
computational chemistry, 372
conformational changes, prediction, 384
fluorescence depolarization, analysis, 373
folding and unfolding process
computational expensive process, 384
FiP35, 385
Theoretical and Computational Biophysics Group, 385
free energy changes, calculation, 386
hydrogen bond exchange, 372
production dynamics
constant pressure dynamics, 380–382
constant temperature dynamics, 380
protein-ligand binding
free energy perturbation methods, 392–393
LIE method, 390–392
MM-PBSA methods, 387–390
thermodynamic integration (TI), 393–394
protein structures, analysis, 372
scheme
computation of properties, 373
Leap-Frog, 375
potential, 374
self-consistent method, 375
SHAKE, 375
Taylor's expansion of position, 374
time dependent behavior, 373
Verlet algorithm, 375
stability of integration, 378–379
system preparation and parametrization
amino acids and nucleic acids, library, 377
comparative protein modeling, 376
minimization role, 377
X-ray crystallography, 376–377
Molecular nanomagnets, ultrafast laser dynamics
clusters, one magnetic center
angular momentum, conservation, 134
Elliot-Yafet mechanism, 134
energy levels, 133
expectation value, energy, 136
laser-induced demagnetization experiments, 134–135
linearly polarized laser pulses, application, 134
time-dependent light-induced polarization, 135
computational methods
adaptive Cash-Karp integration steps, 129
CAS-SCF, 128
DFT, 128
EOM-CC, 128
Hartree-Fock molecular orbitals, 128
nonrelativistic electronic states, calculation, 128
post Hartree-Fock methods, 128
Runge- Kutta method, 129

SAC-CI, 128
SOC, 128–129
time-dependent perturbation theory, 129
finite clusters, more magnetic centers
electronic low-lying states, 137
electronic-population transfer, 137
group-theory analysis, 144
homodinuclear complex, 139
initial and final states, energy differences, 143
ligand-stabilized clusters, 138
low-lying many-body states and spin localization, 139–140
spin flip and spin transfer processes, 137–138, 144
spin switching, partial blocking, 141
time charge-transfer states, 140
transient absorption spectra, 142
Zeeman splitting, 143
functionalization
detuning parameter, 159
ERASE, 158
fidelity vs. detuning, 159
AND gate, 158
logic circuits, 158
low-symmetry geometry, 156
OR gate, 158
spin-density localization, 156
spin flips and transfers, 155, 157
spin, information carrier, 155
spin-transfer possibilities, 155–156
genetic algorithm, 131–132
magnet elements, miniaturization, 127
metallic ferromagnetic sheet, demagnetization, 126
moleculators, 127
phononic degrees of freedom
completeactive- space configuration-interaction (CAS-SCF), 145
direct coupling, 152
electronic levels, splitting, 146
equilibrium interatomic distances, 153
lattice vibrations, 148, 150
low-lying magnetic (triplet) states, calculation, 150–151
second order susceptibility tensor, 146–147
spin density distribution, 150
stretching and shrinking, 151
susceptibility tensor, 145
scattering mechanism, 126
spin-flipping mechanism
laser-induced ED transitions, 130
Λ process, 130
magnetic storing, 129
spin expectation value, 129
substrate
excitation energies and oscillator strengths, 154–155
molecular orbitals, contribution, 153–154
rotational and inversion symmetry, 154
Molecular templates suitable, designing
computational physics and chemistry, 198
fossil fuels, damage, 198
hydrogen storage (see Hydrogen storage materials)
nanoscience, 197
size-dependent properties, 197
Mononuclear phagocyte system (MPS)-pegylation
coating, 14
colloidal particles, 13
immune system, 13
reticulo-endothelial system (RES), 13–14
unmodified nanoparticles, 13–14
Monte Carlo (MC) method
algorithms used, 404
integration technique, 406–408
models, computers

binary copolymerization reaction, 437–438
copolymer composition, polymerization degree, 441
graphs, binary copolymerization, 440
macromolecular complexes, synthesis and formation, 437
reactivity ratios, 438
topological graph, 440
molecular descriptors calculation
acid-catalyzed esterification, 422
algorithm, IRS computer program, 413
alkanes and cycloalkanes, vapor pressure, 409–412
boiling points and surface area, alkanes, 414–416
coefficient of prediction, 416
hard-spheres model, 417
MobyDigs computer program, 416
molecular surface, 413–416
molecular volume, 408–413
packing density, 418
quantitative structureproperty relation (QSPR) models, 413, 421
size and shape, 417–421
steric static integral parameter, 421–429
Taft-type constants, 424
transition states, 423
problem-solving procedures, 404
pseudorandom, 404
random model, 403
random numbers, 404
MPP. *See* Minimum polarizability principle (MPP)
MRT. *See* Maximum residence time (MRT)
Multiwfn program, 177

N

Nanodiamonds (NDs), 46–47
Nanographene oxide (NGO), 48
Nanoparticles
based on rare earths, 53–54
biodistribution studies, 33
β+ radionuclides, half-life, 33
carbon (see Carbon nanoparticles)
classification, 33–34
inorganic (see Inorganic nanoparticles)
lipid (see Lipid nanoparticles)
nanoscale materials, 272
polymeric (see Polymeric nanoparticles)
post mortem nanoparticle determination, 33
quantum dots (see Quantum dots (QDs))
in vivo optical imaging, 33
Nanoporous zeolites, adsorption and catalysis
benzene hydroxylation
Fe-exchange zeolites, 256
N2O decomposition, 257–259
5-T cluster models, 257
extralattice acidic protons, 251
fluid catalytic cracking (FCC) technology, 250
glycine, isomers, 252
hydrogen adsorption and storage, 251–252
hydrogen, storage
Al species, 262–263
Al/ZSM-5 zeolite, 261–262
bicoordination charateristics, 264
effective core potential (ECP), 260
inelastic neutron scattering (INS) experiments, 260
LaIII ion, 264–265
Mulliken charges, 262
n-numbered molecules, adsorption structures, 261

recombination energy barrier, 262
steam reforming, 260
5-T cluster models, 260
25-T cluster models, 260–261
"philosopher's stone," 250
"short-life" intermediates, 251
zwitterion stablization
deprotonated glycine, 254
dianions and monoanion, 252
formation and stabilization, 254–256
glycine, 252–253
nanoporous aluminosilicates, 252
Silicalite-1, 254–255
two-layer ONIOM calculations, 254
ZSM-5 nanopore, 253–254
Nanoscience, 197
Nanosized particles (NPs), 28
Nanotubes and cage-like molecules
activated carbon (AC), 209
BN-nanotubes (BNNTs), 210
carbon nanofibers (CNFs), 209
clathrate hydrates, 209–211
fullerene (C60) molecule, 208
fullerenol derivatives, 210
reaction templates, 209
SiCNTs, 209
storage media, hydrogen, 209
SWCNT, 209
THF, 211–212
Ti-doped CNTs, 210
Nested functionals, 350
NGO. *See* Nanographene oxide (NGO)
N-(2-hydroxypropyl)-methacrylamide (HPMA), 35
NICS. *See* Nucleus-independent chemical shift (NICS)
N_4Li_2 and N_6Ca_2 clusters, 235
4n/4n+2 structures
alternant benzenoid oligomer/polymers, 173
chemistry, CNTs, 174
and Coulomb integration method, 181
electronic analysis, 175
oligomers, 166, 175–177
resonance, 166
Nonlinear optical (NLO) polarizabilities, 315–316
Nonlinear optical properties, computational strategies
computational methodology
ab initio quantum chemical methods, 312–313
Born-Oppenheimer (BO) approximation, 314–315
Hamiltonian, 313–314
nonlinear optical (NLO) polarizabilities, 315–316
Quantum chemistry, 312
fullerenes, 311
nano-systems, 311
nanotubes, 311–312
one dimensional carbon nanotubes
electro-optical properties, 322–324
multiwalled nanotubes (MWNTs), 321
rolling principle, representation, 322
singlewalled nanotubes (SWNTs), 321–322
zigzag nanotubes, 321
two dimensional graphenes
electro-optical properties, 324–328
hexagonal arrangement, 324
zigzag edges, 324
zero dimensional fullerenes
bond distances, calculated and experimental, 318
electron–phonon interaction, 317
electro-optical properties, 319–321
geometrical structures, 317
independent bond types, 318
"isolated pentagon rule," 317
NMR chemical shifts, 317

Nucleus-independent chemical shift (NICS), 214–215

O

oMWCNTs. *See* Oxidized multiwalled carbon nanotubes (oMWCNTs)
"Open-Shell Spherical Aromaticity," 222
Opposite-spin electrons, 183, 349
Oxidized multiwalled carbon nanotubes (oMWCNTs), 47

P

Packing density, 418
PAW. *See* Projector augmented wave (PAW)
PEGylation, 37
Penetrant, 104
Percolation threshold, 97
Perdew-Burke-Ernzerhof (PBE), 215
Permeation, 104–105
PES. *See* Potential energy surface (PES)
P-gp efflux transporter, 17
Philosopher's stone, 250
Photon upconversion, 53
Plasmons, 11
Plateau-like behavior, 159
Polyamidoamine (PAMAM), 5
Pólya-Redfield theorem, 279
Polyethyelene glicol (PEG)
  antigenicity, 38
  biocompatibility, 37
  block copolymers, 39
  in cancer therapy, 38–39
  DOTA, chelating agent, 38
  folate transporters, 38
  SCK-NPs core, 39
  toxicity, 38
Polyethylene oxide (PEO) chains, 39
Poly glycidylmethacrylate, 36
Polymeric nanoparticles
  coated with chitosan, 40
  coated with PEG, 37–40
  dextran, 35–36
  HPMA, 35
  latex, 36–37
Polymer nanocomposites, carbon-based high aspect ratio
  applications and challenges
    electron transport properties, 114
    existence, 115
    filler functionalization, 115
    industrial usage, 115–116
    lithium-ion battery, schematic representation, 113
    in medical applications, 113
    size and geometrical characteristics, 114
  buckyballs (0D), 88
  carbon black (CB), 86
  composition viscosity, 86–87
  dispersion (see Dispersion)
  electrical conductivity
    filler loading conductivity, 97
    nonidealized dispersion, 99
    percolative behavior, schematic representation, 97
    relative rigidity, 99
  gas and liquid permeation, 104–105
  graphene, 87–89
  graphite (3D), 88
  high-density polyethylene (HDPE), 89
  in-situ polymerization
    electrically conductive nanocomposites, 94–95
    mechanical properties, 95–96
    melt mixing, advantages, 93–94
    organic solvents, 93
    phase swelling, 93
  mechanical properties
    absorption conformation, 104
    CNT entanglement effects, 103
    composite films, 103
    elastic strain, 102
    elastomers, 102–103
    hydrogen-bonding interactions, 103
    local deformation, 102
    sample preparation method, 103

thermal fluctuations, 102
morphology and crystallization effects, 108–112
nanoplatelets, 87–88
nanotubes (CNT), 87
natural rubber, 89
polyolefin matrices, 87
rheological measurements, 106–108
solution mixing (see Solution mixing)
thermal and dimensional stability, 106
thermal conductivity, 100–102
Porous inorganic nanoparticles, 13
Positron, maximum energy, 32
Potential energy surface (PES), 215
Projector augmented wave (PAW), 215
Pseudorandom, 404
PW86, 343

Q

QDs. *See* Quantum dots (QDs)
Quantum chemical methods, 310–312
Quantum chemistry, 312
Quantum dots (QDs)
breast cancer marker, 52
drawback, 50–51
functionalization, 50
hydrophilic nature, 52
inorganic dyes, 50
pegylated, 52
PEGylation, 51
semiconductor material, 50
SQDs, 51

R

Radiolabeled nanoparticles
beta particles, 29
as bioprobes, 29
chemotopology
boundary of X, B(X), 69
chemical elements, 63
closure of X, 67
exterior of X, EXT(X), 68–69
graphical representation, 64
HCA, 63–64
interior of X, INT (X), 67–68
for PET imaging, 65–66
positron emitters, classification, 62–63
18F labeled nanoparticles, 29
molecular imaging, 28
nanoparticles, types (see Nanoparticles)
nanosized particles, 28
in PET imaging, 29
purposes, 30
radiochemical purity, 28
radionuclides, in PET Imaging, 30–31
radiopharmaceutical, 28
selection parameters
data decay, 31–32
gamma rays, 32
half life, 32
positron, maximum energy, 32
techniques, diagnostic field, 27
ultrafine particles, 28
utilization, 28
Random number generators, 404, 405
Reduced density gradient, 342

S

SAC-CI. *See* Symmetry-adapted cluster configuration interaction (SAC-CI)
Same-spin electrons, 183
Schrodinger wave equation, 373
SHAKE, 375
Sheet-like networks, hydrogen storage materials
binding capability, 207
3D-CBC3, 208
GODC, 208
GOFs, 207–208
graphene applications, 206
graphyne and graphdiyne, 208
Li adsorption, 207
transmission electron microscopy (TEM), 206
SiCNTs. *See* Si-doped carbon nanotubes (SiCNTs)
Si-doped carbon nanotubes (SiCNTs), 209
Silicon quantum dots (SQDs), 51
Single-wall C-nanocones (SWNCs)

ab initio calculations, 273
BN-hexagonal (h), 273–274
calculation and result
approximation error variance (AEV), 294
cluster distribution size function, 298
concentration dependence, 298
cone angle decays, 288
diffusion coefficient vs. concentration, 301
Gibbs free energies, interaction, 296
graphene hexagonal, pentagons P, cones, angles and covering efficiencies, 289
heat of solution vs. concentration, 299
heat of solution vs. temperature, 300
interaction energy, packing efficiencies/parameters, 289
mean absolute percentage error (MAPE), 294
packing efficiency vs. packing dimension, 292
packing efficiency vs. reciprocal dimension, 293
packing parameters closeness, dimension D and efficiency, 290–292
poly(BN) hexagonal network, 295–296
temperature dependence, solubility, 296
C-NPs, serum-proteins adsorption, 286
computational method
Bundlet model, 275
cluster free energy, 275
droplet cluster model, 277
equilibrium difference, 276
face-centerd cubic (FCC), 276
solubility mechanism, 275
Euler theorem, 273
graphene-sheet symmetry, 273
heterofullerenes, enumeration, 279
heterogeneous photocatalysis
bifunctional photocatalysis, 284–285
dyes, 284
equilibrium adsorption constants, determination, 285
fundamentals, 284
helio-photocatalysis, 284
kinetic parameters, 283
noble metal deposition/ion-doping, 283–284
photonics, 285
water- detoxification photocatalytic devices, 284
hydrocarbon pyrolysis, 273
isolated pentagon rule (IPR), 273
MFM (see Magnetic force microscopy (MFM))
mono- and multi-exponential kinetics, 278–279
photocatalytic degradation
detected p-benzoquinones, 282–283
dichlorophenols (DCPs) and three trichlorophenols (TCPs), 282
synthetic wastewaters, 283
TiO2 photocatalysts, 281–282
quantitative structure–activity/property relationships (QSARs/QSPRs), 286–287
spin transition, 287–288
Single-wall C-nanotube (SWNT) and MWNTs, 273
Single walled carbon nanotubes (SWCNTs), 44–46
SLNs. *See* Solid lipid nanoparticles (SLNs)
SOC. *See* Spin-orbit coupling (SOC)
Solid lipid nanoparticles (SLNs), 43
Solution mixing

CNTs surface functionalization, 92
graphene, synthesis, 92–93
Spin-orbit coupling (SOC), 128–129
Spin-polarized systems, 344
SPIOs. *See* Superparamagnetic iron oxide nanoparticles (SPIOs)
SQDs. *See* Silicon quantum dots (SQDs)
SQUID magnetometer. *See* Supercomputing quantum interface device (SQUID) magnetometer
Stoichiometric threshold, 284, 285
Supercomputing quantum interface device (SQUID) magnetometer, 281
Superparamagnetic iron oxide nanoparticles (SPIOs), 57
Superparamagnetic nanoparticles, 12
SWCNTs. *See* Single walled carbon nanotubes (SWCNTs)
SWNT. *See* Single-wall C-nanotube (SWNT)
Symmetry-adapted cluster configuration interaction (SAC-CI), 128
Synfuel, 202

T

TEER. *See* Transepithelial electrical resistance (TEER)
TEM. *See* Transmission electron microscopy (TEM)
Tetrahydrofuran (THF), 211–212
Thermodynamic integration (TI), 393–394
THF. *See* Tetrahydrofuran (THF)
Time-dependent light-induced polarization, 135
Time-dependent perturbation theory, 129
Transepithelial electrical resistance (TEER), 17
Transmission electron microscopy (TEM), 206
Two dimensional graphenes
electro-optical properties
Conjugation Bridge, 325
donor-graphene nanoribbon-acceptor (D-GNR-A), 325
frequency doubling applications, 325
graphene nanoribbons (GNRs), 325
graphene quantum dots (GQDs), 327
hyperpolarizabilities, 326–327
spin reordering, 325
summation-overstates (SOS) expression, 328
hexagonal arrangement, 324
zigzag edges, 324

U

UCL. *See* Upconversion luminescence (UCL)
UCNPs. *See* Upconverting nanophosphors (UCNPs); Upconverting phosphors (UCNPs)
Ungerade, 148, 154
Upconversion luminescence (UCL), 53
Upconverting nanophosphors (UCNPs), 53–54
Upconverting phosphors (UCNPs), 53

V

Verlet algorithm, 375
Visual Molecular Dynamics (VMD) program, 216
VMD program. *See* Visual Molecular Dynamics (VMD) program

W

Wavefunction analysis
carbon-based molecular nanowires, 183
conductive effect, 183
conjugated benzenoid systems, 182
critical points (CP), 181
electron density (ED), 182
electron localization function (ELF), 182
ELF/ED analysis, 182–183

ring critical points (RCP), 182
Wigner-Seitz radius, 340, 345

Z

Zeeman splitting, 143
Zeolite framework, 257, 260, 264–266
Zero dimensional fullerenes
bond distances, calculated and experimental, 318
electron–phonon interaction, 317
electro-optical properties
cyclopentadienyl rings, 321
heterofullerenes, 319
hyperpolarizability, 321
third-order nonlinear polarizability, 320
"isolated pentagon rule," 317
NMR chemical shifts, 317
Zigzag nanotubes, 321
Zinc oxide nanowires (ZnNWs), 62
ZnNWs. *See* Zinc oxide nanowires (ZnNWs)